Extremes

The Evolution of Human Sociality

Extremes

The Evolution of Human Sociality

Edited by
Kaori Kawai

Kyoto University Press

TRANS PACIFIC PRESS

First published in Japanese by Kyoto University Press in 2020 as *Kyokugen – Jinrui shakai no Shinka.*
This English edition published in 2023 jointly by:

Kyoto University Press
69 Yoshida Konoe-cho
Sakyo-ku, Kyoto 606-8315, Japan
Telephone: +81-75-761-6182
Fax: +81-75-761-6190
Email: sales@kyoto-up.or.jp
Web: http://www.kyoto-up.or.jp

Trans Pacific Press Co., Ltd.
2nd Floor, Hamamatsu-cho Daiya Building
2-2-15 Hamamatsu-cho, Minato-ku, Tokyo
105-0013, Japan
Telephone: +81-50-5371-9475
Email: info@transpacificpress.com
Web: http: / / www.transpacificpress.com

Copyedited by Cathy Edmonds, Kyneton, Vic., Australia
Designed and set by Ryo Kuroda, Tsukuba-city, Ibaraki, Japan
Cover designed by Martyn Dade-Robertson

Distributors

USA and Canada
Independent Publishers Group (IPG)
814 N. Franklin Street
Chicago, IL 60610, USA
Telephone inquiries: +1-312-337-0747
Order placement: 800-888-4741 (domestic only)
Fax: +1-312-337-5985
Email: frontdesk@ipgbook.com
Web: http://www.ipgbook.com

Europe, Oceania, Middle East and Africa
EUROSPAN
Gray's Inn House,
127 Clerkenwell Road
London, EC1R 5DB
United Kingdom
Telephone: +44-(0)20-7240-0856
Email: info@eurospan.co.uk
Web: https://www.eurospangroup.com/

Japan
For purchase orders in Japan, please contact any distributor in Japan.

China
China Publishers Services Ltd.
718, 7/F., Fortune Commercial Building,
362 Sha Tsui Road, Tsuen Wan, N.T.
Hong Kong
Telephone: +852-2491-1436
Email: edwin@cps-hk.com

Southeast Asia
Alkem Company Pte Ltd.
1, Sunview Road #01-27, Eco-Tech@Sunview
Singapore 627615
Telephone: +65 6265 6666
Email: enquiry@alkem.com.sg

The publication of this book was supported by a Grant-in-Aid for Publication of Scientific Research Results (Grant Number 21HP6004), provided by the Japan Society for the Promotion of Science, to which we express our sincere appreciation.

ISBN 978-1-876843-66-3 (hardback)
ISBN 978-1-920850-00-5 (paperback)
ISBN 978-1-920850-01-2 (ebook)

Contents

Figures

Tables

Photographs

Contributors

Kaori Kawai
Professor, Research Institute for Languages and Cultures of Asia and Africa, Tokyo University of Foreign Studies

Kaoru Adachi
Associate Professor, Faculty of Sociology, Kyoto Sangyo University

Noriko Itoh
Researcher, Wildlife Research Center / Center for African Area Studies, Kyoto University

Keiichi Omura
Professor, The Open University of Japan

David S. Sprague
Specialist, National Agriculture and Food Research Organization

Yuji Takenoshita
Professor, School of Nursing and Rehabilitation, Chubu Gakuin University

Michio Nakamura
Associate Professor, Graduate School of Science, Kyoto University

Shunkichi Hanamura
Researcher, Center for African Area Studies, Kyoto University

Hitonaru Nishie
Restart Postdoctoral Fellow, Japan Society for the Promotion of Science / Graduate School of Science, Kyoto University

Ryoko Nishii
Professor, Research Institute for Languages and Cultures of Asia and Africa, Tokyo University of Foreign Studies

Ikuya Tokoro
Professor, Research Institute for Languages and Cultures of Asia and Africa, Tokyo University of Foreign Studies

Toru Soga
Professor, Faculty of Humanities and Social Sciences, Hirosaki University

Takeo Funabiki
Professor Emeritus, The University of Tokyo

Kōji Kitamura
Professor Emeritus, Okayama University

Motomitsu Uchibori
Professor Emeritus, Hitotsubashi University; Professor Emeritus, The Open University of Japan

Masakazu Tanaka
Professor, Professional Institute of International Fashion

Naoki Kasuga
Professor Emeritus, Osaka University; Professor Emeritus, Hitotsubashi University

Suehisa Kuroda
Professor Emeritus, University of Shiga Prefecture

Yuko Sugiyama
Professor, Faculty of Humanities and Social Sciences, Hirosaki University

Naofumi Nakagawa
Professor, Graduate School of Science, Kyoto University

Hideaki Terashima
Professor Emeritus, Faculty of Humanities and Sciences, Kobe Gakuin University

Acknowledgments

This volume is the fourth collection to emerge from Human Society in Evolutionary Perspectives, a joint research project based at the Tokyo University of Foreign Studies (TUFS) Research Institute for Languages and Cultures of Asia and Africa (ILCAA).

The research comprises a series of joint projects that has been ongoing for the past thirteen years, over four terms since 2005. Researchers chiefly concerned with extant wild nonhuman primate societies and researchers chiefly concerned with modern human societies have come together and continued their research while developing themes every three or four years through which to explore the evolution of human society and sociality. This volume collects the results of a three-year joint research project on the theme of Existence, Environment, and Extremity (hereinafter the EENE project) that was carried out between 2015 and 2017. Research from the three preceding terms was published as collections in Japanese and translated into English as *Groups*, *Institutions*, and *Others*, each of which, like this volume, was subtitled *The Evolution of Human Sociality*. The Japanese volumes were published by Kyoto University Press in 2009, 2013, and 2016, while the English translations were jointly published by Kyoto University Press and Trans Pacific Press in 2013, 2016, and 2019, respectively. The Japanese version of this volume was published in 2020. Thus, just as our research is about to enter its next phase, we have the Japanese version of the research collaboration newly completed, along with the English version of the research from the previous phase. In other words, this English version contains our (Japanese researchers') reactions to the previous phase and the Japanese contributors' sense of enthusiasm for the phase to come. If the reader pays attention to the keywords that appear in each of the chapters in this volume, you will be able to see the development of the contributing authors' guiding problematic from 'groups' to 'institutions' and from 'others' to 'extremes' of human survival.

Although our thirteen-year research collaboration has for the time being concluded with the publication of this volume, in 2018, as a follow-up collaboration, we began another ILCAA-based joint research project, 'The Origin of Human Sociality: Towards New Perspectives on Hominization,' and since 2019 activities have been underway on a research collaboration funded by a Ministry of Education, Culture, Sports, Science and Technology (MEXT) of Japan, Grant-in-Aid for Scientific Research entitled 'The Origin and Evolution of Sociality: Developing New Theories of Human Evolution Based on Collaboration between Anthropology and Primatology.' For me, in my capacity as editor, part of the discussion for this volume also helped to establish specific directions for advancing these new topics of research.

Like *Groups, Institutions,* and *Others,* this volume is published with immense support from a great many individuals and institutions to whom I wish to express my gratitude.

Primate sociology and ecology, ecological anthropology, and social / cultural anthropology, the areas in which the contributors to this volume specialize, are all fields of study that regard fieldwork surveys as an indispensable research base and for which researchers' own field data constitute the most important basis for research. Accordingly, first and foremost, and while apologizing for being unable to thank them each by name, I would like to express my gratitude to the local people in each of the contributors' field sites who agreed to their stays and research activities, to those who served as research assistants, and to the primates who always permitted us (sometimes doggedly) to follow them around. This volume would not have been possible without their cooperation in its various forms. Please accept our sincere and heartfelt thanks.

The following organizations and those affiliated with them extended various conveniences that assisted in the conduct of our surveys: Japan Society for the Promotion of Science (JSPS) Nairobi Research Station, Makerere University, the Tanzania Commission for Science and Technology (COSTECH), the Tanzania Wildlife Research Institute (TAWIRI), and the Tanzania National Parks Authority (TANAPA). In addition, the following individuals have given permission for the use of valuable photographs: Ikuko Nishida (for license to use photographs by the late Prof. Toshisada Nishida), Moyo Uehara (for license to use photographs by the late Prof. Shigeo Uehara, teacher), and Kazuhiko Hosaka (for license to use photographs).

The Primate Society of Japan, Jinruigaku Kanren Gakkai Kyōgikai (The Council of Anthropological Societies), and the ILCAA Core Project (Anthropology) provided opportunities to report on the results of this joint research project in the form of various symposia.

Holding the joint research meetings that became the basis for this book caused three years of various headaches for the members of the ILCAA Research Collaboration Division and Joint Research Center. We are indebted to the relevant bodies and individuals at Kyoto University for the joint research meetings held there.

Finally, with the publication of this volume, I wish to acknowledge the Editor-in-Chief of Kyoto University Press, Tetsuya Suzuki, who previously took care of us with *Groups, Institutions,* and *Others* and was involved in every step of this project from the planning stages, throughout the intricate work of editing, to the final correction of the proofs. Despite our habitual tardiness in writing and proofreading manuscripts, he waited patiently and encouraged us, and remained unwavering in his support of an inexperienced editor.

To these various individuals and organizations, I wish to express my wholehearted thanks.

Last but not least, in keeping with current Japanese research practice, I list the grant numbers for the various Grants-in-Aid for Scientific Research awarded by MEXT and JSPS that assisted with the field surveys associated with the writings collected here and with the organization and analysis of the results thereof, as well as with oral presentations and the preparation of written papers. We wish to express our gratitude to all the people concerned.

#16255007, #22221010, #25284171, #6H02737, #12J00004, #14J00963, #15H04429, #15K03034, #16H01969, #16K12445, #17H00948, #17H01648, #17H02436, #17H06341, #17H06381, #17H07232, #17J40059, #17K02014, #17KT0001, #17KT0058, #18H00776, #18K01162, #18K11795, #19H05591, #21HP6004.

Kaori Kawai
February 2023

Introduction

Existence, Environment, and Extremity: Seeking an Evolutionary Historical Basis for Human Sociality

Kaori KAWAI

Living with extinction: problematizing 'extremes' I

We humans, in terms of our biological taxonomy, belong to the species called *Homo sapiens*. Among the extant nonhuman primates, chimpanzees (*Pan troglodytes*) and bonobos (*Pan paniscus*) constitute the species most closely related to humans, but we know that humanity's divergence from the common ancestor that we share with these species took place on the African continent seven million years ago. According to research findings from genomic analysis, which over the past decade or so has yielded a rapid succession of discoveries of human fossils and astonishing technological innovations, the hominins[1] that appeared after this divergence easily number more than twenty species, even going solely by the fossil remains thus far discovered,[2] and there were apparently periods when multiple hominoid species lived at the same time. Nevertheless, all hominins other than *Homo sapiens* are now extinct.

Why is it that only we survived, while these other hominoid species all went extinct? The reason remains a mystery. At the individual level, some would have died out, unable to withstand one harsh situation or another, while others, despite being able to live out their so-called natural lifespan, may have found themselves unable to procreate for some reason, and hence unable to leave behind any offspring. This, perforce, may have led to 'death' at the group level, which is to say extinction. What seems clear, however, is that it would not have been all that strange had we *Homo sapiens* gone extinct before now and that, in any case, we may eventually go extinct in the not-so-distant future.

'The history of biology teaches that 99.9% of those biological species that have appeared on the face of the planet, will in fact go extinct. For the 0.1% who survive, including ourselves, we will in any case go extinct someday—we simply have not done so yet' (Yoshikawa 2014: 6). Over the reportedly four-billion-year history of life on this planet, the biological species that inhabit it have experienced four or five mass extinctions, including the most recent, which claimed the dinosaurs. If we include smaller-scale extinctions, aside from the mass extinction events, we can say that most of the biological species that have existed on this planet have gone extinct. To that extent, extinction is a natural inevitability for living things. Thinking in this way leads us to the conclusion that it is only a matter of time before we *Homo sapiens* go extinct as well.

Perhaps our ancestors, confronted with various situations that threatened them with extinction, may have survived by escaping such situations or overcoming them in some way. If, for the moment, we were to refer to such situations as 'extremes,' it is conceivable that our present circumstances are the culmination of a history of a relationship with such extremes. If existence alongside extinction is a normal mode of being for living things, then it is quite natural to think that our ancestors must have encountered a variety of extremes in the process of human evolution. However, it would probably not be correct to describe *Homo sapiens* as a species that has lived through conditions of extremity. It could just be that we have been lucky. It may be that extant humans and primates have both had repeated near misses with various extremes, which they barely (or luckily) survived, or that these extremes have continually been 'postponed.'

Either way, given that extinction is only a matter of time for a biological species, extremes are undoubtedly an inescapable fate for humans. We—the authors of this volume—have thought about and discussed human society from the perspective of evolution by way of the keywords 'groups,' 'institutions,' and 'others,' summarizing our results in a series of collected volumes (Kawai 2013, 2017, 2019). In terms of a concern with such an evolutionary perspective, 'extremity' could be the ultimate keyword for thinking about human society. In other words, by focusing on states of extremity and the ways in which they are perceived and confronted by humans and primates, we may be able to approach one aspect of the essence of human society.

The aforementioned three predecessors to this volume (*Groups*, *Institutions*, and *Others*, all subtitled *The Evolution of Human Sociality*) dealt with events and phenomena richly tinged with a social coloring, and various modes of sociality were described therein. These, however, are not unrelated to the environments in which each individual or group of individuals exists. In the sense meant here, it goes without saying that 'environment' refers to the natural ecological environment. But because most primates—among whom we include humans—live sympatrically with others as gregarious animals, we are inevitably led to a concept of environment that we could call the 'social environment.' Hence, while the basic concept of this book is to focus more strongly on the environment—a theme that has been a kind of *basso continuo* through the previous three volumes—in order to foreground an awareness of modes of existence, an important keyword in this effort is 'extremes,' which serves as the title of this book.

Illuminating this aspect of extremity constitutes a major step in the construction of a theory about the evolution of human society and sociality. Existence can be conceived of as the act of living as an organism—that is, foremost in biological and ecological terms. However, as noted above, since many primates (including humans) live as gregarious animals, there inevitably is a social element, and therefore survival strategies for humans and primates entail more than individuals simply making 'correct' choices as organisms dictated by their own cranial nerve commands, physiological needs, feeding ecology, or mating behavior. Rather, we can see a tendency toward non-deterministic, flexible, and communicative strategies largely influenced by interactions with others. It is our belief that the discussion of extremes, which are directly bound up with existence, will offer new perspectives on the elucidation of the evolutionary historical basis of sociality.

Recognizing and surviving change: problematizing 'extremes' II

Homo sapiens, which appeared 200,000 years ago on the African continent, underwent an explosive increase in population and spread to almost every corner of the planet within a relatively short period of time (Ōtsuka 2015). By about 10,000 years ago, migration had resulted in the incorporation of almost the entire planet as human habitat. It is estimated that the first migration from the African continent took place approximately 125,000 years ago, and that a more widespread dispersal began to take place approximately 70,000 years ago. Alongside our characteristics as an omnivorous species that consumed both plants and animals, it has been said that the reason our species was able to adapt to temperate, boreal, and even arid zones after leaving our birthplace in the tropics was due in large part to our developed culture. So why did our ancestors begin to move? And why did they continue—even to the point where there was almost no place left to move? Is it perhaps that they aimed for new frontiers when the places where they had been living became too uninhabitable?

As I will explain in just a moment, curiosity would have been an important factor. But curiosity alone does not seem likely to prompt people to move in large groups. Since movement entails a great deal of energy and risk, there would have had to have been a commensurate reason for doing so—one whose ramifications would have been, in a word, existential. It is not hard to imagine that those who became aware of a slight change in the environment they had theretofore inhabited, some sign that threatened to affect their very existence, began to move in search of a safer (or more comfortable) abode. It is quite conceivable that there could have been individual differences (or group differences) in this 'awareness' of environmental changes. Some would have moved upon noticing the changes at an early stage, while others would have been unaware of the changes until the end (by which time they would have already found themselves in straitened circumstances), so that when they began to move they would have done so as though they were being pushed out. In particular, presumably those who continued to move across the sea would only have done so when they were forced, not moving until the very last moment.

Moreover, just because they became aware of a change does not necessarily mean that they would have moved. It could also be that they chose to remain in place even after becoming aware of the fact. This book reports on several cases of people who opted to remain in the midst of danger (see, for example, Nishii's discussion in Chapter 10 and Tokoro's discussion in Chapter 11).

Whether or not a creature becomes aware of (i.e., senses and recognizes) changes in its surrounding environment relates to that creature's behavior. If we regard volcanic eruptions, major earthquakes, tsunamis, and meteorological phenomena such as typhoons and cyclones—all of which could literally be characterized as violent and threatening—as extrinsic factors that cause organisms to change their own behavior (e.g., by making them flee or move), we might be able to refer to the act of perceiving subtle environmental changes, albeit ones that are not as significant as disaster-grade environmental changes, as an intrinsic factor that would also prompt organisms to change their own behavior. Curiosity, noted earlier, is one such intrinsic factor. Perhaps it was not necessarily an 'awareness' of environmental change that prompted those pioneers aiming for new frontiers to begin moving. The very 'idea' that something might happen (someday in the future) is itself important. This is because, conceivably, having acquired the capacity for abstraction, *Homo sapiens* would not necessarily have moved solely based on a desire for material resources.[3] And whether out of curiosity or out of an awareness of change, it is natural to think that in the case of actual movement, humans were constantly being forced to make choices about whether to go or to stay. Assuming that it was their rapid movement and spread across the entirety of the planet that made human survival possible, if we focus on the intrinsic factors that would have been involved, it would seem then, in the case of *Homo sapiens*, that we must consider the conditions of extremity that influenced their existence as something other than extremes in terms of the environments inhabited by living things.

'Disappearance' and extremity, society, and culture

According to the seventh edition of the *Shin meikai kokugo jiten* (New clear-understanding Japanese dictionary) (Yamada et al. 2011), a popular Japanese dictionary noted for its idiosyncratic definitions, the Japanese word *kyokugen*, translated here as 'extremity,' refers to 'the threshold beyond which something's normal state will be lost.' Since 'lost' in this sense implies the 'disappearance' of what had been (i.e., existed) up until that point, to put it more plainly, what we call an 'extremity' is the point immediately prior to a state of affairs in which something that had been functioning would fail, or in which something living would die. If what is meant by a 'normal state' is life lived on the assumption that tomorrow will be the same as today, just as today was the same as yesterday, and that the days of our lives will continue on in the same way indefinitely, then the (end of the) process leading to a state of affairs in which individuals or collectives (e.g., groups, local populations,[4] species, or lineages) are driven to disappear, to experience the collapse of their society (or structure or system), or to suffer a massive emotional trauma (by sudden natural disasters, wars, accidents, pogroms, or other calamities) may be typical of the state defined here as 'extremity.'

When considered thus as a process, extremes can be seen to have 'breadth' in a variety of senses. Of the matters that influence survival, whereas large-scale natural disasters and anthropogenic disasters or accidents are events that assail living creatures all of a sudden, other calamities (such as, for example, the climate change associated with global warming) encroach on a slower time scale—gradually and by degrees. Of course, there is a continuum between what I refer to here as 'all of a sudden' and 'by degrees.' However, we can see that the 'threshold beyond which something's normal state will be lost,' brought on by such an event, itself forms a kind of gradation, and a difference of degree also exists in the manner of its impact.

From the existence of individuals to the survival of societies, there are various levels in the matter of human survival. For example, in contemporary Japan, marginal villages—struggling communities that are barely viable as such—have become a prognosis of the depopulation facing rural villages due to demographic aging, a declining birth rate, and the concentration of the population in Tokyo. Although they do not assume any kind of sudden incident or situation, such villages are a phenomenon in which the resident populations gradually decrease so that their function as communities is eventually lost. Marginal villages on the brink of having their communities disappear—such that, ultimately, one would say, 'and then there was nobody left'—could be described as being at a typical 'extremity' (see Kitamura's discussion in Chapter 14). Similarly, there are other cases in which although individuals go on living, the society they inhabit disappears. Typical examples would be phenomena such as the disappearance of ethnic minority groups and the disappearance of languages that underpin ethnic identity (so-called 'endangered' languages) (see Uchibori's discussion in Chapter 15).

However, even thinking about extremes solely in terms of 'dying out' (or disappearing) is a one-sided approach. Tropical forests are usually considered as environments that are rich in biodiversity. And for the primates that live there, going by the impression of many primatologists, it seems that 'even observing primates in the wild, one rarely gets the sense of their being pressed for time' (see Nakagawa's discussion in Chapter 20). On the other hand, other primatologists have the impression that the great apes, like the gorillas and chimpanzees and colobuses and other old-world monkeys, which live in the tropical forests of Africa, spend all day wandering in search of food, which they have a hard time finding. According to the latter, there is no doubt that the great apes spend a lot of time and energy foraging, which is to say that it seems that they are always living 'on the brink.' But the fact is that the situation is not so desperate that individuals starve to death for want of food. Among wild primates, situations in which, for example, reduced food availability caused by some change in the natural environment leads to the starvation of individuals—and, by extension, the dying out of the group—are extremely rare. In other words, even as they seem to live 'at the brink'—in other words, in extremis—it seems that in reality they live far from any such extremity.[5] In this sense, we could say that the former impression seems to be the more correct of these two opposing impressions. Be that as it may, either way, while it seems that precarity should exist at the individual level, how is it that they are able to overcome crises at this individual level and survive extremes as a collective (e.g., group, local population, or species)? Were they somehow aware of their extreme situation and thus able to successfully sidestep the path to extinction, or did only those individuals that just happened to adapt to the environment survive to become responsible for the subsequent propagation of the species? We can only infer that they must have survived changes in the natural environment and overcome their extreme situation by means of cultural and social coping mechanisms, such as changing their food menu, changing their manner or routes of movement, or changing their grouping patterns.

So what about humans? When Kenya's Turkana pastoralists lost most of their livestock in the great drought of 1980–81, which was described as a once-in-a-century event, they no longer had access to the milk of their livestock, which had been a staple food source. Severe malnutrition led to deaths among both young children and adults. Jun'ichirō Itani, who was conducting field research among the Turkana at the time, made a detailed record of the words and actions of the Turkana people who weathered this drought 'unsure whether they would live or die,' a record preserved in his sublimely written ethnography *Dai kanbatsu: Turukana nikki* (The great drought: A Turkana diary) (Itani 1982). Itani makes frequent use of the term *kyokugen* (extremity). He wrote that the Turkana pastoral people, who made use of their livestock to inhabit extremely arid regions that would otherwise be unavailable to humans, were well aware that they lived on the extreme edge of existence, where it was possible that even the slightest changes in weather phenomena would leave them struggling to survive. And the 'simple life scraped bare of every excess' (Itani 1980), eked out at this extreme edge of existence and tipped over that edge by the great drought, suffered a devastating blow. Livestock and people alike perished in considerable numbers.

The Turkana people recognized that this drought was no normal drought. Despite this, Itani concludes, they were unable to find a way to cope. However, the Turkana people never forgot their pride at being Turkana—at being 'people of cattle' and 'people who live with cattle.' Although they raised their eyes to the heavens and accepted the great drought, saying that everything 'is God's will,' this was another way of saying that the Turkana culture and society knew of no other way to accept the great drought. The Rendille and Gabra, camel-herding pastoralists who live adjacent to the Turkana, were hit by the same great drought at the same time, but did not suffer to the same extent. As the drought progressed, cutting off the milk of their livestock, which was their main food source, the Turkana opted to spare their cattle, which were regarded as more valuable culturally and socially than their camels, which they slaughtered for meat, even though camels are physiologically the most resistant to extreme dryness. But the cattle they had spared proved unable to withstand the drought, and dropped dead in quick succession. As a result, the Turkana ended up losing both their camels and their cattle. And people also began to perish. Meanwhile, among the camel-herding pastoralists of the Rendille and Gabra, people got on with their lives by relying on camels, which were able to supply milk even as the drought progressed, thus averting a catastrophe that would have claimed human lives (Itani 1982).

In the previous section, I mentioned 'extremes' and 'awareness.' Without awareness, an extreme will fail to appear as a process. There will simply be a sudden disappearance (i.e., death or extinction). But the Turkana were aware that the drought was unusually severe. In that respect, an extreme was present, but they were unable to cope. This intense experience on the part of the Turkana is surely a classic example of how cultural values and conventions govern human existence in states of extremity.

The Ik, a group of hunter-gatherers, live at the northeastern tip of Uganda, west of the Turkana's territory. They are an ethnic minority made famous by British anthropologist Colin M. Turnbull (1973) in his ethnography *The Mountain People*. Turnbull scrupulously documents the breakdown of humanity among the Ik, describing how starvation caused by a great famine in the 1960s led to the loss of human sentiment, as illustrated by the apparent absence of even a shred of labor spared for the dying, and of competition for food and a lack of sharing even within families. The Ik at this time could be described as having been in a state of extreme, both biologically (in the sense that they were unsure whether they would live or die) and socially (in the sense of the collapse of family and community relationships) but, even so, they did not die out. Since the territory inhabited by the Ik is adjacent to that of the Dodoth pastoral people, with whom I have conducted field research, I often come across Ik people in my field site. The Ik remain with us today (or at least did so as of 2014). Despite suffering through the terrible famine, it appears that their population did not decrease to the extent that they were no longer able to maintain their cohesion as a people, and they also do not seem to have succumbed to a situation in which this ethnic cohesion collapsed in social terms. Rather, it appears that they have arrived in their current circumstances having maintained their population to some degree and with their identity as Ik intact. Even though the Ik people maintain social and economic relationships with their neighbors, the Dodoth and Turkana pastoralists, they have never been assimilated by them. They have no desire to keep livestock, and continue to live (to survive) as Ik, subsisting primarily by hunting and gathering just as they did before.

Extremes may be avoided or overcome (see Omura's discussion in Chapter 3 and Terashima's discussion in Chapter 21). And while death or extinction await when these circumstances cannot be avoided or overcome, we cannot know which way we will fall. It may, in a sense, be fate—perhaps in the truest sense of the word.

A broad approach for discussing extremes: different perspectives

In this volume, primatology (primate sociology and primate ecology) and anthropology (ecological anthropology and social/cultural anthropology) contributors identify situations that could be called 'extremes' (primarily with reference to various phenomena in their respective fields) to consider in relation to sociality. As a result of these various scholars thinking about modes of extremes and where examples of such extremes might be found in their respective fields, the way that extremes appear has truly expanded. In this book, we intentionally eschew any strict definition of extremity. Although it is sometimes defined in particular chapters, in many chapters it is not. In the book overall, an extreme is loosely considered to be a borderline environment habitable by living things, with borderline environmental conditions in which, if neglected, living things and their societies will perish. We decided to address the questions of how humans, primates, or their respective societies handle extremes, and how they approach extremes, on the basis of concrete examples. In doing so, it became clear that the concept of 'extremity' is enormously broad, and can refer to many different things and be approached in many different ways. It is a wide stage, from individuals (who are related to evolution) to groups (from micro to macro). And yet, although it is wide, and although it is varied, it is not simply a motley collection, but one of mutual correlation. Although I will also touch on this in the next section ('A conceptual schema (ontology) of extremes'), I point out here that the breadth of what we call extremes and their mutual interrelationships have been recognized.

Below, we will look at extremes from the aspect of how they appear, since the ways in which extremes appear, and the living things that inhabit them, are indeed truly varied.

Extremes as invisible externalities

One way of conceiving of extremes is to think of them as marginal or borderline situations that are enormously inhospitable for sustaining life—situations in which individual organisms or groups of organisms (e.g., clusters, groups, local populations, species, or phylogenetic groups) struggle to eke out an existence even as they are exposed to crises that threaten their survival due to the additional presence of some kind of powerful stressor. Here, a stressor is envisioned as a force that is external from the point of view of the living being(s) in question. That is, extremes in this sense are extreme natural ecological environments produced by some shift in the natural environment, such as a large-scale natural disaster or slow but steadily encroaching climate change. Such stresses have the potential to become extrinsic factors, driving not only individuals but also local populations and biological species to extinction. Extrinsic factors, although they are in most cases visible and apparent, can sometimes be invisible to the naked eye, odorless and colorless, as with the radioactive contamination caused by the nuclear accidents at Chernobyl and Fukushima. While it could be argued that radioactive contamination differs from a natural disaster in that it is caused by human error (i.e., a nuclear accident), it can still be treated as a 'risk or hazard' and as a force that brings about an extrinsic environment that plunges living beings into a condition of extremity, just as in the case of a natural disaster (Hoffman and Oliver-Smith 2002). In this way, extremes can be discussed as being extrinsically caused by external environmental factors. Yet although this is what the term 'extremity' typically brings to mind, in fact, of the extremes dealt with in this volume, there are only two cases that touch on extrinsic factors purely as 'risks or hazards' and these are both limited to the issue of radiation damage in Fukushima (see Nishii's discussion in Chapter 10 and Tokoro's discussion in Chapter 11). By contrast, most of the chapters in this volume deal with extremes brought about as a result of factors that are intrinsic to human and primate society and individual psychology and spirit. It might be said that these are attempts to interrogate the 'internality' of extremes.

Extremes as a chronic disease

Many of the chapters in this volume present situations of extremes that humans and primates inevitably end up facing simply by virtue of existing. Since primates (including *Homo sapiens*), as gregarious animals, live in social groups composed of multiple individuals, conditions of what we might call 'social extremity' are always near at hand.

In this context, I want to highlight Kōji Kitamura's discussion in Chapter 14 as a classic argument about the relationship between social extremes and evolution. The field site in question is a marginal village on one of the remote islands of the Seto Inland Sea. Kitamura situates the phenomena that are actually taking place in the confrontation with the extreme—a question of the life or death of island society—in the form of a chronic problem in the context of human history, namely the reproduction (i.e., survival) of society. In doing so, he is able to discuss marginal villages in terms of an issue bearing on the evolution of human sociality. He poses the question of why extreme situations that have arisen somewhere along the process of human evolution should recur in the midst of everyday practice in extant human and nonhuman primate societies.

Various social extremes are considered as chronic problems for humans and primates living in the company of others. They seem to give the impression of a tightrope stretched between two cliff edges that seems likely to fray at any moment. Humans and primates will resist, combat, confront, overcome, tame, or flee from extremes. And if they flee, it may be that they do so only to come into contact once again with others (or other groups) and become exposed to new social extremes.

The disappearance of categories

A wide array of situations can be envisioned for social extremes, from those that occur at the individual level to those that occur at the level of the collective (i.e., cluster, group, local population, species, or phylogenetic group). Moreover, at these various levels of extremity, situations will sometimes occur with reference to substantive objects that can be specifically singled out, such as an individual, local population, or species (e.g., the death of an individual, the disappearance of a local population, or the extinction of a species); they can also occur in terms of the disappearance of a group as a category. Despite the ostensibly distinct difference between these two ideas as objects and categories, they are not in fact clearly distinguishable. For example, the death of an individual, the disappearance of a local population, and the extinction of a species are most certainly real problems. They could be the death of individual a, the disappearance of local population X, and the extinction of species Z. Conversely, the process of a group's disappearance as a category is slightly more complicated. Ethnic group A or species Z both refer to a collective of individuals named A or Z. Even if a, a', and a''—the individuals who make up A—should all die, the category A can still remain as an empty set (Uchibori 1989). This is because even if a, a', and a'' all die, so that none of the people who made up A remain, the ethnonym A will still remain in the memories and records of those who once 'knew' ethnic group A and the people who belonged to it. At this point, we could say that A continued to exist as an empty set. The disappearance of A as a category is a situation that will only arrive when the ethnonym A is lost (or expunged) from the memories and the records of all people. It is worth noting here, as well, that although a (biological) species in one sense seems to have a physical reality, in another sense it exists as a category and, as such, its boundaries can be ambiguous. The extinction of a species is essentially the disappearance of a category, and in that sense it can be thought of as having something in common with the disappearance of a human group, even if it cannot be said to be identical. We will see this in more detail below when we discuss the extinction of *Homo neanderthalensis*.

It hardly needs to be said that category-based behaviors such as domination and subordination, majority and minority, selfishness and altruism, and tolerance and intolerance, among others, overlap with our discussions of groups, institutions, and others that were taken up in the preceding three volumes in this series.

Society is not immutable. Its members, structures, organizations, and systems are continually in flux. What could we say is maintained with the continuation of society? For example, in the marginal village question mentioned earlier, the disappearance at issue was not the physical death of all the people living in the villages in question. Rather, it is generally the case that a marginal village becomes so when people leave it to move somewhere else. Although no one remains in that place, those who were born and lived there will still be living somewhere else. This is a phenomenon whereby the category named for the people who live in that place disappears (becomes an empty set), while in biological terms the DNA (deoxyribonucleic acid) of the people who were born and lived there does not disappear. The disappearance of the people named X who lived in the land of Y is simply the disappearance of a social category. From the perspective of people living in marginal villages, it is the disappearance of the collective 'we'-ness that had distinguished them from 'others.'

Of course, this is a serious problem in the context of contemporary Japan. But can it be appreciated in biological terms? For example, when it comes to the Neanderthals (who in systematic terms are a cousin of *Homo sapiens*) as a human fossil, the issue of 'disappearance' becomes more subtle. It has been presumed that the Neanderthals went extinct. However, from the latest results of genomic analysis, we know that Neanderthal-derived DNA accounts for between 1.5% and 2.1% of the DNA of extant humans (*Homo sapiens*) living outside the African continent (Reich 2018). If so, does this mean that, in biological terms, the Neanderthals have not disappeared (i.e., have not gone extinct)? How might we then characterize the 'extinction' of the Neanderthals? Considered in this way, it must be said that no clear distinction can be drawn between the death of a social category (e.g., the disappearance of a village or ethnic group) and the death of a biological species (e.g., the extinction of an endangered species). Or, since the word 'Neanderthal' refers to a category that we named, it may be better to think that such situations could occur naturally. Here lies the issue of whether it is at all possible to distinguish between the disappearance solely of a category or name and a disappearance in substantive terms. And it can also be said that without some way (like genomic analysis) of scientifically visualizing substances that would otherwise be 'invisible,' there will be no logically decisive means of interpretation.

However, there is still the question of whether the death (or disappearance) of categories can be considered by expanding it to include nonhuman primates who lack language. This immediately calls to mind relations between unit groups of chimpanzees. Groups of wild chimpanzees (unit groups) are groups with fixed membership (although females tend to change group allegiance around the time of their sexual maturity), and relations between groups are known to be hostile. Chimpanzees are likely also aware of these 'groups' and are conceivably aware of which group a given individual belongs to. Of course, since we cannot interview chimpanzees, we cannot know whether chimpanzee unit groups are simply categories for observers (i.e., humans) or whether chimpanzees also inhabit a society that is predicated on these categories. However, a well-known case revealed by a field survey of wild chimpanzees conducted over long years offers a (perhaps excessively) famous example of the disappearance of a group in the wake of a conflict (i.e., the avoidance of hostile relationships and encounters) observed between neighboring groups of chimpanzees in Mahale Mountains National Park (e.g., Kawanaka and Nishida 1977; Nishida 1994; Takahata 2015). Similar cases have been observed in Gombe, Tanzania, and in Ngogo, Uganda. In fact, even though what was observed between groups in Mahale was not even a skirmish, let alone a battle or slaughter, this case is apparently often cited incorrectly as an example of the disappearance of a group as the result of 'killing' (slaughter), owing in part to the strength of the impact of the group's disappearance (Nakamura 2015). What is certain, however, is that one of the two adjacent groups did 'disappear.' The question here is whether what disappeared was the adjacent unit group as a 'category,' or the unit group as a collection of individuals.

With the problem of categories, what is essential is the operation of naming the category. That is, the presence or absence of language is considered to be decisive for the conceptual formation of the category, and thus it is also a fact that the difference between humans and nonhuman primates will be all too conspicuous. But is the presence or absence of language actually decisive? It seems to us that there is still room for discussion with regard to the continuity and discontinuity of the notion of categories.

Routinization: extremes that lurk in everyday life

As mentioned above, an important aspect of extremes is the act of 'being aware' of them. There can be no extreme without the subject being aware of it. Extremes are not always visible, and some—such as radioactivity—are invisible (and tasteless and odorless). Similarly, and as also mentioned, extremes that develop gradually—those of a type that encroaches by slow degrees, as with environmental changes that take place over an extended period—are rarely characterized by visible or noticeable changes that can be easily spotted, but are more often invisible situations in which extremes progress without anyone's awareness of them. In the case of social extremes, as well, although they may be visible, as in the case of riots or terrorism, they generally seem to be invisible, similar to the type that encroaches by slow degrees. Where conditions present extremes like natural disasters and accidents caused by human error, which could be called acute and visible risks or hazards, extremes will not take long to detect. However, invisible extremes, extremes that encroach by slow degrees, and intrinsic extremes are all difficult to see. How (or on what basis) can these extremes, so difficult to grasp with the five senses, be perceived? How might they be brought into our consciousness (given structure)? How do humans or primates become aware of extremes and change their behavior to cope? It also seems quite possible that some scope of time may be necessary before recognition of an extreme dawns into awareness. There may also be individual differences or differences between groups as to whether such awareness comes immediately or with great difficulty. A wide range of responses was observed following the radioactive contamination resulting from the Fukushima nuclear accident; some people evacuated immediately, while others were less easily moved (see Nishii's discussion in Chapter 10). And in evolutionary terms, all we can say is that we do not know which response was correct. Further, people's recognition of an extreme may be misguided. Nevertheless, it seems certain that the forms that society takes will differ significantly on the basis of that perception.

Where the perception, recognition, and conscious awareness of an extremity (even in dire conditions that are extremely adverse to making a living) continue over a long time, and where people are able to continue living as they did before, we can no longer call such a situation an extremity. Although we do not think of risks and hazards such as natural disasters and accidents caused by human error as the kind of thing one can 'get used to,' in the case of social extremes, it may happen that the extreme could become entrenched (see Kawai's discussion in Chapter 9). However, at such times, it would be correct to say that the extreme has already been absorbed into the rhythms of daily life.

For Japan's marginal villages and ethnic minority groups around the world in danger of disappearing, it seems unlikely that their states of social extreme will have gone completely unnoticed. A sense of crisis, the feeling that 'if things continue on like this, we will end up disappearing' or that 'we will be no more,' is usually something that is clearly recognized by local residents (see Funabiki's discussion in Chapter 13 and Kitamura's discussion in Chapter 14). With regard to people's behavior, as well, changes are generally recognized at various levels. Conversely, in the context of situations that are invisible, and where people feel hardly any sense of crisis, it is sometimes possible to purposefully incite the

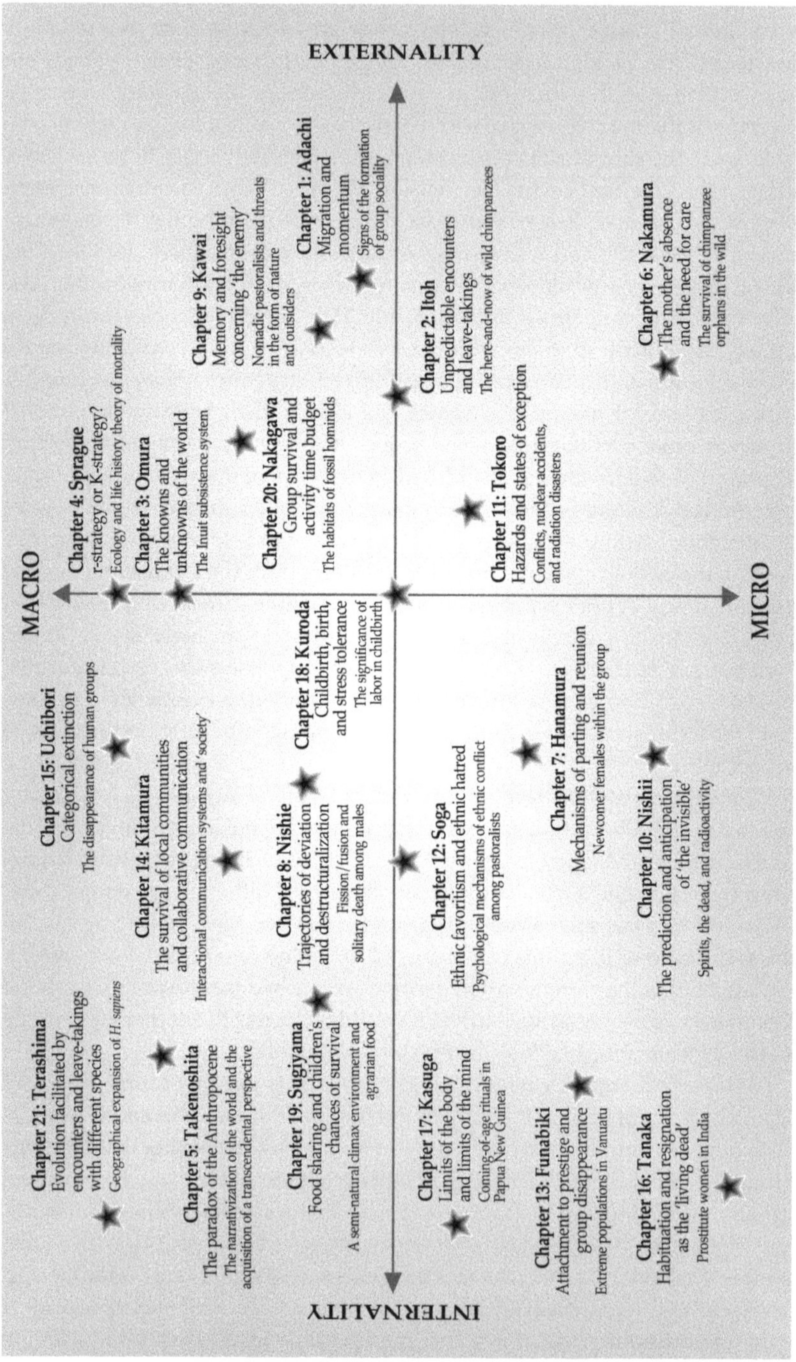

Figure 0.1: Various phases of extremes

sense that a crisis is imminent, thereby intentionally (arbitrarily) bringing the extreme condition into view.

Looking back over history, one need hardly mention Germany's Nazis to find that there have been many attempts to exploit extreme conditions to suit political ends. A similar tendency might also be seen in the way that the Anthropocene has been taken up and handled in recent years (see Takenoshita's discussion in Chapter 5). It could also be said that statements like those decrying that 'everything is fake news' are aimed at the same thing. The visualization of extremes has a significant influence over people's behaviors and ideas about how society should be. Bringing the 'invisible enemy' into view and marshalling people to fight can take place even when these are not political tactics on the part of politicians. Depending on category and (the manipulation of) identity, even among a normally 'peace-loving people' someone might be a mass murderer or a martyr who offers up his or her own life (see Soga's discussion in Chapter 12). Or, frightened by invisible enemies, it is sometimes the case that, having brought an enemy into view through various social and cultural procedures, someone will repeatedly undertake an (essentially unfounded) movement in order to escape being victimized by an (imaginary) enemy (see Kawai's discussion in Chapter 9).

A conceptual schema (ontology) of extremes

As will be clear from what I have written thus far, this book is not actually about 'extremes' as such. Nor is it a book intended to answer the question of what extremes are. We find phenomena and events that could be conceived of as extremes in our respective fields, carefully observe how these extremes are treated therein, and attempt to explore how these extremes are received and inhabited by humans and primates. In doing so, our intention is to drive a wedge between (modes of) sociality and human and primate societies. Extremes are handled and received in different ways by different individuals and different societies (or groups). There are too many different ways of perceiving, recognizing, and becoming aware of extremes to feasibly arrive at any simple pattern or formulation. I stated earlier that this book defines 'extremity' only loosely. It is because of this that various forms of extremes have come to light. Moreover, in addition to realizing the unexpected breadth of extremes, we came to grasp their interrelatedness, as I have also already mentioned.

Therefore, in Figure 0.1, I have endeavored, perhaps somewhat forcibly, to plot all the extremes presented in the twenty-one chapters collected in this volume into a conceptually drawn two-dimensional space (having first generalized them with a greater degree of abstraction). Structures sometimes become visible as more ethnographic facts are accumulated. And this sometimes engenders expectations for the emergence of a grand theory. I would like the reader to understand this schematization (i.e., the formulation of Figure 0.1) as the preparation of a sketch for understanding various extremes with various approaches, as well as a tentative effort toward conceptualizing extremes and illuminating the outlines of their ontological mode.

In terms of coordinates, I have arranged the externality–internality spectrum on the horizontal axis (x-axis) and the micro–macro spectrum on the vertical axis (y-axis). The x-axis signifies the level of the natural ecological environment (= externality) and the level of the social and psycho-spiritual environment. This axis could also be said to capture the physical and material level (= externality) and the non-physical and non-material level (= internality). The y-axis signifies the level of individual everyday life (= micro) and the level of groups, species, and evolution (= macro).

The figure is classified into four quadrants, across which I have plotted the extremes discussed in each chapter. However, some extremes have been plotted on the vertical or horizontal axis—situated on the 'borderline,' so to speak. Although each extreme is marked with a star, each is actually wider than an isolated point. But as they would overlap with adjacent stars, which is difficult to adequately represent on the figure, I would like the reader to think of the stars as radiating outward in all directions. It might be best to picture a central point from which something (a mode) spreads radially outward. Each extreme can be extended in any direction from its central point, and can also be influenced from any direction. The outward spread radiating from a star overlaps in part with the spread of the surrounding stars. The areas of overlap are areas of mutual influence. In that sense, each extreme is represented as being interrelated with the others. Or perhaps we should say this tentative partition itself appears as being intertwined in a hybrid fashion.

Not offering a precise definition for 'extreme' is only one way that this volume elects not to present the reader with any sort of framework. Although it is important to show an initial framework in this introduction in the sense that it prepares readers to make their way through the book, on this occasion I have intentionally chosen not to do so. Instead, I hope that this figure, which shows the state (positionality) of extremes that the authors have constructed from the actual sites of extremes in their respective fields will constitute a kind of navigation aid. Readers will be able to read each chapter and do their own mapping.

Another point is that this conceptual schema of extremes does not include a temporal axis (z-axis). Extremes most certainly have a time axis (history and speed). With a little thought, it is easy to see how quickly the things that confer or bring about extremes (e.g., external factors) approach. Some, like natural disasters such as tsunamis and typhoons, descend in an instant, while others, such as climate change, are of a type that encroaches by degrees, occurring slowly over a span of years, decades, or even centuries. The stars that represent the (central points of the) extremes actually have their own time scale and speed.

Although I have simplified matters by omitting a temporal axis here, since it would have been difficult to illustrate a three-dimensional space that included it, I would nevertheless like the reader to try to keep the temporal axis in mind.

Structure of this book

Following this introduction, this volume contains four parts consisting of twenty-one chapters. This four-part structure, however, has no correspondence with the quadrants sketched out above. Recalling that the quadrants are intertwined in a hybrid fashion so that they are not independent, and that the stars representing the individual extremes plotted in the quadrants each have their own spread that radiates outward, so that they interact with each other, it will be easily understood that it would be impossible to deal with the four quadrants and the extremes contained therein by dividing them into four parts as mutually exclusive and separate objects. 'Extreme' is an enormously broad concept. In this volume, which clarifies and discusses extremes in relation to various topics and from various approaches, as I have already mentioned, hardly any consideration is given to external extremes—which can be summarized as so-called 'risks and hazards' such as natural disasters, accidents stemming from human error, and wars—in favor of an almost exclusive focus on social extremes inherent in human and primate societies. In other words, the object of our discussions and considerations consists of the extreme situations that intrinsically arise when humans or primates live in groups (i.e., coexist or cohabit sympatrically with others). By focusing on aspects such as the character of extremes in such socially bounded environments and how humans and primates inhabit such extremes, the whole has been divided into the four parts that comprise this volume. As with the previous three volumes in the series (*Groups*, *Institutions*, and *Others*), the synopsis of each chapter can be grasped by the conceptual schema and explanatory text provided at the beginning of each chapter. Here, then, I explain the intention for each of the four parts.

Part I: Livable Extremes: Between Nature and Society

In Part I, while managing the natural ecological foundation from a natural scientific perspective, we focus on various situations of extremes for humans and primates who inhabit the distance between existence as organisms and existence as social beings; we also focus on ways of confronting such situations, demonstrating the semantic breadth of the extremes discussed in this volume and using this as a point of entry. Focusing more on theoretical statements than observational cases or ethnographic descriptions, this part brings together discussions that posit bold hypotheses about what might constitute conditions of extremity for living things and develops discussions that make us think about whether such things can in fact be called extremes. These discussions think logically about the evolutionary mechanisms of sociality with respect to extremes.

Part II: Realizable Extremes I—Existence and Social Environments

Part II and Part III both bring together papers that take up specific examples, all based on ethnographic descriptions and the records of direct observations, to discuss social extremes that are produced when living things form groups and coexist or cohabit with others. Part II contains three papers dealing with wild chimpanzees and three papers dealing with human societies. By carefully following how individuals act and how groups respond in each case to differing extremes, the authors clarify the contrivances by which primates and humans encounter social extremes, how they confront the extremes they encounter, and how they try to maintain their existence, while also touching on assumptions about how conditions of extremity arise or are produced in the first place.

Part III: Realizable Extremes II—The Power of Memory and Imagination

Continuing on from Part II, Part III brings together discussions about social extremes produced by the act of living (coexisting or cohabiting) sympatrically with others, especially for cases that are characteristic of human societies. Because humans have language, we are able to 'name' various social categories. Although named social categories foster corresponding identities and provide foundations for the lives of those who belong to them, they can also produce social extremes, such as by directing people toward the eradication of people who belong to other social categories. A situation like the disappearance of a category can become an empty set while shifting or overlapping with the actual disappearance of a human group (of living individuals), so that eventually they will be erased from both memory and written records.

Part IV: Extremes Engraved in the History of Human Evolution

As mentioned earlier, consistent with the previous three volumes in this series, the contributors to this volume are concerned with the evolution of human sociality. Part IV takes up events and phenomena that could be conceptualized as extremes from among events that have occurred in society and the human body (i.e., form) in the course of the history of human evolution, bringing together papers that pursue the process of human evolution more realistically, such as by presenting the names of fossil humans (i.e., fossil species of the genus *Homo*) and showing ancient periods of history, in order to situate and discuss extremes in the context of concrete evolutionary processes. The human species must have encountered various types of extremes over the seven million years of evolution since we diverged from the common ancestors we share with chimpanzees and bonobos. With the exception of *Homo sapiens*, all the hominins that have arisen on this planet have gone extinct, a fact that may have been the consequence of being unable to acquire the skills to overcome conditions of extremity.

■■■

As described above, this volume is the fruit of debates that have involved open-minded discussions about where human society has been and where it is going, discussions that have taken place steadily, based on concrete observational records and ethnographic descriptions, while drawing on bold ideas to blaze a trail into fields with scarce evidence. As Hideaki Terashima points out in the final chapter (Chapter 21), 'movement, encounters, and interactions created the world of today's Sapiens.' The 'encounters' between different types of people could have represented one type of extremity. While unfortunate events may have occurred (such as the extinction of one of the parties to the encounter), it is highly likely that the extreme of catastrophe was avoided by drawing fully on our negotiating capabilities as *Homo communicans* and our locomotive faculties as *Homo mobilitas*. Could we not say that it was our superb agility at making use of extremes that became the driving mental and physical force behind the geographical expansion of Sapiens, who would have repeatedly engaged in a 'swing-by' that saw them approach one another and then cohabit for a time after the encounter before separating once more?

Notes

1. The term 'hominin' refers to the genetic lineage that includes modern humans. In terms of official taxonomic nomenclature, hominins constitute the tribe Hominini, which is the classificatory category directly above genus *Homo*.
2. Ways of counting the number of human species differ according to researchers' interpretations. Moreover, given that the fossils discovered thus far are considered to be only a small subset of the fossil humans that lived in the past, the actual number of species is likely to be higher (Sarashina 2018: 18).
3. In this respect, *Homo sapiens* and nonhuman primates likely differ, even among extant primate species. The latest view among primatologists is that the imaginative faculties with which humans are equipped are not possessed even by chimpanzees, the great apes most closely related to humans. Chimpanzees inhabit the 'here and now' and do not feel hope or worries in anticipation of the future (Suddendorf 2013).
4. A 'local population' refers to a group of one species of organism that lives within a prescribed area.
5. Organisms inhabit an Umwelt—a self-referential world of their own in which they employ only a fraction of environmental resources in accordance with their own perceptual faculties (Uexküll and Kriszat 2010).

References

Hoffman, S. M. and A. Oliver-Smith (eds) (2002) *Catastrophe & Culture: The Anthropology of Disaster*. School of American Research Press.

Itani, J. (1980) *Turukana no shizen-shi – kashaku naki hitobito* (A natural history of Turkana: A people without remorse). Yūzankaku.

Itani, J. (1982) *Dai kanbatsu: Turukana nikki* (The great drought: A Turkana diary). Shincho-sha.

Kawai, K. (ed.) (2013) *Groups: The Evolution of Human Sociality*. Kyoto University Press and Trans Pacific Press.

Kawai, K. (ed.) (2017) *Institutions: The Evolution of Human Sociality*. Kyoto University Press and Trans Pacific Press.

Kawai, K. (ed.) (2019) *Others: The Evolution of Human Sociality*. Kyoto University Press and Trans Pacific Press.

Kawanaka, K. and T. Nishida (1977) 'Mahare sankai no chinpanjī (2) shūdan-kan kankei' (Chimpanzees of the Mahale Mountains (2): Inter-group relations). In J. Itani (ed.), *Chinpanjī-ki* (The chimpanzees). Kodansha, 639–694.

Nakamura, M. (2015) *'Saru-gaku' no keifu: Hito to chinpanjī no 50-nen* (Genealogy of Japanese primatology: 50 years of people and chimpanzees). Chuokoron-Shinsha Inc.

Nishida, T. (1994) *Chinpanjī omoshiro kansatsu-ki* (Thirty-four stories of chimpanzees). Kinokuniya.

Ōtsuka, R. (2015) *Hito wa kōshite fuete kita: 20 man-nen no jinkō hensen-shi* (The proliferation of *Homo sapiens*: A 200,000-year history of population transition). Shincho-sha.

Reich, D. (2018) *Who We Are and How We Got Here: Ancient DNA and the New Science of the Human Past*. Oxford University Press.

Sarashina, I. (2018) *Zetsumetsu no jinrui-shi: Naze 'watashitachi' ga ikinobita no ka* (A history of human extinction: Why did 'we' survive?). NHK Publishing Inc.

Suddendorf, T. (2013) *The Gap: The Science of What Separates Us from Other Animals*. Basic Books.

Takahata, Y. (2015) 'Disappearance of K-group male chimpanzees: Re-examination of group extinction'. In M. Nakamura, K. Hosaka, N. Itoh and K. Zamma (eds), *Mahale Chimpanzees: 50 Years of Research*. Cambridge University Press, 119–127.

Turnbull, C. M. (1973) *The Mountain People*. Jonathan Cape.

Uchibori, M. (1989) 'Minzoku-ron memorandamu' (Memorandum on ethnicity discourse). In S. Tanabe (ed.), *Jinruigakuteki ninshiki no bōken: Ideorogī to purakutisu* (Adventures in anthropological awareness: Ideology and practice). Dobunkan.

Uexküll, J. and G. Kriszat (2010) *A Foray into the Worlds of Animals and Humans, with A Theory of Meaning*, Joseph D. O'Neil (trans.). University of Minnesota Press. (Original German edition published 1934).

Yamada, T., T. Shibata, K. Sakai, Y. Kuramochi, A. Yamada, Z. Uwano, M. Ijima and H. Sasahara (2011) *Shin meikai kokugo jiten* (New clear-understanding Japanese dictionary), seventh edition. Sanseidō.

Yoshikawa, H. (2014) *Rifujin'na shinka: Idenshi to un no aida* (Outrageous evolution: Between genes and destiny). Asahi Press.

PART • ONE

Livable Extremes

Between Nature and Society

CHAPTER 1

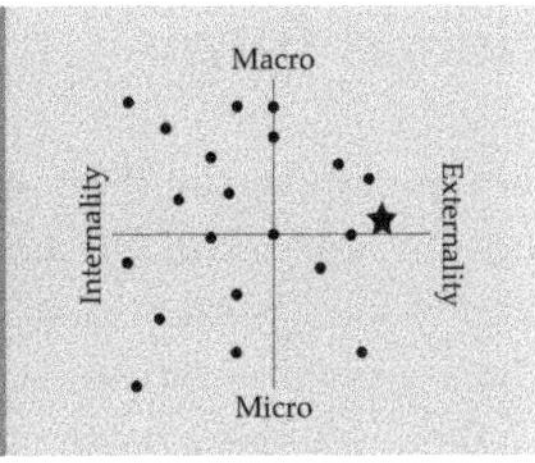

Momentum as an Extreme:

The Sociality of Moving Groups

Kaoru ADACHI

Keywords: movement, environment, momentum, future destination, anticipation, anti-structure

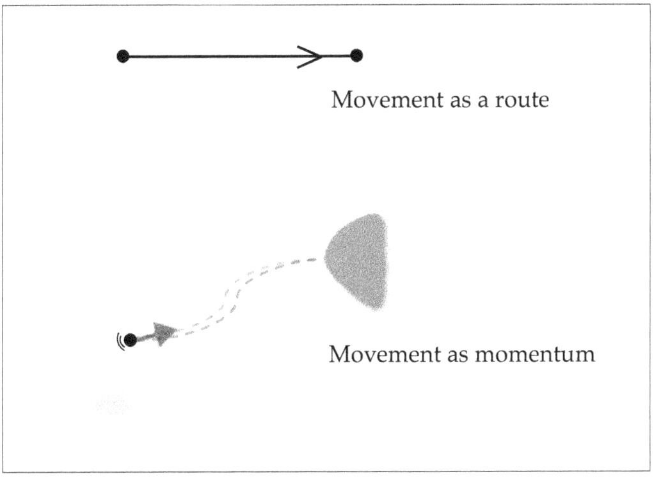

If we consider a group's movement in terms of a route, we can then predict a single destination and so gain a bird's-eye view of that movement. Yet doing so offers no insight into the social aspects of that group. By contrast, when movement is regarded as *momentum,* the destination can be predicted as an area with a certain range through a momentum that appears inseparably in both the individual and the group. Because the intended destination is not limited to a single point and the success or failure of coexistence is not guaranteed, a primordial sociality will emerge in the context of movement as an extreme.

Social forms from the perspective of movement

When we consider the evolutionary basis of human sociality, two different positions are available. One is to consider human sociality as something special that is not possessed by any other living things, and thus to focus on the quantum leap from zero to one. The second is to begin by seeking out a sociality that humans share with other living things, and from there to come to grips with the manner of evolution that brought it into being. This chapter aims at the latter position. It seeks to capture the sociality of both humans and other living things from within the universal qualities that they share in common. In other words, rather than seeking to understand people merely as beings who, as living things that are a step beyond other living things, possess sociality as an added extra, this position seeks to clarify an underlying sociality in order to understand the sorts of diversity that have developed on that basis.

For most animals, including humans, movement behavior is regarded as an essential condition for survival. Moving makes effective feeding possible, as well as survival by escaping from predators. When we focus on movement behavior, we can see how sociality appears in the context of a basic behavior that supports day-to-day survival, one not conventionally regarded as a social behavior. In this chapter, I would like to confirm that the social behavior by which groups are constituted, rather than being a behavior that could be clearly distinguished as either affiliative or antagonistic, is instead something that appears primordially in the context of behaviors that have been considered to be only indirectly related to social formation itself, such as movement and feeding. By focusing on movement, which for many animals is a behavior that is essential for survival, I explore the possibility that *sociality as an extreme*—which is to say the moment when social behavior occurs in relation to others—appears in the organization of movement in a group. Here my use of the term 'extreme' does not indicate a marginal or excessively harsh situation but, rather, an observation of repeated daily group movements from a minimal point of view. I would like to investigate how sociality as it relates to movement is realized through societies in which primate groups, especially mixed species associations, maintain loose connections while undergoing repeated encounters and separations from each other.

Grasping the moving environment

The movements of animals, including humans, have been widely studied in the fields of ethology and ecology and have yielded a variety of findings. Modes of movement vary by the animal species in question, and the nature of the ecological significance one finds depends on a combination of those animals' characteristics and the environments they inhabit. Moreover, even within the same animal species, the significance of movement will be utterly different depending on factors such as habitat, age, or sex.

Typically, background factors raised in the context of animal movement research include population dynamics; spatial redistribution; home ranges, territories, and groups; group movement and dynamics; informed dispersal and prospecting; memory; individual condition; food provision; energy balance; and encounter rates and patterns (Hooten et al. 2017). These various factors are intricately intertwined, exerting mutual influence on each other to determine animals' movement behavior. As is evident at a glance, these movement-related topics straddle several temporal and spatial scales. Movement behaviors vary in scale from the viewer's perspective, with characteristics that are established at various levels from the macro to the micro, such as, for example, global movements, movement within populations, group movements, home ranges and territories, individual movements, or the mental processes that movement entails. For example, with regard to movements such as bird migration, these routes attract attention at the global level, often on the long-term time scale of an entire year. Conversely, when considering the day-to-day behavior of individual small mammals at the neurophysiological level, an observer may be interested in movements that take place in a maze placed in a cage on a time scale of only tens of minutes.

The development of movement behavior research, similar to the study of other behaviors and perhaps to a greater extent, has been supported by methodological developments. Traditionally, an observer wishing to learn the behavior of a given animal would have had to follow it constantly in order to directly observe the movements of an individual. Animals make steady progress, traversing high mountains and deep valleys while occasionally slipping through densely tangled undergrowth. And with animals that move at high speeds over long periods of time, humans simply lack the physical abilities necessary for direct pursuit. In addition, the observable range for animals that move through the sky or sea is, in the first place, limited. Since the 1960s, observers have made active use of radio telemetry to follow the movement behavior of animals (primarily medium- and large-sized mammals). The position of a given individual can be measured by attaching a small transmitter with something like a collar and then receiving transmitted signals remotely. Another option that can be used is biotelemetry, which involves implanting a transmitter in the animal's body to continually measure biological data such as body temperature and heart rate. This method was actively used for primates by a research team lead by Masao Kawai (Mizuno et al. 1976).[1]

Developments in global positioning system (GPS) telemetry in recent years have substantially improved effectiveness in estimating positions through the use of radio signals. With radio telemetry, observers could only estimate the position of a given individual within a certain broadly defined range. Newer GPS techniques, however, give a target's position as coordinates with a relatively high degree of accuracy. And as well as being compact and lightweight, GPS devices also offer many other advantages, such as the potential to be equipped with onboard data storage or the ability to obtain large amounts of recorded data by collecting it remotely or retrospectively. Some aspects of in vivo implantations of transmitters for the collection of biological data, particularly 'biologging,' have contributed significantly to the fields of ethology and physiology. In parallel with measuring position using GPS telemetry, observers are able to collect data at desired regular intervals concerning the body's internal environment (e.g., body temperature, heart rate, and blood pressure) and the extracorporeal environment (e.g., temperature, atmospheric pressure, and illuminance), which can be recorded and stored in a compact device. By installing an accelerometer, it is also possible to perceive and quantify phenomena such as the way that birds move their wings. This has proven to be particularly effective with marine animals that had previously been difficult to observe. It has also yielded many findings that have helped to elucidate mechanisms such as diving behavior among penguins and fluctuations of body temperature over long-distance movements by migratory birds (Yoda 2018). In terms of the environment that surrounds living things, the outside world can be captured from an animal's 'viewpoint' by attaching a camera to the biologger. Doing so has made it possible to learn about a variety of ocean-based behaviors on the part of marine animals for which detailed observation had not previously been possible.

In addition, advances are also being made in drone-based data collection, greatly expanding a world that until now has not been available to direct observation. Furthermore, in recent years, movement behavior has also been observed using camera traps to estimate population sizes and densities. Camera traps are a tool for investigating the behavior of animals passing by a fixed location (O'Connell et al. 2018). Observations made using multiple camera traps will, to a certain extent, clarify how specific animals make use of their habitat areas. In other words, fixed cameras allow us to learn how animals move; instead of actively pursuing the animal in question, as in the case of telemetry, this technique involves passively perceiving the state of an animal at a predetermined point. Moreover, the development of analytical methods that use simulation models is also attracting increased attention. The movements of large groups made up of birds, insects, or fish is developing into a group-based information science by enlisting not only ethologists, but also physicists, neurologists, and computer scientists. In this context, a group's movements is thought of as being formed as the result of some kind of information transmission between its constituent individuals.[2]

Such diverse analytical studies of movement behavior tell us how animals perceive environments and make choices about the environments they inhabit. Studies of movement behavior represent studies of environmental adaptation, and various methods are used in order to capture interactions between environments and living things. And the fact that living things interact with their environment is basically synonymous with the fact that they are alive (see Itoh's discussion in Chapter 2). Arguably, the development of methods in the study of movement behavior could also be said to represent an effort to somehow isolate the very fact of an animal's being alive. Detailed trajectories of animal movement captured with GPS telemetry and records of muscle movements obtained by biologging are records of interactions that living things have had with situated environments. Movement behavior data for understanding the mechanisms by which an individual selects its environment will naturally also include an animal's internal state alongside elements from the external environment. This is because movement data will also include environmental data such as wind speed, water depth, or incline, and information about what exists in a given location.

Although movement is a question of how living things perceive environments and choose environments to inhabit, for animals that belong to a group there is yet another important element in the environment to be selected; namely, other individuals in the group. Just as elements such as sunshine, terrain, and ocean depth are inextricable from the behavior of living things that move through that environment, to the eyes of a moving individual, other members of the group will appear as yet another environmental element. If we regard the transmission of information with other individuals as an interaction that influences the shape of a moving group, we could say that movement is deeply implicated in the generation of sociality at the most primordial level of animal life. This is a sociality that is contiguous with how living things perceive their environment and move about while inextricably engaged in interactions with that environment (see Kawai's discussion in Chapter 9).

The organization of movement

Focusing on primate groups, I would first like to consider movement in principle as a mode of animal behavior. What I am interested in here is not movement as data obtained from GPS or camera traps, but movement as a specific action that people can get a real sense of through observation or firsthand experience. In order to give equal treatment to movement for many living things, I extract common elements shared by a variety of movements. The constituents of movement behavior extracted here are by no means the 'shadows' of abstract movement; rather, I want to consider them as being the minimum necessary elements when observing movement itself, elements that can be seen when observing movement with as few preconceptions as possible.

Movement as a route: position and timing

The most basic way to describe how animals move is to mark the beginning and the end of that movement. The beginning and end constitute two points, and the route connecting those points represents the distance over which movement occurs. Groups are not always moving, but after staying at a certain location for a certain amount of time, they will move on to the next location. In order to know a group's movement, we have to specify when and where that movement begins, and when and where that movement stops. What is revealed thereby is movement as a route that links two points, and when and where this movement occurs is important, as is its *position* and *timing*. In addition, as I shall describe later, animals' basic survival activities consist of their interactions with the environment through movement, which is reflected in this movement as a route.

When observing primate groups, just as with humans like ourselves, behavioral patterns—what might be called daily rhythms—become evident. Monkeys settle on relatively safe spots to take shelter and sleep for the night. An ideal observation would involve waiting until the monkeys arrive at their sleeping site and finish their day (Anderson 1999). By returning to the same place early the next morning, a lucky observer would be able to encounter the same group under observation once again.[3] For example, among the Cercopithecines of Côte d'Ivoire, groups setting out from their sleeping site early in the morning end up feeding along the way. Sometimes they will eat a variety of fruits in a variety of places, and at other times they may gather together and consume a large amount of ripe fruit in one single tree. During these fruit feeding periods, they will also eat leaves and insects, and roam relatively actively. Later the group will settle down to rest, and individuals in the group will begin grooming here and there in a relaxed manner, or lie on their stomachs on a thick tree branch with all four limbs dangling—to all appearances, sleeping soundly. In this state, the entire group abides quietly, not moving. During this time, observers might organize their notebooks, eat sandwiches they packed that day, chat with each other, or doze off in the company of the monkeys (while, of course, continuing to collect data at predetermined intervals). After a certain amount of time spent resting, the monkeys will begin to show the slightest signs of moving off. Then, all of a sudden, several monkeys will begin moving in the same direction, signaling to the observers that the movement of the group is underway. Continuing the observation to keep up with the monkeys, as their pace increases or slackens, the monkeys will be seen to break up into loose clusters and change locations before the entire group comes to rest somewhere, once again not moving. After several repetitions of such intermittent bursts of movement, the troop will eventually arrive at the sleeping site.

In this way, a typical daily activity pattern often constitutes a chain of moving – feeding – moving – resting – moving that is repeated in a constant rhythm. This pattern and rhythm consists of movement bookended by activities that are essential for survival; namely, feeding and resting. Although movement does not produce any benefit to survival in and of itself, interactions with the environment that are achieved as a result of movement occasion energy consumption and the avoidance of danger and fatigue as essential benefits to individual survival.

Of the reasons why animals must move, feeding has a significant impact on survival. Not only must animals eat, but in order to eat effectively they must move. Movement behavior is strongly associated with foraging behavior. Therefore, the route that determines position will be deeply implicated in movement, especially in the timing of when that movement starts and stops (Boinski and Garber 2000).

Given the distribution of food in the environment and of the living things that eat it, it will be possible to estimate optimal distance in terms of how many places to forage in over the course of a day. For animals, efficiency will be maximized when moving over a distance at which the benefit obtained by foraging outweighs the cost of movement. Likewise, the same efficiency may be calculated with regards to the duration time for movement. How long animals will stay in a single food patch can be predicted based on the potential benefit to be gained there and the cost expended by the stay. With regard to the timing of when movement will resume after stopping, we can estimate that the troop will stay put for as long as the benefits obtained foraging exceed the costs of movement, and then begin moving again once the benefits fall below the costs. Many detailed observations and theories have been advanced with regard to the optimal movement associated with the benefits of feeding from the viewpoint of feeding ecology (e.g., Nakagawa 1994). Choices that influence behavior, such as how far to travel in a day, how long to stay in any one place, and how frequently to repeat movements, will not always have the same value for all individuals in the group. Larger individuals and pregnant or lactating females naturally will not be able to ingest enough calories for survival without eating a lot. Conversely, younger and smaller individuals may not need as much food. Those at the top of the dominance hierarchy will have priority access to food, and so be able to eat a lot at each food patch, even if not many patches can be found. Where subordinate members end up having to compete for food, however, they will not be able to eat enough without going in search of the next food patch. In other words, for monkeys living in groups, the stakes of movement at the level of the group will not always match those for particular individuals.

The choice of a route is also influenced by elements other than feeding behavior itself. Territory for protecting resources from neighboring groups also has a powerful impact on movement behavior. For example, in the case of a territory that contains food resources, food patches within the home range (especially those situated near the territorial border) are sites that must be visited and defended on a regular basis. When a neighboring group crosses the border into another group's territory, members of the defending group will confront the intruders to make their presence known and use display behavior and raised voices to drive them off.

Another important activity for survival is rest. It is important to be able to get adequate rest safely. Food contributes to survival only when you have time to digest it and convert it into energy. For example, for colobuses, which subsist primarily on leaves largely made up of digestion-inhibiting components, rest to assist digestion is an important part of daily activity. For other monkeys, as well, digesting the food they have eaten to convert it into energy and sitting out the hottest hours of the day by suppressing wasteful expenditures of energy are behaviors directly linked to survival. Nighttime rest is particularly important for diurnal monkeys. Resting is not achieved simply because the monkeys happen to stop other activities at a given location; rather, it is made possible by moving to specific resting sites. Just as they search for feeding sites, monkeys must also move in search of suitable resting sites. Such sites must be free from predator attacks and, moreover, must allow the monkeys to protect themselves from inorganic natural threats such as high temperatures, precipitation, and strong winds.

Nor do movement-related factors such as those described above operate independently of one another. The location of a resting site depends heavily on previous feeding sites, and avoiding predators—which is crucial for rest—must also be accomplished while feeding. Even while patrolling territorial borders, individual group members will continue feeding, or else interrupt their feeding and hurry over from a great distance in order to address an emergency posed by an incursion by a neighboring group. All these factors work in combination to determine a route.

To think about movement behavior comprehensively from the interplay of these various factors is also to understand it using a model that can be calculated retrospectively based on the costs and benefits of movement. We might also say that an understanding of the kind of benefits that can be gained by movement and the sorts of opportunities that are potentially and practically lost is something that can only be gained through a bird's-eye view on position and timing. This cannot be linked with a perspective that seeks to capture how movement actually occurs, when and where movement begins, which direction it takes, and where it stops as phenomena in and of themselves. Also, we could say that movement understood by way of a route is only partial in the sense that if focuses only on the start and end points and is not concerned with the route taken or events that occur along the way. The average movement between the starting point and the destination is re-examined afterwards. Although movement as a route or distance can be an index for evaluating the environment in relation to other activities, when focusing on movement as behavior in and of itself, it means that we must limit ourselves to what is ultimately only a partial approach. And, as mentioned earlier, the results of interactions with the environment that give rise to movement behavior are not necessarily produced equally for all members of the group individually. Group movement, which is influenced and determined by multiple factors, is realized at a level that transcends the respective interests of individual group members. Here we come to the great universal mystery concerning group movement: namely, the question of how the behavior of disparate individuals is controlled in the context of group movement. This question probes the integration of parts with the whole, and is deeply implicated in the problem of the nature of sociality in the context of group, especially the problem of the synthesis of the two levels of the individual and the whole.

Movement as momentum: speed and direction

How might we conceptualize movement behavior holistically without using the idea of 'route' as an approach? When observing animals' movement behavior, two other factors that seem important, aside from position and timing, are those of *speed* and *direction*. Indeed, in the very middle of a moving group, it is the manner of individuals moving at a high speed in a certain direction which indicates the fact of that movement. We could say that it is by virtue of speed and direction that a substantive depiction of how an animal moves becomes possible.

When the starting point and ending point are determined as a route, speed and direction are calculated retrospectively as the average value from start to finish. Conversely, the speed and direction of movement at a given moment constitute not an averaged movement, but movement as it appears in the instant it is seen, when the observed span of time has been minimized (i.e., reduced to an extreme point). When we try to understand movement with a route linking start to finish, that movement can only be grasped retrospectively, which only partially explains the movement itself. In contrast, the entirety of movement behavior appears in an instant as speed and direction. In the moment that moving animals or groups are seen, the entirety of their movement is established as a mode of behavior that expresses where they are going and how they are trying to get there. When pursuing a group, even if the entire group of individuals cannot be seen as a whole, the speed and direction shown by an individual or a limited number of individuals at any given moment will inform the observer that the group is moving.

Speed is the average distance traveled per unit time when changing position from one point to the next. Speed as an average of distance traveled is calculated irrespective of how one actually moves from one point to the next. Conversely, when the time span in this calculation is minimized, instantaneous speed is obtained as a tangent at a certain point on the trajectory. At this minimal extreme, the tangent vector represents instantaneous speed. Along with expressing this kind of physical formula, the moving individual indicates speed and direction, allowing the observer to understand that the individual is moving at the moment it is seen (see Nishie's discussion in Chapter 8). Here I would like to refer to the instantaneous speed and direction that provide the observer with clues to movement as the 'momentum' of that movement. Momentum can be seen for an individual or could also be perceived for the entire group. The entirety of the movement is perceived as a momentum as it appears at this minimal extreme. While this momentum will be perceived by the people observing the group's movement, is it conceivable that it might also be sensed by the individual monkeys (as social animals sharing the same environment) being observed?

As we saw in the previous section, one of the great mysteries of group-based movement is the issue of how individuals capable of free movement come to terms with the group to establish movement on the part of the whole group. If each individual were to begin moving according to its own schedule in a direction of its own choosing, it would be impossible to maintain group cohesion. While movement is the result of a decision-making process about the kind of environment to choose and how to live therein, in the context of interactions with the environment, there is another process wherein individual decisions are somehow coordinated into a single movement on the part of the entire group. In the context of movement, this process is regarded as a form of collective decision making, and various mechanisms for structuring various positional relationships have been considered (Boinski and Garber 2000; Fisher 2009). When we look at the interactions within these mechanisms, we find that it is the social interactions between individuals that connect the two different levels of individual and group. These interactions between individuals manifest as an information-sharing process that is perceived as momentum.

One of the oldest mechanisms that has been considered for overcoming the contradiction between individuals and groups in the context of movement involves group members who follow decisions made by a leader. The roles of leader and follower are differentiated within a group, and the direction of movement and when to start and stop are determined by individual behavior on the part of the leader. Other individuals pay attention to and synchronize their actions with the leader's behavior, which has the effect of determining the movement of the entire group. Leaders may be determined by virtue of their dominance or sex within the group. One role of the leader is literally to lead the movement of the group. Although individual members of the group may think in different directions, as long as the other individuals follow the direction of the one who is leader, the group will be able to move as one. The leader determines the timing of the start of the movement, and the other individuals will begin moving to follow the leader's movement. When the leader stops moving, then, at that point, the other individuals will also come to rest. However, this kind of leader-driven integration process is only rarely seen in the scientific literature on animal behavior. Among primates, it is said that 'the alpha individual' does not exist, and the idea of top-down processes of decision making with regards to movement tends to be rejected.

Another, more democratic, alternative to a leader is that of a majority decision. Even if some individuals begin moving, movement on the part of the entire group will not take place unless many others also begin moving. Accordingly, the first individual to begin moving will have no choice but to return to where it started and stay put, giving up on the idea of moving. If, after several repetitions of such exchanges, more and more individuals take the cue from the individual starting the move, a majority of the group will begin moving, and when the number of individuals not following this lead becomes a minority, we can say that the group has started to move. When a minority leads by starting, determining a direction, and finishing, it has been observed that 'back and forth' interactions will take place and gradually converge into a majority before the individual members will follow that lead.

In the case of movement in which other individuals follow a leader, or of movement in which the timing and route are determined by the majority, a variety of information relating to movement will be exchanged in interactions between individuals. For example, the posture of an individual's body or the noises it makes when it begins moving and the location and speed of movement all comprise valuable information for other individuals. Furthermore, there have also been observed cases of peculiar vocalizations emitted only when moving, and of changes in the frequency of contact calls to maintain proximity. These are elements of social information that individuals use to move in concert with the group as a whole. In the context of movement, these social exchanges between individuals who are adjacent to one another constitute a collective decision-making process.

In addition, we also find examples in which this social process has been studied by taking up specific modes of movement for particular species. A famous example in the context of primate research, especially studies of chimpanzees, is the study of processions. When crossing roads (often artificial roads) running through their habitat area, individuals in a group have been observed to form rows and move in turns (Itani 2008; Nishida 1994; Hockings et al. 2006). The composition of such processions is sometimes said to have a structure that reflects the group in terms of a periphery and center: in the open environment of a road, where a group would tend to be defenseless, young individuals make their appearance first, followed by adult males, then females and offspring positioned in the middle so that they will be protected by the adult males, and then finally the young adults of the group. A spatial arrangement that reflects this orderly social structure cannot be separated from the environment of the road, which compels them to form a line. A group that forms a procession to move, when not moving along a road, will often have a more ambiguous structure, with individuals each moving on their own in the woods. Baboons are another example of primates that adopt complexly structured processions. Baboons are known to form composite groups consisting of multiple one-male units in which a lone male is followed by several females. The spatial arrangements involved in this complex movement can become very complicated, but it has been confirmed to be made possible through social interactions with nearby individuals and involves following relatively simple rules (Farine et al. 2016).

For a group of individuals to move as a loosely integrated unit that is constantly changing its shape in various ways is not something that is limited to primates. Photograph 1.1 shows a flock of ancient murrelets (*Synthliboramphus antiquus*) flying together while maintaining overall control as a group. In such vast groups of birds or fish, various modeling studies have attempted to discover the basis of their collective unity, which has led to the creation of an artificial life simulation program known as 'boids'—a term coined by Craig Reynolds (1987), who referred to 'simulated bird-like, "bird-oid" objects generically as "boids" even when they represent other sorts of [organisms] such as schooling fish.' Boid simulation shows how disjointed individuals can flock together and move like a natural group using only three rules of operation: namely, separation, alignment, and cohesion (Gunji 2013). Since its proposal, the boid approach has undergone various improvements and, as well as being used to depict crowds in entertainment media such as films and computer graphics, has also become the basis for model-bases studies of animal groups.

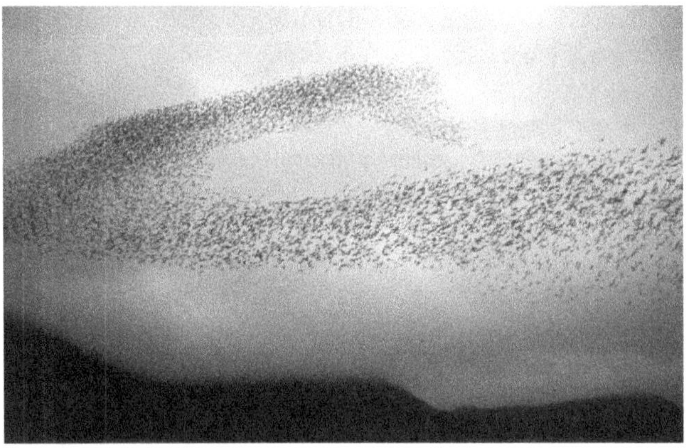

Photograph 1.1: Ancient murrelets flocking in a murmuration, in which a vast number of individual birds move while maintaining control of the whole. The boid simulation attempts to reproduce this kind of group movement solely with the three movement rules of separation, alignment, and cohesion.

Yukio-Pegio Gunji (2013) uses the terms 'prediction' and 'anticipation' to express how groups become possible, based on group simulation model-based studies such as boid. According to Gunji, whereas prediction is the calculation of a single future point or value, anticipation is a future-oriented calculation that cannot be determined as a single point or value, but only as an area or span. Contrasting things in this way, we can regard movement-as-route as being a prediction calculated based on the history of movement thus far and the knowledge and memory of the habitat as a whole. Here we can calculate the costs and benefits of movement, and a given point is indicated as a destination as one of two choices to either hit or miss. Conversely, anticipation does not indicate a single point, but appears as a span or area with a certain degree of possibility. Anticipation cannot point to a destination; all that is indicated is an ambiguous and broadly defined future destination of movement. The anticipated future destination of the group appears as a bundle of possibilities defined by 'going in that general direction.'

Both route and future destination indicate the mode of movement. Both are calculated on the accumulation and memory of interactions with the environment realized through past movement. However, the future that lies beyond that calculation is vastly different. While one might grasp the route given complete information and access to a bird's-eye view of the environment, these things are not attained in interactions with the environment in the context of normal animal behavior. Accordingly, it is thought that movement of a group in the context of movement is something that can only be realized by anticipation, rather than by prediction. The individuals that comprise the group each exchange social information with nearby individuals, using cues such as posture at a given point in time, the movements of each part of the body, and vocalizations. Based on the information and memories obtained from these, the optimal possible behavior for the future will be anticipated as something that leaves open a variety of possibilities rather than predicting by converging on a single given option. This process involves the anticipation of a broad bundle of possibilities that are triggered directly and instantaneously through social interactions. That is, although individuals that interact with others within their range of sight will make choices that are available to them individually, the fact that these are grounded in anticipation means there is no full guarantee of the instantaneous achievement of harmonious or concerted action. Social information perceived between nearby individuals in these momentary exchanges is the momentum of movement as a vector that has a direction at its minimum extreme. Surely we could consider the site to which this points as the group's future destination.

Extremes of movement and mixed species associations

The phenomena of mixed-species groups known as 'mixed species associations' occur when groups of animals from different species come together to form one large group that moves and feeds together. I have observed mixed species associations of Cercopithecines in Taï National Park in Côte d'Ivoire, West Africa (Photograph 1.2; Adachi 2013: 31).

As mentioned above (in the section 'Grasping the moving environment'), the nomadic pattern of a group of primates over the course of a day is determined to some extent by behavioral rhythms, and mixed species associations are no exception. With the mixed species associations of Cercopithecines in Taï, as well, the character of their daily nomadic range could be said to be more or less fixed. These species which form associations are often of a similar size and shape, and their diets overlap to a great extent. Consequently, they interact with the environment in a similar manner. In an elaboration of his own theory of biological society, the Japanese primatologist Kinji Imanishi coined the term 'specia' to refer to a species society that includes all the individuals who share the same life form, and adopted the ecological term 'synusia' for multiple species society with overlapping and similar life forms (Imanishi 1949; Niwa 1993; Adachi 2013). The life form is a concept that refers to the inseparable connection between a subject and its environment, indicating the whole that encompasses both the individual that uses the environment and the environment that is used. Mixed species associations consist of species with overlapping life forms, corresponding to what Imanishi called 'synusiae.' In this type of synusia, individuals of different species form a group and move

Photograph 1.2: An example of a mixed species association among Cercopithecines

as one according to their life rhythms, coordinating with each other in their exploitation of environments. When observed continuously for a certain period, the life rhythms of a mixed species group will be observed to stabilize, reflecting the stability of the environment and shaping the pattern of movement as a result. Mixed species associations appear to be so consistently and coherently established that observers can predict their future movement patterns based on the accumulation of data regarding past behavior.

In order to think about mixed species associations as a meaningful phenomenon, and as a mechanism for dismissing the possibility that mixed species associations are (only) together by chance, and possess no social significance, a null hypothesis model for mixed species associations has been proposed (Waser 1982). Assuming that groups composed of different species will each move randomly, this model calculates the probability that their ranges will overlap by chance, and compares this with empirical data on the rate with which mixed species associations occur in reality. Here the analogy of gas molecules in their ideal state is used to model the random movement of groups; in addition to the molecules' size and velocity, the model calls for a container entirely filled with gas. These correspond respectively to the size of a single-species group, the speed of the group's movement, and the overall size of its home range. The null hypothesis model, having specified in advance the range within which movement is possible, takes a bird's-eye view of movement by conceptualizing a group's movement by analogy with molecules that move about freely. In the case of mixed species associations of Cercopithecines in Taï National Park, the actual prevalence of mixed species associations is much higher than the rate obtained from this kind of modeling (Adachi 2013). In other words, this means that rather than moving freely like gas molecules, these groups cooperate and move together in an integrated manner at a rate beyond that which could be put down to chance. While here it is groups, rather than individuals, that are likened to molecules, the same argument could be applied to relationships between individuals. In the null hypothesis model, a simple simulation calculation is performed by assuming individuals (or single-species groups in the case of mixed species associations) to be randomly moving molecules. This is a similar approach to group movement as in models typified by boid simulations. Through a comparison with the model, the phenomenon of mixed species associations as they actually occur can be grasped as non-random, meaningful movement.

In mixed species associations as social groups, as well as the two different levels of individuals and groups, there is also a third level comprised by the multi-species group. The movement of a mixed species association, as well as producing a disjuncture between the interests of individuals and the interests of the group, also achieves integrated movement as a whole, as though it were a single group, while addressing conflicts of interest between different species with respectively differing characteristics. In other words, it seems that mixed species associations form beyond the level of differences between individuals or even of differences between species, as well as beyond the level of differences between these two levels. When analyzing movement on the part of mixed species associations, which is accomplished by overcoming these three different levels of individual, group, and species, two orientations are available to the observer, depending on how he or she perceives the gaps between levels (Heymann 2011).

One approach is to disregard the species level. Generally, mixed species associations are known to form among sympatric species that have similar ecological requirements. Thus considered, individuals of different species in the context of a mixed species association can be thought of in terms of having a somewhat extended version of the differences that occur between individuals within a single-species group. That is, they are able to forage for food as a single group precisely because they have similar ecological needs. Here the interests of individuals should be integrated into the group, regardless of species. Another approach is to disregard the level of the individual by placing the emphasis on competition for benefit between species and options for avoiding the same. This is a way of looking at mixed species associations as a form of behavior in which, although competition arises between similar species for the same type of food, a single-species group will avoid competition with the different species group as a whole to gain an even greater benefit. This perspective is the basis for the attitude of seeing a single-species group as an individual. Regardless of which of these perspectives one adopts, they are alike in that individual differences are integrated into a single moving whole. In the case of mixed species associations, there are two different levels of the individual, and the composite complexity of individuals is evident.

Here I would like to confirm what the observer actually sees when observing a troop of monkeys. For example, what I did when observing mixed species associations was as follows. When observing moving groups, I would wander around inside the group to determine the center of the group and then observe movement by plotting this as a point (or a certain area) on a map. That is, I would walk around to ascertain which individuals represented the periphery of the group as a whole, and then perform a mental calculation to work out the center point of the area bounded by those individuals. The path linking the points drawn on the map at regular time intervals (e.g., every ten minutes) would thus represent the group's route. When actually observed, single-species clusters will be present to some degree, and the center point of these various single-species groups will be the center point of the mixed species association as a whole.

When recording position at ten-minute intervals, each time I completed the task of checking the periphery to calculate the center meant that I had to repeat the same task ten minutes later; however, it was not always the case that I had to walk around to confirm the group's periphery by actual observation. For example, in the heat of the day, if the individuals within my visual range were stretched out and dozing in the treetop branches, and when I no longer heard any sound of leaves or branches rustling as the monkeys moved around in the treetops, then, as the observer, I would also settle in and mark the same spot that I had entered into my field notebook ten minutes earlier. In other words, the observer senses that the group is not moving. A group that has been resting quietly will also resume its activities at some point. As soon as several individuals in the surrounding area begin shaking the branches, several animals within visual range will begin moving in the same direction, and when it is time to record, the observer will stand up and begin moving closer to the periphery of the group that is producing sounds. If, by walking around, the observer determines that a majority of the group has begun moving in the same direction, then he or she will have to walk the periphery again to confirm that the center point of the group has moved. However, even without always walking around the entire group like this, an observer will often be able to make a conjecture about the direction and timing in which movement will begin with some degree of accuracy. And while this conjecture occasionally misses the mark, it proves correct more often than not.

How can an observer make conjectures about a group's future destination? What signs intimate that the group will soon begin moving, that it will almost certainly go in a particular direction, or that it will come to rest at a particular spot? Confronting a group of monkeys, the observer watches to see which types of individuals are moving with which posture at what speed in which parts of the trees with which other individuals. The momentum of the behavior of individuals will be conveyed the moment they are seen. Although this is something that is in principle perceived separately from whether or not the entire group is moving, in reality it is deeply connected. The observer accomplishes his or her perception of movement behavior by alternating freely between the two scales of individuals and the whole. A similar perceptual process seems to occur in the exchange of information among the individuals that constitute the group. This exchange of information that takes place among individuals is what makes the group's movement possible, and it is the momentum of movement that constitutes this shared information. This is momentum at an instant that we might call the minimal extreme of sociality, one that fluidly connects scales that exist at different levels to make normal movement behavior possible. Although this momentum is primarily observed at the level of each individual, in the sense that this momentum is linked to the whole and cannot exist independently of the group, it is also the momentum of the group.

Various differences are evident in momentum as it is perceived in groups or individuals, depending on time, place, and class of the individual observed.

Numerous patterns can be extracted from observations of mixed species associations, and observers have found different momentums in the respective movements of each pattern. One such pattern is a predictable everyday pattern. As the group moves over a certain range, there will be some individuals who move and others who linger, each being seen to move back and forth. This is a pattern in which individuals in a group, while apparently making continuous and repeated detours and seeming to go off in arbitrary directions, are gradually corrected in their interactions with one another so that they finally converge in a single direction, exhibiting the semblance of a kind of majority vote. Here, various bits of information are perceived, such as individuals' vocalizations, postures, and bodily movements, what they are doing, and who they are with, and from these an observer will be able to predict what will happen next. In a very simple example, in the case that only young animals begin moving in a certain direction away from their mothers, who remain stationary, then we can conjecture that even if several individuals begin moving, they will usually come back. Here the momentum created by the combination of the site and individuals is perceived alongside the accumulation of past behaviors, making it possible to conjecture about the future. Of course, such conjectures will sometimes miss the mark, and it may be that before long the mothers will rouse themselves to begin moving in the same direction as their young—but, at the very least, the observer is able to make a conjecture. In most cases, momentum will be formed over time by many individuals, gradually determining the direction of movement.

However, there is also a special kind of momentum that occurs only occasionally. For example, a group that has been proceeding straight in a particular direction might change its direction sharply in a short amount of time (sometimes even returning to its original position), or might begin moving rapidly as a whole by suddenly accelerating from a resting or static position so that its speed changes drastically. When plotting this trajectory on a map, the resulting pattern will be an extremely sharp angle. In Figure 1.1, which shows the movement of a mixed species association over a four-day period, this corresponds to the movement that follows the first rest stop of the day (large black circle) along the route on the first day (Figure 1.1a) and the way that the dots are spaced relatively far apart, so that the trajectory line makes a sharp turn at the end of the second day (Figure 1.1b). At such times, there will not always be any convincing reason for such a sharp turn or, if there is, it may become discovered after the fact.

From a bird's-eye view, the former type of movement (i.e., the everyday pattern) traces a gentle curve, while the latter (i.e., sudden changes in direction) traces a sharp corner. As the parties actually performing this movement, how is such movement perceived by the monkeys themselves? Or from the perspective of the observers who are following them? And how would animals lacking in such a bird's-eye view in the first place even perceive such differences in movement? From the perspective of the individuals that make up the group, the only information that is perceived is that which is available on the spot. When we look at the phenomenon of movement from the perspective of a moving individual, it will appear as momentum as expressed by another individual (or several other individuals). This is not movement as a route, but momentum that appears as a tangent at one instant on a route. Momentum is momentary; it is a sign that appears only when the temporal frame of reference is reduced to its minimal extreme. The tangent obtained by calculating a differential is the most primitive form of movement as a route, representing the entirety of movement as it appears only in that instant. Is this not a bundle of potential future behaviors that we could call the group's 'future destination'? A future possibility that should be anticipated? For individuals who live in groups, this bundle of potential behaviors represents a foothold guaranteeing sociality in its most primordial form. As the observer of such a group gains experience, he or she will learn to perceive this momentary momentum as a node between the past and future and thereby predict a future destination so that he or she will become implicated into the sociality of the group.[4]

In the perception of 'momentum' apparent in the movements of mixed species associations, we can recognize a primordial sociality, in that all behaviors occur only in the context of relationships. In the moment when momentum is captured, there is no distinction between the individual and the group; coexistence as a group will be apparent even when looking at a single individual, just as separation and isolation will be apparent when looking at coexistence among multiple individuals. Even when an animal is off on its own, its isolation occurs in the midst of a potential society, and when many individuals come together, this also shows the potential that they will scatter apart in the next moment. Among mixed species associations in Taï National Park, single-species groups form loose connections with each other, repeatedly separating and coming together again (Adachi 2013). The dissolution and

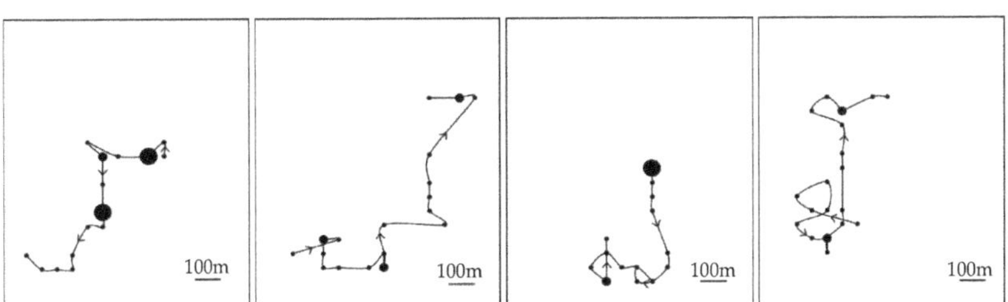

a) 1993/11/4 b) 1993/11/5 c) 1993/11/6 d) 1993/11/7

Figure 1.1: Daily movement trajectories for Diana monkeys (*Cercopithecus diana*) and their mixed species associations (the observation period for all four days lasts from 7:00 a.m. to 5:30 p.m., with the size of the dot representing the time spent at rest)

formation of mixed species associations in most cases takes place in association with changes in movement behavior, with movement serving as a trigger for instances of fission and fusion.[5] In mixed species associations, in particular, because group composition is constantly changing due to these loose connections, signs of the formation of groups can be seen within momentum. A characteristic of momentum is that it is inseparable from the future destination. The future destination is not a simple future, but something anticipated as a bundle of possibilities. And while it is not a fixed or predictable point, the future destination is nevertheless deeply implicated in *some* future. We might say that momentum supports momentary individual existences with a future as a hoped-for future destination. So momentum is established by implicating both the individual and the whole at an extreme point between a selection from a store of past behaviors and an anticipatory future. The momentum of the instant when animals' behavior is brought to an extreme is indispensable for perceiving a group's future behavior. While an extreme is the task of bringing the finite asymptotically close to the infinite, in the context of movement behavior it is instantaneously perceived as momentum and constitutes an important component of the social environment. That is, individuals in a moving group interact with their environment by sharing and exchanging information about momentum. This extreme determines the mutual social embeddedness of living things as they continue to live in groups and interact with their environment.

Inhabiting extremes: niche, anti-structure, and time

Niches are considered to be elements of an environment that influence the choices made by living things. Aside from the inorganic environment, the presence of conspecific and heterospecific individuals are indispensable for the process of niche formation. When the behavior of an animal modifies the environment so that it becomes a new niche that will determine its own behavioral choices, as well as those of other animals, this process is known as 'niche construction' (Odling-Smee et al. 2003). In particular, when this means modifying a social environment—that is, when it involves other organisms and individuals that will potentially interact with one another—it is called 'social niche construction.'

In the context of a group's movement, the future destination will be perceived at the same time as momentum. The form of the momentum is also the shape of how animals select and make use of their environment together with other individuals. The choice of the environment and the way it is used overlap with the area known in traditional ecology as a niche. Although living things move in actual spaces, there are virtual 'niche spaces' that consist of mutually constructed environments, and the path that is taken is determined not only by the material environment but also by social factors. Like water flowing along a narrow and winding waterway, living things slide easily through these niche spaces, sometimes slowly, sometimes as a torrent. Only social beings are qualified to pass through this virtual niche space. Movement patterns that might seem incomprehensible and unpredictable in real space (e.g., sharp changes in direction) may also exist as smooth flows in virtual niche spaces. Movement in groups that transcend individuals seem to be established by using each other's momentum as an element of their environment. And these relationships with the environment, or virtual niche spaces, are supported by a primordial sociality as a momentum visible in a single instant.

In order to survive, all living things have no choice but to interact socially with their environment. This being the case, could we say that all living things inhabit conditions of extremity? Momentum appears for both individuals and the group. We cannot make any distinction between momentum on the part of an individual and momentum on the part of the group. In terms of momentum, individuals and the group are joined together in a single whole through a primordial sociality. At the instant momentum is perceived, in the sense that it is not connected to anything at all, a single animal is neither an individual nor a group. Although clothed in potential coexistence in the form of a future destination, in that instant it is solitary. An animal tossed away at that instant, while being alone as an individual, is connected to society in that it has momentum. Even if that individual will subsequently spend a long period on its own, separated from others for a while, the fact that it is a being with perceptible momentum points to a primordial sociality. In that respect, it can never be fully solitary.

The state that appears at the instant that a span of time is narrowed to its extreme represents the state of inhabiting perfect solitude as an individual with a difference that cannot be recovered by anything, from which all links to society have been lost. At the same time, momentum is a faintly sensed sign of sociality that appears as a fleeting hope. It is a state of primordial sociality in the sense that even if its structural parts—the group or species—were to be lost, the individual would retain the sensation of still being connected to something far away. The anti-structure that Jun'ichirō Itani found in the sociality of mixed species associations and solitary monkeys among Japanese macaques may be a term for expressing this primordial sociality (Adachi 2013). The solitary state of lone monkeys separated from the larger group, when it is clothed in the momentum of being a 'lone' monkey, points to a connection that will take place with other monkeys somewhere as a future destination. The sociality of anti-structure is precisely this faint sign of connection that can be seen ahead as a future destination. Anti-structure expresses coexistence as a potential. This is why loose connections such as those in mixed species associations are regarded as anti-structure. Further to an anti-structural line of thinking, solitude is a pattern of connection, and being solitary at a distance is itself already a kind of society. Ecological niches loosely point to the functional role they play in the environment, and they could also be said to be a means of showing relatedness as potentiality (Adachi 2017).

The acts of coming loosely together and moving loosely apart are bound together by momentum as an apparent position at a given moment in time. This looseness expresses the fragility of (possible) breakage and the perfectly flat and equal relationship free from hierarchy in the context of the particular moment. In mixed species associations, relationships with other species that are similar but different appear as an even and flat moment of connection. Such characteristics may be why Itani regarded the 'communitas' of the liminal phase of Victor W. Turner's (1970) theory of ritual as being identical to anti-structure. The liminal phase refers to the phase in which a lack of connection provides assurance of future connection, as the light of re-aggregation that is to follow afterwards. Anti-structure is 'disconnectedness' as momentum, a peculiar and unstable connection that is observable in periods of ritual transition. The aspect of anti-structure that appears as a sign of primordial sociality rather than a rule-bound structure is captured therein.[6]

Animals are ephemeral. They lack the capacity to reflect on the past or look into the future, living only in the present of the here-and-now. Humans are said to be the only exception. Humans are able to look back over their past history, pass on that history as stories, and make forecasts by looking to the future. And yet we know that considerable diversity can be seen with regard to how humans interact with their social and physical environments when faced with the need to look to the future in the practice of daily life (see Kawai's discussion in Chapter 9). This chapter is a discussion of movement, but at the same time it is also a treatise on how animals living in the moment inhabit the present as a node between past and future. Moreover, it is also an attempt to explore how we can talk about time and history—concepts that have been regarded as unique to humans—from the perspective of animal society. Rather than depicting the societies that animals inhabit as a debased form of human societies, by perceiving the momentum of movement behavior that appears in extremes my aim has been to pioneer an approach to the primordial sociality that is common to both.

Notes

1. In recent years, telemetry has also been widely used in the context of environmental conservation and countermeasures against wildlife damage (e.g., Ueda et al. 2018).
2. A new field of study is developing in which research into mechanisms of information transmission in animals' movement behavior is linked with neural networks and examples of so-called 'big data' networks through the accumulation of human-centered digital data (examples are known of the recording of behavior using information services such as smartphones) (Fukuyama and Hato 2015). This field is also associated with the development of methods of deep learning and pattern recognition, and is involved in advances in robotics.
3. Among mixed species associations of Cercopithecines in Côte d'Ivoire, monkeys would move about on clear nights when the moon was full, often reuniting in the morning at a site far removed from the site where they had parted ways the previous day.
4. The discussion of group formation in Gunji (2013) demonstrates that what is apparent at the level of the group as a whole and what is apparent locally, while differing decisively, are at the same time smoothly connected. This is a stance that conceives of an individual's independence (freedom) as a thing and the totality recognized in the group as an event. Here the thing and event exist simultaneously without contradiction, engendering the sociality of the group.
5. Among chimpanzees, for whom fission and fusion are a normal state, movement plays an important role when members of a group come together or separate (see Hanamura's discussion in Chapter 7).
6. Imanishi's argument, which conceived of a biological society from the concept of a life form in terms of an organism's being inseparable from its environment, overlaps precisely with the argument of anti-structure, which is established as this kind of individual momentum (Adachi 2013).

References

Adachi, K. (2013) 'The sociology of anti-structure: Toward a climax of groups'. In K. Kawai (ed.), *Groups: The Evolution of Human Sociality*. Kyoto University Press and Trans Pacific Press, 21–42.

Adachi, K. (2017) 'Living one's role under institution: Ecological niches and animal societies'. In K. Kawai (ed.), *Institutions: The Evolution of Human Sociality*. Kyoto University Press and Trans Pacific Press, 265–286.

Anderson, J. R. (1999) 'Sleep, sleeping sites, and sleep-related activities: Awakening to their significance'. *American Journal of Primatology*, 46: 63–75.

Boinski, S. and P. A. Garber (2000) *On the Move: How and Why Animals Travel in Groups*. The University of Chicago Press.

Farine, D. R., A. Strandburg-Peshkin, T. Berger-Wolf, B. Ziebart, I. Brugere, J. Li and M. C. Crofoot (2016) 'Both nearest neighbours and long-term affiliates predict individual locations during collective movement in wild baboons'. *Scientific Reports*, 6, article number 27704.

Fisher, L. (2009) *The Perfect Swarm: The Science of Complexity in Everyday Life*. Basic Books.

Fukuyama, S. and E. Hato (2015) 'Observation and behavior models for spatial planning in networks'. *Journal of Japan Society of Civil Engineers*, Ser. D3.

Gunji, Yukio-Pegio (2013) *Mure wa ishiki o motsu: Kono jiyū to shūdan no chitsujo* (The swarm is conscious: Individual freedom and group order). PHP Institute.

Heymann, E. W. (2011) 'Coordination in primate polyspecific associations'. In M. Boos, M. Kolbe, P. M. Kappeler and T. Ellwart (eds), *Coordination in Human and Primate Groups*. Springer, 263–281.

Hockings, K. J., J. R. Anderson and T. Matsuzawa (2006) 'Road crossing in chimpanzees: A risky business'. *Current Biology*, 16(17): 668–670.

Hooten M. B., D. S. Johnson, B. T. McClintock and J. M. Morales (2017) *Animal Movement: Statistical Models for Telemetry Data*. Routledge.

Imanishi, K. (1949) *Seibutsu shakai no ronri* (The logic of a society of living things). Mainichi Shimbun.

Itani, J. (2008) 'Chinpanjī o otte' (Tracking chimpanzees). In *Itani Jun'ichirō chosaku-shū dai 2-kan* (Collected writings of Jun'ichirō Itani, vol. 2). Heibonsha, 203-366.

Mizuno, A., M. Kawai and S. Ando (1976) 'Ecological studies of forest-living monkeys in the Kibale Forest of Uganda'. *Kyoto University African Studies*, 10: 1–35.

Nakagawa, N. (1994) *Saru no shokutaku—saishoku seitaigaku nyūmon* (The monkey's dinner table: Introduction to foraging ecology). Heibonsha.

Nishida, T. (1994) *Chinpanjī omoshiro kansatsu-ki* (Thirty-four stories of chimpanzees). Kinokuniya.

Niwa, F. (1993) *Nipponteki jinenkan no hōhō: Imanishi seitaigaku no imi suru mono* (A Japanese way of viewing nature: Ecology in the Imanishian sense). Rural Culture Association Japan.

O'Connell, A. F., J. D. Nichols and K. U. Karanth (2011) *Camera Traps in Animal Ecology: Methods and Analyses*. Springer.

Odling-Smee, J. F., K. N. Laland and M. W. Feldman (2003) *Niche Construction: The Neglected Process in Evolution*. Monographs in Population Biology 37, Princeton University Press.

Reynolds, C. (1987) 'Flocks, herds, and schools: A distributed behavioral model'. *Computer Graphics*, 21(4): 1.

Turner, V. W. (1970). *The Ritual Process: Structure and Anti-structure*. Aldine.

Ueda, Y., M. Kiyono, T. Nagano, S. Mochizuki and T. Murakami (2018) 'Damage control strategies affecting crop-raiding Japanese macaque behaviors in a farming community'. *Human Ecology*, 46: 259–268.

Waser, P. M. (1982) 'Polyspecific associations: Do they occur by chance?' *Animal Behaviour*, 30: 1–8.

Yoda, K. (2018) 'Bio-logging and navigation of marine animals'. *Japanese Journal of Animal Psychology*, 68: 49–56.

CHAPTER 2

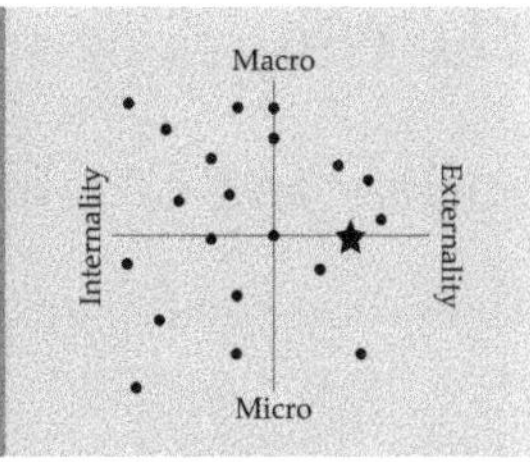

At the Fringes of Society:

The World Brought into Being by the Extreme of the Present

Noriko ITOH

Keywords: extremities = here-and-now, imbalance, instability, incompleteness, environment (surroundings), fission and fusion, nature, chimpanzees

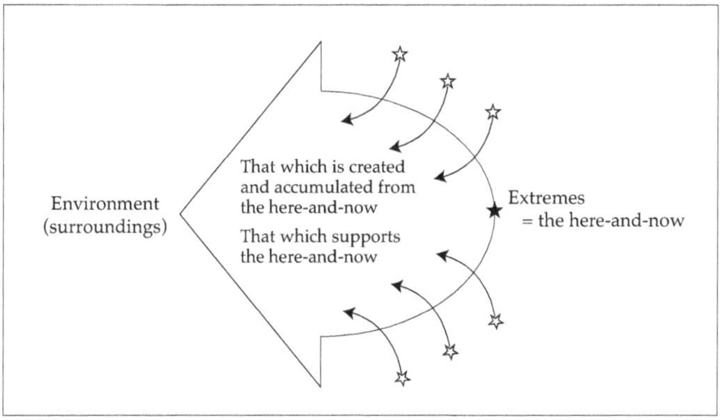

This present moment is an extremity beyond which there is nothing *yet*. We somehow manage to while away and accumulate these moments one by one in the creation of our own manifold environments (surroundings). But environments cannot be created ex nihilo in a single here-and-now. They are unstable and incomplete, produced with the benefit of diverse and continuous aid. At the same time, in turn, these environments—whether social or natural—also support the here-and-now. Though we could not survive without our environment, we cannot survey its entirety, nor are we usually aware of it. It is something elicited only partially in a type of subtle imbalance—a gradual sense of discomfort, a difficulty breathing, or else feeling a little flustered or, more simply, warmer or colder—as though one has lost track of where one is *in the now*.

The fringes of society

Society and extremities

As also noted in the introduction to this volume, a variety of views are possible as to the relationship between extremities—the topic of this collection—and society. Therefore, I would like to begin by sketching out to some degree how we will be handling extremities in this chapter. The main actors in this chapter are chimpanzees of *a particular group*—not chimpanzees in general.

The Japanese word *kyokugen*—which is the original title of this book—has been translated as 'extremity' in this English edition. The nuances of *kyokugen* evoke harshness and misery—imbued with certain evaluations, except in its mathematical usage—which might also be true for the English usage of extremes. So, from the beginning, when I prepared my paper in Japanese, the more neutral English word 'extremity' was in my mind for imagining *kyokugen*, which is more affiliative to the Japanese word *hashi* (fringe, edge). I have made considerable modifications to this section from the original book, as it would be very strange if these background explanations remained in their original form.

The dictionary definition of 'extremity' in English indicates the farthest part of an object. The hands, feet, and limbs, as the fringes of the body, are referred to as one's 'extremities.' Beyond objects, the word can also be used in relation to quantity or degree. So what does the 'fringes of society' mean? This will depend on our view of society.

A typical image of society is that it spreads outward in a circle from a central core. If we think of what lies in the center as being individuals, or else organizations and groups, then the fringes will also be populated by individuals or organizations. With chimpanzee society, while there certainly exists a (human) view of society that regards adult males as the core, this does not necessarily mean that all the other and various individuals could be placed in a gradation that proceeds from the core to the fringes. More importantly, although this view might be useful when considering society as a structure, it serves as an irrelevant starting point if we focus on the anti-structural aspect (Adachi 2009) that operates at the root of the structure.

What about society in terms of quantity and degree? Let us consider it in terms of emotions, for example. When our feelings such as anger or sadness reach extreme levels, we become trapped in those emotions, so that we cease to be able to make any other judgments, which we might easily make under normal circumstances. Moderate laughter is good for one's health, but if it goes too far, it can lead to breathing difficulties, or even to death. That is, there are quantitative extremes in the emotions of individuals that it might seem we are able to measure. However, when a society turns in a certain direction, flourishing (or fading) in strength or quantity, then we must extract comparable elements that represent only a part or a particular aspect of that society. Moreover, to make comparisons, we will need a sufficient amount of comparable data covering quite a long time. Surveys of wild chimpanzees have only been carried out for fifty or so years worldwide. This may seem like a long time, but it is less than the life expectancy of a single individual chimpanzee, let alone that of a society. Thus, some day in the distant future, there may be a comparative historical study of a chimpanzee society, but this is not the time.

A third presumable fringe lies in the temporal dimension. Although we cannot specify beginnings, their opposite extremity—namely, the present moment—can be dealt with in a concrete manner.

The extremity of the here-and-now

In terms of the 'here-and-now,' it should suffice to say that this consists solely of having lived and continuing to live still. Yet describing the 'here-and-now' (seemingly inevitable when trying to convey this to others) is not this simple, as the indistinct and incoherent scales of time and space are intermingled and we cannot grasp their entirety. So it is easy to imagine that despite my most careful intentions, my ventured descriptions will be quite awkward—either too vast or too small, too much or too little, both temporally and spatially. On the other hand, the here-and-now of the actual living is such a chaotic place that any attempt to describe it would seem to be ill-advised in its own right. However, as we shall see later, I have endeavored to avoid this problem as much as possible by focusing not on the whole, but on certain types of imbalances found therein.

In this present moment, I do not exist on my own. My existence in this present moment is supported by my various 'surroundings' (see Takenoshita's discussion in Chapter 5). These surroundings, including myself, constitute the here-and-now against the background of respectively different spatiotemporal scales (these are not necessarily limited to a single scale in each case). Here—the here-and-now that I inhabit—society and nature are equally important; it would make no sense in the present moment, for example, to separate what represents a social problem from what does not. Usually, research would involve performing such a separation or extraction for analysis or to discuss each arbitrary area, and this prior work would seem essential for deepening our understanding. And yet such processing is distant from our sense of living in reality.

Accordingly, here, in keeping with everyday sensibilities, I would like to think about the here-and-now for chimpanzees without drawing any a priori distinctions between social and non-social aspects. As Kawai stated in the introduction, extremes also lurk in the context of our everyday lives. Some might wonder why I use the term 'extremity' to refer to a *now* that we have been able to manage (competently or otherwise), and the cases dealt with herein do not entail that the group has collapsed or is even in any danger of doing so.

Aside from my own slight misgivings about how long the chimpanzee group in question here, the M group, or even the whole species, might persist, the group remains extant as of the time of this writing. While M group chimpanzees live their lives somehow the 'same as it ever was,' we can say so merely retrospectively—it is already the past as seen from the present moment of *now*. A person walking breezily along the street could quite plausibly be hit by an object falling suddenly from the sky. Just so, this moment of here-and-now at the end of time, when nothing has yet happened, will always remain open to other possibilities.

Life among the chimpanzees

In the following sections, we will take a concrete look at the here-and-now for chimpanzees, but no matter how important the moment is, it may seem difficult to imagine without having some inkling of the background. In fact, as I make my own observations, I also mobilize a variety of background knowledge, just as the chimpanzees also mobilize a variety of empirical knowledge in the pursuit of their daily lives. With that in mind, here I would like to sketch an outline of the 'surroundings,' including their society, that have been created by the accretion of moments in the here-and-now.

The chimpanzees we will meet in this chapter live in Tanzania's Mahale Mountains National Park, just like those featured in Chapters 6, 7, and 8 (by Nakamura, Hanamura, and Nishie). More precisely, they live in Kasoge (or Kasoje) Forest, which extends across the western part of the national park. Kasoge Forest, to borrow a phrase from Toshisada Nishida, is 'a "forest island" in the midst of the vast miombo woodland' (Nishida 1990: 15). This is largely due to the geographical conditions, with Lake Tanganyika in the west and the Mahale Mountains massif in the east (Photograph 2.1). Kasoge can be classified as a semideciduous lowland tropical forest. Walking through the forest, one finds a variety of mosaic patches in terms of vegetation types and forest maturity. These patchy environments stem from abiotic disturbances caused by the complex topography and numerous water systems that characterize the local area, and another major effect is the disturbance caused by the past activities of the local Tongwe people, slash-and-burn cultivators who inhabited the area. According to Reice (1994), who proposed the nonequilibrium hypothesis that ecosystems are constantly in the process of recovering from earlier disturbance events, both the disturbances and the patches created thereby (which have their own impact on disturbances) are dynamic processes that foster diversity and promote ecosystem maintenance. Chimpanzees use a variety of plants in pursuing basic activities like eating and sleeping.[1] Forests are constantly changing from moment to moment as abiotic dynamics intermingle with the activities of living things on various temporal and spatial scales. The sustained and worldwide phenomenon of global climate change has also had an impact. It should also be noted that this chapter features many plant species, most of which have local Tongwe names.[2] In what follows, I note the scientific name only when a species is first mentioned, and then use the local name thereafter.

As presented in other chapters, although chimpanzees form social groups, they accomplish this through a process of mutual fission and fusion with each other. In their society, precisely where and when they meet or part is presumed to be unpredictable. The here-and-now in a chimpanzee fission-fusion society appears anytime and anywhere in different space and time. To perceive the *now* in various places while still grasping the whole at the same time would be impossible—both for me and for the chimpanzees in question.

In this chapter, to secure a minimum connection with the main issue of this book—sociality—I focused on settings where chimpanzees encounter or separate from each other. With regard to dealing with these encounters and separations, my discussion overlaps with those found in Chapter 7 (Hanamura) and Chapter 8 (Nishie). However, as I describe in the final section, it is not only society that withdraws from the here-and-now but nature as well, and in that sense this chapter also has connection with the discussions in Chapter 5 (Takenoshita) and Chapter 21 (Terashima). For living individuals, the here-and-now is the end point of their world, beyond which no society or living space has yet been established; at the same time, it is also a starting point of their world.

Photograph 2.1: The Mahale Mountains and Kasoge Forest

The Mahale Mountains, photographed from Lake Tanganyika. This massif transforms the moist air from the lake into rainfall, nurturing the lush Kasoge Forest that spreads out at the foot of the mountains. Since the departure of the Tongwe from the area, the forest has grown even larger. However, the question of whether or not the ecosystem established here, including human activity, has been beneficial to the other creatures with whom humans coexisted is not easily answered, given the simultaneous impact of ongoing climate change (photographed October 2007).

The observations presented below are from my fieldwork in 2018 between the end of June and the middle of October. On arrival, local field assistants told me that the chimpanzees were staying in the high area of the massif. The frequency of encounters and separations (hereinafter referred to as 'fluidity') of Mahale M group chimpanzees are known to seasonally fluctuate considerably (Itoh and Nishida 2007). From around July, when they begin eating the fruit of a woody vine called ilombo (*Saba comorensis* [Bojer ex A.DC.] Pichon, Apocynaceae; see Photograph 2.2), their most important food source, this fluidity increases and results in the formation of a large nomadic party, with most of the group members participating in repeated fission and fusion with each other (Itoh and Nishida 2007; Itoh 2017). This year, however, despite the fact that I was eagerly awaiting it, the sight of the ripened yellow fruits in bunches that appear here and there never appeared. Since the blooms of the flowers were apparent instead, the fruiting season may have been delayed.

Because of this, most of the chimpanzees stayed up in the mountains until the end of August, although we did find that some of the females, albeit only a few, made their way down to the lowlands, either on their own or with their young. Since these were all either elderly females or females carrying infants, the highlands may have been a somewhat difficult place to live, even for chimpanzees. These females also interacted with each other, although not as frequently as in the ilombo fruiting season. Locating these chimpanzees in

Photograph 2.2: Flowers and fruit of the ilombo plant

Left: Ilombo in bloom (photographed July 2018). Right: Fruit half-eaten by a chimpanzee (photographed November 2015). The size is that of an almost fully grown fruit, although not yet ripe. When ripe, the outer skin turns yellow. The inside of the fruit is filled with yellow pulp that sticks to the fruit's relatively large seeds. Chimpanzees, as they eat, will sometimes spit out seeds that still have some pulp attached but will often swallow the seeds whole. In the ilombo season, when the ilombo fruits are their main food, they leave behind feces filled with clumps of these seeds, thereby transporting the ilombo to various places where it will germinate and grow.

the lowlands was almost always by sheer luck for me since they were sitting quietly in the treetops and bushes—that is, they were hard to find. So, I was desperate to know how they find each other. At the same time, unlike in times of high fluidity—when it required everything I could do just to keep up with who was coming and who was going—I was able to closely observe instances of encounters and separations. Of these, let us take a closer look at three cases that I was able to observe under relatively good conditions.

The present as beginning and end

In the course of fission and fusion, chimpanzees form temporary groups (hereinafter 'parties'), with certain individuals repeatedly coming or going in a manner that is unpredictable (referred to as 'face-to-face parties' in Itoh and Nishida 2007). Hence, a situation in which the same members continue to stay together over a long period, other than in the case of a mother and her child / children, is in some sense a rare sight. Staying together for a prolonged time seems difficult not just because other chimpanzees come and go. Rather, once formed, the continuation of a party—perhaps precisely because it is not inevitable in the context of chimpanzee society—seems to require considerable effort. Though it might seem paradoxical, this can be guessed from instances where this togetherness ends up being interrupted.

In a case on August 27, Puffy and her son Peace (Photograph 2.3) ended up separating from two other families that they had been with for more than three hours. The participants of the two other families included Devota and her daughter, as well as Koopy and her son. Devota is about the same age as Puffy and her still-breastfed daughter turned two years old that year. Koopy had recently transferred from a different group and seemed to be slightly younger than the others. She delivered her first baby—as far as we know—in the year of this field study.

That day, at 08:41, I came across Devota and her daughter picking and eating fruit in an ihambwa tree canopy (*Ficus vallis-choudae* Delile, a species of fig) along a dry gully located slightly downstream from the confluence of two rivers. At 09:32, Devota began moving from tree to tree back up the valley toward the main stream, where she met Puffy's and Koopy's families resting in separate trees. Several ihambwa trees were growing here, too. It seemed likely that the chimpanzees had been eating the fruit, as the fresh remains of these fruits were found on the ground. Perhaps they had slept around here. I could see chimpanzee beds, likely from the night before, in a nearby flowering tree, known locally as mukibhughwakakamba, or kakamba for short (*Cordia africana* Lam., Boraginaceae). Since it was about a four-minute walk from where Devota had been, she may also have slept here.

This party lasted for a long time after Devota joined, but ended up being *dissolved* around 14:00. Since it is normal for such a party not to stay together, it may seem strange to put it this way. However, that they were able to stay with each other for more than three hours was due to the fact that they kept mutually adjusting the timing and direction of their movements, and it seemed to me more than anything that Puffy was still actively trying to stay with the others, particularly with Devota.

I believe the party started to dissolve when the chimpanzees began eating ilombo fruit around 13:21. Leaving my rationale for thinking so until later, I should make a note about this woody vine. Ilombo fruit has been recognized since the beginning of chimpanzee surveys here in 1965 as one of the most important foods (Nishida 1974; Itoh and Nakamura 2015a). The fruit is also edible by humans, and some Tongwe informants now in their late sixties recall having their parents squeeze the juice of ilombo fruit for them as a treat when they were children. As mentioned earlier, however, there were hardly any ripe fruits to be found this time. Here, too, they were searching for and eating a bare handful of unripe fruits. As a result, the party had spread out to the widest extent since coming together that day. From the ground, the figures of the chimpanzees in the canopy overhead were almost indiscernible, and I followed the positions of the three families by relying on the rustling of leaves and the faint thuds of peeled fruit rinds as they dropped to the ground.

Photograph 2.3: Puffy and Peace

Typically, a female will leave her natal group, transferring to another group when she reaches sexual maturity. Puffy is a female who lost her mother at an early age. She was raised by another female named Gwekulo, a friend of her mother's, and ended up having offspring in her natal group. She has a tranquil personality, much like her mother. Her first son, Peace, although he was by then four years old and had started his weaning period, was often observed riding his mother's back whenever she moved—quite a heavy burden for Puffy (photographed June 2016).

About twenty minutes after they began eating the ilombo fruit, while the others were still eating—as I could tell from sounds above—at 13:41, Puffy and her son descended to the ground—maybe they had had difficulty finding more fruit, or were perhaps already sated—and started moving to the western edge relative to the party, where they began eating ants (or ant larvae) in some dead branches. As my field assistant appeared to be following Devota, I decided to keep an eye on Puffy and Peace, on whom I had a direct line of sight.

In a few minutes, Puffy and Peace moved further northwest in the bush and onto the observation trail, but continued foraging for ants. At 13:47, while they were moving, Devota, to the southeast, was following them and Koopy was still feeding in the northeastern canopy, judging from the sounds they were making. Although five minutes passed, Devota failed to appear, but seemed to still be in the same area as the field assistant. Puffy made her way slowly westward, with occasional interruptions, as when she had to stop along the way to pull her son up onto her back. However, at 13:56, she sat down heavily and looked back in a southeastern direction, where Devota's family should have been (although I could see neither Devota nor the field assistant), and made a small vocalization, 'nnnn…' It was unclear exactly for whom this sound was made, or what it might have meant, though it did not appear to have been directed at her son.

With no particular movement (or sound) from Devota, Puffy turned southward and at 13:57 she polished off a fallen ilombo fruit on the ground, then climbed into the trees, where she began once more to eat ilombo fruit. Puffy had to wait and then ushered her son Peace into the treetop ahead of her before she was able to start eating, as he made a short whimpering sound (indicating dissatisfaction or anxiety). During this time, Puffy was watching in Devota's assumed direction (as we had moved to the south, Devota's location was now east of Puffy).

At 14:10, a hush suddenly settled over the nearby forest, and since I could not even hear my field assistant, I became slightly uneasy. Two minutes later, Puffy finally climbed down to the ground (and once again, she had to stop her descent halfway to sit and wait for her son, who was following after her and giving a small whimper), whereupon she moved a little south and then west in the bush. At 14:16, she began to sniff at some plants growing close to the ground. Rendezvousing with my field assistant, I asked the whereabouts of Devota and was told that she had already moved off. Thanks to my assistant, I now knew that the place where Puffy had been sniffing, as well as the route that she had ultimately followed to the west, was where Devota had passed by. Since I could hear Koopy still staying in the northern canopy, where she had been all along (as Puffy also probably knew), she must have been searching for Devota.

If she could keep tracking her spoor further west, she might be able to meet up with Devota again. However, at 14:18, after two minutes of close scrutiny that consisted of sniffing and looking over her surroundings (and listening for sounds, though I cannot know for sure), Puffy turned in the opposite direction, hurrying back to the spot where she had originally stopped eating ilombo and come down out of the trees, where she once again started to sniff the ground and look around. There was not a sound, let alone anyone to see. Perhaps because she could not know where Devota was heading from her scent, at 14:20, she went back the way she had come and emerged onto the trail, where she uttered a soft grunt, 'nfu,' whereupon she moved a little farther to the west. There she began ant-fishing with her son from a low spot on a large bhuhono tree (*Pseudospondias microcarpa* [A.Rich.] Engl., Anacardiaceae). In the end, even though no one appeared, as she fished for ants, Puffy continued for a long time to repeatedly check the direction in which Devota had been eating ilombo fruits and the direction she had herself been sniffing. This lasted at least until around 15:00.

Puffy and Peace did not subsequently rejoin Devota's family or Koopy's family before we stopped our observations for the day at 16:30. Although Koopy could be heard making intermittent sounds in the forest canopy until at least 15:10, Puffy did not seem to care. When she had finished ant-fishing, at around 16:00, Puffy moved off to the southwest. I paused my observation between 15:31 and 16:09 to search the area, during which Puffy and Peace started to travel. We tracked them by following glimpses of them and the tracks of their footsteps on the ground vegetation from far behind. By 16:31, we had no luck catching up with them and gave up trying to follow.

Although I had the impression that Puffy had 'wanted' to stay with Devota rather than with Koopy, my rationale for thinking so was that Puffy basically did not seem to pay any heed to Koopy's movements, despite the sounds she was making, and conversely that Puffy seemed to be looking back for Devota and searching for her by sniffing the ground when she could no longer be seen or heard. I did not smell anything at the time, but I have repeatedly observed similar sniffing behavior before. In some cases, they were strongly assumed to be tracking the movements of specific individuals by their scent. Presumably, as well as other individuals, they may smell leopards and other animals. Chimpanzees sometimes give off quite a powerful body odor, likely depending on their physical condition, and on occasion even I, walking upright on two legs, have been able to smell the scent of chimpanzee wafting up from the ground. Accordingly, a chimpanzee could hardly miss it. In some cases, chimpanzees have been able to follow scents to reunite with partners.

On this occasion, Puffy could not, or at least did not while we were following, meet Devota again. But in contrast to a lack of obvious reaction to Koopy's movements and sounds, Puffy kept searching for scents or other clues and continued afterwards (and beforehand) to check the place where Devota still was, as well as the places where Devota had been, or should have been (though it is not certain whether this would have been in the past tense for

Puffy). From these facts, Devota was presumably a source of concern for Puffy. However, my impression might have been somewhat biased by the nearly two hours (09:38~11:17) of enthusiastic reciprocal grooming we observed after they met up with each other. Let's now rewind a bit.

Movement up to this point, which was far removed from the point where they had come together, was basically prompted (not to say led) by the actions of Koopy, who was the youngest and retained something of a childish restlessness (Photograph 2.4). Even in the midst of the enthusiasm of the grooming that began to take place as soon as they encountered each other, Koopy 'stepped away' between 10:35 and 10:48, and for about a minute from 10:51, to snack on some nearby kakamba flowers, while deftly avoiding their buds. Puffy and Devota's children engrossed themselves in play, chasing each other while occasionally snacking on nearby ihambwa fruits or picking at an ilombo fruit, which happened to be in the vicinity. Koopy's and Devota's children, who still depended on breast milk, also found time to breastfeed when they had a chance.

It may have been around this time that Koopy was the first to get itchy feet. From 11:14 to 11:16, she clambered out to sit on the tip of a branch, then suddenly moved back, very much as though she was inviting the group to move. At 11:17, when she moved at last to another tree, Puffy and then Devota followed her to a fig tree called ikubhila (*Ficus sur* Forssk.) several trees away. They engaged in a variety of activities in this single tree, including eating fruit (they chose half-ripe fruit, as no ripe fruits were available) and grooming on the part of the mothers and breastfeeding and playing on the part of the children. The fruit of this fig species is important, but in a different sense than the ilombo, as the fruiting cycle of individual ikubhila trees varies and thus, taken as a whole, the fruits are available throughout the year.

Around 12:40, although we were unable to confirm the instigator, the entire party began moving along the ground, with Puffy at the rear. At 12:47, they finally came out of the bush onto the trail, where they continued moving while taking time here and there to groom or play. They had moved roughly southwest for about 250 meters through the bushes to reach the east–west trail, whereupon they began heading west along the path. Since the children, aside from Koopy's nursing baby, were always stopping to play, their mothers were constantly left waiting for them to catch up. Devota, herself, also often stopped to play with the children. Thus the positional relationship on the part of the three mothers was in a constant state of disarray, somewhere between movement and rest (Photograph 2.5).

At 12:57, Koopy began to eat a fallen ripe ilombo fruit on the ground that she found. Both Devota and Puffy quickly approached and gazed at it for almost a full minute. During this time, it could be that their minds were preoccupied with the ilombo fruit. When Koopy finished eating, she started moving again (12:59). Puffy, Peace, and then Devota with her daughter on her back followed her. At 13:00, Koopy came to an intersection with a north–south trail. She paused, turned around, then traveled west for about a minute, and then sat down. Once again, the chimpanzees began repeatedly moving a little further and then stopping, while taking time to engage in grooming.

Around 13:05, Puffy began heading for a tall mukamba tree (*Milicia excelsa* [Welw.] C.C.Berg, Moraceae) in the bush just south of the path. Koopy followed

Photograph 2.4: Koopy

Koopy transferred from another group in 2014 (she was estimated to be about fourteen years old when this picture was taken in 2017). She is very playful, but also tends to frequently stir up a commotion. Although she bore a child in 2018, she continued, with a nursing baby on her belly, to put on displays that could be quite nerve-racking to watch. This photo was taken in December 2017, so she would have been pregnant at the time.

Photograph 2.5: Resting? Moving?

It is not always easy to tell whether chimpanzees are on the move or not. Sometimes they will take a few steps and then stop, before taking a few more steps, as though remaining in place by seeming to move in a vague way. In this photo, since the chimpanzees are grooming, it may look like they are resting. However, they are in fact moving along slowly but steadily. One of these chimpanzees is waiting earnestly for the return of her daughter, who has just snatched her baby sister (born that year) and gone off to eat in the trees in the bush to the west (left from the photo). They stop to move for a variety of reasons, to search for fruit in the trees or because a baby begins to fuss, etc. When multiple individuals are present, as shown here, this kind of intermittent movement occurs frequently, giving an overall appearance that is akin to a mixture between movement and rest. This same state of betwixt and between holds true for individuals as well. While even they sometimes seem to be waiting for some other individual(s), in the next moment they may simply wander away, as though they were not waiting for anything at all (photographed August 2018).

immediately after, followed by Devota and her daughter, who lagged behind since Devota was grooming her daughter. By 13:07 they had all begun to eat galls formed on the fresh mukamba leaves. This was to be the end of their activities together on a single tree. Inspired, perhaps, by Koopy's ilombo snack, a little more than ten minutes later the party spread out as each family moved off in search of ilombo that were spread throughout the nearby canopy. Although I had completely missed seeing any ilombo fruits in the canopy from the ground when they began eating the galls, they had probably been eyeing them since well beforehand.

Earlier, I mentioned my suspicion that it was the activity of ilombo eating (coupled with the scarcity of the fruit at the time) that occasioned their separation. This was because until this point, the members of the party had remained in places where they were in each other's view—in neighboring trees at most—and so would not have completely lost sight of each other's positions. Of course, their movement in a state in which they were able to see each other was accomplished based on the mutual coordination over an extended period of time—no one going off ahead alone, stopping any number of times and paying attention to their rear. If no one moved, then it would be easy for them to remain together, but since this will never happen, so long as they remain alive, they have no choice but to move along while harboring diverse desires at the individual level.

Changing the present

Thus far we have been looking at the aspect of sustained encounters. From the perspective of social extremities, this could be characterized as a continuing effort to change the moment of the *now*. However, it seems there may be another way of transforming the preceding phase. We can see this in the following case, which seems to have involved seeking out and encountering Nkombo, the group's most venerable female. This took place on August 6, three weeks prior to the previous case.

Since August 5 was a day off for all researchers and assistants, we had no idea where chimpanzees were. Most were supposed to be still up in the mountains, but I took the chance to see any individuals in the lowlands and departed our camp site at 08:00 that morning. Luckily, we soon discovered Koopy and Genie with their babies. Like Koopy, Genie delivered her daughter earlier that year. They were sitting in an ikubhila tree (which I had spied, days before, at the side of the trail), eating figs (08:53).

Around 09:00, as I followed Genie, she shifted her location slightly and started eating young leaves from a woody vine known locally as itambuka or ifumiya (*Dalbergia malangensis* Sousa, Leguminosae-Papilionoideae). At this time, Juju and her newly weaned daughter Jura came into my sight, climbing another tree nearby to start eating young leaves from another vine known as ijubhila (likely *Baphia capparidifolia* Baker, also in the Leguminosae-Papilionoideae family). A few moments later, all three families were stuffing their mouths with young itambuka leaves and chewing (Photograph **2.6).**

After about one-and-a-half hours of feeding and postprandial rest, the chimpanzees returned to the trail and began moving (two hours after we had first spotted them). However, they soon stopped along the path and all at once began staring fixedly into the bushes to the north. Since they did not have the air of looking for food, but rather the atmosphere was a little tense, I also began searching, but did not sense any signs or scents. Juju stood motionless on her four limbs for almost a minute, wary of the interior of the bushes. Her daughter Jura also brought her face close to nearby trees and the ground to smell and then gaze into the bushes, and she did this sequence of movements repeatedly. Afterward, although Juju and her daughter occupied themselves with grooming and the like, not really making any further progress, they stayed within sight of Genie and Koopy and their children, who had moved ahead.

Eventually, fourteen minutes after returning to the trail, Juju and Jura finally shifted (11:11) when Genie began moving west along the trail. They proceeded to the west, plucking young leaves from the side of the trail as they moved. At 11:21, Juju and Genie both began to look around with a slightly tense expression, as they had earlier. One minute later, Juju, who had been taking up the rear, suddenly began moving in a sprint, quickly taking the lead. At 11:32, she moved into the bush on the north side of the road, and three minutes later they all

Photograph 2.6: Fresh itambuka leaves

During the 2018 dry season, female chimpanzees were ravenously eating young leaves of itambuka, which might be a catastrophe for the plants. However, being eaten is something that the plants take into account—they sometimes show a positive response (compensatory growth). Even so, the reaction of these harvested itambuka plants is still unknown (photographed September 2016).

became somewhat excited as they started eating young itambuka leaves, which I thought they had been gorging themselves on earlier. It seemed that there was another chimpanzee up in the tree.

After about five minutes, I finally made out the figure of Nkombo, the group's oldest female, who was also eating young leaves of itambuka (Photograph 2.7). In addition to my poor footing, the visibility was terrible from the ground up to the treetops, so, as was the case in the previous episode, I confirmed approximate positions from the sounds of rustling movements and the wadges (Photograph 2.8) that occasionally dropped down from on high. Nevertheless, I was able to ascertain that no particular interaction took place between the females. The only thing that I was able to glean was that they were somehow excited. It is certainly not impossible that this itambuka was more delicious than the one they had been feeding on earlier. However, considering the two instances of tense and excited visual and olfactory exploration and Juju's sudden sprint ahead, Nkombo seemed to be the key. As I continued my observation with this suspicion in mind, it seemed to me that Juju and her daughter, at least, had wanted to spend time with Nkombo. This inkling stemmed from the following episode.

Genie's and Koopy's families were slowly making their way through the trees in a northerly direction, but by 12:15, I lost track of their position utterly as I could not hear their sounds. To the north of our location was a swamp that

Photograph 2.7: Nkombo

The oldest female in the group, as of 2018. She transferred from a now extinct neighboring group in the 1980s. By the time of this photo, all the other females who had transferred with her had died. Although this photo was taken in June 2016, she appears to have hardly changed since the old days, probably thanks to her youthful face. She befriended an elderly male named Kalunde as she grew older, but he also disappeared (and is presumed dead). Since I had seen them moving and playing together all the time, except when she was in estrus, I was a little worried, but it seems that she began to spend time in the company of another male—her junior, but the oldest male in the group.

the Tongwe had once used for rice cultivation. Chimpanzees visit such swamps in all seasons to eat various types of leaf and pith. Even if it is an important place for chimpanzees, it can be a treacherous locale for humans, as one's feet can become completely mired in the mud and be stuck without the help of others. Hoping sincerely that the chimpanzees would give up any plans to visit the swamp, I kept waiting for the remaining females, relying on my ears to guide me.

Nkombo, who finally descended to the understory at 12:22, stopped along the way to defecate before reaching the ground and proceeding south on foot. Since this was the opposite direction from Genie and the others, I followed this eldest chimpanzee, wondering suspensefully if the party was already breaking up. At 12:27, she began climbing an ikubhila tree (mentioned earlier about the species) and began to eat figs that were mature in size but still unripe.

Jura, who seemed to have been following Nkombo's movements, appeared in the background at 12:31 and instantly clambered up beside her to join her in eating figs. Nkombo, who kept eating without paying any particular heed to Jura, began climbing down the tree at 12:36, and when she retraced her route a little northward along the ground, I found Jura's mother, Juju, sitting and facing toward us. When Juju saw Nkombo coming, she turned around and began heading north, with Jura following suit in the trees.

Photograph 2.8: Wadges made by chimpanzees

Unless swallowed, chimpanzees make wadges in their mouths to spit out fibers and small seeds, rather than spitting them out one by one. The photo on the left, taken in June 2016, shows a wadge of lulyolwakape (*Psychotria peduncularis* (Salisb.) Steyerm., Rubiaceae). Each of the small fruit covered with a thin blue skin has fresh, white, sweet pulp surrounding a seed approximately 5 millimeters in diameter. As they fruit in clusters (the center photo, taken in January 2015, shows immature fruits), where colonies of the plants are found, it is relatively easy to fill the stomach. In addition, these shrubby herbs only grow to approximately 50 centimeters in height, so there is no need to climb the trees. The photo on the right, taken in July 2018, shows a desiccated wadge seemingly of lujongololo (*Uvaria angolensis* Welw. ex Oliv., Annonaceae). These fruits, which contain multiple seeds approximately 10 millimeters long by 6–7 millimeters wide, can be formed into clumps. This plant can grow as a liana or a shrub or in a mixture of the two lifestyles.

After two minutes' walk with Juju in the lead, the party re-encountered Genie's and Koopy's families ahead of them on the ground. Koopy, who had been resting at the very front, immediately began grooming Juju on her arrival, while Genie, making greeting vocalizations (a sound called a pant-grunt) ran up to Nkombo, the next to arrive, in a bit of a fluster. A minute later, Koopy began grooming Nkombo, but before another minute had passed, Koopy and then all the others began heading for the swamp. They all continued eating various kinds of foods in the swamp until I started home at 15:27 with my assistant.

Although Genie's and Koopy's families went out of sight once again, Juju and Jura continued foraging near Nkombo until observation ceased for the day. Curiously, this does not necessarily mean that Juju and her daughter always wanted to be near Nkombo. As we shall see, it seems that there are also times when, despite a fortuitous encounter, they seem less than enthusiastic about her company.

The present bound to the past

Changing one's situation in the *now*, while it may constitute a major change of course from what has gone before, is nevertheless not free of that past. The following case on July 19 left me with the strong impression that chimpanzees also have 'grown-up relationships.'

The mukibhughwakakamba (or kakamba), mentioned earlier, is a relatively large tree that is found scattered across a relatively dry area in the southern half of M group's home range. Wood obtained from this tree represented an important timber resource for the Tongwe people in the past, as did that of *Cordia millenii* Baker, which belongs to the same genus (the local name, like the species name, appends ghwakakamba to mukibhu in the case of the former and ghwesimbwa in the latter, with ghwa/ghwe being an infix corresponding to the English 'of'). Although these two species look similar, save for their flowers and fruits, Tongwe people say they differ recognizably in terms of the quality of their wood. Kakamba, which is the stronger wood, is suitable for making drums, and so is also sometimes known as mufunangoma (ngoma meaning 'drum').

The fruits of this tree, which the chimpanzees eat, peak in production at the beginning of the dry season, roughly between July and August each year, with some annual variation (Itoh and Muramatsu 2015). Individual variation may also be significant, as we saw with the trees that were flowering in the first case discussed above. When the chimpanzees were difficult to find, we guessed that they would likely be found in the fruiting trees that they usually eat in that season. On this occasion, as well, we sometimes took roundabout routes to search for kakamba fruiting trees, but it seemed as though it was a poor harvest that year. In these circumstances, on July 19, when I did find some chimpanzees eating these fruits, the only tree bearing much fruit at all was the one in question.

At 09:37, in a fruit tree in the bush a little west of the trail, we came across a party of eight individuals consisting of three families—namely Ua, with her daughter Upepo (age seven) and son born the year prior, Vera and her son, also born the previous year, and Genie and her daughter—as well as the elderly female Nkombo. At 09:50 some individuals began to engage in grooming and other rest activities, and by 10:15 the entire party had gone into rest mode. Even the children were relaxing quietly, without quarreling. Although the calls of chimpanzees could be heard from the eastern mountains on two occasions prior to 10:41, nobody made any particular reaction.

However, at 11:06, when a chimpanzee child was heard screaming from the trail, a distance of about a minute or two away on foot, the chimpanzees instantly climbed down the trees to hurry toward the sound. I was not able to navigate the bush as nimbly as they could, so unfortunately could not reach the scene of the encounter in time to witness the excitement. I could only hear greeting voices from ahead, the location of the initial child vocalization.

What I want to focus on is what came next. Joining the party were Juju and her daughter Jura (Photograph 2.9). Apparently, it seems they had only happened to be heading this way (likely aiming for the kakamba) when Jura raised her voice, attracting the attention of the other chimpanzees, who had

Photograph 2.9: Juju and Jura

The photo on the left shows Juju in 2014 with her daughter Jura. The photo on the right shows Jura in 2018, all grown up. Although Juju is a little on the small side, she now seems quite robust, perhaps from carrying Jura around for four years. As of the time of writing, she is doing well raising her second offspring, whom she gave birth to in 2018 (photo © Michio Nakamura).

been enjoying a leisurely rest. As described later, Juju and Jura stayed with the others for about an hour-and-a-half, which was partly due to Jura following them closely at first, and Juju had to follow her daughter. Afterward, from the movements of Juju in contrast to Jura, I got the sense that the encounter had not been intended and Juju had no choice but to keep company with the others for a while.

Once they all finished meeting on the trail, they began moving east along the path at 11:10. Juju did not move right away—or rather it seemed that she would have preferred to go somewhere else—but her daughter Jura completely ignored her mother and promptly followed the other females, and Juju also eventually followed.

Later, having moved slightly further east along the trail, they all climbed up into an ihambwa tree in the bush to the south (around 11:20). This fig is commonly found growing in bright spaces along rivers, as we saw in the first case, but although it was humid here, the light was dim, with no river nearby. The chimpanzees all began to eat unripe fruits that they found here and there, picking out the best they could find. After about five or six minutes, as though starting over, some individuals began to rest, and by 11:50 everyone was resting, except possibly Genie and her daughter, who could not be seen in the dim canopy.

After a short break, the entire group began to stir, and just after 12:00 descended to the ground to move further east. At 12:11, Genie began eating the stiff leaves of a fig (shrub) known locally as kankolonkombe (*Ficus asperifolia* Miq.), which she came across at the side of the trail. This prompted Nkombo to begin eating the pith of the young stalks of a woody vine of the Apocynaceae family known as mpila (*Landolphia owariensis* P.Beauv.); Vera, the new leaves of a young tree called kakubhabholo (*Sterculia tragacantha* Lindl. Sterculiaceae); Jura, the pith of the leaf stalks of an isakama (*Myrianthus arboreus* P.Beauv. Cecropiaceae); and Juju, the young leaves of a fig tree (unidentified). With such a variety of options available in such a small area, each individual began helping themselves to handfuls of their own preferred food.

I thought that they would all continue to stay together for a little while more. But after thirty minutes or so of feeding, when they began heading southeast, Juju and Jura remained. All the others climbed up and started to eat in a large mukamba tree after a few minutes' walk. Mukamba grow conspicuously very tall in the Kasoge Forest, and the one in which the chimpanzees were now eating the galls that had formed on young leaves and flowers in the canopy was extraordinarily high.

When we left Juju, she and her daughter Jura were near each other, eating the same young fig leaves as before. I looked back for a little while, wondering whether they would belatedly join the others, but they did not appear. For some reason, whether waiting for Juju and Jura or simply to rest, Nkombo did not climb the mukamba tree immediately, but lay down at the side of the trail and watched behind her for a little while, just like I did. According to T. Matsumoto, who tracked Juju and her daughter until later in the evening, they subsequently headed west (not stopping by the kakamba mentioned above), without ever rejoining Nkombo or the others.

That which is birthed by the present

Fission and fusion society and the here-and-now: that which is created

When the fluidity of Mahale M group is high, someone is always arriving or departing from one moment to the next, mostly unpredictably. By contrast, however, it may give an impression of stability and predictability for all the above cases when seen as a whole. Even so, when we take a detailed look at the here-and-now, as in the case of Puffy when she noticed that Devota was no longer present (in the section 'The present as beginning and end') or Juju when she was accidentally discovered (in 'The present bound to the past'), not knowing what will happen next holds true even when fluidity is low. The same applies to other situations, such as when sustaining encounters, when not doing so, or when catching the scent of another individual that one hopes to meet.

Taking into account that these chimpanzees basically are not forced or need to stay together, every moment of here-and-now of a party for the participants has equally and always both the possibility of continuity and discontinuity. Although I did not put the emphasis on this point, all of the cases include situations in which individuals staying together continues, even though the length of time and the intensity of the 'intention' may vary. The behaviors that led to this continuation included staying put to engage in social interactions such as grooming or play, stopping to eat or rest, moving incrementally, or looking back. None of these behaviors are very special taken as such, but when multiple individuals stay together, they serve a coordinating function in terms of mutual timing and position, as when waiting or adjusting the pace of movement.

In particular, the act of looking back suggests the possibility that chimpanzees identify branching points in the here-and-now among participants. Such branching points are not only spatial but also temporal. For example, when an individual decides to move ahead, the other participant might go that way instead of this way (spatial) or the other participant might still be resting in that spot (temporal). In such situations, it is immaterial whether they can read correctly or not the action of the others. What is important in the here-and-now is the attempt to find a branching point at which the movements of other individuals deviate from one's own, including mistakes or misunderstandings.

The act of looking back, in this sense, indicates that they do monitor the other's movements, including those yet to be realized. Hence, as well as the looking back, depending on the time or occasion, there may be other behaviors that work in a similar manner.

In the case of Juju, this branching point was sought from behind the party (which might be more difficult than from ahead). Getting away from an oblivious partner is surprisingly difficult. In the case of something momentous, like fear, escape would conceivably be easy for humans and chimpanzees alike. However, although we cannot know if the chimpanzees find it difficult, after the encounter with the others, Juju did not immediately go her own way. Rather, she stayed at the rear, ultimately turning to leave only when the others were clearly headed further for the second time. Contrary to the moments of staying together, Juju here may have been soughing for a branching point of going 'this way' instead of the others' 'that way' in the transitory *now* as it passes by from moment to moment.

This could perhaps be rephrased as 'continuing to wait.' For example, when we saw Juju in the section 'Changing the present,' she seemed to have been waiting for something even before her encounter with the elderly female, Nkombo (though the same could perhaps be said for other individuals, as well). From the series of behaviors that consisted of her careful sniffing, visual exploration, and finally sprinting ahead, it seems that her 'hunt' for Nkombo had been ongoing for quite a while. While it would be easier to draw this conclusion if she had done something like running up to embrace Nkombo, would this have been chimpanzee-like behavior? They are not that easy to fully understand for us humans.

I do not mean that chimpanzees spend every waking moment apprehensive about what is going to happen next. Certainly, they may be a little disconcerted when an individual they were expecting to be there is not, or surprised when someone is unexpectedly nearby at an unexpected time. Even so, rather than being overwhelmed in such instances, they will try to react to the prevailing situation by seeking first of all to find out what is happening or to gradually distance themselves from it. Then they will revisit the situation in the form of reacting to the result of that reaction, and so on.

Not knowing what will happen next could be said to be the quintessence of their fission and fusion society. In other words, this instability is a prerequisite. The full membership of the unit-group (or community) is almost never in the here-and-now; rather, it is the case that they are always only partially involved in their own unit-group. Nevertheless, it is by repeating such encounters and partings with each other that an identifiable unit-group is created (even though its constituent membership will change with the deaths and births of individuals and with the intergroup transfers of females). In this sense—although it is doubtful that there will be anything like a complete society—the group will continue to exist, albeit incompletely.

We cannot have a bird's-eye view of the whole fission and fusion society that is accumulated in the here-and-now. Its relation to the here-and-now will only ever be partial. Perhaps, however, it could be said to appear precisely in the subtle malfunction of the here-and-now. In the here-and-now, branching points for various individuals appear one after another as possibilities in accordance with the passage of time and shifting spatial positions, and these individuals will sometimes become separated or come together in practice, as well. Encounters and partings may be unexpected or even surprising. Or else, in the sense that it is not something that an individual can control, we could say that even the encounters and partings that are expected are ultimately realized only fortuitously. Such encounters might be scents and similar traces found by happenstance, or the call of distant voices. In sensing the encounter or parting, reunion or absence arises with the unexpected actualization of the here-and-now under sensibilities working toward these branching points. It is in this way that a discrete society is brought into being, if only partially. Though it is not an exaggerated question such as life or death, the realization of an unexpected here-and-now nevertheless brings with it a subtle malfunction.

Is this malfunction brought about when fluidity is low, which makes it difficult to meet up with other individuals? In situations where other individuals are always coming and going, as though through alternation or replacement, it can feel like individual encounters are passing by like a flowing current. And yet, although this will require further study, the same kind of thing, not necessarily being exactly the same, may actually be occurring.

The fact that chimpanzee society is unstable and discontinuous may be important for their habitation and creation of an unstable natural environment, as we shall see next. For chimpanzees, the act of eating constitutes predation from the standpoint of plants. However, that this occurs in a discontinuous fashion keeps intensive damage to a minimum, and the fact that scattered feeding and defecation (i.e., transportation of seeds) takes place at the same time is likely to have a nontrivial effect on the structure of the forest.

One more creation

In the cases discussed here, several plants have appeared as important actors, just like the other individuals who have played a part in establishing events in the various aspects of encounters and leave-takings, including the commensal event of eating together. The way that fruits grow bears a direct relationship to encounters and leave-takings in the form of commensal feeding and separations that occur while eating. Food also plays an important role when considering the evolution of human society, such as in terms of food sharing. Though it tends to be obscured in the shadow of more illustrious topics, the routine act of eating together on a daily basis is of fundamental importance to the cumulative creation of the 'natural' environments in which individuals live.

Plants emerged on Earth long before mammals like ourselves, and together with the microorganisms that eat and decompose their detritus, worked over the long years to create the rich soil that covers our planet. Over geological spans of time, plants have cultivated relationships with their various abiotic and biotic surroundings, thereby providing habitats for many living things, including humans. The diverse creatures that have appeared since that time have created a planetary ecosystem while interacting with these surroundings on their own respective time scales. Chimpanzees are one of the member species that form a part of this web of relationships.

Chimpanzees are fructivorous omnivores, and also eat insects and other mammals (Itoh and Nakamura 2015a). When eating fruit, they swallow tiny seeds such as those found in figs, and sometimes swallow larger seeds whole that are stuck tightly with the pulp of the fruit, like most of the wild fruits. In Kibale National Park in Uganda, chimpanzees reportedly swallowed whole seeds in 77% of their fruit-eating bouts (Lambert 1999). In other cases, the juice may be squeezed out in the mouth, while the seeds and the leftover pulp are discarded as a wadge (Photograph 2.8).

By eating fruits, chimpanzees also nurture their living forest. Plants produce fruits, which are then eaten by animals, and thus the seeds are carried far from the parent tree, thereby spreading the living area of its offspring. In the case of chimpanzees, the swallowed seeds are extracted in their feces after about a day of travel. In several experiments, it has been shown that seeds that have passed through the intestinal tracts of chimpanzees have a higher germination rate than those that have not. Seeds that have been discarded as wadges also germinate, and one species examined in Nyungwe National Park in Rwanda showed a higher survival rate for such sprouts than even those from seeds spread in feces (Gross-Camp and Kaplin 2005). Fully understanding the role of chimpanzees in nurturing forests through seed dispersal still requires future study in the context of various species (for a study of seed swallowing and seed dispersal by chimpanzees, see the review in Nakamura and Itoh 2015).

Kasoge Forest is not nurtured solely through the efforts of chimpanzees. Various living thing in the sky, on and under the ground, and in the water have all played an active part. Humans are no exception. Nishida (2002) states plainly that chimpanzees were able to live in Mahale until the late 1960s simply thanks to the way Tongwe people lived; continuing to live in low density and maintaining their humble, traditional lifestyle.

The Tongwe inhabited a vast area (called Tongweland) that encompassed about 20,000 square kilometers, including what is now the national park (Kakeya 2017). Although the population density was relatively high in Kasoge, it still amounted only to five people per square kilometer (Nishida 2002). By the mid-1970s, as a result of the government's resettlement policy and, moreover, the creation of the national park in 1985, the Tongwe, who had been knowledgeable about the forest, were no longer able to live there.

Kakeya (2017) characterized the people's livelihoods in Kasoge Forest as having a rich variety. Aside from rice cultivation in the swamps (nowadays you can see suggestions of what may once have been raised levees between rice paddies in the swamps), the Tongwe used slash-and-burn techniques to clear small-scale swiddens. The fields were abandoned shortly as they moved and left until the recovery of vegetation. Traces of this practice remain in the patchwork distribution of secondary growth forests of various ages.

It seems that these small-scale disturbances may also have had the effect of maintaining food variation for chimpanzees living sympatrically in the area. For example, the grassy herb known as iswe (likely *Pennisetum purpureum* Schumach., Gramineae), which appears shortly after the abandonment of a swidden, was still present to some degree even in 1995, when I started my fieldwork. The chimpanzees eat the moist pith of this plant, relishing its flavor as their surroundings were pervaded by its refreshing perfume. It is rarely found nowadays, growing only sparsely in limited locales. As the succession progresses, a Guttiferae shrub known as mutunu (*Harungana madagascariensis* Lam. ex Poir) will appear, its fruits also part of the chimpanzees' diet. Like the iswe, however, it is now seen only rarely.

Kasoge Forest is gradually growing larger as it transitions, with the suppression of human activity to the maximum possible extent and the relocation of the Tongwe out of the national park (Itoh and Nakamura 2015b). This change in the scale of the forest is thought to be due to 'nature' reasserting itself, but the fact of the matter does not seem quite so simple.

One factor, as touched on earlier, has been the effect of the elimination of disturbances caused by human activities. But the disappearance of human activity does not immediately make it 'natural.' In East Africa, including Mahale, the environment is constantly changing due not only to the effects of geology and climate, but also as a result of human activities, to the extent that it is difficult to imagine what might constitute pristine 'natural vegetation' (Hamilton 1974: 205). This is not limited to East Africa. As described in Terashima's discussion in this volume (Chapter 21), in their long history of moving across the planet, humans must have been involved in the formation of ecosystems through their repeated disturbances in the various places they have gone. In addition, because ecosystems are affected by things like random chance and the state of other neighboring ecosystems, the changes they experience are never uniform. Accordingly, although Kasoge Forest has and will follow a unique 'Kasoge-style' trajectory of change, it does not make change easier to predict.

Nishida (1990) hypothesized the succession process by referring to vegetation found at sacred sites where the village chiefs are buried, called isigho lyetabhami, which would not have been affected—or minimally affected—by human activities such as the traditional slash-and-burn cultivation. This prediction is supported by the proliferation of some of the tree species that he speculated would remain as a climax forest (notably an Annonaceae species called kahwibhili (*Xylopia*)[3] and the bhuhono). On the other hand, there is a robust presence on the part of pioneer species such as ibhonobhono (a species of the genus *Croton*, Euphorbiaceae)[4] and muhehu (*Trema orientalis* [L.] Blume, Ulmaceae), indicating that the forest is the site of ongoing disturbances other than those caused by direct human activity.

Global climate change represents another complicating factor. As I mentioned in the cases above, in 2018 we were waiting and waiting for the ilombo fruit, which should have been ubiquitous, but never really appeared. The chimpanzees ate the ones they found, even when they were few in number or still unripe, and if lucky enough to happen upon even a single ripe fruit would devour it with gusto. Such disturbances in the fruiting pattern have been observed since around 2005 not only in connection with the ilombo, but also in other fruit species relied upon by chimpanzees (Itoh and Muramatsu 2015). One cause of this phenomenon is thought to be the impact of climate change. Long-term meteorological data from Kasoge suggests the effects of climate change include declining rainfall, a shift in the annual rainfall distribution pattern, and daily maximum temperatures exhibiting large variabilities with daily minimum temperatures that conversely remain stable (Itoh 2015a).

The impact of climate change is also being felt elsewhere, such as in the turnover rate for trees[5] and the increasing trends of lianas. These trends have been apparent around the world. It is known that the turnover rate has been on the rise since the 1950s, with concerns about the direct and indirect effects of climate change (Phillips and Gentry 1994).[6] This may bear some relation to the ongoing presence of pioneer species, as noted above. The increasing trend of lianas, which has been found in Kasoge, may be influencing both the turnover rate of trees and the presence of pioneer tree species. Since 1983, densities of vines and lianas have been found to be increasing in tropical forests around the world (Phillips and Gentry 1994; Phillips et al. 2002; Swaine and Grace 2007; Chave et al. 2008; Ingwell et al. 2010), suggesting the effects of global climate change—that is, the increasing concentration of carbon dioxide in the atmosphere (Sasek and Strain 1988; Phillips et al. 1994; Granados and Körner 2002). In Kasoge, the increasing rate of ilombo density has been particularly notable, even among other lianas. Although also part of the diet of redtail monkeys, this hard-shelled fruit is mainly eaten by chimpanzees, so the proliferation may stem from current environmental conditions and dispersal of the seeds as a result of the chimpanzees enthusiastically eating the fruits.

Though I am glad to see the proliferation of the ilombo that provides the fruits that the chimpanzees love, we cannot simply rejoice. As the canopy becomes covered with lianas, the trees that they cling to may not be able to carry out their normal production activities, beyond which they may break or topple, unable to bear the burden of the weight (Photograph 2.10). Since lianas often spread through the canopy by traversing several individual trees, if one tree was to fall, it may drag others down with it. While some lianas are able to adjust their life form from lianas to shrubs, depending on the habitat conditions, the ilombo seems unable to change. Accordingly, as the lianas proliferate, the trees that support them will be lost, so that their own survival may be threatened in turn. This would not be a pleasant prospect for the chimpanzees, either, given the current situation in which there is no food source that might serve as an alternative to the ilombo fruit.

From the above description, although clearly not exhaustive, 'opaque' and 'instable' are two words that may well capture the characteristics of the

Photograph 2.10: The suffering of the trees

This may appear to be a tree, but the tree itself is invisible, being shrouded from tip to toe in vines and lianas. Most of the leaves that we can see belong to the ilombo plant, a woody vine frequently mentioned in this chapter (though there are probably other species of vine and liana, as well). When the tree is covered to this extent, not only does photosynthesis become difficult, but its branches may also break or come crashing down from the sheer weight of the vines (photographed July 2018).

chimpanzees' environment, as with most environments. This opacity, beyond the vastness of its full scope with respect to the here-and-now, is also related to the instability resulting from continuous and ongoing disturbances and changes, including those that are essential for a healthy ecosystem—in which case, perhaps these should be referred to as 'fundamental instabilities.' In other words, the environment can exist as long as it stays incomplete due to the changes occurring on various time scales, such as seasonality and repeated recoveries from disturbances.

The extremity of the here-and-now is the edge of this incompleteness. Chimpanzees, by actually acting at this edge, are partly involved in the generation of the incomplete nature as a whole on which they themselves depend. It may be that this holistic picture is not usually consciously perceived or, to be precise, there may be no need to do so. But could we say that we can sense a part of it as a subtle malfunction? For example, contrary to our expectation of finding ripe ilombo fruits, by finding still unripe ilombo, only one single ilombo fruit, or a lack of fruit? To this partially emergent situation, chimpanzees, rather than giving up entirely, will respond partially with partial countermeasures, such as eating other things while seeking out ilombo fruits, accepting unripe fruits as fine for now, and recognizing that though there is only one, still, I can eat one!

Unstable and incomplete

The surroundings, both social and natural, that should support the here-and-now for individual chimpanzees are unstable. They are also incomplete in the sense that they exist in an ongoing present. Individuals are only partially involved in the here-and-now. The whole picture of their surroundings remains unseen, but there is no need for it to be seen. Their surroundings are called forth, albeit only partially, as an awareness of subtle malfunction in the here-and-now.

Meanwhile, when we shift our perspective from individuals to their surroundings, the here-and-now builds up unstable and incomplete surroundings as its own background for better or for worse. In recent years, many people living in what are called 'developed' societies have rediscovered nature as something problematic. In Chapter 5 of this volume, Yuji Takenoshita discusses this in terms of a conflict between narrative and reality. Here, I prefer to see it as a confusion over the manifestation of the environment as a problem that we have brought upon ourselves, how we have ended up rediscovering that nature is not something external, but rather something within which we exist.

Various environmental problems, including that of global climate change, are anthropogenic. And while this is hardly good news, it is thus that we have proven ourselves beyond reasonable doubt to be constituents of our own environment. My emphasis on indubitability is intended here as a rebuttal to humanity's romantic aesthetic toward (the concept of) pristine nature (Ichikawa

2003). Although the idea of nature (or the environment) as the object of an external gaze has engendered management efforts that have sought to support and maintain environmental values, human efforts to manage and stabilize ecosystems have only resulted in the loss of the biodiversity that was its goal (Reice 1994). The romantic view of nature itself is in the first place only partial, and also quite recent, and not a view shared by humanity in general. As Morton (2007: 21) points out, 'indigenous societies often refer to nature as a shape-shifting trickster rather than as a firm basis.'

No matter how important disturbances are to the environment, as Takenoshita notes in Chapter 5, the disturbances we have caused on a global scale have already caused damage that may be irrecoverable and cannot simply be dealt with in the same way as other disturbances. But what can we do about this now? The universal characteristic of living creatures that Adachi (2019: 386) calls 'dealing with the surrounding environment "on the whole" well' is congruent with something like the 'swing-by' that Hideaki Terashima speculates for humankind in Chapter 21. Humans have never been able to change these constantly shifting surroundings all at once, and nor is this the nature of those surroundings. Rather, us ordinary people commit to the here-and-now on usual bases to inhabit that extremity 'moderately well'—and it is perhaps by virtue of the accumulation that accrues as a result that we are able to play a part in the creation of new surroundings. Of course, this is not something that can be brought into being by humans alone. Needless to say, we must draw on the aid of innumerable other living things, from the largest organisms to invisible microorganisms.

Notes

1. Since chimpanzees make beds in the trees every day to sleep in, trees are essential for a good night's sleep (Zamma 2016).
2. I should note with regard to plant identification, I isolated and compared plant characteristics in the field based on the list in Itoh (2015b), which itself represents an update of the list in Nishida and Uehara (1981). I include the name of the naming authority with the species name in order to avoid confusion with the same name used by other naming authorities—that is, for separate species. For local names, I have revised the notation where there were significant deviations from the pronunciations provided by local informants (unpublished material).
3. This was recently reported as a new species unique to this area.
4. Note that the same name refers to a different tree species even within the Tongweland.
5. This is the sum of the mortality rate and the proportion of newly introduced individuals.
6. Kasoge values compare favorably with the average of old world and new world turnover rates (1.61%, 1.81±0.19%, and 1.80±0.15%, respectively) (Itoh and Nakamura 2015b).

References

Adachi, K. (2009) 'Hikouzou no shakaigaku: Syudan no kyokuso e' (The sociology of anti-structure: Toward a climax of groups). In K. Kawai (ed.), *Shudan—Jinrui shakai no shinka* (Groups: The evolution of human sociality). Kyoto University Press, 3–21.

Adachi, K. (2019) 'Toward the environmental others: An ethological essay on equilibrium and coexistence'. In K. Kawai (ed.), *Others: The Evolution of Human Sociality*. Kyoto University Press and Trans Pacific Press, 365–386.

Chave, J., J. Olivier, F. Bongers, P. Châtelet, P.-M. Forget, P. van der Meer, N. Norden, B. Riéra and P. Charles-Dominique (2008) 'Above-ground biomass and productivity in a rain forest of eastern South America'. *Journal of Tropical Ecology*, 24: 355.

Granados, J. and C. Körner (2002) 'In deep shade, elevated CO_2 increases the vigor of tropical climbing plants'. *Global Change Biology*, 8(11): 1109–1117.

Gross-Camp, N. and B. A. Kaplin (2005) 'Chimpanzee (*Pan troglodytes*) seed dispersal in an afromontane forest: Microhabitat influences on the postdispersal fate of large seeds'. *Biotropica*, 37(4): 641–649.

Hamilton, A. C. (1974) 'The history of the vegetation'. In E. M. Lind and M. E. S. Morrison, with a contribution by A. C. Hamilton, *East African Vegetation*. Longman, 188–209.

Ichikawa, M. (2003) 'Kankyō mondai ni taisuru mitsu no seitaigaku' (Three ecological approaches to environment problems). In K. Ikeya (ed.), *Chikyū kankyō mondai no jinruigaku* (The anthropology of global environmental problems). Sekai Shishosha, 44–64.

Ingwell, L. L., S. Joseph Wright, K. K. Becklund, S. P. Hubbell and S. A. Schnitzer (2010) 'The impact of lianas on 10 years of tree growth and mortality on Barro Colorado Island, Panama'. *Journal of Ecology*, 98: 879–887.

Itoh, N. (2015a) 'Climate and climatological trends in the Kasoje forest'. In M. Nakamura, K. Hosaka, N. Itoh and K. Zamma (eds), *Mahale Chimpanzees: 50 Years of Research*. Cambridge University Press, 143–149.

Itoh, N. (2015b) 'Appendix I: Floral list'. In M. Nakamura, K. Hosaka, N. Itoh and K. Zamma (eds), *Mahale Chimpanzees: 50 Years of Research*. Cambridge University Press, 691–716.

Itoh, N. (2017) 'Duality of the mode of coexistence and action selection: Groups and the emergence of "institutions" in chimpanzees'. In K. Kawai (ed.), *Institutions: The Evolution of Human Sociality*. Kyoto University Press and Trans Pacific Press, 141–163.

Itoh, N. and D. Muramatsu (2015) 'Patterns and trends in fruiting phenology: Some important implications for chimpanzee diet'. In M. Nakamura, K. Hosaka, N. Itoh and K. Zamma (eds), *Mahale Chimpanzees: 50 Years of Research*. Cambridge University Press, 174–194.

Itoh, N. and M. Nakamura (2015a) 'Diet and feeding behavior'. In M. Nakamura, K. Hosaka, N. Itoh and K. Zamma (eds), *Mahale Chimpanzees: 50 Years of Research*. Cambridge University Press, 227–245.

Itoh, N. and M. Nakamura (2015b) 'Mahale flora: Its historical background and long-term changes'. In M. Nakamura, K. Hosaka, N. Itoh and K. Zamma (eds), *Mahale Chimpanzees: 50 Years of Research*. Cambridge University Press, 150–173.

Itoh, N. and T. Nishida (2007) 'Chimpanzee grouping patterns and food availability in Mahale Mountains National Park, Tanzania'. *Primates*, 48: 87–96.

Kakeya, M. (2017) *Kakeya Makoto chosaku-shū dai 1-kan: Hito to shizen no seitaigaku* (The collected writings of Makoto Kakeya, volume 1: The ecology of nature and humanity). Kyoto University Press.

Lambert, J. E. (1999) 'Seed handling in chimpanzees (*Pan troglodytes*) and redtail monkeys (*Cercopithecus ascanius*): Implications for understanding hominoid and cercopithecine fruit-processing strategies and seed dispersal'. *American Journal of Physical Anthropology*, 109(3): 365–386.

Morton, T. (2007) *Ecology without Nature: Rethinking Environmental Aesthetics*, Harvard University Press.

Nakamura, M. and N. Itoh (2015) 'Seeds from feces: Implications for seed dispersal and fecal analyses'. In M. Nakamura, K. Hosaka, N. Itoh and K. Zamma (eds), *Mahale Chimpanzees: 50 Years of Research*. Cambridge University Press, 259–273.

Nishida, T. (1974) 'Yasei chinpanjī no seitai' (The ecology of wild chimpanzees). In R. Otsuka, J. Tanaka and T. Nishida (eds), *Jinrui no seitai* (Human ecology). Kyoritsu Shuppan, 15–60.

Nishida, T. (1990) 'A quarter century of research in the Mahale Mountains: An overview'. In T. Nishida (ed.), *The Chimpanzees of the Mahale Mountains: Sexual and Life History Strategies*. University of Tokyo Press, 3–35.

Nishida, T. (2002) 'Mahare chōsa shōshi' (A brief history of the Mahale survey). In T. Nishida, S. Uehara and K. Kawanaka (eds), *Mahare no chinpanjī: 'Pansuroporojī' no 37-nen* (Mahale chimpanzees: 37 years of 'Panthropology'). Kyoto University Press, 5–28.

Nishida, T. and S. Uehara (1981) 'Kitongwe name of plants: A preliminary listing'. *African Study Monographs*, 1: 109–131.

Phillips, O. L. and A. H. Gentry (1994) 'Increasing turnover through time in tropical forests'. *Science*, 263: 954–958.

Phillips, O. L., P. Hall, A. Gentry, S. Sawyer and R. Vasquez (1994) 'Dynamics and species richness of tropical rain forests'. *Proceedings of the National Academy of Sciences of the USA*, 91: 2805–2809.

Phillips, O. L., Y. Malhi, B. Vinceti, T. Baker, S. L. Lewis, N. Higuchi, W. F. Laurance, P. Núñez Vargas, R. Vásquez Martinez, S. V. Laurance, L. V. Ferreira, M. Stern, S. Brown and J. Grace (2002) 'Changes in growth of tropical forests: Evaluating potential biases'. *Ecological Applications*, 12(2): 576–587.

Reice, S. R. (1994) 'Nonequilibrium determinants of biological community structure'. *American Scientist*, 82(5): 424–435.

Sasek, T. W. and B. R. Strain (1988) 'Effects of carbon dioxide enrichment on the growth and morphology of kudzu (*Pueraria lobata*)'. *Weed Science*, 36(1): 28–36.

Swaine, M. D. and J. Grace (2007) 'Lianas may be favoured by low rainfall: Evidence from Ghana'. *Plant Ecology*, 192(2): 271–276.

Zamma, K. (2016) *Chinpanjī wa 365-nichi beddo o tsukuru: Nemuri no jinrui shinkaron* (Chimpanzees make their beds 365 days a year: A sleep-based approach to human evolution). POPLAR Publishing.

CHAPTER 3

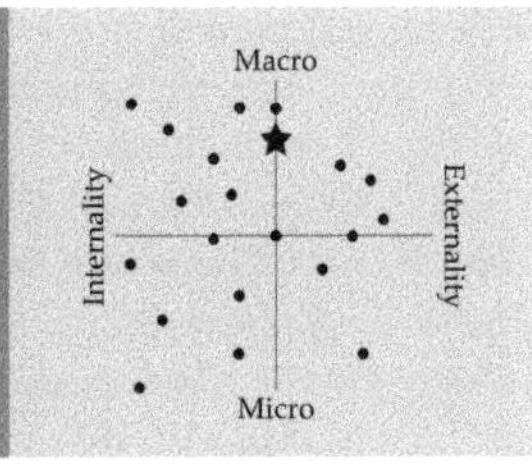

The Ontology of the Extreme:

Seeking an Evolutionary Basis for Human Sociality in the Systems of Inuit Subsistence and Modernity

Keiichi OMURA

Keywords: Canadian Inuit, subsistence system, 'land' (*nuna*), modernity, global networks

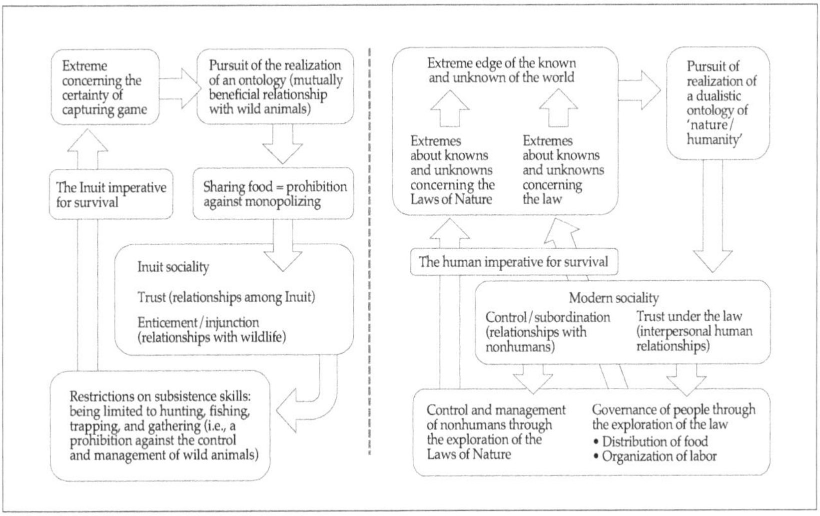

The extreme as a 'limit-defining/horizon-opening tension-relation' that gives direction to humanity's imperative for survival, thereby transforming it into the driving force of a system that encompasses humans and nonhumans alike, is inextricably bound up with human sociality. For example, an 'extreme concerning the certainty of capturing game,' which gives direction to the Inuit imperative for survival and transforms it into the driving force of a subsistence system that generates and sustains a 'land' (*nuna*) comprised of both humans and nonhumans (wild animals), is inextricably bound up with Inuit sociality, which involves a relationship of 'enticement/injunction' between the Inuit and animals, as well as relationships of trust among the Inuit themselves. Moreover, the modern extreme, which concerns 'what is known and what is not known about the world,' arises from the entanglement of two other extremes—those that concern 'what is known and what is not known about the law' and 'what is known and what is not known about the Laws of Nature.' This modern extreme gives direction to the survival imperative of the inhabitants of modernity and transforms it into the driving force of the modern system that creates and sustains global networks that encompass humans and nonhumans alike and is inextricably bound up with modern sociality, which involves a relationship of 'control/subordination' between the human and the nonhuman, as well as relationships of 'trust under the law' among humans themselves. The construction of systems that engender diverse socialities by giving direction through these various extremes to our survival imperative—which for humans is our most fundamental strength—constitutes an evolutionary basis for human sociality.

Introduction

> [T]he extreme simultaneously promises dangerous unknown territory (whether physical or epistemological) and the possibility for naming and sustaining control over that territory. This dual meaning is evident in its etymological history. While in modernity 'extreme' is about states and conditions which can endlessly be overcome, earlier meanings denoted an absolute limit beyond which nothing could pass...[The] extreme's limit-defining/horizon-opening character, far from contradictory, is in fact key to its work as a mediating trope, site, and figure in late modernity, which brings limits and horizons into relation in complex and often unexpected ways. [In the deployment of] cultural stagings of the extreme (from violence, temporalities, and weather, to finance, corporate brands, and embodiments)...their representation is only possible within the tension-relation of limit/horizon. Taking epistemological and ontological relationality as its central characteristic, extreme's extensiveness enables a productive meeting of the known and unknown, the present and future, the here and now and the as-yet-not. As such, the extreme offers an explicit mode of action and meaning-making as an organizing resource for contemporary subjects grappling with the disjunctures of late modernity. It is both symptomatic of and responsive to a historical moment when modernity's promises have faded, even as they are renewed by technological, scientific, economic, and social change. (Valentine et al. 2012: 1008–09)

The trope of the 'extreme' is ubiquitous in today's world. As pointed out by David Valentine and other anthropologists of science who track the practices of scientists at the forefront of confronting the extremes of the human world, such as outer space and the ocean depths, the trope of the extreme is frequently invoked across a broad spectrum of politics, economics, society, and culture—one that includes climate change and other global environmental shifts, war and other forms of violence, finance, sports, cosmetic surgery, corporate brands, and religious movements—and is tied up with 'the problem of what humans, human practices, and human environments have become and are becoming' (Valentine et al. 2012: 1008).

The reason that the trope of the extreme seems so current that it could be said to be fashionable, as Valentine et al. (2012) note, is that the extreme, which condenses the tension-relation between the definition of the limits of the known world and the opening of the horizons of the unknown world, could be said to constitute a vivid reflection of our ambivalent situation in the world of late modernity. As the inhabitants of late modernity, we feel a sense of impending crisis in having come face to face with the dangers of unknown territories beyond our control, and which lie at the limits of the safe and known world. At the same time, we have the expectation that it may be possible to open up new horizons that transcend this known world (which, although safe, is increasingly obsolete) by overcoming even the dangers of these unknown territories, and so bring them under our own control. In this sense, the circumstances of the late modern world that we inhabit could be said to be condensed in their purest form within the practices of scientists at the forefront of extreme conditions such as outer space and the ocean depths—scholars who literally inhabit the tension-relation of the extreme as they seek to open up new horizons while defining the limits of the human world. As in the attempts made by Valentine et al. (2012), it is by tracking the practices of the scientists who confront these extremes that we may be able to cast light on the current state of our species as inhabitants of the world of late modernity and, in doing so, illuminate our future.

Projects like those of Valentine and his colleagues pose a question that resonates not only in the world of late modernity, but also in anthropology, a field that explores humanity's limits and potential in the past, present, and future. Is the extreme a state of affairs that is exclusively characteristic of the world of late modernity? Certainly, the dramatic emphasis on the extreme as a trope, as Valentine and his colleagues indicate, is a phenomenon peculiar to late modernity, and one that may not have existed in the past. However, even if this trope was not present as such within the span of past human history, can it be true that humanity has never experienced the extreme as a site to open up new horizons while defining our own limits? Can it be true that, prior to the world of late modernity, humanity sought simply to define its limits, without ever attempting to open up new horizons and transcend them?

In this chapter, I think about these questions by analyzing the mechanisms by which the world of the Inuit,[1] the indigenous people of the Canadian Arctic, is generated and sustained through a subsistence system, which I then compare with the mechanisms by which the modern world is generated and sustained through the modern system.

To that end, I first briefly examine the reality of subsistence activities for Inuit in the Canadian Arctic. In doing so, I clarify the uncertainty involved in catching wild animals with subsistence techniques such as hunting, fishing, trapping, and gathering in the context of the Arctic's ecological environment, and how, because of this, Inuit who engage in such subsistence activities are always driven to the extreme as a tension-relation between defining the limits of and expanding the horizons of the inherent uncertainty of these subsistence activities. Then, I analyze the mechanism of the Inuit subsistence system under the conditions of this reality, revealing how the extreme to which the Inuit are driven by their subsistence system is effectively used in the mechanism by which that subsistence system operates and by which the Inuit lifeworld, called *nuna* (land), is continuously generated. Finally, based on Bruno Latour's (1993) analysis of the modern world, I compare this mechanism of the Inuit subsistence system with the mechanism by which the modern world is generated and sustained through the modern system, while structurally transforming and translating the Inuit subsistence system into the modern system. In doing so, I am able to identify commonalities and differences between the two, providing a foundation for considering an evolutionary basis for human sociality.

The extreme in the reality of Inuit subsistence activities

Inuit subsistence activities, which consist of hunting, fishing, trapping, and gathering, often provide milieux for the unforeseen. This is due to the fact that predicting what might happen in the future, while possible to a degree, can never be perfect in such activities. This means that subsistence activities cannot always follow a detailed plan that has been worked out in advance. Instead, what is needed, while allowing for some degree of prediction about the future, is the versatility to respond flexibly to sudden contingencies.

That the Inuit emphasize the importance of versatility toward such contingencies—not only in subsistence activities but throughout their lives—is a common refrain highlighted in the writings of many ethnographers such as Jean Briggs (1968, 1970) and Hugh Brody (1975), among many others (e.g., Morrow 1990; Omura 2013a; Willmott 1960). In Inuit society, people do not insist on sticking to preconceptions or prior plans or worry about unforeseen circumstances. Rather, those who are able to actively accept a situation and deal with it flexibly, taking advantage of the potential hidden therein, are regarded as *inirnit* (adults) possessed of *ihuma-* (wisdom), while *qaplunaat* (white people) and scientists who attempt to rigidly follow detailed plans are viewed as *nutaraqpaluktuq* (childish) (Briggs 1968; Omura 2013a).

But, of course, this does not mean that Inuit are indiscriminate in how they engage in their subsistence activities. As the findings of Arctic anthropology have shown thus far (cf. Omura 2013a, 2018), Inuit are well-versed in knowledge about things like seasonal movement patterns and the behavioral habits of wildlife, and they make full use of this knowledge in the practice of their subsistence activities. For example, when fishing, anglers will set ambushes by reading the patterns of the movements of trout and arctic char, which migrate seasonally in large shoals from the ocean to the lakes or from one lake to another, taking advantage of their habits to catch them with weirs or bait (Stewart 1992, 1993). Also, in the case of seals that move unseen beneath the frozen sea ice, groups of hunters will guess at the animals' behavioral patterns in order to ambush and harpoon those that surface to breathe at one of dozens of breathing holes that the seals have opened in the ice (Balikci 1989). Furthermore, in the case of caribou on the land, since the animals repeat their seasonal migrations over distances of hundreds of kilometers, hunters will ambush and catch their quarry by reading these migration routes (Balikci 1989; Stewart 1990). The same is true for the narwhals and bowhead whales that range over vast expanses of ocean (Omura 2013a).

In fact, while staying in the homes of hunters almost every summer and winter since 1989, I have routinely witnessed exchanges of information among Inuit hunters on matters pertaining to the distribution, migration status, and behavioral habits of wild animals (Omura 2013a, 2018). Hunters returning from their outings will almost always gather in their parents' households to discuss the state of the day's subsistence activities, exchanging and updating information concerning these wild animals on a daily basis. Moreover, for hunters and elders, visiting each other's households to spend hours in discussion over meals or tea is part of their daily routine. In such conversations, information on trends pertaining to wild animals is frequently exchanged and will then be put to use in subsequent subsistence activities. Now that the telephone and email through internet have enabled the possibility of long-distance communication, it is not at all uncommon for hunters to share information about the distribution and movements of caribou over the phone or by email with family members in settlements located hundreds of kilometers away.

Nevertheless, no matter how much insight one might glean about the tendencies of wild animals through these daily exchanges of information, the fact that they move freely across the vast environment of the far north, and the impossibility of tracking all their movements, means that the reality of the situation is that they remain difficult to encounter. And even though remote sensing technologies that employ artificial satellites and the like might work with caribou, which move across the land, it remains almost impossible to track the movements of fish and marine mammals like whales and seals that move through the oceans. Moreover, although there are patterns in the seasonal migrations of wild animals, they do not always follow the same route every year, and the timing of their journey can fluctuate by days or weeks. Accordingly, one will not always encounter the animals in accordance with predictions. In fact, in my own experiences accompanying Inuit during their subsistence activities, it was not uncommon that we never came across the wild animals that were our quarry, even after spending close to twelve hours searching (cf. Omura 2013a). Other times, herds of wild animals that had appeared the previous year would suddenly disappear the following year, and it was often related to me how famines that had occurred in the past had been due to changes in the migration routes of these wild animals. No matter how much one attempts to grasp the distribution and migration routes of wild animals, there will be limits, and in the practice of subsistence activities, the basic principle is to shoot on sight, in the sense that if you catch your quarry, you will not have let the opportunity pass.

In seal hunts conducted on the sea ice in the winter and spring, especially, the fact that the seals' movements under the ice are invisible to the eye makes them extremely difficult to catch. In the seal hunt, groups of hunters lie in ambush to harpoon seals that have risen to breathe at one of the dozens of breathing holes the seals make in the ice. On such occasions, hunters will first explore the state of the sea ice by piercing it with their harpoons near pressure ridges or cracks in the ice, seeking sites where seals are likely to open breathing holes. Once a grouping of breathing holes is found, the hunters will examine their state and distribution and, based on a knowledge of the behavioral habits of seals and relying on their dogs' sense of smell, will select several holes as being most likely to be in use at that time, at which point several hunters will lie in ambush near the holes. Even so, it is difficult to identify at which of the dozens of holes a seal will choose to make its appearance. In addition, because seals are sensitive to changes in conditions on sea ice (such as noises and the condition of the hole) and will slip away if they detect any disturbance, it is always uncertain whether a seal will emerge at a hole where an ambush has been laid, and some things will always be left to fate.

In the context of such subsistence activities, there will always be limits, no matter how familiar an Inuk is with the distribution, migration routes, or behavioral habits of wild animals. There will always be some uncertainty not only with regard to whether the wild animals sought as quarry will be caught, but whether they will even be encountered at all. In that sense, the element of luck is ever present in Inuit subsistence activities. However, as Inuit do not hunt for sport, as in trophy hunting, but are engaging in subsistence activities to secure their daily food and resources for living, their survival depends on the elimination of these vagaries of fortune as much as possible.

This is why, as we have seen, Inuit hunters endeavor so diligently day after day to get a sense of animals' movements, continually accumulating and updating a store of knowledge that will ensure they are able to encounter wild animals as reliably as possible. Moreover, they hone their versatility to respond flexibly to sudden contingencies so that they are better able to manage when predictions based on that knowledge miss the mark. At the very limits of certainty with respect to their livelihoods, where the element of luck inevitably remains as a factor, Inuit hunters who engage in subsistence activities as a matter of survival strive to extend the horizons of certainty while minimizing the uncertainty associated with these activities. In this sense, the experience of Inuit who practice subsistence activities is nothing other than the tension of an extreme that seeks to expand horizons while defining the limits of certainty.

If an extreme can thus be seen even in Inuit subsistence activities, this means that the extreme is not a phenomenon peculiar to the world of late modernity. Of course, the extreme in the context of Inuit subsistence activities and the extreme in the context of the world of late modernity, although alike in that they are both limit-defining and horizon-opening tension-relations, will still differ in terms of what those limits and horizons pertain to. This is because, whereas as noted at the outset in the passage quoted from Valentine et al. (2012), the extreme in the world of late modernity is a 'limit-defining/horizon-opening' tension-relation pertaining to the known and unknown worlds, the extreme in Inuit subsistence activities pertains to the uncertainty of those subsistence activities. So why does such a difference arise between the extreme in Inuit subsistence activities and the extreme in the world of late modernity? And regardless of this difference, why is it that the extreme appears as a limit-defining and horizon-opening tension-relation in both instances?

To think about this problem, I want to consider how the extreme as it pertains to uncertainty in subsistence activities functions within the subsistence system by which the Inuit world of *nuna* is continuously generated. This is because the subsistence activities by which the Inuit experience the extreme as it pertains to uncertainty are bound up in their social–cultural–economic system called 'subsistence,' which means that this problem cannot be solved unless we consider these activities in the context of this system.

The Inuit subsistence system: a mechanism that continuously generates the world (*nuna*)

Arctic anthropologists have pointed out that the existence of a social–cultural–economic system called 'subsistence' is commonly observed among the Inuit and Yupik peoples who are distributed widely from the Arctic zone of easternmost Siberia through North America to Greenland (e.g., Bodenhorn 1989; Fienup-Riordan 1983; Kishigami 2007; Nuttall 1992; Stewart 1991, 1995; Wenzel 1991). This subsistence system connects Inuit with wild animals and with each other through processes of food acquisition, distribution, and consumption, which are regulated by an ontology expressed as the world as it should be, thus working continuously to generate the Inuit lifeworld, called *nuna* (land), and the extended family group, which is the basic unit of Inuit sociality.[2] The mechanism by which this Inuit subsistence system continuously produces their lifeworld as well as the extended family groups embedded therein can be modeled as a circulation system, as follows (Omura 2012, 2013b).

First, by employing subsistence techniques such as hunting, fishing, and trapping, an Inuk enters into a food supplier/recipient relationship with the animal. At the same time, it is when the resulting food or other resources for living (things that are received) are shared out among the Inuit that the extended family group, which constitutes the foundation of everyday social relationships among the Inuit, is generated. The important thing here is that this food sharing is normalized among the Inuit, since in Inuit ontology, which shows the world as it should be, the ideal relationship to be achieved in subsistence between Inuit and wildlife is a reciprocal relationship, as follows.

In Inuit ontology, animals are regarded as possessing a *tarniq* (spirit), so that even when the body perishes, the spirit persists. However, for this animal spirit to be reincarnated in a new body, the Inuit must share and consume its flesh completely. To this end, the animal spirit presents its own body to the Inuit to become shared food so that it may be reincarnated. From the Inuit point of view, this is akin to receiving assistance from the animal, since it amounts to a gift of the resources needed for survival. In other words, in the Inuit world as it should be, the aim is to realize a reciprocal relationship whereby animals help Inuit survive by offering their bodies as food, while the Inuit, by sharing that food among themselves, help the animals reincarnate into new bodies.

As a result of the guiding principles provided by this ontology, the Inuit, situated in an inferior position with respect to the animals as the 'receivers of food,' must always share the food presented to them by the animals among themselves, so that sharing among Inuit becomes a norm. This is because, unless the food is shared among the Inuit, the animal spirits cannot reincarnate, and the animals will stop offering themselves to the Inuit as food. What is important here is the convention by which this sharing norm is imposed on the Inuit by the animals, rather than by the Inuit themselves. Consequently, a cooperative relationship is established among the Inuit in which everyone follows the same norm and shares food without anyone enjoining anyone else to do so.

Thus, within the context of this equal sharing of food, relationships of trust arise among the Inuit whereby they rely on each other's willingness to cooperate in the same commensal act of eating in the mutual expectation that there will be no betrayals or usurpation of food. By standing in a permanently inferior relation to animals as the recipients of food and attributing the injunction to impose food-sharing norms on themselves to the animals, the Inuit abjure any hierarchical relationships of dominance and subordination among themselves, instead establishing relationships of trust based on egalitarian cooperation.

The cost of this, however, is that the Inuit are proscribed from taming (domesticating) animals. If the Inuit tamed animals, it would no longer be the animals, but rather the Inuit that had tamed the animals, that impose the sharing norm on the Inuit. In this case, Inuit would be placing injunctions on other Inuit, reimposing relationships of dominance and subordination that they should have abjured. In order for equitable relationships to be established among the Inuit, animals must be placed in a superior position relative to any Inuit person. As a result, the Inuit are unable to employ methods that lead to domination or control of animals, such as raising livestock. Instead, they must devote their energies to hunting, fishing, trapping, and gathering, tactics that involve mastery of techniques of enticement from a subordinate position—tactics that are, in other words, the weapons of the weak.

This normalization of sharing encourages not only the sharing of food, but also cooperation and the sharing of subsistence techniques and knowledge. Because the norm ensures that Inuit always share food obtained as a result of these subsistence activities, it allows them to cooperate in subsistence activities without any fear of usurpation or betrayal. If food acquired as a result of subsistence cannot be monopolized and must always be shared with others, then hunting or fishing alone or monopolizing techniques or knowledge will no longer have any positive meaning, whereas working together by sharing techniques and knowledge will.

As sharing and cooperation thus become commonplace, these practices facilitate the enrichment and refinement of the knowledge and skills used for subsistence, thereby increasing the chances that the Inuit will re-engage in relationships with new individual wild animals as receivers of food (subject to the injunction that it be shared) relative to the animals who bestow it (and who enjoin the food-sharing norm). Then, once this relationship is realized in practice, everything reverts to the starting point of subsistence for the same cycle to be repeated once more. In this way, the rebirth of the animal spirits in new bodies is achieved, just as in Inuit ontology, in the form of the reproduction of the relationship between Inuit and wildlife in the cyclical process of subsistence. That is, the spirits of the animals that are endlessly reborn indicate the relationship between the Inuit and the animals.

Thus the relationship of enticement/injunction between the Inuit and the animals becomes entangled with relationships of mutual cooperation and trust among the Inuit, and clearly differentiates the Inuit (and their extended families). As this cycle progresses, two categories are differentiated and brought to the surface, namely the Inuit (and their extended families) as people who (for the Inuit) are to be trusted and cooperated with, and the animals as objects of enticement whose injunctions must be followed. Of course, the Inuit/animal relationships that are generated and renewed through this cyclical process are not limited to a particular species of wildlife, but are formed with many different animal species. For this reason, the Inuit extended family finds its genesis in the articulation of various relationships that are cyclically generated with multiple animal species through subsistence practices, emerging as one node within a network formed by a myriad of interconnected nodes of various animal populations. It is precisely this network that constitutes the Inuit lifeworld—the *nuna* (land).

This Inuit subsistence system could be said to articulate and relate the various extended family groups of the Inuit and herds of wild animal species out of an undifferentiated chaos without either humans or nature, thereby creating order for the entire lifeworld of the Inuit and thus generating their *nuna*. Nor is this practice of subsistence solely an economic activity for acquiring and circulating the resources necessary for survival; it is also a sociopolitical, ethical, and cultural practice that realizes an ontology represented as a social intercourse that forges ecological relationships with wild animals and social relationships among the Inuit themselves—as the world that should be. In this sense the Inuit subsistence system binds together all the realms of politics, economics, society, and culture as the core of Inuit society. It is a system that establishes the order of their lifeworld in its entirety all at once. As the Inuit themselves frequently emphasize, it merits being called a way of life, one that constitutes the cornerstone of one's entire life as an Inuk.

The extreme in the context of the Inuit subsistence system

When we attempt to examine the extreme edge of certainty that Inuit experience in their subsistence activities under the conditions of the Inuit subsistence system discussed above, we find that this extreme edge of certainty is cleverly staged to facilitate the continuous generation and maintenance of equal relationships of trust among the Inuit. In this subsistence system, the sharing of food is continuously normalized through the system's cyclical operation. And it is in the course of this process that equal relationships of trust among the Inuit are generated and sustained. Yet the system's continuous operation owes to the fact that the Inuit, driven by the system to an extreme over certainty in their subsistence activities, must work frantically to expand its horizons in order to survive at the limits of the certainty of catching their quarry, effectively ensuring the continuous operation of the subsistence system, as follows.

As we saw above (in the section 'The extreme in the reality of Inuit subsistence activities'), Inuit hunters, driven to this extreme, stake their lives on frantic efforts to extend the horizons of the certainty of catching their quarry because there are limits to their ability to grasp the movement of wild animals—try as they might, the uncertainty of luck is ever present. It follows that if they were somehow able to eliminate that uncertainty, they would be freed from this extreme and thus no longer need to strive to extend the horizons demarcating the limits of the certainty of catching their quarry. The only feasible way would be to herd the animals together in one place so that they could not move freely, thus bringing them all under Inuit control, because it is impossible to completely grasp the movement of animals in the wild, even using remote sensing technologies. But as we saw earlier, within this subsistence system, the Inuit never stand in a dominant position relative either to wild animals or themselves: the domination and control of wildlife by the Inuit is prohibited in order that the sharing of food is normalized among the Inuit without any hierarchical relationships of dominance and subordination. They are barred from eliminating the extreme by herding, dominating, or controlling animals in a single place, and have no choice but to devote themselves to enticing the animals from a subordinate position through hunting, fishing, trapping, and gathering. Within the limits of the certainty of catching wild animals, the inevitable consequence of the subsistence system's generation and maintenance of equal relationships of trust among the Inuit, and its repudiation of hierarchical relationships of domination and subordination, is that the Inuit are driven to the extreme of staking their own survival on their frantic efforts to expand these horizons under this restriction of subsistence techniques.

Once the Inuit are driven to this extreme edge of certainty in their subsistence activities, while still continuing to refine their subsistence techniques in order to make up for the persistent uncertainty that cannot be overcome, they are forced by their ontology to practice the food sharing that is the premise of acquiring their quarry, so that this sharing becomes normalized. Were the acquisition of quarry certain, it would be taken regardless of whether or not food was shared and the imperative to share would lose its coercive force. The uncertainty whereby a successful hunt on one occasion offers no guarantees for the next hunt is precisely what makes it difficult for people to ignore what is indicated as the condition of a successful hunt by their ontology; that is, 'the sharing of food is the prerequisite of a successful hunt'—in other words, 'unless you share food, you will no longer be able to catch it.' Whether or not this is true, and whether or not one believes it, when one does catch some quarry under this condition, the safest thing is to share it out regardless. Thus, with normalization of food sharing among Inuit, their subsistence system begins to circulate and thereafter operates continuously, the cycle repeating with each successful catch, with the result that relationships of trust among the Inuit are continuously generated.

Of course, regardless of being driven to this extreme edge of certainty in their subsistence activities, the Inuit would still be willing to share their quarry with each other to bring their ontology into being. Since an ontology shows the world to be realized, they would consider it their mission to make that ontology a literal reality, and to that end would share the food bestowed upon them by the wild animals in the attempt to help the spirits of the animals be reborn into new bodies. In that case, after sharing the spoils of their hunt, they can realize their ontology in truth by taking new quarry as they devote themselves to hunting, fishing, trapping, and gathering, the subsistence techniques used from a subordinate position to entice animals, without trying to herd, dominate, or control them. Certainly, there is no way to directly confirm whether the ontology has actually been realized as a literal reality by sharing food, or to ascertain whether or not the animal spirits have indeed been reborn. However, at the very least, so long as the Inuit always share the food bestowed by the animals and devote themselves to their subsistence techniques, then when they are given food by new animals, they may regard themselves as having assisted the rebirth of the spirits of the animals that had previously bestowed food upon them.

Moreover, Inuit need not always consider making their ontology a literal reality to be their mission in order to end up aspiring to do so. This is because, for Inuit who are not hunters,[3] realizing this ontology can have positive meaning, and there are reasons to act toward doing so themselves. Because non-hunting Inuit are not directly involved in subsistence activities and do not receive their food directly from wild animals, were this ontology not realized, they would cease to be able to obtain food without worrying about being subordinated to hunters. This is because, in that case, non-hunting Inuit would receive the food as a one-way gift from the hunters, in which they would be unilaterally indebted to the hunters for food and thus would be exposed to the fear that they would be subordinated to the hunters. However, if the ontology was realized, then it would be possible for non-hunting Inuit to secure food and survive without worrying about being subordinated to hunters, under the condition that they should never try to hoard nor monopolize but always share the food received from animals.[4] This is because, if the ontology was realized, then non-hunting Inuit would also participate in subsistence activities in the form of aiding the rebirth of the wild animal spirits by simply sharing and eating the food received from the animals and setting up the premise that animals give food to Inuit. At the same time, in this case the animals would be giving food not only to the hunters but also to the Inuit who had once shared food as a whole people.

However, in the case of the hunters who actually take part in subsistence activities and are given food by the animals directly (unlike for Inuit who do not hunt), the realization of this ontology in practice has no positive meaning. For hunters, the realization of the ontology amounts to a unilateral relinquishing of food that they have worked so hard to catch at the extreme edge of certainty in their subsistence activities without any individual reward. However, even in this case, when hunters are driven to this extreme edge of certainty in their subsistence activities, a positive reason for realizing their ontology will emerge for the hunters. Try as they might, their subsistence activities will never be free of uncertainty. So if there is always a chance that they may not capture their quarry, hunters will always be exposed to the possibility of finding themselves in the same position as Inuit who do not hunt. In that case, were this ontology to be realized, even if they failed to take their own quarry, they would still be able to survive by eating food shared by other hunters, without worrying about being subordinated to those more successful hunters. For hunters, as long as they are driven to the extreme edge of certainty in their subsistence activities, the realization of their ontology is insurance for times when their own subsistence activities fail and they are not blessed with quarry.

Considered in this way, we find that the extreme edge of certainty to which Inuit are inevitably driven by their subsistence system functions as an apparatus that does not so much work actively to get Inuit to share their food, but rather deters the monopolization of food by creating a precarious situation in which food would be unavailable if it were not shared. Especially in the case of hunters, not everyone is willing to adhere to the sharing norm, since sharing food means relinquishing food that they have obtained themselves in the practice of their own subsistence activities. Quite the opposite, rather than sharing with other Inuit the food they have obtained for the common pot, so to speak, for them to decide instead to lay a claim to the food as their own, and then to distribute any surplus, can give rise to the possibility that they are trying to dominate other Inuit by making them indebted to the hunters for food. This is precisely why a way has been devised to drive hunters into the precarious situation of being unable to obtain food unless it is shared—so that they cannot help but adhere to the norms of sharing.

This has important implications for iterations of the subsistence system's cyclical process. To start with, unless the hunters who are originally bestowed with food by the animals relinquish that food to share it with other Inuit, the system's cyclical process will not engage. Moreover, no matter how much it is said that ontology shows the world as it should be, because hunters have no positive reason to realize that ontology in terms of their own survival, they will not always share their food. Conversely, Inuit who do not hunt do have a positive reason to realize that ontology in terms of their own survival. Even so, no matter how much they desire to realize that ontology by sharing their food in order to survive, they cannot even try to do so unless the hunters are willing to relinquish and share with non-hunting Inuit the food that they have obtained by their own efforts. In other words, the continuous operation of this subsistence system is predicated first and foremost on ensuring that when hunters catch their quarry, they always share the spoils of the hunt. To that end, it is not enough to rely on hunters voluntarily accepting the realization of the Inuit ontology as their own mission, which is precisely why the sharing of food by hunters is ensured by driving hunters into a precarious situation in which they may be unable to obtain food unless it is shared.

What is important here is that the extreme edge of certainty in the subsistence activities that drive the continuous operation of the subsistence system (by deterring the monopolization of food by the Inuit) is an inevitable consequence of that subsistence system. This suggests that the Inuit survival imperative is directed by the subsistence system to render it useable in the system's cyclical operation, as follows.

There is no need for Inuit to adhere to the subsistence techniques of hunting, fishing, trapping, and gathering for their continued survival. Rather, if they wanted to ensure their survival, it would be better for them to round up and manage wild animals to ensure the ability to catch their quarry. However, to do so would be to shatter the equal relationships of trust among the Inuit, which are generated and sustained by their subsistence system. To prevent this from happening, their subsistence techniques are limited to hunting, fishing, trapping, and gathering. By endeavoring to survive under this limitation, the Inuit are driven to the extreme of trying to expand their horizons while defining the limits of the certainty of catching their quarry. As a result, the Inuit survival imperative is converted into the power to extend the horizons of the certainty of catching their quarry. It is precisely because they are motivated by this survival imperative that when the prerequisite of a successful hunt is suggested to them by their ontology under this condition, regardless of whether or not it is true, and whether or not they believe it, a precarious situation is brought into being in which they can no longer obtain food unless they share it. Under these circumstances, the Inuit's desperate efforts to extend the horizons of certainty lead them to actually follow a norm of food sharing without exception, which leads to the continuous generation of relationships of trust among them. Thereafter, in order that these relationships of trust not be broken, the domination and control of wild animals is forbidden and subsistence techniques limited, thereby bringing the cycle back to the beginning.

In this way, in the Inuit subsistence system, the imperative to survive, at the extreme edge of the certainty of catching their quarry, is converted into the power to extend (even a little) the horizons of that certainty. Then, when the prerequisite of a successful hunt is suggested by their ontology under this condition, this power to extend the horizons is converted into a force that binds the Inuit to a norm of food sharing. The cyclical operation of this system is sustained by the driving force of these two powers that are successively converted from the Inuit survival imperative by the system, and it is in this process that relationships of trust are generated and sustained among the Inuit themselves.

The extreme in the context of late modernity: a mechanism that continuously expands global networks

It follows from what we have considered thus far that the extreme edge of certainty in Inuit subsistence activities can be understood to constitute an engine that drives the subsistence system by using the Inuit survival imperative as fuel to generate and sustain relationships of trust among the Inuit. From this mechanism, we can perhaps derive the reason that the extremes of late modernity pointed out by Valentine et al. (2012) take the form of the extreme of the 'knowns and unknowns' of the world. For it is by structurally transforming and translating the Inuit subsistence system that the extreme edge of the knowns and unknowns of the world in late modernity can be derived from the extreme edge of certainty within Inuit subsistence activities.

To this end, let us first posit that, in place of the Inuit ontology we have considered thus far, the world that the Inuit should seek to achieve would be suggested by a modern dualistic ontology. As Latour (1993) has shown, the aim of modernity has been to achieve a dualistic world of 'humanity / nature.' Here, it is held that nonhumans in the realm of 'nature' follow the Laws of Nature while humans in the realm of 'humanity' make their own laws based on reason, which they agree upon and follow in accordance with that reason (cf. Omura 2010). If we go by this dualistic ontology, the relationships to be achieved in subsistence between Inuit and wildlife as well as among themselves in the Inuit ontology could be rewritten as follows.

First, since wild animals follow natural laws, they have no spirits, and since there is no spirit to be reborn, they will not be reincarnated even once the animals are shared and consumed by the Inuit. On the other hand, the Inuit will formulate laws based on reason in order to share their food and organize their own labor to obtain food, and will accordingly share their food while organizing their labor in agreement with those laws. In the case of the Inuit ontology, because the sharing of food not only encourages cooperation and the sharing of subsistence techniques and knowledge but also aids the rebirth of animal spirits and sets up the premise for catching quarry, the practice of sharing prescribes collaboration in the performance of subsistence activities. But in the case of the modern ontology, the sharing of food bears no relation to collaboration, and the law that the Inuit have made based on reason must simultaneously prescribe not only the sharing of food but also the collaboration by which that food is obtained.

To summarize, Inuit ontology, now rewritten according to the modern dualistic ontology of humanity / nature, would state that 'animals have no spirits; they behave and breed in accordance with natural laws, and perish when they are eaten by Inuit. For their part, the Inuit make laws according to reason, share food, and organize their labor in subsistence activities in accordance with those laws.' Here, Inuit and animals respectively follow the different regimes of laws based on reason and the Laws of Nature. As a result, the Inuit, without being forced by any other Inuit, voluntarily share food and organize their labor in accordance with laws made on the basis of reason and consensus, and relations of trust will be generated and sustained under Inuit laws independent of their relationship with animals. Thus the way has now been opened for Inuit to round up and manage wild animals, which has been prohibited in order not to shatter the equal relationships of trust among the Inuit in the case of the Inuit ontology.

Of course, even if this is the case, simply being freed of the constraints of subsistence techniques does not mean that the Inuit will now be able to round up and manage animals, because these wild animals will not necessarily abide by the Inuit's intentions. Since wild animal species already move freely across the environment of the far north, where they live according to their own varied circumstances, they cannot so easily be rounded up and managed by the Inuit. The key point here is that the modern dualistic ontology provides the Inuit with a solution to this problem. In this ontology, wild animals are supposed to abide by the Laws of Nature. Thus, were this ontology to be realized, it would become possible to control wild animals by dominating them through the use of the Laws of Nature. Since wild animals cannot escape the Laws of Nature, even were they to attempt to bolt from the Inuit's enclosures and escape their control, as long as the Inuit enclose and control the animals in accordance with nature's laws, the animals will be unable to escape.

However, for this to be possible, it is not enough simply to *say* that animals obey the Laws of Nature; a dualistic ontology must be realized by making it clear that it is actually the case that they do so. To this end, having first observed the behavior and breeding of wild animals and derived a hypothesis about natural laws based on these observations, that hypothesis must be compared with the actual behavior and breeding of animals to confirm that the animals actually abide by those laws. Now the enclosure and management of wild animals comes itself to constitute the realization of this ontology through the experimental model made up of hypothesis and verification. If it was possible to observe animals under identical conditions at all times by rounding up and managing wildlife all in one place, it would be possible to derive more effective and accurate hypotheses about natural laws that govern those animals than it would be otherwise. If animals could be successfully managed based on the laws derived by such hypotheses, then it would be confirmed through experimentation that the animals are governed by those laws. In other words, by deriving and making use of natural laws, the very process of continuing to round up and manage animals would itself realize this ontology through a process of experimentation consisting of hypothesis and verification.

Thus, by rounding them up, the Inuit maintain control over the animals at all times. This ensures that they can catch animals at will, which means that the extreme edge of the certainty of catching animals disappears. Even so, a new extreme emerges in its place. Certainly, the natural laws that govern animals can be used to ensure that they do not escape their enclosures or Inuit control. However, whether or not they are truly governed by the natural laws derived by observing their behavior and breeding can only be confirmed retrospectively; that is, by means of experiments to compare those hypothetical laws with the empirical facts of animal behavior and breeding and then using those laws to determine whether or not the animals are being well managed. Therefore, knowledge of the Laws of Nature will not come all at once, but only tentatively and gradually, leaving an ever-present boundary between what is known and what is not known about the Laws of Nature. What is more, because the knowledge of these natural laws facilitates the acquisition of food through the containment and management of animals, the Inuit will end up striving to extend the horizons of the unknown at the limits of knowledge, and so be driven to the extreme edge of the knowns and unknowns of the Laws of Nature.

The same applies to the formulation of laws in accordance with reason. With the realization of this ontology, as shown above, both the sharing of food and the ideal way to cooperate in subsistence activities will be prescribed by law, and all Inuit will obey this law based on reason, so that relationships of trust will be established. However, such rational and universally agreed upon laws are not self-evident from the start. Certainly, in this ontology everyone agrees to obey the law, because it is based on reason, and everyone follows that reason. However, just as in the case of natural laws, what constitutes rationality and reason in this ontology is not prescribed, so there is actually no alternative but to think that a law that is universally obeyed actually holds true and that both the law and those who obey it adhere to reason. In other words, as in the case of gaining knowledge about natural laws, without creating a hypothetical law and testing it through experiments that put it into operation, we cannot know what manner of laws are rational or universally agreed upon or anything about the reason to which that law adheres.

Moreover, in the subsistence system converted in accordance with the modern dualistic ontology, since relationships of trust are established under such laws, Inuit are free of the obligations to share food and cooperate with everyone, and so long as they are under a universally agreed law, they will have the freedom to distribute and consume food and to organize their labor in subsistence activities any way they like. Of course, there are any number of ways to distribute and consume food and ways to organize labor, as well as ways to combine and link the two, so it is most difficult to find a law that would prescribe them in a way that would be universally agreed upon. Moreover, as we saw earlier, because neither rationality nor reason have been made clear in advance, the only option is to actually make and apply a law, and then to confirm by case-by-case experiment whether or not everyone agrees. Accordingly, knowledge about the reason to which the Inuit and the law adhere does not come all at once, but can only be gleaned tentatively and gradually, so the boundary between the known and the unknown will always end up arising here. What is more, because prescribing the distribution and consumption of food pertains directly to their individual survival, the Inuit will have to work to extend the horizons of the unknown at the limits of their knowledge of laws based on reason, and so be driven to the extreme edge of the knowns and unknowns of the law.

In this way, in the subsistence system converted in accordance with the modern dualistic ontology, the Inuit would end up being driven to a double extreme.[5] One is the extreme edge of the knowns and unknowns of the Laws of Nature, while the other is the extreme edge of the knowns and unknowns of laws based on reason. What is important here is that once the Inuit begin to multiply animals within this system, these two extremes will begin to cycle endlessly and, in the context of that circulation, the Inuit and animals will end up being linked even more closely than before.

In the context of the extreme edge of the knowns and unknowns of the Laws of Nature, if one could manage animals while further clarifying the natural laws that govern the behavior and breeding of those animals, then by rounding up and containing those animals, more than simply stabilizing food provision, it would also become possible to augment food supply by multiplying the number of animals by way of intervening in their behavior and reproduction in accordance with natural laws. Once food supply increases, the Inuit population will gradually increase in turn. As a result, in order to feed this larger population, it will be necessary to further multiply the animals to continue to augment the food supply in a stable fashion. Increasing the food supply by multiplying animals in this way will gradually lead to further increases in the Inuit population, which will necessitate more animals to support that population, so that the increase in the number of animals and the Inuit population will accelerate in a cyclical fashion. Once the animal population has been increased in this way, it will no longer be possible for Inuit to stop multiplying animals, so the number of animals and the Inuit population will only continue to increase as the cycle continues.

In the course of this circulation, efforts to extend the horizons of the unknown at the limits of the knowledge of the Laws of Nature will become endless. In this system, once the animal population is increased, it becomes impossible to stop multiplying animals in husbandry to feed the burgeoning population of Inuit that eventuates, with the result that efforts to extend the horizons of the unknown at the limits of the knowledge of natural laws also become unstoppable. Moreover, in keeping with increases in food supply and population, the Inuit will have to adjust their methods of distribution and consumption to balance the burgeoning food supply with the increasing population in a way that is universally agreed upon. Furthermore, the fact that it becomes impossible to stop multiplying animals means that the Inuit will have no choice but to improve and develop the way to organize their labor in order to facilitate that increase by extending, even if only a little, the horizons of the unknown about natural laws. Thus, in order to nourish and ensure the survival of their burgeoning population, the Inuit would have to work ceaselessly to increase the stock of animals by expanding the horizons of the unknown at the limits of their knowledge of natural laws. And the more they strove, the more they would be forced to strive ceaselessly to extend the horizons of the unknown at the limits of their knowledge of the laws that organize their labor and the distribution and consumption of food.

As this endless process of circulation progressed, the Inuit and animals would grow ever more intimately linked as they adhered respectively to the different orders of reason and natural laws. In order to nourish and ensure the survival of their burgeoning population, the Inuit would be driven to ceaseless efforts to extend the horizons of the unknown at the limits of their knowledge of the Laws of Nature and the laws that organize their labor and the distribution and consumption of their food. Through these efforts, the animals would be increasingly husbanded by the Inuit, and the Inuit would be increasingly mobilized to husband the animals. So, by working to develop a closer relationship with these animals in this ongoing and endless effort, the horizon of their ignorance of the natural laws that govern those animals will be endlessly extended, just as the horizons of their ignorance of the laws that organize their labor and the distribution and consumption of their food will be endlessly extended toward a universally agreed upon law in accordance with reason. Of course, since the number of animals and Inuit alike both increase in this process, the Inuit mobilized for labor and the animals being husbanded will continue endlessly to expand spatially along a temporal axis, creating yet another boundary between known and unknown. Thus the Inuit will be driven to the extreme of trying to extend the horizons of the unknown world by staking their own survival at the limits of the known world that comprises themselves and their animals.

Here, by extending 'the Inuit' to 'humanity' writ large and 'the animals' to the entirety of 'the nonhuman,' we find that the Inuit subsistence system converted according to the modern dualistic ontology is what generates the global networks of modernity elucidated by Latour (1993).[6] Within these networks, the more we strive to realize the dualistic world of humanity/nature while drawing distinctions between a humanity that obeys reason and a nature that obeys natural laws, the more the human and the nonhuman will intermingle. The more the human and the nonhuman intermingle, the clearer the reason obeyed by the human and the natural laws obeyed by the nonhuman will become. Thus the effort to realize the dualistic ontology of humanity/nature accelerates the extension of the network comprising the human and the nonhuman, a network that now is extending all over the surface of the entire planet and even into outer space. Having started to manage the nonhuman, intervene in its activities, and increase the amount of resources exploited thereby, humans, to survive, will be driven to the extreme edge of the realm of nature, where they will attempt to expand the horizons of the unknown at the limits of their knowledge of the Laws of Nature, while at the same time also being driven to the extreme edge of the realm of humanity (i.e., society, politics, and economics), where they will attempt to expand the horizons of the unknown at the limits of their knowledge of laws based on reason. And in humanity's frantic efforts to open new horizons beyond the limits of these two extremes, the human and the nonhuman will become ever more closely intertwined and the network created through this intertwining will continue endlessly to expand spatially along the temporal axis.[7]

Thus what is constantly occurring at the outer edge of the human–nonhuman network—which, driven by the engine of humanity's effort to stake its survival on the realization of a dualistic ontology of humanity / nature, is expanding ever faster—is nothing other than the extreme of late modernity concerning the known and the unknown of the world pointed out by Valentine et al. (2012). In the course of an unlimited circulation of the increase in resources and population growth that begins with an increase in the resources exploited by managing the nonhuman and intervening in its activities, the human–nonhuman network, fueled by humanity's survival imperative and driven by an engine consisting of extremes in the realms of both nature and humanity, is driven to the extreme of attempting to open up the horizons of the unknown world while defining the limits of the known. These efforts to extend the horizons of the unknown world at the limits of the known world are part of efforts to realize the dualistic world of humanity / nature. Apportioned to the two engines constituted by the extreme in the realm of nature and the extreme in the realm of humanity, and fueled by humanity's survival imperative, these efforts work to further expand the human–nonhuman network and, moreover, to produce an extreme that concerns the known and unknown world.[8]

Considered thus, we find that the reason that the extreme of late modernity indicated by Valentine et al. (2012) takes the form of an extreme that concerns the knowns and unknowns of the world is that by seeking to realize a dualistic world separated into a humanity that obeys laws based on reason and a nature that obeys natural laws, humanity is liberated from the proscription on dominating and managing the nonhuman—as well as, at the same time, from the obligation to cooperate with everyone and to share food among themselves.

By aspiring to a dualistic world of humanity / nature, humans become free to dominate and control the nonhuman while pursuing natural laws. More than this, they also acquire the freedom to distribute and consume food among themselves and to organize their labor in subsistence activities as they please, on the condition that this is done under a law to which everyone consents on the basis of reason. And while the freedom to dominate and manage the nonhuman produces an extreme in the realm of nature while extinguishing the extreme that concerns certainty in the Inuit subsistence system, the freedom to distribute and consume food and organize labor under universally agreed laws produces an extreme in the realm of humanity. Furthermore, the expansion of the human–nonhuman network, driven by extremes in the realms of nature and humanity that have been produced by these two freedoms, in turn produces an extreme that concerns the knowns and unknowns of the world.

Important here is that in the Inuit subsistence system, the extreme that constitutes the engine of the system functions to constrain the Inuit and to prevent the extension of the world of *nuna* comprised by the relationships between the Inuit and wildlife and among themselves. In contrast, the twin extremes as engines of the modern system are produced with the liberation of humanity from the constraints that have been placed on the Inuit, and function as an engine that expands the world of the human–nonhuman network. In the Inuit subsistence system, the extreme edge of certainty in their subsistence activities that was produced as the engine of the system continuously drives the system by enjoining the Inuit to share while prohibiting their enclosure or management of wild animals, and thus generates equal relationships of trust among the Inuit while prescribing that Inuit labor should be employed for collective benefit. Accordingly, from the very outset, increasing the number of animals by intervening in their behavior or breeding was never an option, so neither the Inuit population nor their food supply could increase. Conversely, in the modern system, humanity is free of any prohibition on the domination or control of the nonhuman. At the same time, they are also liberated from the obligation to cooperate with everyone and to share food among themselves. This leads to the emergence of extremes in the two realms of nature and humanity, and these two extremes serve to drive the ever-accelerating expansion of the human–nonhuman network.

Accordingly, the extreme of late modernity pointed out by Valentine et al. (2012) and the extreme of the Inuit subsistence system—though they share a common function as engines (fueled by humanity's survival imperative) of systems that generate and sustain equal relationships of trust among humans while generating and sustaining their lifeworlds—may nevertheless be considered to be diametrically opposed to each other, as follows. When generating and sustaining equal relationships of trust among the Inuit, the extreme in the Inuit subsistence system, by prohibiting the domination or control of animals while simultaneously restricting the distribution and consumption of food, as well as how labor is to be employed, functions to prevent the expansion of the world generated and sustained by that system. Conversely, the extreme in the context of late modernity, by opening the way to the unrestricted domination and control of animals and, at the same time, to the unrestricted distribution and consumption of resources and organization of labor, drives the expansion of the world generated and sustained by that system. These two extremes are alike in that they both give direction to humanity's survival imperative to drive their respective systems, but differ in terms of the direction in which that force is turned. In the subsistence system, this is a direction that imposes limits on the Inuit to prevent the expansion of their world; in the modern system, it is the opposite direction, which grants humanity freedom to bring about the expansion of the world.

Ontology of the extreme: an evolutionary basis for human sociality

As I have described thus far, using the extreme as an axis to compare the Inuit subsistence system with the modern system can teach us several things about the evolutionary basis for human sociality.

One is the possibility that the extreme as a limit-defining / horizon-opening tension-relation is universally present in human societies as an indispensable engine for systems that, fueled by humanity's survival imperative, generate and sustain worlds that encompass human and nonhuman alike. As discussed above, extremes emerge not only in the modern system that generates and sustains the global networks of our contemporary world, but also in the Inuit subsistence system. In both systems, unless the imperative to survive is steered by the system to the extreme and converted into a force that will drive the system's continuous operation, the circulation of the system will be disrupted. In this sense, the extreme represents an indispensable engine for the continuous operation of either system. Of course, whether this applies universally to all human societies is a matter for future empirical study. Nevertheless, the fact that the extreme functions in this way not only in the modern system but also in the context of the Inuit subsistence system could be said to indicate the possibility that the extreme is not a phenomenon peculiar to modernity, but something that appears universally in systems for generating and maintaining humanity's various lifeworlds and plays a key role in the respective systems.

What must be noted here is that it is not only human societies, but human–nonhuman worlds that are generated and sustained by the systems in which these extremes function as engines. Human / nonhuman relationships are inextricably incorporated as part of the cyclical processes of both systems, and human social groups are only ever generated and sustained in the ways in which these are intertwined with the generation and maintenance of human / nonhuman relationships. This indicates that human sociality inevitably includes not only relationships among humans, but also relationships with the nonhumans. As we have discussed in this chapter, in either system, the kinds of relationships built between humans are cyclically linked to and inseparable from the kinds of relationships built with the nonhumans around them. In this sense, the sociality generated by the engine of the extreme could be considered as semio-materially generating and sustaining a human–nonhuman world, establishing order in relationships among humans as well as in their relationships with the nonhumans. This is the second thing that our comparison of the Inuit subsistence system and modern system teaches us.

However, although the extreme might have a universal function for humans as the engine for systems by which humanity generates and sustains worlds while generating a human sociality that engages with the human and nonhuman alike, the way that it manifests will differ significantly for each system. As discussed, in the Inuit subsistence system, the extreme manifests as the extreme edge of the certainty of catching their quarry, while in the modern system it manifests as the extreme edge of the knowns and unknowns of the world that emerges through the association of the extreme edge of the knowns and unknowns of the law and the extreme edge of the knowns and unknowns of the Laws of Nature. Also, whereas sociality in the former is a combination of the relationships of enticement/injunction between Inuit and wild animals and relationships of trust and cooperation among the Inuit, in the latter it is a combination of relationships of control/subordination between the human and the nonhuman and relationships of trust between humans under the law. Moreover, these manifestations of sociality and the extreme are linked in that their respective extremes are produced as engines to drive the operation of the systems that produce their respective socialities. This suggests that there can be various types of extremes, and the possibility that different extremes may be called for and produced depending on the type of sociality and the kind of systems that will be built for it. This is the third thing that our comparison teaches us.

The important thing here is that depending on the extreme produced as the engine of a system that produces each sociality, the human survival imperative will be converted into a force that drives the continuous operation of the system that produces the corresponding sociality. This shows us that the construction of systems that engender diverse socialities through these various extremes by directing and controlling our survival imperative—which for humans is our most fundamental strength—constitutes an evolutionary basis for human sociality. The imperative to survive is universally apparent as the strength to continue being oneself, not only for humans but for all creatures, and all manner of beings.[9] Likewise, as an interaction with all creatures and all manner of beings in one's environment, including other members of one's own species, sociality is not something that is restricted to humankind. If this is the case, then what kind of system has humanity built to provide a solution to the universal problems concerning sociality: that is, the problems of how to direct the universal survival imperative shared by all beings, humanity included, in order to interact not only with other members of our own species but also with the other beings all around us? To ask these questions is to think about human sociality not only in the context of human history, but within the evolutionary history of all beings, including living creatures. This is the last thing that our comparison has to teach us.

And if, just as in the Inuit subsistence system and the modern system considered in this chapter, the extreme functions as an engine to convert the universal survival imperative into a power that drives the operation of systems to produce sociality not only for humans but for all creatures and all manner of beings, then a focus on the extremes that appear in each type of human society, for each type of creature, and for each manner of being will offer a starting point for thinking about the evolutionary history of those various socialities. What sorts of extremes are various creatures and various beings, including the people of human societies, driven to? In each case, what sort of power is derived from their imperative to survive: that is, their inherent strength to continue being as they are? And how does this power drive the operation of systems that produce their own unique forms of sociality? We must elucidate the mechanisms that produce the power to drive the operation of the various systems that engender diverse forms of sociality from the universal strength by which all beings strive to remain themselves. This is the ontology of the extreme, which will constitute a foundation for thinking about the evolutionary basis of sociality not only for humanity, but for all creatures and all manner of beings.

Notes

1. Since their transition from a seasonally migratory lifestyle to sedentary life under the Canadian federal government's assimilation policies from the 1950s to the 1960s, all aspects of Inuit life have been buffeted by waves of rapid change. As a result, Inuit today inhabit a high-consumption society no different from those of any other advanced nation. However, even under these circumstances, their traditional subsistence activities have not lost their importance as a foundation that provides support for their lifestyle and identity. Certainly, the ways such activities are carried out today have changed significantly, with many hunters engaging in subsistence in parallel with wage labor. This is because subsistence activities have become highly mechanized, with equipment such as high-powered rifles, snowmobiles, all-terrain vehicles, and metal boats with outboard motors. Cash is needed to procure and maintain this equipment, as well as to cover the costs of gasoline and ammunition. Nevertheless, subsistence is enthusiastically practiced, to the extent that people say that 'Inuit who do not engage in subsistence are not Inuit' (Omura 2013a). Moreover, despite the fact that purchasing processed foods with cash income is common, animal meat obtained through subsistence activities is preferred as *niqinmarik* (real food), and is considered essential for maintaining ethnic identity. Furthermore, the distribution of such meat still functions to sustain social relationships (Kishigami 1996, 2007; Stewart 1995; Wenzel 1991).
2. Social groups that represent the basis for Inuit social relations are the kinship groups known as *ilagiit* (kindred), among which the extended family group, called *ilagiimariktut* (true *ilagiit*), serves as the basic unit for everyday social relations. The *ilagiit* is a group of 'related people who may go away but then come back and then share food, help each other, and stay together' (Balikci 1989: 112). Moreover, the *ilagiimariktut* 'refers to members of an extended family group who live in the same place and have close relations in economic and other activities, i.e., those who form a concrete social group. As a result, [this group] is generally comprised of one's parents, siblings, wife, children, grandchildren, uncles, aunties, grandparents, and cousins' (Kishigami and Stewart 1994: 421).

With this extended family group at the core, fictive or quasi-kin relationships (such as adoption) and voluntary associations (such as namesake relationships and avoidance relationships) that transcend kin relationships are sometimes formed, resulting in the production of complex social relationships.

3. References to Inuit who are today no longer directly involved in subsistence activities as a result of social change generally indicate women.
4. Herein lies a weakness of the Inuit subsistence system, which at the same time allows for the division of labor among Inuit. As I have pointed out elsewhere (Omura 2019), in this system it is impossible to eliminate freeloaders who consume food without making any contribution of their own. At the same time, however, as Bodenhorn (1990) has pointed out, it is by virtue of this fact that women who do not directly go out to take part in subsistence activities are nevertheless able to participate, thereby enabling a sexual division of labor.
5. In this sense, we could say that our planet's situation as a laboratory, as noted by Suzuki et al. (2016), is something that is not limited to today, but began with the advent of the modern world. Compared with the Inuit subsistence system, we might even say that the modern system is characterized by the fact that it employs the realms of both humanity and nature as sites of experimentation.
6. While it is certainly true that the modern world constitutes networks, as a phenomenon, it is also true that the generation and extension of those networks is supported by a system. This is the same as the relationship between the Inuit world and their subsistence system. As I demonstrate in this chapter, the Inuit world, called *nuna*, also constitutes a network, as a phenomenon, but one whose generation and maintenance is supported by their subsistence system.
7. Herein we find the basis for the 'progress' mindset characteristic of modernity.
8. Certainly, the industrial capitalist system and the nation-state regime, historically, were integral elements of modernity. However, they are not *inevitable* aspects of modernity, essentially speaking. As shown here, the essential element of modernity is the performance of experiments in the realms of both humanity and nature, and, in the realm of humanity specifically, the performance of experiments for establishing systems that will generate and sustain equal relationships of trust under the law. Industrial capitalism and the nation-state regime are transitory institutions that have arisen as a result of the accumulation of historical contingencies as these experiments have progressed. If further experiments are conducted in the future, it may be that other institutions will be created, for better or for worse.
9. This power can be rephrased as what Spinoza (2018) called 'conatus.' Valuable insights may be gleaned by considering this chapter's discussion of extremes and sociality in relation to Spinoza's treatise on politics (Spinoza 2007; cf. Ueno 2014).

References

Balikci, A. (1989) *The Netsilik Eskimo*. Waveland Press.

Bodenhorn, B. (1989) *The Animals Come to Me, They Know I Share: Inupiaq Kinship, Changing Economic Relations and Enduring World Views on Alaska's North Slope*. PhD thesis, Cambridge University.

Bodenhorn, B. (1990) 'I Am Not the Great Hunter, My Wife Is'. *Études/Inuit/Studies*, 14(2): 55–74.

Briggs, J. L. (1968) *Utkuhikhalingmiut Eskimo Emotional Expression*. Department of Indian Affairs and Northern Development, Northern Science Research Group.

Briggs, J. L. (1970) *Never in Anger: Portrait of an Eskimo Family*. Harvard University Press.

Brody, H. (1975) *The People's Land: Whites and the Eastern Arctic*. Penguin Books.

Fienup-Riordan, A. (1983) *The Nelson Island Eskimo: Social Structure and Ritual Distribution*. Alaska Pacific University Press.

Kishigami, N. (1996) 'Kanada kyokuhoku chiiki ni okeru shakai henka no tokushitsu ni tsuite' (Characteristics of social change in Canada's far north). In H. Stewart (ed.), *Saishūshuryōmin no genzai: Seigyou bunka no hen'yō to saisei* (Hunter-gatherers today: The transformation and renewal of subsistence cultures). Gensousha, 13–52.

Kishigami, N. (2007) *Kanada Inuitto no shoku bunka to shakai henka* (Food culture and social change among Canada's Inuit). Sekaishisosha.

Kishigami, N. and H. Stewart (1994) 'Gendai Netsurikku Inuitto shakai ni okeru shakai kankei ni tsuite' (Indigenous social relations in a contemporary Canadian Inuit society: A case study from Pelly Bay, Northwest Territories, Canada). *Kokuritsu minzokugaku hakubutsukan kenkyū hōkoku* (Bulletin of the National Museum of Ethnology), 19(3): 405–448.

Latour, B. (1993) *We Have Never Been Modern*. Harvard University Press.

Morrow, P. (1990) 'Symbolic actions, indirect expressions: Limits to interpretations of Yupik society'. *Études/Inuit/Studies*, 14(1–2): 141–158.

Nuttall, M. (1992) *Arctic Homeland*. University of Toronto Press.

Omura, K. (2010) 'Shizen = bunka sōtai shugi ni mukete: Inuito no senjūmin undō kara miru gurōbarizēshon no mirai' (Toward natural-cultural relativism: Considering the future of globalization from the standpoint of the Inuit indigenous movement). *Bunka jinruigaku* (Japanese journal of cultural anthropology), 75(1): 54–72.

Omura, K. (2012) 'Gijutsu no ontorogī: Inuito no gijutsu fukugō shisutemu o tōshite miru shizen = bunka jinrui-gaku no kanōsei' (The ontology of technology: Considering the potentiality of natural-cultural anthropology through an analysis of the Inuit technological-complex system). *Bunka jinruigaku* (Japanese journal of cultural anthropology), 77(1): 105–27.

Omura, K. (2013a) *Kanada inuito no minzoku-shi: Nichijō-teki jissen no dainamikusu* (Ethnography of Canadian Inuit: Dynamics of everyday practices). Osaka University Press.

Omura, K. (2013b) 'The ontology of sociality: "Sharing" and subsistence mechanisms'. In K. Kawai (ed.), *Groups: The Evolution of Human Sociality*. Kyoto University Press and Trans Pacific Press, 123–142.

Omura, K. (2018) 'Maps in action: Quotidian politics through boundary translational matrix for world multiple in contemporary Inuit everyday life'. In K. Omura, A. Morita, S. Satsuka and G. Otsuki (eds), *The World Multiple: The Quotidian Politics of Knowing and Generating Entangled Worlds*. Routledge, 68–82.

Omura, K. (2019) 'The ontology of the other: The evolutionary basis of human sociality and ethics in the formation and continuation of Inuit society'. In K. Kawai (ed.), *Others: The Evolution of Human Sociality*. Kyoto University Press and Trans Pacific Press, 225–244.

Spinoza, B. (2007) *Theological–Political Treatise*, Jonathan Israel and Michael Silverthorne (eds). Cambridge Texts in the History of Philosophy, Cambridge University Press.

Spinoza, B. (2018) *Ethics: Proved in Geometrical Order*, Matthew J. Kisner (ed.), Michael Silverthorne (trans.). Cambridge Texts in the History of Philosophy, Cambridge University Press.

Stewart, H. (1990) 'Dentō Netsurikku Inuitto no inukushuku ni yoru karibū ryō: Minzokugaku to kōkogaku no hazama' (Traditional Netsilik Inuit caribou hunting using Inuksuit). *Minzokugaku kenkyū* (Japanese journal of ethnology), 55(1): 75–86.

Stewart. H. (1991) 'Shokuryō bunpai ni okeru danjo no yakuwari buntan ni tsuite: Netsurikku Inuitto shakai ni okeru emono bunpai sekai-kan' (Gender roles in food distribution: Game, distribution, and worldview in the Netsilik Inuit society). *Shakai jinruigaku nenpō* (Annual report of social anthropology), 17: 45–56.

Stewart, H. (1992) 'Netsurikku Inuitto no gyorō: Natsu no yanaryō o chūshin ni' (Netsilik Inuit fishing: Focus on summer weir fishing). *Hokkaidō-ritsu Hoppō Minzoku Hakubutsukan kenkyū kiyō* (Bulletin of the Hokkaido Museum of Northern Peoples), 1: 31–52.

Stewart, H. (1993) 'Netsurikku Inuitto shakai ni okeru haru no seigyou: 5/6-gatsu no karibū ryō to gyorō o chūshin ni' (Springtime subsistence in Netsilik Inuit society: Focus on fishing and caribou hunting in May and June). *Hokkaidō-ritsu Hoppō Minzoku Hakubutsukan kenkyū kiyō* (Bulletin of the Hokkaido Museum of Northern Peoples), 2: 13–36.

Stewart, H. (1995) 'Gendai no Netsurikku Inuitto shakai ni okeru seigyou katsudō: Seizon kara bunka-teki sabaibaru e' (Subsistence activities in modern-day Netsilik Inuit society). *Dai 9-kai Hoppō minzoku bunka shinpojiumu hōkokusho* (People and cultures of the Tundra—Proceedings of the 9th International Abashiri Symposium), Hokkaido Museum of Northern Peoples, 37–67.

Suzuki, W., A. Morita and L. Nyland Krause (2016) 'Jinshinsei no jidai ni okeru jikken shisutemu: Ningen to hoka no seimei to no kankei no saikō e mukete' (Experimental systems in the Anthropocene: Toward a rethinking of the relationship between humans and other forms of life). *Gendai shisō*, 44(5): 202–213.

Ueno, O. (2014) *Spinoza 'Shingaku seiji-ron' o yomu* (Reading Spinoza's *Theological–Political Treatise*). Chikumashobo.

Valentine, D., V. Olson and D. Battaglia (eds) (2012) 'Introduction: Extreme: Limits and horizons in the once and future cosmos'. *Anthropological Quarterly*, 85(4): 1007–1026.

Wenzel, G. (1991) *Animal Rights, Human Rights*. University of Toronto Press.

Willmott, W. E. (1960) 'The flexibility of Eskimo social organization'. *Anthropologica*, 2: 48–59.

CHAPTER 4

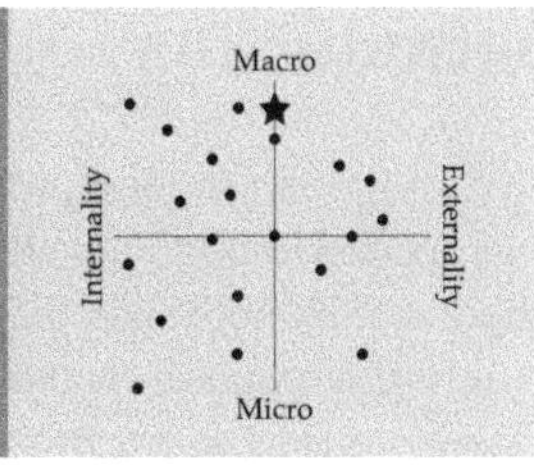

Mortality:

Ecological and Life History Interpretations of the Mortality Rate in Human Evolution

David S. SPRAGUE

Keywords: sustainability, mortality rate, demography, ecology, life history

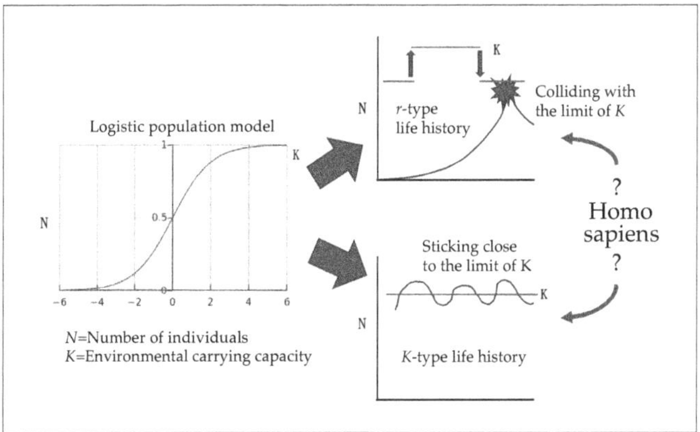

According to the theory of *r*/*K* selection, *r*-selected species are those with a high mortality rate and short life span that nevertheless attempt to leave behind offspring by overcoming fluctuations in their habitats with their vigorous reproductive capacity (*r*). Meanwhile, *K*-selected species are those with a low mortality rate and longer life span that raise few offspring and live at the brink of their habitats' environmental carrying capacities (*K*). Analyzing the life history of *Homo sapiens* in terms of *r*/*K* selection may be able to tell us whether our own species tends toward an *r*-strategy or a *K*-strategy, or else expresses both selectively, and enables us to assess growth and decline in human populations, perhaps even up to modern fertility decline.

Introduction

Reflecting on the long path of human evolution, should we consider the life of our species to be sustainable? In view of the fact that we have survived thus far, there can be no doubt that the life of our species has persisted in the past. But surely this persistence was never guaranteed. In the deep past, many species closely related to our own have met with extinction during the long passage of evolution after branching away from the evolutionary lineages of chimpanzees and gorillas to forge a path of their own. The sole surviving human lineage just happens to be that of our own species, *Homo sapiens*. Moreover, it is likely that many peoples died in obscurity at every step along the pathways by which *Homo sapiens* dispersed to every corner of the globe. Those who perished in this way may have fallen into a state of extremity, in some sense. It may be that having been unable to withstand these extreme conditions, they were unable to leave any descendants behind.

What kind of state would this extreme have been? In everyday terminology, what we mean by an 'extreme' condition is 'an extremely harsh environment.' Often, it seems, this description is used to refer to environments where life is difficult for humans as biological organisms. In this case it could refer, for example, to a very cold or hot region of the planet. However, harsh living conditions may have a subliminal presence in our day-to-day lives. In our contemporary daily experience, in which food, clothing, shelter, and medical care are in ample supply for many of us, we may only rarely have to worry about the hidden extremes that lurk just out of sight. However, just as we are susceptible to being hit by earthquakes, typhoons, or major natural disasters, the dangers of encountering the extremes that accompany illnesses, accidents, and other sudden emergencies are submerged in our daily lives. The possibility of falling into a state of extremity that cannot be surmounted is always present. The ultimate proof of this is death, the inevitable fate of all living things.

To persist, in other words, is to survive. Death, which interrupts survival, is both a key concern for society and a fundamental phenomenon in the context of the biological world. Death represents the end of individual existence. For a society, a mortality rate that remains high will have ramifications for the survival of that society. In biology, observing the death of an individual is one of the most important challenges for research. Death, alongside birth, is one of the most important events that all living things will necessarily experience. By contrast, reproduction is not necessarily experienced by all individuals. In fact, most individual organisms do not experience reproduction—they die before they have the opportunity to do so.

In biological theory, mortality represents an important explanatory factor at the level of populations. In evolutionary theory, mortality is a driving force of natural selection. Mortality rates that vary due to differences in the characteristics of individual organisms lead to differences in reproductive success. In ecology, mortality is a major determinant of population size. And mortality rates are arguably also a basic factor for evaluating the sustainability of an entire ecosystem. To elucidate the kind of organisms that humans have survived *as*, is, moreover, one of the basic tasks of ecological anthropology. In other words, when evaluating the evolution and ecology of living organisms, we need to consider mortality alongside—or even prior to—reproductive strategies. In recent years, biological research has tended to focus on organisms' reproductive strategies, but there is still a need, first, to summarize discussions on mortality rates as their counterpart.

Mortality in ecology and life history theory

Briefly, we must note that what the term 'mortality rate' signifies is not constant across the various theories that constitute the field of biology. Broadly conceived, theories for evaluating the sustainability of biological species can be divided into the two approaches of ecology and life history. Hence I now summarize the role that mortality plays in the respective discussions leading from these two approaches.

Ecology

Ecology is a field of research that seeks to explain the number and distribution of biological organisms. Ecologists focus on populations of organisms, the fluctuations in their size and distribution, and the environmental factors that cause such fluctuations. An ecologist's most basic task is to count the number of individuals and record the geographical extent of the population under study. Mortality constitutes a key factor for population fluctuations in ecology, since populations will increase or decrease in the balance between reproduction and death. When the number of deaths exceeds the number of births, the total number of individuals that make up a population may decrease, the distribution of that population may narrow, and the population itself may disappear. It follows that mortality rate is a variable of the population, which constitutes the most basic object of ecological research.

How do mortality rates relate to evolution in the ecological context? The focus of 'everyday' ecology is the study of immediately observable populations for the conservation and preservation of wild plant and animal species, population control, or resource management. The primary goal of this approach is to observe and explain population fluctuations in real time (i.e., in the short term). Thus the pursuit of ordinary ecological research presupposes a time scale short enough to obviate any need to consider evolution. However, the mortality rates elucidated in these ecological studies are an important factor that determines 'life history,' which seeks to offer a more comprehensive explanation of the life of biological organisms.

Life history

The concept of life history refers to an approach that seeks to comprehend and study the entire life span of organisms as a single whole. This is sometimes referred to in the social sciences as 'life course' or 'life cycle.' The study of life history in biology is a set of theories that seeks to explain the survival and reproductive strategies of organisms over the course of their entire life. The phenomena studied include the sequence of life events in which birth is followed by growth, maturity, reproduction, and aging, as well as the timing by which these occur (i.e., their order and spacing, as well as duration). Within life history theory, mortality plays an important role as a factor that limits other life events. However, the mortality that is important here is not the mortality rate of the population as a whole, but the age-specific mortality rate of individuals starting at birth. In other words, the idea is that the lifelong 'plan' of dying affects the schedule of other life history events, particularly the timing for reproduction.

Figure 4.1 is a schematic depiction of the age-specific survival curve and its inverse, the age-specific mortality curve, that represent this schedule of mortality. From an individual standpoint, the survival curve represents the probability that the individual will survive to a certain age after birth. From a group standpoint, this curve can also be read as the proportion of a cohort of individuals born at the same time that will survive to a certain age. As a general pattern, for most organisms, the survival curve trends fairly steeply downward from the point of birth. The reason for this is easier to understand when we look at the mortality curve. For many species, the mortality rate is very high for organisms immediately after birth. Although the mortality rate decreases with subsequent growth, it will eventually begin to increase again with age. In the end, the survival curve approaches 0%, and the mortality curve 100%. Naturally, the actual shape of the survival curve will differ depending on the species, but while the curve shows enormous diversity when we survey the

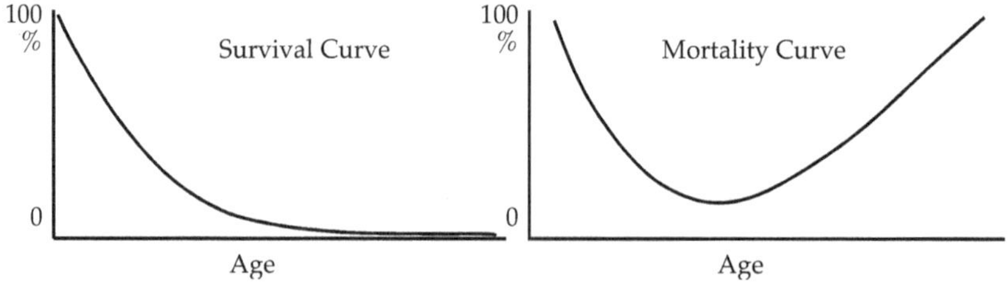

Figure 4.1: Schematic curves for survival and its inverse, mortality, of biological organisms

biological world as a whole, the basic principle of life history holds that each individual must live out its life by finding opportunities to grow and reproduce within the survival and mortality schedules of the species into which it is born.

To leave behind offspring before death is the prime objective for biological organisms. The time allotted to a newborn individual varies greatly depending on whether it will die relatively soon after birth or many years afterward. In some species, individuals will die within mere days of birth; in other species, they may live as long as thousands of years. Life history theory attempts to explain the life plan of these species with varying survival schedules. For example, it predicts that organisms that die relatively quickly may reproduce early, maybe only once, while longer-lived organisms will have opportunities to reproduce repeatedly.

Life history and ecology are linked in an odd relationship with each other. Typical studies in life history or ecology treat each other as preconditions, to be treated as fixed explanatory factors. In ecology, the life histories of a species under study are the preconditions entered into models of population fluctuation. For example, considerable variation in the scale and speed of population fluctuation exists between species with long life spans versus those with short life spans. However, there is no need for the life history of organisms itself to evolve when calculating a population fluctuation model. Conversely, ecology takes the role of specifying extrinsic factors that determine life history. Ecological phenomena, such as climate, vegetation (forests and grasslands), food, and predators, that govern the lives of organisms are treated as explanatory factors for life history.

However, even while ecology and life history are sometimes treated as mutually extrinsic, causal, or prerequisite factors, we must also be vigilant that theorists without warning may merge ecology and life history into a single theoretical framework. A good example of this is the theory of r/K selection.

r-strategy and K-strategy life histories

The theoretical system known as r/K selection seeks to combine ecology and life history in the attempt to explain the lives of many biological species at once. Although this theory has been widely invoked since the 1970s, it has fallen out of favor after many years of study and criticism (Parry 1981). That said, it remains heavily cited, notably in the context of human evolution. Of course, a great deal of criticism has been leveled against r/K selection due to the obvious fact that not all organisms fit into such a simple scheme. Nevertheless, given that many researchers have used it as an analytic tool with which to approach the analysis of the life history of many organisms, here, too, I similarly would like to use it as a tool to investigate the sustainability of human life and ecology.

The symbols r and K signify constants used in the formula of the logistic population model devised to represent a growing population of organisms

(Figure 4.2). The value of r sets the natural rate of increase for a population, and K sets the environmental carrying capacity. In the logistic population model, the size of the population traces the iconic S-curve. This curve expresses the basic principle that a population starting with a small number of individuals increases in size according to the rate of increase r for as long as environmental conditions allow. However, because an increase in the number of individuals entails an increase in density, as the population approaches the carrying capacity K, the rate of increase will fall due to some type of density effect until, finally, the number of individuals will balance at a level conforming precisely to the carrying capacity (Sibly et al. 2005).

When this population model was first proposed, it was intended to address the theory of island biogeography (MacArthur and Wilson 1967). For example, consider the case of an animal species that is brought by ocean currents (or people) to a Pacific island where it has not previously existed. With abundant food resources and no natural enemies, the number of individuals in this newly introduced species may increase explosively. In the limited space of the island, however, they will presumably eventually reach the limits of the environmental carrying capacity. In this case, we can easily imagine that the number of individuals trace the S-curve of the logistic model. Thus inspired by the logistic model that succinctly expresses the relationship between population increase and carrying capacity, biologists sought a general explanation for the life history of organisms with the relationship between r and K (Pianka 1970). As a result, to determine where actual plant and animal species lived along the S-curve became one major goal of life history studies. Do organisms normally live close to K, or do they move up and down the curve?

Against the backdrop of this logistic population model, r/K selection theorists classified species into r-selected species that possess large reproductive capacity and K-selected species that live tenaciously near their environment's carrying capacity (Parry 1981). The habitat environment for r-strategists is assumed to be extremely unstable. That is to say, although r-selected species have higher mortality rates and shorter life spans, they attempt to leave behind

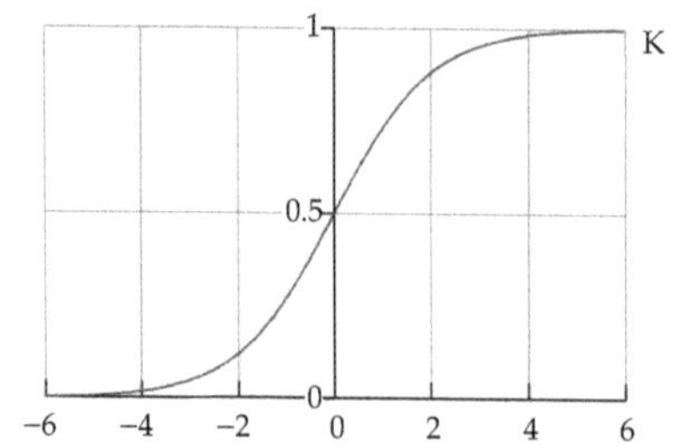

$$\frac{dN}{dt} = rN\left(1 - \frac{N}{K}\right)$$

N=Number of individuals
K=Environmental carrying capacity
r = natural rate of increase

Figure 4.2: Logistic population model

offspring by weathering environmental fluctuations in their habitat with their vigorous reproductive capacity (*r*). Conversely, *K*-selected species, which have lower mortality rates, live longer, and raise fewer offspring, can increase the size of their populations until they approach the brink of a habitat's carrying capacity (*K*). At this point, they will live in relative stability, enhancing their competitive capacity with sophisticated adaptive strategies.

Here we face the problem of the 'density effect' that should control the number of individuals which has increased into the neighborhood of *K*. Researchers have frequently pointed out that the meanings of 'density effect' and 'carrying capacity' are blind spots in the logistic population model (e.g., Dewar 1984; Moore 1983; Parry 1981). The population model is merely a mathematical formula, and *K* is a simple constant defining the curve with a rather vague meaning in biological terms. It simply envisions that when a population increases in size, the per capita availability of food resources decreases by some mechanism, resulting in a decline in the rate of population increase. The model says nothing specific, for either *r*-strategists or *K*-strategists, about what actually befalls organisms facing higher densities. In order to specifically observe the impact of the density effect, we need to conduct field-based surveys to investigate the ecology of each individual biological species. Moreover, in the real world, *K* itself is likely to fluctuate. Will organisms exposed to the density effect die of starvation? Or will insufficient nutrition undermine growth and cause more sickness, and would a population die out when its members no longer have the energy to reproduce and leave behind offspring?

The theory of *r*/*K*-selection offers a certain solution to this challenge. Namely, the life histories of *r*- and *K*-selected species propose different adaptations to the density effect in accordance with fluctuations in their respective populations. As organisms assumed to live close to the environmental carrying capacity, *K*-selected species are highly adapted to living tenaciously in their particular habitats in order to prevail in competition with other members of the same species with which they live, while curbing their reproductive rate, having fewer offspring, and enhancing their survival rate by growing larger bodies. Large mammals offer typical examples of organisms that fit this image, including primates that have evolved advanced intelligence and sociality (Lovejoy 1981). In recent years, after *r*/*K* selection theory fell out of vogue, life histories have been categorized as either *fast* or *slow*, with primate life histories regarded as the latter (Charnov and Berrigan 1993; Jones 2011). By contrast, the assumption for *r*-strategists is that they do not attain very high densities due to the extreme instability of their habitat environments. In other words, the number of individuals is assumed to repeatedly rise or fall due to extreme predator pressure or large environmental fluctuations not directly attributable to population density. Or perhaps it is the case that *r*-strategists, which are vulnerable to density effects, proliferate rapidly and deplete their resources, or modify their habitat to the point that they can no longer inhabit it, and end up dispersing to seek out new frontiers.

Plants that may be classified as secondary vegetation are among the organisms that correspond to this latter scenario. 'Secondary pioneer species vegetation' refers to the first plants to colonize and begin to grow when part of a mature forest is opened up by an event such as fire, landslide, or a fallen tree. Many such species are known as 'light-demanding' trees. Light-demanding trees germinate all at once and grow rapidly while strong sunlight is available before the canopy is re-established. When such growth is successful, however, a dilemma arises. As their own growth results in the establishment of a canopy, the forest floor becomes darker, creating a difficult environment for light-demanding plant species to reproduce. Moreover, as shade tree species slowly begin to become established, the forest will grow darker and darker, so that shade trees with a higher competitive capacity will eventually shut out the light-demanding trees. Therefore, light-demanding trees produce a large volume of seeds in a limited time for the wind and wildlife to carry off and disperse across a wide area. While most animal and plant species show some dispersal capability, in *r*-strategy theory the light-demanding trees of secondary vegetation are introduced in textbooks as examples of species that demonstrate life histories with both great reproductive capacity and the ability to disperse seeds over long distances.

Are humans *r*-strategists or *K*-strategists?

Should we think of the human species as *r*-selected or *K*-selected? Comparing our life history with that of other animals, *Homo sapiens* is a fairly large-sized mammal that enjoys a high survival rate and bears few children, suggesting that we should be considered as *K*-strategists (Martin 1983) or as animals with a slow life history (Smith and Tompkins 1995). A wealth of demographic data exists that would support this perspective. In the case of our own modern society, however, our medical care and food situations have been secured in the relentless pursuit of improvements to our lives. This has led to a very high survival rate, and in recent years we have grown increasingly heavy, our physiques more robust. Nevertheless, our fertility is in decline, and as an organism our situation has become quite extreme. As anthropologists, we prefer, where possible, to obtain information on the population dynamics and ecology of people living in a state that reflects the environment in which humans evolved. To that end, anthropologists have embarked on long-term surveys of hunter-gatherer peoples around the world.

Whereas Figure 4.1 shows a schematic survival curve for organisms, Figure 4.3 plots actual survival curves for human populations. The survival rates in Figure 4.3 have been constructed from data, collected by anthropologists, on age at death for various hunter-gatherer peoples. For comparison, the figure also plots survival curves for chimpanzees (*Pan troglodytes*), the ape species closest to humans, and for Japanese people in the Meiji era (1868 to 1912). First, comparing chimpanzees with humans, we see that the human survival curves progress through a much higher region than that for chimpanzees. Whereas the steeply sloping survival curve for chimpanzees terminates at about age fifty-five, those for humans exhibit much more gradual slopes, extending beyond age seventy. Focusing on the human survival curves, the main apparent difference between ethnic groups seems to be the infant mortality rate. Among hunter-gatherer peoples and Meiji era Japanese, we find that between 25% and 45% of children died before the age of ten. However, focusing on the curves for adults from the age of ten onward, we find that the slopes of the survival curves are broadly similar for all ethnic groups. The curves for hunter-gatherer peoples show a gentle transition, on a level with that of Meiji era Japanese. In fact, the curve for elderly Meiji era Japanese appears to slope rather steeply.

The gentle slope of the adult survival curve may be said to trace a typical survival curve for organisms that reproduce repeatedly over the course of a long life. That is, hunter-gatherer peoples are quite in keeping with the image of a *K*-selected species by skillfully adapting to their environments and maintaining fairly high rates of survival, exploiting their diverse tool technologies and cooperative social relationships. Moreover, it is possible to construct a narrative in which humans, as the ultimate *K*-type organism, developed diverse modes of

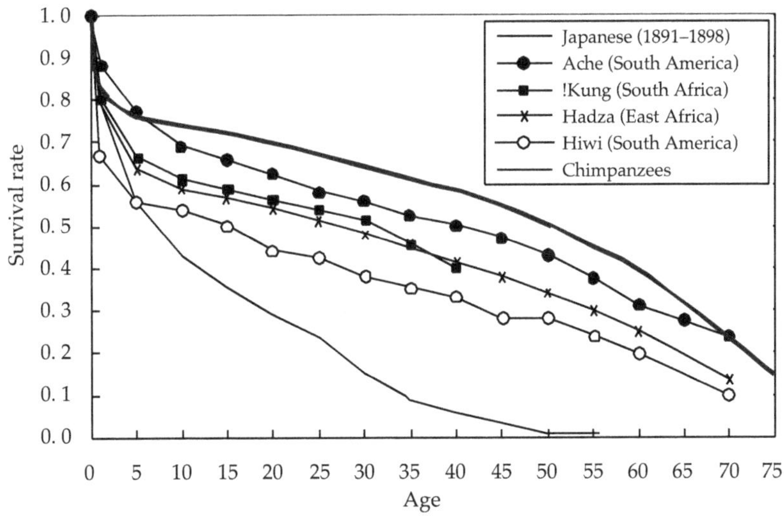

Figure 4.3: Survival curves for Meiji era Japanese, hunter-gatherer peoples, and chimpanzees (Ministry of Health, Labor, and Welfare 2001; Kaplan et al. 2000)

subsistence, such as agriculture and pastoralism, that have brought them success in repeatedly expanding the carrying capacity of their own environments.

Considered from a different perspective, however, the life history of *Homo sapiens* also seems to have many aspects of an *r*-strategist species. First, examining their life history in detail shows that among great apes, humans enjoy fairly high rates not of only survival, but also reproduction. Furthermore, after setting out from southern Africa, over the course of about 200,000 years, humans have rapidly expanded their habitat to encompass the whole world as they sought new frontiers, and their population continued to increase (Ōtsuka 2015). From this standpoint, the human species could also be described as a powerful *r*-strategist.

Table 4.1 compares characteristics of the life history of *Homo sapiens* with those of other apes. Looking at the table, one has the impression that the life history of the human species comprises a mix of both *r*-type and *K*-type characteristics. First, *Homo sapiens* is relatively heavy and has the longest life span. That the species has the latest age at first birth suggests a long growth period. Furthermore, the species has the longest gestation period and the heaviest newborns, as well as the highest newborn weight as a proportion of

	Orangutan (*Pongo pygmaeus*)	Gorilla (*Gorilla gorilla*)	Bonobo (*Pan paniscus*)	Chimpanzee (*Pan troglo-dytes*)	Human (*Homo sapiens*)
Average female life span (years)	58.7	54.0	50.0	53.4	85.0
Age at first birth (years)	15.6	10.0	14.2	13.3	19.5
Adult female body weight (kg)	36.0	84.5	33.0	35.0	47.0
Gestation period (days)	260	255	244	225	270
Newborn infant weight (kg)	1.56	1.95	1.38	1.90	3.00
Newborn weight as % of mother's weight	4.3	2.3	4.2	5.4	5.9
Weaning age (years)	7.0	2.8	–	4.5	2.8
Interbirth interval (years)	8.05	4.40	6.25	5.46	3.69
Age at last birth (years)	>41	–	–	42	45

Table 4.1: A comparison of the life histories of humans and great apes (Robson et al. 2006)

the mother's weight. These characteristics could be said to represent a *K*-type life history. Given that infants pose such a heavy burden on mothers, it is tempting to speculate that the interbirth interval is long as well. But, in fact, the opposite is true.

Among apes, the human interbirth interval is quite short. Anthropologists have proposed the evolution of 'childhood' as a theory of how we were able to achieve the feat of accommodating the heavy burden of a shorter interval between births. *Homo sapiens* has successfully incorporated into its life history a childhood phase not found in other ape species, weaning children early and shortening the interbirth interval (Bogin 1997, 2015; Gibbons 2008).

In other primate species, offspring need to forage for and eat their own food while they are being weaned, and the digestive and masticatory organs will develop to the point where they will be able to do so by the time weaning is complete. Anthropologists frequently make note of the eruption of an individual's first adult molar as a morphological indicator of growth. Among the young of nonhuman primate species, the molars will begin to emerge by the time they are weaned, allowing them to chew and digest the same food as adults. If humans continued to breastfeed their young until the same stage of growth as other primates, they would likely do so until the age of about five years. Yet human infants begin to be weaned from a stage at which they are still unable to walk, and adults give the child food by which to grow before the milk teeth are replaced, thus securing the mother an earlier release from breastfeeding.

In other words, the theory of childhood proposes that humans raised their reproductive capacity by encouraging food distribution from adults to children and giving birth successively while raising children communally (Gibbons 2008; Kaplan et al. 2000; Robson et al. 2006; Smith and Tompkins 1995). Then, while maintaining a high rate of reproduction with advanced technology and organizational power, humans dispersed across the planet.

However, continuous dispersal and expansion may not have been the case everywhere. If we assume that humans have *r*-strategic affinities relative to other apes, we need to consider the danger of humans breaching the limits of carrying capacity in cases where their population increased in a particular area. What are the mechanisms that control human populations? Might they be the Four Horsemen of the Apocalypse, who feature in the Christian Bible—Pestilence, War, Famine, and Death?

Carrying capacity for *Homo sapiens*

Our species has been visited by large-scale famines, plagues, and wars on many occasions. Each features in numerous historical chronicles, wherein historians frequently introduce the attendant misery. Even the great civilizations of

China, India, the Middle East, and Europe have not escaped these tragedies. And in some eras such large-scale tragedies have actually led to widespread population decline.

In Europe, especially in the fourteenth century, the Four Horsemen of the Apocalypse ran amok. First, a long period of climatic warmth gave way to a rather cool climate known as the Little Ice Age, with the result that population growth that had persisted until that point entered a difficult phase. With the Little Ice Age, what is now known as the Great Famine embroiled the whole of Europe, striking most heavily over the three years from 1315 to 1317. It was only around 1322 that the food situation and population began to recover to some extent. When this period of food shortages had finally eased, a pestilence that had been carried along the trade routes from Asia arrived in Italy in 1347. This plague, arguably the most well-known example of infectious disease, was transmitted by fleas and mice, spreading rapidly throughout Europe in an epidemic that lasted until 1353. Dreaded as the Black Death, due to the symptoms of the disease, it is generally estimated that the population declined by anywhere between 30% and 60%, depending on the region (Bramanti et al. 2016; DeWitte 2015; Grada 2010; Hays 2005). Regardless of the famine and pestilence, political conflicts persisted. Central Europe became embroiled in the Hundred Years War (1337–1453), with Britain and France as the principal belligerents. Agricultural turmoil caused by the war exacerbated food shortages, while the movements of soldiers and refugees over long distances spurred the spread of the epidemic.

While the fury of the Four Horsemen has been passed down to us as historical fact, thanks to recent progress in the study of climate change, we have seen the development of new methods for analyzing historical records together with quantitative environmental information. With the development of databases matching continuous long-term records of meteorological data with population numbers and other social information, it has become possible to statistically demonstrate in considerable detail the relationship between climate and population from around 1000 AD. In the European context, for example, Lima (2014) has found a relationship between long-term temperature fluctuations and population growth rates. Figure 4.4 presents Lima's data for the period from 800 AD to around 1800 AD, showing that even during periods when the Western European population trended upwards, as temperatures rose, so did the rate of population increase, and when temperatures dropped, the rate of population increase slowed. Then, at more or less the same time as the extreme temperature drop that took place around 1300 AD, the rate of population growth fell into negative territory across Western Europe. Also, low temperatures in the sixteenth and seventeenth centuries coincided with more localized Western European population declines (mainly in France, Flanders, Germany, and Italy).

In a study focused on the latter period, Zhang et al. (2011) analyzed an array of socioeconomic factors in the context of the cool period from 1560 to 1660. They found that factors that tended to decrease in line with decreasing temperature included the width of annular rings in trees, grain production and indices of agricultural production, workers' wages, and the average height of Europeans. They also point out that factors which tended to increase with lower temperature were grain prices, famine, plague, migration, and the number of wars, as well as the number of deaths attributable to them. Furthermore, inferring a causal relationship from the time-series analysis of these factors, they conclude that the famine, plague, and wars that broke out due to food shortages caused by lower temperatures drastically reduced the rate of population increase from +0.4% to –0.3%, prompting the collapse of the European population, which in 1650 fell to its lowest point (105 million people) during the period of analysis. Moreover, beyond Europe, Zhang et al. (2007) report that China also experienced many famines and wars that broke out after the decline in food production during the Little Ice Age.

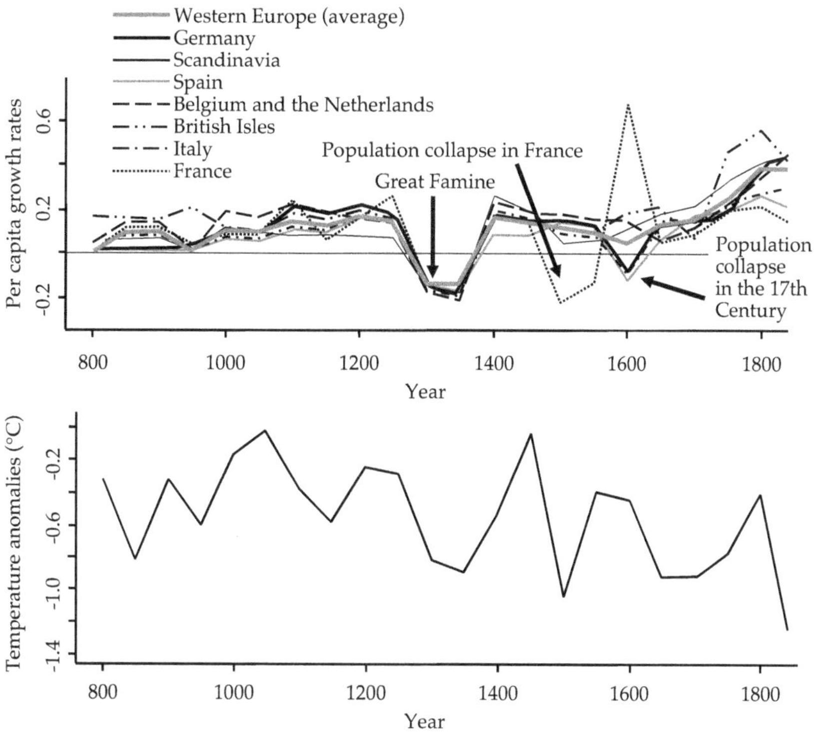

Figure 4.4: Population and temperature changes in Western Europe from the Middle Ages to the early modern period (adapted from Lima 2014: Figure 1)

Another point of interest—from the perspective of this chapter—highlighted in Lima (2014) and Zhang et al. (2007, 2011) is that these studies are based on carrying capacity. These authors build their arguments on the ecological premise that climate change affected populations supported by the prevailing environment and the contemporary level of agricultural technology. They argue that their longitudinal time-series analyses of climate and population demonstrate that as the climate changed and temperatures fell, a corresponding decrease in agricultural production impaired the population's ability to sustain itself.

While the agroecological theories underpinned by these time-series statistical analyses show broad historical trends, I want to ask how life in particular societies might have been carried out over long periods at the brink of the carrying capacity of their respective environments. A concrete example of this is Japan in the early modern period during the Edo era (1603–1868). Historical demographers have been able to estimate trends for the Japanese population in considerable detail, allowing in-depth analyses of the background of population

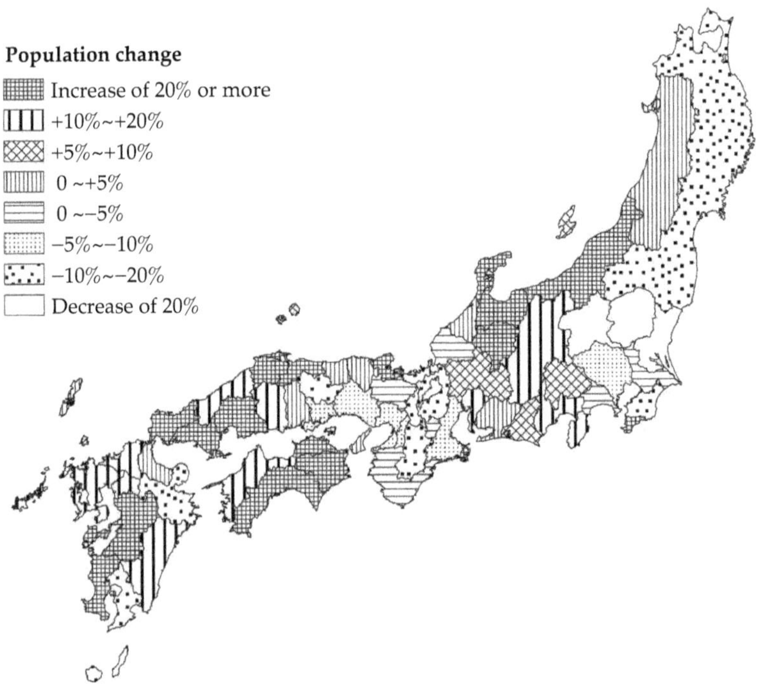

Figure 4.5: Rates of population fluctuation for Japanese provinces in the latter half of the Tokugawa period (1721–1846) (adapted from Hayami 2008: Figure 1; cartographic data from Tatuoka 2018)

change (e.g., Kitō 2000a). Although early modern Japan could be said to have prospered thanks to skillful agricultural and forestry strategies that allowed it to remain completely self-sufficient in terms of natural resources, Japan nevertheless experienced several famines, and population levels remained static during the middle of this period. From the latter half of the eighteenth century to the first half of the nineteenth, population declines were experienced, notably in eastern Japan and the Tōkai and Kinki regions (Hayami 2008; Figure 4.5). Nawata (2006) cites three phenomena as underlying the stagnation of the population in the latter half of the Edo era. First, he points out that there were limits to the expansion of agricultural production in the pre-industrial era, and that the plateauing of the expansion of land under cultivation could have been a limiting population factor in an agriculturally centered society. Moreover, the period of the Little Ice Age more or less includes the period of the Kyōhō famine (1732), the Tenmei famine (1782–87), and the Tenpō famine (1833–39), known as the 'Three Great Famines' of the Edo era, that led to widespread death due to starvation and disease. Furthermore, childbirth restrictions, including infanticide (referred to euphemistically as *mabiki*, or 'thinning out'), are thought to have been practiced in many areas, an inference that is suggested by biases in sex ratios and the express prohibition of *mabiki* by certain clans and the shogunate, as well as the fact that funding was made available for *akago yōiku* (the care of infants) (Takahashi 2007).

Further, Kitō (2000b) offers an ecological perspective while pointing to a contemporary decline in fertility that was characterized by social trends toward fewer births and later marriages. He speculates that one reason for the decline in fertility, alongside a significant reduction in the area of farmland under cultivation by farm households as a result of the increased population, was the impact of increasingly stringent restrictions on the supply of resources such as fuel, fertilizer, and water. Totman (1986: 468), who expresses an even clearer ecological perspective, uses the term 'carrying capacity' in an article entitled 'Tokugawa peasants: Win, lose, or draw?' wherein he advances the view that 'by the early eighteenth century, the archipelago's human population had consumed most of the accumulated biomass that was accessible.' Many researchers thus believe that early modern agriculture, in its dependence on the biological productivity of natural resources in the hills and fields near farming villages, had reached its ecological limits.

Hunter-gatherer lifestyles and carrying capacity

The idea of the Four Horsemen of the Apocalypse belongs to the period of recorded history in which agriculture and pastoralism were already highly developed. In this period, societies were established in which people lived at higher densities than had been possible before the advent of agriculture, and it is in this context that religious literature portrayed emergencies in which climate change and war gave rise to famine and plague. However, although famines and plagues have had a significant impact on populations as ecological facts, they are in fact a very recent development in the context of our evolutionary history, and do not bear directly on the evolution of *Homo sapiens* as a species. To think about our evolution as humans, we need to reflect on the hunter-gatherer lifestyle in which we spent most of the long sweep of our history as a species. The people who were the first to rapidly disperse across the face of the planet were all hunter-gatherers.

Research has shown that hunter-gatherer peoples are less susceptible to famine than agriculturalists (Berbesque et al. 2014). Lacking crops, hunter-gatherer peoples are never troubled by poor harvests, and may thus be able to maintain a well-balanced diet by gathering a variety of foods. Moreover, the population dynamics of hunter-gather peoples are moderate relative to those of agriculturalists, with longer interbirth intervals. While infant mortality rates can be quite high, their low-density and dispersed living patterns mitigate the spread of infectious disease, and the survival rates for adults are comparable with those of agriculturalists (Larsen 2006). In a way, we might say that hunter-gatherer peoples pursue a more typical *K*-strategist lifestyle than their agriculturalist counterparts.

However, even though hunter-gatherer population density is low relative to that of agriculturalists, the relationship between their population and the natural resources that support their subsistence lifestyle is quite another issue, and there is a possibility that they are also susceptible to a type of density effect. Based on a geographical analysis that found population densities among the world's hunter-gatherer peoples are strongly correlated with net primary productivity, biodiversity, and environmental pathogen stress, Tallavaara et al. (2018) concluded that subtropical and temperate regions have the highest carrying capacities for hunter-gatherer peoples. Walker and Hamilton (2008) have also argued that population density for hunter-gatherer peoples correlates with body size; drawing on results from studies of population dynamics among multiple hunter-gatherer peoples, they present the results of a statistical analysis showing that more densely populated hunter-gatherer peoples tend to be smaller in stature, and they claim that this is the result of a density effect by some mechanism. However, anthropologists who have conducted field studies of the lives of hunter-gatherer peoples remain divided over the question of whether hunter-gatherer populations actually approach the carrying capacity of their environments.

Many researchers hold to the position that hunter-gatherer peoples do not reach their environmental carrying capacity. From his fieldwork among the !Kung San people of the semi-arid Kalahari Desert in southern Africa, Richard Lee (1980) argued that the fertility of local hunter-gatherers was so low that they could not attain a population that would deplete the flora and fauna that support their subsistence. He made particular note of the length of the interbirth interval as a reason for this low fertility. One characteristic of !Kung child-rearing customs is that infants have a long breastfeeding period, and may continue breastfeeding until around three years of age. Lactation has the effect of suppressing women's menstruation, thus contributing to wider intervals between pregnancies. In addition, the 'gathering' part of hunter-gatherer subsistence falls primarily to women, who routinely walk long distances to reach their foraging sites. In some cases, the total distance walked, carrying their pickings on their return journeys, can amount to around 2400 kilometers per year. They also carry household goods when shifting the location of their camps. What is more, mothers carry their children when traveling, so for mothers of infants under the age of two or so, although they employ a kind of harness strap, this means that they also carry their children over these 2400 kilometers each year. Although the distance over which a child is carried will decrease as he or she gains the ability to walk independently, over the four years in which a child is dependent on her, a mother can be calculated to carry her child over a total of 7800 kilometers. And while their lives are for the most part extremely healthy, these women's energy expenditure—to keep up with activities that include foraging, housework, child-rearing, and breastfeeding—is also extremely high, and it is conceivable that this may have the effect of suppressing menstruation, as in the case of some athletes (Bentley 1985). Because fertility had not increased in tandem with the physiological adaptations to this active lifestyle on the part of these women, Lee (1980) felt that a mild form of population control must be in effect.

Nevertheless, many other researchers believe that the hunter-gatherer lifestyle may be limited by the availability of resources. Based on his many years of field research among the Hadza, a hunter-gatherer group indigenous to East Africa, Nicholas Blurton Jones (2016) argues that the number of animals to be hunted and the volume of plant-based resources to be gathered will naturally fluctuate according to time and circumstance, and that we cannot therefore discount the possibility that even hunter-gatherer peoples have experienced famines caused by climate change-induced fluctuations in the available animal and plant resources.

Yet Blurton Jones falls victim to a dilemma, to which he himself points, that ensnares researchers studying the population dynamics of the hunter-gatherer peoples. This dilemma, known as the 'hunter-gatherer paradox,' holds that it is unlikely that hunter-gatherer populations will be precisely balanced at the brink of their environmental carrying capacity due to some density effect. This is because the populations of several hunter-gatherer peoples surveyed by anthropologists, including the Hadza, have continued to rise, albeit sometimes moderately. For a population to stabilize at a given level without increasing or decreasing would require much higher mortality and lower fertility than have been observed, and a life grim enough to have fallen into such a plight has almost never been observed despite long-term research. Although Blurton Jones (2016) carries out a detailed analysis of the potential for diseases, recourse to violence, and resource depletion that are present in Hadza life, he finds nothing that would be sufficient to exert a density effect that would limit the population. Therefore, he draws on a study proposing the idea of a 'sawtooth'-style of population dynamics—characterized by a pattern by which a population, while steadily increasing under good environmental conditions, will repeatedly experience rapid decreases in the event of droughts and other catastrophes—to arrive (somewhat convolutedly) at the suggestion that hunter-gatherers such as the Hadza also conform to these sawtooth-style population dynamics (Blurton Jones 2016).

In his discussion, Blurton Jones makes no mention of *r*/*K*-selection theory. What I find interesting here, however, is that by way of his proposal of sawtooth-style population dynamics, he is in effect proposing that the life history of *Homo sapiens* is considerably closer to *r*-selection than not. It seems as though as a result of his analysis of the possibility that the human species has a *K*-type life history, he has arrived at the opposite conclusion. Of course, Blurton Jones never gives up on the possibility that the influence of carrying capacity will be exerted in some instances. Moreover, a somewhat ambiguous aspect of *r*/*K* selection theory is that it allows for fluctuations in carrying capacity to some degree, and we cannot therefore reject the possibility that the population of *K*-type organisms will also fluctuate in step with carrying capacity. However, even though Blurton Jones was seeking the presence of a density effect as a population dynamic close to the carrying capacity, it seems as though he concluded that he could find no clear confirmation of it. What is relevant for this chapter are the examples of hunter-gatherers who have demonstrated mild population dynamics that are able to respond adaptably while making skillful use of the resources at hand.

Life history theory and fertility decline

Homo sapiens has a higher reproductive capacity than other apes, and while developing new forms of subsistence one after another and repeatedly dispersing across the planet, we have enjoyed success at settling some areas at high densities. Looking back over the history of our species, it also appears as though *Homo sapiens* has applied both *r*-type and *K*-type life strategies in different times and places. One cannot help but consider that hunter-gatherer subsistence conferred many people with the potential for *r*-type capacities for reproduction and dispersal. Moreover, by applying our advanced intelligence and adaptability commensurate with *K*-strategists, we have increased food production through the development of new modes of subsistence such as pastoralism and agriculture. We could perhaps go so far as to say that *Homo sapiens* is a species that has succeeded in combining the best of both strategies, which would certainly feel like a satisfying conclusion.

However, our species in recent years has once again begun to exhibit a new development in our life history. Namely, fertility decline. Moreover, this fertility decline seems to be becoming more acute precisely in developed countries, which are regarded as economically affluent, with excellent medical care and diets. Whereas the population should continue to increase as mortality falls (i.e., as the survival rate increases), more and more countries are finding that their populations are actually in decline. This kind of contemporary fertility decline poses a somewhat paradoxical challenge for the traditional social sciences and natural sciences alike (Stulp and Barrett 2016). While considering this issue in light of an evaluation of the sustainability of life for *Homo sapiens*, I would like to reconsider the conditions of that sustainability more clearly, while also confirming the conditions under which *r*- and *K*-type life history strategies are exerted.

My reasons for attempting this here, beyond the existence of biologists who have tried explicitly to apply *r*/*K*-selection theory to the evolution of humans as a species, is that I feel that the premises and logic of *r*/*K*-selection theory are included in the background of the demographic theories by many thinkers up to the present day. Thomas Malthus, the forefather of demographic theory, offers a good example. Malthus (1798) developed a theory preoccupied with the tendency of populations to grow exponentially. The basis of this exponential population increase is the same *r*-coefficient used in the logistic population model. A deeper reading of Malthus's ideas reveals that he considered humans to be *r*-strategists, and was thus concerned about the dismal situation that would occur when their endlessly increasing population reached the limits of *K*. It is also well documented that Malthus was deeply pessimistic about the prospect of agricultural production expanding to keep up with population growth, since he felt that food production would only increase arithmetically. In other words, this deeper reading of Malthus might yield the interpretation that he did not consider humans to be *K*-strategists, and thus underestimated their potential ability to enhance their competitive capacity. While it is thus possible to find elements of *r*/*K*-selection theory in demographic theories like those of Malthus and others, what *r*/*K*-selection theory teaches us is that for both *r*-strategists and *K*-strategists extrapolating the apparent biological success of population growth has the potential to generate their own respective forms of affluence or poverty.

With their small bodies and short life spans, *r*-strategists may evoke the slightly pitiful image of being easily swayed by an unstable environment. However, an implicit tenet of *r*-strategy theory is that the habitat for *r*-type organisms is not always limited. An *r*-strategy would be pointless unless it afforded opportunities for the population to increase. Thinking back to our logistic population model, the bottom of the curve is supposed to depict the presence of an increasing population. While the population continues to increase within the limits allowed by the environment, the density effect will remain weak, so that the population can expand freely. Contrary to the originally pitiful image, in some cases the life actually experienced by individual *r*-strategists may be a kind of (at least temporary) affluence.

However, the *r*-strategy comes with a time limit. Inherent within *r*-strategy theory is the premise that population growth is difficult to sustain. There are two reasons for this. The first is none other than the sudden catastrophes that destabilize habitat environments, which is a major premise for the *r*-strategy. Extreme seasonal changes and rapid shifts in the climate, such as droughts, storms, wildfires, floods, or an influx of predators, have the potential to cause populations to go into steep decline. There is even the possibility that a population will go extinct. In theory, the reason why *r*-strategists need a life history characterized by short life spans, small bodies, and numerous offspring is that it is difficult to predict when and where their offspring will find opportunities to survive. Sustainability for *r*-strategists relies on an expectation that their vigorous reproductive and dispersal capacities will ensure that some of their offspring, somewhere, somehow, will be able to survive. Another reason is that stable environments hospitable for *r*-strategists might paradoxically somehow persist. The doom of *r*-strategists is that they can despoil their own habitats or approach the limits of the carrying capacity. At that point they must either disperse far away to seek out new frontiers or else transition to a *K*-strategy.

The hunting-and-gathering subsistence systems must have allowed many peoples to exert reproductive and dispersal capacities similar to elements of an *r*-strategy. And in the case of humans, we have augmented our food production by various means, including technological innovation, tool use, and the development of new subsistence patterns. Considered this way, the increase in the human population over the past few tens of thousands of years may have been a fleeting *r*-strategy phase. Many researchers and experts are concerned about our future, wondering what difficult times we will face when humanity's ever-increasing population someday confronts the limits of the planet's carrying capacity. Some continue to sound the alarm about global shortages even as science fiction-inspired proposals calling for humanity's dispersal into the cosmos are taken increasingly seriously.

By contrast, *K*-strategists, who are expected to maintain stable populations by dint of their highly evolved environmental adaptations, enjoy a rather more elegant image. Animals such as elephants, whales, and primates—who are large in size, long-lived, and care for their offspring within complex social systems—are quite easy for us to empathize with. However, a more thorough consideration of *K*-strategy theory suggests that this elegant image may not be entirely accurate. By definition, *K*-strategists must always live at the brink of their environmental carrying capacity. Life at the brink is also hard. Even if the habitat environment is 'abundant' and affords the stable maintenance of a dense population, the subjective experience of individual *K*-strategists may be quite severe due to all the density effects. For example, tropical rainforests constitute the emblematic example of a rich natural environment, boasting the most abundant biomass and higher biodiversity than anywhere on earth, and fascinating our human sensibilities with the fantastic adaptations that each individual species has made to this environment. Considered from an ecological standpoint, however, rainforests are the site of fierce competition between and within innumerable plant and animal species for limited space and resources. In other words, when we probe the logic of the *K*-strategy, we find that one of its premises is that a society 'sustained' by sympatrically living individuals of the same species at high densities is a fiercely competitive society, and that each individual must be equipped with the competitive capacity to compete successfully. Considered in this way, the elegant image of *K*-strategy life quite possibly belies what is in fact an inherently harsh reality.

While I admit that *K*-strategy theory is much too simple to explain contemporary human fertility decline solely in its own terms, I am certain that it offers valuable insights for the analysis of population dynamics. Since lower fertility is postulated to be one of the important characteristics of *K*-strategy life, the question here is whether it is possible to think of recent fertility decline as the consequence of some kind of density effect associated with life near an environment's carrying capacity.

Researchers are divided over the degree to which population density might explain fertility decline. Recent scholarship in the field of demography has been quite negative about the possibility that population density might itself be a direct cause of fertility decline. To counter this argument, Lutz et al. (2007) report the existence of a statistical correlation between population and density at the national level based on a survey of demographics from 159 countries. However, even here, the authors' discussion goes no further than proposing possible mechanisms regarding the causes of decreased fertility. Moreover, given that migration from rural villages to cities has historically been associated with fertility decline, the direction of the causal relationship between population density and fertility remains unclear. Until the early modern period, high population densities sometimes invited the spread of infectious disease. Kitō (2000a) and Hayami (2008) have used the term *ari jigoku* (antlion pit; i.e., a sand trap created by the antlion insect to capture its prey) to refer to the great cities of early modern Japan, which were unable to reproduce their own populations due to several factors, including the spread of disease. However, in our contemporary era, characterized by improved hygiene and advanced medical care, it seems unlikely that we can explain fertility decline solely in terms of population density.

In the modern context, fertility decline began in Europe in the nineteenth century with the development of the industrial economy. So has a new density effect also emerged with the advent of this industrial economy? Even as the industrial economy has brought us, on the whole, an affluent life, it may also be a surprisingly hard life in terms of how we actually experience it as biological organisms. I suspect we may find insights into contemporary fertility decline by exploring the possibility of this being an expression of some kind of density effect mediated by the monetary economy. In fact, Malthus himself was a pioneer of the idea that the monetary economy could act as a mechanism for population control. He pointed out that as the population increases, the soaring price of food places strain on farmers and laborers who are unable to afford it. This is a kind of concrete ecological argument, and there may well have been some societies or communities in which fertility levels actually declined due to the rising cost of living. However, fertility decline is also occurring in societies where food resources are abundant and where food costs do not account for such a high proportion of household expenses.

Therefore, the point on which I would like to focus our attention is the question of what is meant by 'density effect.' Essentially, 'density effect' is an ecological term that refers to situations in which the number of individual organisms is large relative to the amount of resources necessary for those organisms to live, and, as the term implies, the concept is based on the premise that the amount of resources depends on a given area. However, the amount of resources per unit area can rise or fall, or these resources may not even be derived from land at all. Failing to take these points into consideration runs the risk that they become blind spots for simplistic demographic explanations. For example, one reason that Malthus believed food production only increased arithmetically was that he considered the acreage of farmland to be a limiting factor. His belief that agricultural production could not support population increase was premised on the idea that any increase in food production would be achieved mainly by expanding the amount of land under cultivation. Fortunately, humanity has been able to vastly increase the productivity of crop harvests per unit area thanks to developments in agricultural technology (Boserup 1965). From the standpoint of human ecology, I want to emphasize the need to conduct research with attention to the specific resources necessary for life among people engaged in hunting and gathering, pastoralism, agriculture, and the industrial economy as modes of subsistence, as well as to the population sizes and mechanisms of resource distribution for each resource.

Another point worthy of attention when investigating density effects is to pay closer attention to the perspective of life history, since density effects are exerted throughout life history. The life history factor that first comes to mind as being directly associated with modern fertility decline is, of course, the cost of raising children. Most people are aware that parenting costs are frequently mentioned in the various fields of the natural and social sciences that seek explanations for fertility decline. However, in life history theory, the degree to which parents pay the cost of raising children is regarded as an option. Life history theory assumes that the problems organisms confront over the course of their lives involve various trade-offs, and that life history evolves depending on which available trade-off option organisms choose. And the costs expended by parents on raising children also seem to be a manifestation of complex trade-offs that involve many important factors associated with reproduction.

The most basic trade-off associated with child-rearing is the quantity versus quality of offspring. For parents allocating energy they should spend on child-rearing, the cost of bearing and raising a large number of high-quality offspring is a difficult prospect. Hence parents are forced to choose between raising a small number of high-quality children or raising more children at a much lower cost. The observation that humans, when placed at this juncture, tend to prioritize the 'quality' of children is gradually becoming an established theory in the various fields of biology and the social sciences (Ōyama 2004; Lawson and Mulder 2016). In zoology, the quality of offspring is often measured in terms of size, and among mammals the number of offspring borne by a mother in a single litter will be inversely proportional to the weight of each newborn as a percentage of its mother's weight (Walker and Gurven 2008). Primate mothers typically give birth to a single offspring at a time, but the number of offspring that a mother bears in her lifetime will still be inversely proportional to her weight. In the human context, as noted earlier, the weight of a newborn relative to its mother's weight is the highest among all the apes. By this standard, the cost of raising offspring for primates (and especially for *Homo sapiens*) is already quite significant by the time the offspring is born, and it is for this reason that they have traditionally been regarded as *K*-strategist animals.

In this way, life history theory attempts to explain the course of an organism's life by means of these trade-offs. One textbook introduces forty-five such trade-offs (Stearns 1992). Life history theory represents the idea that individuals, from the moment of fertilization in the mother's womb, are faced with new life trade-offs one after another throughout their lifetimes and that the resulting chain of trade-offs will steer the life course of the organism. Besides quantity versus quality, many trade-offs concerning offspring have been proposed. First, to introduce another trade-off directly related to reproduction, 'growth versus reproduction' is regarded in zoology as an important trade-off to determine the age of adolescence. It is difficult for growth to coincide with reproduction. For young and growing individuals, the choice of whether to continue to concentrate energy on growth or, alternatively, to wind up growth and pivot to reproduction is one of the most important crossroads in an organism's life. Next, a fully grown adult will face new trade-offs. An urgent problem for long-lived animals is that of the maintenance of life versus reproduction. Reproduction involves the expenditure of a great deal of energy and is accompanied by the risk of injury. For *K*-strategist animals—for whom improving the rate of survival is of paramount importance—to take a gamble and meet an early death means to lose the entire investment in themselves. Concentrating for the moment on feeding and avoiding danger in order to survive leaves open the possibility of having offspring in the longer term. Moreover, once a child has been born, the trade-off between 'the child already born versus the next child' may influence the interbirth interval, particularly in species where the cost of raising offspring is high.

How do these trade-offs relate to the density effect? In conventional life history theory, unlike ecology, the density effect is not regarded as a central issue. Although an overall tendency toward enhanced competitive capacity to deal with density effects is highlighted in the ecologically informed context of *K*-strategy theory, the specific mechanisms of an organism's competitive capacity are extremely diverse and, as such, *K*-strategy theory does not always have specific proposals for individual species. Deriving any specific hypothesis for some ambiguous density effect in the context of contemporary human society seems even more difficult. What I would like to propose, however, is that life history theory provides a point that links trade-offs with the density effect.

When we strip life history theory down to its very essence, it will be realized that the ultimate limiting factor of life is time. Think back to the survival curve introduced at the outset of this chapter. Organisms need to live out their lives within the time allotted to them by the age-specific mortality rate. Within this limited time, they will search for food, grow, attempt reproduction, and raise their young. Furthermore, *K*-strategy organisms will spend a large part of their lives acquiring competitive capacity to successfully overcome fierce competition. The long period of growth necessary to develop a large body, a chief characteristic of *K*-strategists, stands as a prime example. Then, in being pressed to make difficult choices at successive crossroads in their lives, *K*-strategist animals have chosen ever more time-consuming and costly life histories as a way of enhancing their competitive capacity. What is more, they may also end up facing a trade-off in which, having expended too much time or cost to accumulate these conditions for reproduction, they risk running out of time and so losing the opportunity to reproduce altogether.

Of all the primates, the life history of *Homo sapiens* is extremely time-consuming. First, the period of physiological growth is long, and those who pursue some livelihood or occupation will spend long years of study as youths before they can become self-supporting. And the more complex the knowledge and skills required for pursuing a livelihood, and the longer the training or period of apprenticeship after starting an occupation, the longer that marriage and raising children may be postponed for young people not yet able to fully support themselves. Also, if a couple prioritized enhancing a family's financial strength after marriage, then the birth of a first child may be delayed until the household has sufficient capacity to bring up children without lowering their standard of living. Subsequently, if the birth of the next child waits until the child already born has grown sufficiently, the interbirth interval will widen. It is no doubt rare for couples to intentionally lower their standard of living to have more children. In other words, in an industrial society, competition is quite fierce to gain employment in an occupation by which one can earn sufficient income to cover the anticipated costs of child-rearing. In this context, parents and children alike may find themselves in situations where they must expend all their time and money on the acquisition and maintenance of resources necessary for life, such as education and employment.

Discussion: the telos of life history evolution

As a mammal, *Homo sapiens* is certainly an animal that leans heavily toward being a *K*-strategist, which is why the comprehensive child-rearing, learning skills, and high intelligence of humans have been emphasized not only in anthropology but also in many other fields of study. However, a final point that I want to raise is the suggestion that any pride humans might feel at being the ultimate *K*-strategist may be mere self-satisfaction. Humanity combines both the potential to fall victim to mass death caused by classic famine and pestilence from aspects of being an *r*-strategist species, and the potential to expose themselves to the extreme competition characteristic of *K*-strategists. In other words, what the theory of *r*/*K*-selection seems to be teaching us is that we may end up producing extreme states characteristic of both the *r*-strategy and *K*-strategy, as the consequence of pursuing are own lives.

When considering the sustainability of the human species, needless to say, the obvious necessity to remain vigilant against the risks of food shortages or disease means our efforts to ensure public hygiene and food production must continue relentlessly. Moreover, the sustainability of our life histories is also important for humanity. The key to assessing the sustainability of life history is to examine whether a specific life history is viable under the complex series of interacting trade-offs.

This can be easily understood by imagining a life history that would not be viable. For example, the trade-off between the size and number of offspring makes it hard to establish a life history that involves giving birth to many large offspring or only a few small offspring. In the former case, the parents would be unable to bear the strain, and in the latter case the lineage would die out because the survival rate would be too low. As a result, the life histories often observed in the animal world show an inverse relationship between the size and number of offspring. Therefore, analyzing the viability of our everyday life history is one key to assessing the sustainability of contemporary society. To use the tropical rainforest as an analogy, although modern industrial society has successfully amassed enormous wealth, and we are now free of the need to worry day to day about shortages of food, clothing, or shelter, the lives of the people who live in modern society are not always easy. As vividly expressed in the phrase 'work-life balance,' researchers and the public alike are concerned that the more severe the trade-off between work and family, the less viable will be life history in contemporary society. I would like to suggest that among researchers in the various fields of scholarship concerned with fertility decline, a consensus view is gradually emerging that we may have fallen into a paradox in the sense that industrial society represents a very difficult society in which to fulfill our life history as humans.

However, in the end I have come to believe that there is no necessity to take a dim view of fertility decline itself. From an ecological perspective, populations are quite dynamic, and it is not at all surprising that population size rises or falls at different times. Rather, as descendants of hunter-gatherer peoples, the inhabitants of developed countries seem to have successfully reduced the number of children they raise, almost as though they were intentionally seeking to avoid the density effects. In a certain sense, these are ideal population dynamics that will eventually sidestep famines and plagues. If we can accept this optimistic vision of the future, one based on faith in the wisdom of a humanity that has applied both *r*- and *K*-strategies to its own benefit, then I hope that in years to come, *Homo sapiens* will continue to build better lives while successfully controlling its population through a variety of socioeconomic mechanisms.

References

Bentley, G. R. (1985) 'Hunter-gatherer energetics and fertility: A reassessment of the !Kung San'. *Human Ecology*, 13(1): 79–109.

Berbesque, J. C., F. W. Marlowe, P. Shaw and P. Thompson (2014) 'Hunter-gatherers have less famine than agriculturalists'. *Biology Letters*, 10(1): 20130853–20130853.

Blurton Jones, N. (2016) *Demography and Evolutionary Ecology of the Hadza Hunter-gatherers*. Cambridge University Press.

Bogin, B. (1997) 'Evolutionary hypotheses for human childhood'. *Yearbook of Physical Anthropology*, 40: 63–89.

Bogin, B. (2015) 'Human growth and development'. In M. Meuhlenbein (ed.), *Basics in Human Evolution*. Academic Press, 285–293.

Boserup, E. (1965) *The Conditions of Agricultural Growth: The Economics of Agrarian Change Under Population Pressure*. Allen & Unwin.

Bramanti, B., N. C. Stenseth, L. Walløe and X. Lei (2016) 'Plague: A disease which changed the path of human civilization'. *Advances in Experimental Medicine and Biology*, 918(2): 1–26.

Charnov, E. L. and D. Berrigan (1993) 'Why do female primates have such long lifespans and so few babies? Or life in the slow lane'. *Evolutionary Anthropology*, 1(6): 191–194.

Dewar, R. E. (1984) 'Environmental productivity, population regulation, and carrying capacity'. *American Anthropologist*, 86(3): 601–614.

DeWitte, S. N. (2015) 'Setting the stage for medieval plague: Pre-black death trends in survival and mortality'. *American Journal of Physical Anthropology*, 158: 441–451.

Gibbons, A. (2008) 'The birth of childhood'. *Science*, 322(5904): 1040–1043.

Grada, C. O. (2010) *Famine: A Short History*. Princeton University Press.

Hayami, Y. (2008) 'Kinsei Nihon no jinkō kōzō to hendō' (Population of Tokugawa Japan through macro-data). *Nihon gakushi'in kiyō* (Transactions of the Japan Academy), 62(3): 285–309.

Hays, J. N. (2005) *Epidemics and Pandemics: Their Impacts on Human History*. ABC-CLIO.

Jones, J. H. (2011) 'Primates and the evolution of long, slow life histories'. *Current Biology*, 21(18), R708–R717.

Kaplan, H., K. Hill, J. Lancaster and A. Hurtado (2000) 'A theory of human life history evolution: Diet, intelligence, and longevity'. *Evolutionary Anthropology*, 9(4): 156–185.

Kitō, H. (2000a) *Jinkō kara yomu Nihon no rekishi* (A demographic reading of Japanese history). Kodansha.

Kitō, H. (2000b) 'Jinsei 40-nen no sekai — Edo jidai no shussei to shibō' (The 40-year life expectancy world: Birth and death in the Edo period). Presented at the Public Symposium of the Japanese Society for Anthropology, Tokyo.

Larsen, C. S. (2006) 'The agricultural revolution as environmental catastrophe: Implications for health and lifestyle in the Holocene'. *Quaternary International*, 150(1): 12–20.

Lawson, D. W. and M. B. Mulder (2016) 'The offspring quantity–quality trade-off and human fertility variation'. *Philosophical Transactions of the Royal Society B: Biological Sciences*, 371(1692), 20150145.

Lee, R. (1980) 'Lactation, ovulation, infanticide and women's work'. In M. Cohen, R. Malpass and H. Klein (eds), *Biosocial Mechanisms of Population Regulation*. Yale University Press, 321–348.

Lima, M. (2014) 'Climate change and the population collapse during the "Great Famine" in pre-industrial Europe'. *Ecology and Evolution*, 4(3): 284–291.

Lovejoy, C. O. (1981) 'The origin of man'. *Science*, 211(4480): 341–350.

Lutz, W., M. R. Testa and D. J. Penn (2007) 'Population density is a key factor in declining human fertility'. *Population and Environment*, 28(2): 69–81.

MacArthur, R. and E. Wilson (1967) *The Theory of Island Biogeography*. Princeton University Press.

Malthus, T. (1798) *An Essay on the Principle of Population*. J. Johnson.

Martin, R. D. (1983) 'Human brain evolution in an ecological context'. James Arthur Lecture on the Evolution of the Human Brain, no. 52, 1982. American Museum of Natural History.

Ministry of Health, Labor, and Welfare (2001) *Abridged Life Table for Japan 2001*. Government of Japan.

Moore, J. (1983) 'Carrying capacity, cycles, and culture'. *Journal of Human Evolution*, 12(6): 505–514.

Nawata, Y. (2006) 'Rekishi-teki ni mita Nihon no jinkō to Kazoku' (A historical view of Japanese demographics and families). *Rippō to chōsa* (Legislation and inquiry) 260. Planning and Coordination Office, Secretariat of the House of Councillors.

Ōtsuka, R. (2015) *Hito wa kōshite fuete kita: 20 man-nen no jinkō hensen-shi* (How humans have proliferated: A 200,000-year history of population transition). Shinchosha Publishing.

Ōyama, M. (2004) 'Kodomo no yōiku kyōiku hiyō to shusshō-ritsu teika' (Cost of raising children and the recent fertility decline in Japan). *Jinkōgaku Kenkyu* (Demographic research), 35: 45–58.

Parry, G. D. (1981) 'The meanings of *r*- and *K*-selection'. *Oecologia*, 48(2): 260–264.

Pianka, E. R. (1970) 'On *r*-and *K*-selection'. *American Naturalist*, 104(940): 592–597.

Robson, S. L., C. P. van Schaik and K. Hawkes (2006) 'The derived features of human life history'. In K. Hawkes and R. R. Paine (eds), *The Evolution of Human Life History*. School of American Research Press, 155–196.

Sibly, R. M., D. Barker, M. C. Denham, J. Hone and M. Pagel (2005) 'On the regulation of populations of mammals, birds, fish, and insects'. *Science*, 309(5734): 607–610.

Smith, B. H. and R. L. Tompkins (1995) 'Toward a life history of the Hominidae'. *Annual Review of Anthropology*, 24(1): 257–279.

Stearns, S. C. (1992) *The Evolution of Life Histories*. Oxford University Press.

Stulp, G. and L. Barrett (2016) 'Wealth, fertility and adaptive behaviour in industrial populations'. *Philosophical Transactions of the Royal Society B: Biological Sciences*, 371(1692): 20150153.

Takahashi, M. (2007) 'Jinkō genshō to shōshika shakai: Kinsei chūki no jinkō genshō to shōshika taisaku (Tokushū: Rekishi wa ni-do kurikaesu?)' (Population decline and the social impact of a falling birthrate: Coping measures in the early modern period (Special issue: Is history repeating itself?)). *Nihon rōdō kenkyū zasshi* (Japanese journal of labor studies), 562: 3–12.

Tallavaara, M., J. T. Eronen and M. Luoto (2018) 'Productivity, biodiversity, and pathogens influence the global hunter-gatherer population density'. *Proceedings of the National Academy of Sciences*, 115(6): 1232–1237.

Tatuoka, Y. (2018) Ryōseikoku no chizu (Map of the provinces of Ritsuryō Japan). ArcGIS Hub, https://hub.arcgis.com (accessed August 9, 2019).

Totman, C. (1986) 'Tokugawa peasants: Win, lose, or draw?' *Monumenta Nipponica*, 41(4): 457–476.

Walker, R. S. and M. Gurven (2008) 'The trade-off between number and size of offspring in humans and other primates'. *Proceedings of the Royal Society of London, Series B: Biological Sciences*, 275: 827–834.

Walker, R. S. and M. J. Hamilton (2008) 'Life-history consequences of density dependence and the evolution of human body size'. *Current Anthropology*, 49(1): 115–122.

Zhang, D. D., P. Brecke, H. F. Lee, Y.-Q. He and J. Zhang (2007) 'Global climate change, war, and population decline in recent human history'. *Proceedings of the National Academy of Science*, 104(49): 19214–19219.

Zhang, D. D., H. F. Lee, C. Wang, B. Li, Q. Pei, J. Zhang and Y. An (2011) 'The causality analysis of climate change and large-scale human crisis'. *Proceedings of the National Academy of Science*, 108(42): 17296–17301.

CHAPTER 5

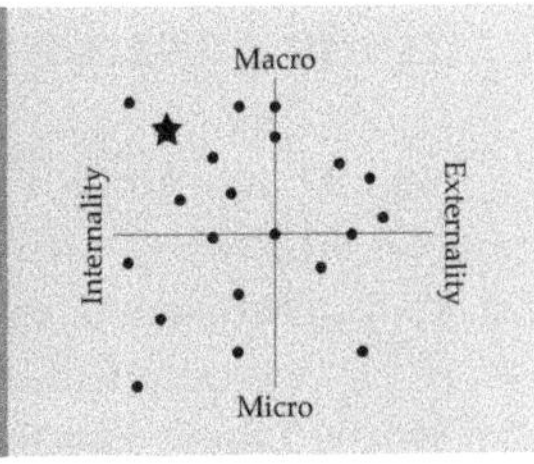

The Anthropocene as Extreme

Yuji TAKENOSHITA

Keywords: Anthropocene, stories, narrative

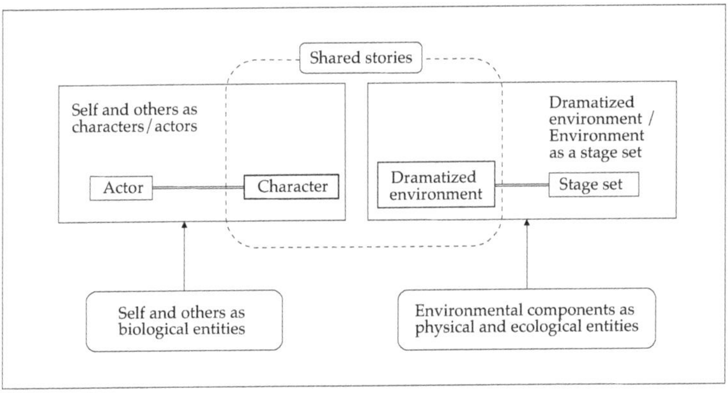

In my essay 'Society as a "story"' (Takenoshita 2019), which I advanced in a previous volume, *Others*, I argued that the self and others possessed a duality whereby they are characters in 'stories' that they simultaneously perform as actors—not as biological entities, but as fictions spun therein. To extend this insight to the environment, the constituents of an environment would also be understood to possess the duality of being dramatized environments within stories spun from bare reality, while at the same time being the mise-en-scène (stage set) that forms the architecture of those stories.

Extreme environments that support human survival

The aim of this volume is to discuss extreme environments that support human survival. But what exactly are these extremes? I would like to start from this question. There are four terms included in this phrase—namely, 'human,' 'survival,' 'environment,' and 'extreme.' Delimiting the semantic scope of these four words in their own terms is not as easy as it might seem. Let us look at each in turn.

First, what do we mean by 'human'? Does this refer to extant *Homo sapiens*? Or should it include fossil humans, as well? If so, then where do we say that 'humanity' begins? Do we start with Sahelanthropus, the oldest known hominin? Alternatively, do we trace humans back to *Ardipithecus ramidus* as a direct ancestor of *Homo sapiens*? Or do we regard humanity as beginning only with the genus *Homo*? In that case, what about those who diverged from the lineage that connects to modern humans, such as *Paranthropus*, Flores Man (*Homo floresiensis*), and Neanderthals (*Homo neanderthalensis*)? Or, conversely, do we expand the scope to include apes classified as hominids in the most recent classification?

Next, what is 'survival,' precisely? We could say 'to live,' but then there are various levels of life. It exists at the individual level, at the group and population levels, at the species level and at the lineage level. If the antonym of life is death, then death is on the same level as life. Just as we have death at the individual level, so too do we have the formation and extinction of (social) groups and populations, and the extinction of species and higher-order classificatory groupings (lineages). Hence the problem arises of the level at which we should conceive of 'survival.' This also pertains to what we mean by the word 'human.'

Death at a lower level of survival is encompassed by life at higher levels. That is, the death of the individual is part of life, one of the changes that takes place in the process of survival at the level of the group or population. Likewise, the formation and extinction of groups is also part of life, one of the changes that takes place in the state of a species in the context of the process of its continuance. Even when we consider levels below the individual, a relationship is established whereby both life and death are encompassed within life at higher levels. The life of an individual is supported by life and death at the cellular level, for example. In other words, the equation 'life = life + death' holds true. Accordingly, our way of thinking will differ depending on whether survival is understood as life on the left side of this formula (i.e., something that encompasses both life and death) or as life on the right side (i.e., something that is opposed to death).

As to 'environment,' since this is one of the main themes discussed in this chapter, I will simply point out that it is a relative concept, in a double sense, and that an 'absolute' environment does not exist. The etymology of 'environment' derives from the Greek περιβάλλον, meaning 'what surround.' This means that an environment manifests only referentially, as in an 'environment for whom (or what)?' Simply put, environment is relative in the sense that it varies according to the individual. And the 'whom (or what)' that manifests an environment this way itself exists as something 'surround' for yet another 'whom (or what)'—namely, as an environment in its own right. So environments are also relative in the sense that they exist as such for each other.

Finally, we come to 'extreme.' If we consider 'survival' as being opposed to death, then an extreme would mean conditions at the very edge of survival, where death lies waiting even one step beyond. However, if we consider 'survival' as encompassing both life and death, then we must also think of the 'extreme' from a commensurate perspective. The apposite image, rather than as a boundary with death, would be of the limit of life's potential, which is intrinsic to life.

In this chapter, first, I regard the 'human' as the lineage leading to extant *Homo sapiens* following its divergence from our common ancestor with the extant great apes, including also collateral lines such as *Paranthropus*. And I regard 'survival' as a concept that encompasses both life and death. It is in these senses that I position 'extreme environments that support human survival' as the limits of the possibilities of environments that humans can survive. Having done so, I discuss the Anthropocene as a concrete example of just such an extreme environment that supports human survival.

The Anthropocene as extreme environment

Although the Anthropocene is a concept that has been mooted as a geological epoch, its first proponents were not geologists but the climatologists Eugene F. Stoermer and Paul J. Crutzen. Considering the epoch known as the Holocene to have ended in the mid-eighteenth century, these thinkers felt that the modern era had moved into a new geological period that they proposed calling the Anthropocene. They argued that the impact of human activity since the Industrial Revolution had spread to the far reaches of the globe and that human activity was not only changing global material circulation systems, but also bringing about the fifth era of biological mass extinction in the history of the planet (Crutzen and Stoermer 2000; Crutzen 2002).

Crutzen and Stoermer's (2000) article had a brief but powerful impact in having proposed a new geological epoch. In fact, however, their article was more provocative that empirical. In the first place, they were not geologists, and some critics have shown skepticism as to whether human activity since the Industrial Revolution would actually have left its mark on the planet as a geological stratum. Although the Anthropocene Working Group of the International Union of Geological Sciences, a body responsible for defining geological epochs, is currently discussing whether to acknowledge the Anthropocene as a geological epoch and, if so, when to mark its beginning, such recognition was not yet forthcoming at the time of writing. Furthermore, all organisms live in constant interaction with their environment. And while they are influenced by their respective habitat environments through these interactions, it is also a fact that they have their own impact on their habitat environments, which they modify to a greater or lesser extent. Thus, that humans are modifying the environment by their own activities is not to say that they are doing anything out of order as biological organisms. To start with, the Holocene, which began about 11,000 years ago, was already understood as an era marked by the spread of *Homo sapiens* and the advent of farming and livestock husbandry (i.e., a deviation from natural ecosystems due to artificial modifications of the environment), without the need to intentionally build up a new geological epoch as a human era. The Pleistocene, which in turn began 2.58 million years earlier, was positioned as the epoch that had seen the appearance of the genus *Homo*, as well as its subsequent evolution and diffusion.

In most organisms, however, the impact of their survival on their habitat environment is trifling. Rather, they are almost unilaterally subject to the influence of the environment, and in most cases build a stable system of interactions with the environment within ecological and evolutionary historical time scales. In contrast, environmental modifications by modern humans are unique in that changes have been applied to the habitat environment that are greater by orders of magnitude within a short period of time, and these are more or less global in scale. Their tremendous impact is said to be on a level with the major restructuring of our planet's atmospheric composition by aerobic cyanobacteria in the Proterozoic Eon and the accompanying mass extinction of biological organisms. However, owing to the talismanic power of the 'Anthropocene' as an idea, it is also a fact that a sense of crisis—one engendered by the idea that, as a result of having magnified the impact of human activity on the natural world and repeating irreversible changes to the planetary environment, humans may have passed the point of no return—has taken on an urgency shared not only among researchers in the natural sciences but also by many researchers in the humanities and social sciences (Steffen et al. 2011). As a symbol of this sense of crisis, the Anthropocene has transcended (or perhaps deviated from) the framework of geological epochs to become a buzzword among researchers in a wide range of disciplines, including geoscience, ecology, sociology, economics, and philosophy, and is gradually spreading into popular usage (Shinohara 2018).

However, this chapter does not focus on the Anthropocene as a crisis situation in this sense, which would only be extreme in the context of a mindset in which survival is opposed to death. Rather, I approach the Anthropocene as the limit of possibilities for environments that support survival as something encompassing life *and* death. It is a situation in which the impact of human activity has spread to the far reaches of the earth, so that most environments that now surround us are somehow subject to human influence.

Human characteristics that gave rise to the Anthropocene

The major alteration of the global environment by cyanobacteria was accomplished by their acquisition of a new capability for aerobic respiration—that is, the ability to produce oxygen. So, then, what sort of capability did we humans—or, in other words, *Homo sapiens*—acquire that enabled the large-scale anthropogenic environmental changes that we are now experiencing?

The 'official' inception of the Anthropocene took place in the 1800s, when the Industrial Revolution took place in Europe (Crutzen and Stoermer 2000; Crutzen 2002; Steffen et al. 2007, 2011, 2015). That being said, Steffen et al. (2011) describe the 1800s (which they regard as the beginning of the Anthropocene) not as a period when humans began to exert significant impact on the global environment, but as the period when the pace of the expansion of that impact began to accelerate to a point where it could be distinguished from what came before. In other words, humans prior to the Anthropocene were not necessarily living an environmentally friendly life; humans had 'become a global geophysical force' much earlier (Steffen et al. 2007: 1). Steffen et al. (2007) trace this back to the period when *Homo erectus* first began using fire.

However, this is quite a superficial way of looking at things. First, it only considers the human relationship with the environment in terms of the destruction of the environment through technology. Moreover, it considers the evolutionary history of humanity solely through the simple lens of the linear increase of intelligence, and equates this increased intelligence with technological progress. The thing is, humans firstly invented stone tools 2.58 million years ago, then mastered fire between 1 and 1.40 million years ago, then pioneered cultivation 10,000 years ago, and finally entered the Anthropocene with the Industrial Revolution.[1] Progress in terms of intelligence does not equate to technological progress, and technological progress is not necessarily tied to environmental destruction. As to why humans began making modifications to their environment, this is not something that can be explained by the evolution of intelligence in any quantitative sense.

Crutzen and Stormer (2000) cite the concept of the 'noösphere' proposed by Vladimir Vernadsky, a Soviet geologist, and Pierre Teilhard de Chardin, a French philosopher and Jesuit priest, claiming it to be a 'world of thought' based on consciousness and thought that humans had acquired in the process of evolution. Having obtained this world of thought, humans began to modify the environment around them in order to make this world of thought a reality through their 'brainpower and technological talents,' while at the same time changing their own future (Crutzen and Stoermer 2000: 17).

This way of thinking is consistent with the discussion I advanced in our previous volume, *Others: The Evolution of Human Sociality* (Kawai, K. 2019), regarding the evolution of alterity in the context of the human species. I argued that, unlike other animals, we humans do not live in the world of reality as such; rather, we perform 'stories' that we create based on that world. By integrating the stories created by the self with those of others, we would live out the 'shared story of the group' in the company of others; this, I felt, was the model for human sociality (Takenoshita 2019).

In *Others*, my discussion focused on the integration of scenarios between the self and others who live out these stories together. This was because the object of the previous book (as its title suggests) was to discuss the theme of 'others.' Thus I was not able to fully discuss the environments that provided sites where the self and others would interact. However, my discussion in *Others* owes much to Thomas Suddendorf's 'theater metaphor' in his theory of mental time travel (Suddendorf 2013; Suddendorf and Corballis 2007), wherein he likens not only the self and others to actors, but also the constituent elements of the environment to those of the theater, such as the stage and props. Hence, by extending the discussion from our previous volume, we should be able to discuss the environment, as well.

In human societies, which have learned to inhabit stories rather than bare reality, the self and others possess a duality whereby they are characters who appear in *stories* as fictions rather than as biological entities, while at the same time being actors who perform those stories (in *Others*, I described the self and others as 'characters/actors'). Extending this theater metaphor to the environment, just as the self and others possess the duality of being characters/actors, the constituents of the environment are also understood as entities possessing the duality of being both dramatized environments within stories spun from bare reality, while at the same time being the stage set that forms the architecture of those stories (see the schematic diagram at the opening of this chapter). Stories are spun according to the environment of entities in the here-and-now. Meanwhile, once stories are created and shared, they begin to prescribe the environment as a stage set. Should any awkwardness arise in the setting of this stage while performing these stories, it is only natural to modify it so that it will better serve those stories. In other words, it is by virtue of this 'narrativization' of the world that humans are able to embrace the will to modify their environment.

The transcendental perspective and narrativization of the world

For the 'narrativization of the world' to occur, we need to assume the perspective of 'the self as actor' discussed in the previous volume. The actor's perspective might be rephrased as a transcendental perspective that breaks away from the here-and-now. In other words, it is a perspective that, even while becoming immersed in the role of a character in a drama to perform a story, still objectively acknowledges self, others, and the dramatic environment while surveying the story's overall scenario.

Humans reputedly only acquired this kind of transcendental perspective after the emergence of modern humans, or *Homo sapiens*. During this period, the brain of *Homo sapiens* developed as anatomically identical to that of humans today, and many products (such as paintings and ornaments) have been preserved as evidence of the fact that such *Homo sapiens* possessed a transcendental perspective and engaged in symbolic thought. Klein (1989) considers that the evolution of cognitive ability in humans occurred rapidly over a relatively short time scale between 50,000 and 30,000 years ago.

In his bestselling *Sapiens: A Brief History of Humankind*, historian Yuval Noah Harari (2011) says that while it was the Industrial Revolution that brought about the Anthropocene, humanity had by that time already undergone three other revolutions—namely, the Scientific Revolution during the Renaissance of the sixteenth century, the Agricultural Revolution at the inception of cultivation about 10,000 years ago, and the Cognitive Revolution, in which the evolution of cognitive abilities took place between 70,000 and 30,000 years ago. And he positions the fact that the Cognitive Revolution made it possible to relate and share 'fictions' with each other using language—that is, the fact that humanity began to inhabit its own stories—as the starting point of the revolutions that followed thereafter (Harari 2011). The key point of this argument is that, as opposed to attempting to grasp the evolution of intelligence in quantitative terms, in the manner of Steffen et al. (2011), it focuses instead on qualitative transformations. Harari (2011) even states that because the brain is an energy-intensive organ, its enlargement cannot necessarily be said to have offered humans an edge for survival prior to the Cognitive Revolution.[2]

What Harari describes as the ability to build fictions is synonymous with Crutzen and Stoermer's (2000) 'world of thought' and my own 'story-telling ability' (Takenoshita 2019: 394). If so, might we be able to conclude that it was the transcendental perspective and linguistic ability acquired as a result of the Cognitive Revolution that brought about the narrativization of the world?

And yet there seems to be some room to question whether the transcendental perspective that humanity possesses is unique to us (as *Homo sapiens*), and whether it could even be said to have first appeared with humanity. Harari's

(2011) *Sapiens* is targeted to a mass readership, errs on the side of emphasizing the peculiarities of *Homo sapiens*, and depicts its evolutionary history in a dramatic fashion. Moreover, his mode of understanding the process of human evolution is progressive. In other words, he regards modern society as a kind of evolutionary telos, and looks back on human history from the viewpoint of when and at what stage the abilities and technologies that made it possible emerged. Accordingly, prehistoric humans and nonhuman primates are dismissed as evolutionary sidelines.

In fact, it has been found from newly discovered remains of early *Homo sapiens* in Africa that the evolution of cognitive abilities in *Homo sapiens* is not quite so groundbreaking, and likely evolved gradually from the time of ancestral species that preceded *Homo sapiens* (Mcbrearty and Brooks 2000). It is not necessarily apposite to think that the various cognitive traits possessed by modern humans evolved all at once as a set. Individual cognitive traits likely evolved slowly and independently of each other, and it may have been only when they finally came together that they occasioned behavioral changes that would merit being called 'revolutionary.' And what Harari is actually emphasizing by saying 'cognitive evolution' is the emergence of the possibility of humans giving precise expression to fictions by language (i.e., telling stories) and, moreover, sharing them. We could not definitively say that earlier human species lacked a transcendental perspective that broke away from the here-and-now. Therefore, in the next section, I would like to probe the origins of this transcendental perspective by tracing back from *Homo sapiens*.

Origins of the transcendental perspective

Fossil humans

The absence of evidence that human species prior to *Homo sapiens* ever acquired a transcendental perspective means that tracing its origins requires some creative thinking. Therefore, as a clue, I focus on the empirical morphology of human evolution, with particular reference to the evolution of brain capacity and discrepancies in the timing of behavioral, ecological, and social evolution.

First, to return to the story of the Anthropocene, its inception is regarded as having taken place around two centuries ago, from the time of the beginnings of the Industrial Revolution in the eighteenth century. Even if the origins of humans' intentional modification of their environment dates back to the start of cultivation, this would still only be 10,000 years ago, long after the emergence of *Homo sapiens* (Ōtsuka 2015). However, for all that it was a gradual evolution, humans should have acquired a brain anatomically identical to that of modern humans by about 30,000 years ago. In other words, for the first several tens of millennia after *Homo sapiens* acquired a brain identical to that of modern humans, they carried out environmental modifications only slowly, until more active environmental modifications began in various places about 10,000 years ago. When we look at human species prior to *Homo sapiens*, this kind of discrepancy becomes even more apparent. Let us observe as we move backward in time.

Homo erectus is known to have carried out systematic hunts in the pursuit of large game. Moreover, according to Richard Wrangham (2009), *Homo erectus* employed a sexual division of labor and cooked with fire. The Acheulean stone tools these hominids fashioned were not haphazardly improvised, but rather objects that could not have been created without having imagined the finished form in advance. In other words, *Homo erectus* possessed the power to imagine a finished form that did not yet exist in the here-and-now.

This behavioral and sociological evolution can essentially be explained by an increase in brain capacity. With *Homo erectus*, humans began to increase in physical size. The increase in brain capacity also accelerated. By the time of *Homo erectus*, between about 1.5 to 1 million years ago, brain capacity had reached around 1000 cubic centimeters (Saito et al. 2006).

However, things are not quite so simple. *Homo floresiensis*, discovered on the Indonesian island of Flores in 2003, is a descendant of *Homo erectus* that lived from about one million years ago to around 10,000 years ago. It had attained a smaller stature, possibly due to the island effect (Foster's Rule), and had a reduced brain capacity of only about 400 cubic centimeters (on par with chimpanzees). Nevertheless, we know that *Homo floresiensis* maintained techniques for crafting elaborate flake tools and hunted elephants and other megafauna that lived on the island (Kawai, N. 2010). As this would naturally have required collaborative work, it seems clear that although the small-bodied *Homo floresiensis* possessed only a small brain, it had high cognitive abilities and communality.

Going back further, we find that *Homo habilis* (the first species to belong to the genus *Homo*), in terms of its morphology, is not significantly different from *Australopithecus africanus*. However, it was *Homo habilis* that formed the Oldowan stone tool culture, which is the earliest stone tool culture for humans of any kind. For stone tool crafting techniques using largely the same production method to spread widely through East Africa within a relatively short period of time, it would be necessary that these techniques be taught (and learned).

Morgan et al. (2015) carried out an experiment pertaining to the transmission of the manufacturing techniques used to produce Oldowan stone tools that involved 184 modern research participants. After first teaching a subset of the participants how to make Oldowan stone tools, they investigated the efficiency and accuracy of the transmission of stone tool manufacturing techniques to others by five different methods: (1) reproducing the object by looking at one manufactured by someone else, (2) reproducing the object after observing the manufacture of one by someone else, (3) reproducing the object by having someone else explicitly demonstrate how it is manufactured, (4) having someone else teach the production method through gestures while manufacturing one of their own, and (5) having someone else teach the production method with language while manufacturing one of their own. Doing so, they found that even modern people were unable to produce serviceable Oldowan stone tools simply by imitating an object or observing its production by someone else. In other words, they found that teaching was necessary for transmitting the techniques to make Oldowan stone tools and, moreover, that teaching by means of language was very important. In addition, when participants learned in sequence, such as when they learned from someone who had learned from someone else, it was found that in the absence of gestures or language, the stone tool manufacturing techniques deteriorated steadily.

From this result, Morgan et al. (2015) considered that active teaching and an ability to manipulate symbols (such as those that constitute the basis of language) were crucial for the spread via social learning of the knowledge of how to fashion Oldowan stone tools. Initially, the abilities to teach and manipulate symbols were underdeveloped, so that the transmission of the Oldowan culture was slow. However, once this stone tool culture was established, a positive selection pressure began to be placed on these abilities, resulting in the evolution of the ability to manipulate symbols. Morgan et al. (2015) proposed a cultural evolutionary hypothesis positing that this process facilitated the transmission of complex and sophisticated techniques for crafting stone tools that had previously been exceedingly difficult and hard to pass on, thereby laying the groundwork for the emergence of the Acheulean stone tool culture of the *Homo erectus* period. We will revisit this theory of cultural evolution later, but what is important here is the possibility that properties such as the ability to manipulate symbols may already have appeared in the *Homo habilis* period. Moreover, exercising the communality that would be involved in actively teaching things to others would also necessitate the ability to fathom the minds of others, which is to say a theory of mind. And this theory of mind is nothing other than a departure from the self-centered view of the world.

Going back even further, Owen Lovejoy (2009, 2014) raises the possibility that among *Ardipithecus ramidus* (4.4 million years ago), males may have carried and distributed food for and to females. If this were the case, then at the stage of Ardipithecus, a transcendental perspective that had broken away from the here-and-now had become evident (even if only incipiently). Although chimpanzees and gorillas also distribute food among themselves, in their case food distribution is basically food sharing that takes place in the here-and-now—that is, only so long as it follows the natural flow of events as they occur. However, for a male to carry and distribute food for and to a female who is not there at the time means that he must act with a future time or other place in mind, either by casting his thoughts to an individual who is not present in the here-and-now and carrying food that he will distribute to her or, conversely, understanding that the individual in question will be waiting in anticipation of his delivery of food to her.

Seen in this way, communality and a transcendental perspective that breaks away from the here-and-now do not seem to be as correlated with brain capacity as once thought. If we consider that the acquisition of a transcendental perspective is indispensable for creating a society that involves considerations for others and actual collaboration and cooperation, then its origins may be traced back well before the emergence of *Homo sapiens*.

Extant great apes

If this is the case, do nonhuman primates, especially the great apes, really lack a transcendental perspective? In our previous volume, with reference to Suddendorf (2013), I suggested that the difference between great apes and humans lay in whether they acknowledged that others were, like the self, simultaneously characters who appeared in stories (others as characters) and actors performing the role of those characters (others as actors), which is to say whether or not they had a transcendental perspective (Takenoshita 2019). Since the logical development in *Others* suffered from being somewhat overly concerned with typology, I would like to revisit this here.

The debate as to whether humans are special or exist on a spectrum with monkeys is a longstanding and ongoing debate, though a barren one (Premack 1986). And the history of the study of great apes has also revealed that many of the psychosocial abilities and traits once thought to have been unique to humans are found in great apes as well, if to varying degrees and frequencies. While Suddendorf (2013) made something of an impact with his argument of a clear gap between humans and apes at a time when the wider discourse placed more stock on 'the discovery of commonality,' what are the facts of the matter?

Musgrave et al. (2016) report a case of 'teaching' by chimpanzees at the Goualougo Triangle in the Republic of the Congo. The Goualougo chimpanzees feed on termites (*Macrotermes* genus) and make use of tools when doing so. After inserting the stem of a terrestrial monocotyledonous plant into a hole in the termite mound, they catch and eat the soldier termites that have bitten into it. When doing so, even better results can be obtained by biting and chewing the tip of the stem to transform it into a brush-like shape. So when a female chimpanzee fishes for termites, she will chew on the end of the stem she is using and give it to her infant beside her, who is also hungry for termites. This means that the infant, which has not yet been able to make a fishing implement on her own, will have been given a sample fishing implement and, moreover, an opportunity to experience termite fishing, which satisfies Caro and Hauser's (1992) definition of animal teaching. Though it may sound trivial, it is something that chimpanzees do not usually do, even to this extent. However, that instances of teaching have thus been observed, if rarely, indicates not that they *cannot* do it, but only that they *do not always* do it.

Nishida (1994) relates an excellent episode showing how chimpanzees have the power to imagine things that do not exist in the here-and-now. In Tanzania's Mahale Mountains National Park, an alpha male chimpanzee, while ranging, suddenly broke into a run to seek out and upset a group of rival adult males that had gathered surreptitiously for grooming. Nishida was certain that the alpha male, while wandering about, had eventually noticed that his rivals were nowhere to be seen, at which point, imagining that they must have gathered together somewhere to groom each other out of his sight, he decided to go and disturb them. No less a figure than Nishida himself imagined that 'the alpha male must have gone to bother them after surmising that they must be off somewhere grooming each other in secret' (Nishida 1994: 165). This story is too good to be disregarded or cast aside as an example of anthropomorphism. This, too, is another example of how great apes can sometimes, though not always, demonstrate a transcendental perspective

Habitus as a transcendental perspective

Seen in this way, a transcendental perspective that has broken away from the here-and-now cannot be said to be the same ability acquired by *Homo sapiens* (at the time of the Cognitive Revolution). In terms of ability and habit, chimpanzees can exercise such a perspective, and occasionally do. Thus it is likely that fossil humans prior to *Homo sapiens* could also do so, and likely did. If so, the difference between humans, apes, and fossil humans may only be a quantitative difference in terms of how often they made use of the mental abilities afforded by a transcendental perspective, and to what ends.

This can be expressed as follows, in line with the discussion of others that I advanced in our previous volume. In that volume, I stated that 'Great apes are always on the stage and do not direct the story from the wings or watch it play out from the gallery' (Takenoshita 2019: 397). I meant that great apes are always playing 'characters,' but never take on the perspective of 'actors.' However, the example above shows that chimpanzees can sometimes adopt an actor's point of view and cast their thoughts to the story they are themselves engaged in performing at a holistic level. However, they do so only rarely, and the use of insights gained from an actor's perspective is limited to realizing short-term benefits for oneself and one's immediate kin. And, most likely, they do not anticipate a future many years hence or the existence of chimpanzees in distant areas that they have never encountered.

By contrast, modern people, at least, frequently adopt an actor's perspective. They are in fact more likely to do so by default. And by so doing, they reinvest the insights they gain to benefit society as a whole. They also labor with a view to a future several years off, and behave in such a way as to benefit society as a whole, including those who are strangers to them.

Thinking thus, it would seem inappropriate to regard the acquisition of a transcendental perspective as a first step toward the Anthropocene. It might be more accurate to consider the Anthropocene to have been brought about not by a transcendental perspective as an evolutionarily unique property of humans, but more likely by a human *habitus* (Bourdieu 1977) that had gained familiarity with a transcendental perspective that had been inherited from the common ancestor of humans and great apes and which had come to be used more frequently. Of course, cognitive abilities of modern humans and chimpanzees are different in terms of the quality of the transcendental perspective. But it does not suffice to discuss only differences of quality. Rather, we must also consider the qualitative evolution of cognitive ability as well as the transition of *habitus* in terms of when, where, and how to exercise the transcendental perspective. To that end, rather than relying solely on the framework of biological evolution, a cultural evolutionary approach (Mesoudi 2011) such as that advocated by Morgan et al. (2015) with regard to tool production may be more effective.

The paradox and future-orientation of the Anthropocene

Incidentally, it is something of a paradox to think that humans' capability to alter the environment and superlative ability to predict the future based on a transcendental perspective would have ushered in the arrival of the Anthropocene. Humanity must have been trying to make the environment suitable for itself by making the most of its ability to predict the future. But with regard to our current planetary environment, we have created a variety of unintended inconveniences, as a result of which it is becoming less optimal for humanity's survival. While we might feel that a safer and more stable environment is one in which people can rest easy, the numerous environmental modifications we have undertaken to that end have ended up producing an unstable and unsustainable environment almost unknown in the history of our planet. The more subject an environment is to human influence, the more likely conditions will arise that are physically, chemically, socially, or psychologically ill-suited to human survival. How might this paradox be best understood?

The simplest answer is that there are limits to humanity's ability to predict the future, and it is impossible to predict the future completely accurately. Just so. But if our environmental alterations produce unexpected or unfavorable results, then we should also be able to stop and put things back the way they were, or else abandon our approach and switch to a different method.

However, cases of a technology, once attained, being abandoned are rare in human history, and attempts to redress the negative impacts of environmental modifications have usually tended to involve further environmental modifications and social changes.[3] Thus we could say that what brought about the Anthropocene was not so much the increased frequency of our use of a transcendental perspective (world of thought) that breaks away from the here-and-now, but a headlong orientation toward the future in terms of our attempt to make full use of this perspective to break away and create a new future. In other words, Crutzen and Stoermer's (2000) argument that it was humanity's acquisition of a world of thought that led to the Anthropocene is insufficient; it is important to add that we acquired an urge to modify reality in order to match that world of thought.

In fact, as a biological species, *Homo sapiens* is extremely future oriented and frontier oriented compared to its predecessors. Humans have even attempted sea journeys aiming for lands that could not be seen from the here-and-now, which seems to be something that earlier human species were not capable of (Kaifu 2016). And while it is tempting to speculate about the circumstances by which the kind of future orientation we see in humanity was acquired, I personally have the sense that the term 'future orientation' is itself too obscure, and it does not seem feasible to move the discussion forward.

Self-propelled stories

While acknowledging that the following approaches the level of fancy, and so becomes a story in itself, I would like to think about the relationship between the fact that humans are beings who inhabit stories and the fact that they are future-oriented beings, the temporal axis of evolutionary history notwithstanding.

The reality-ascendant world

In our previous volume, *Others*, I considered the narrativization process to have four layers (Takenoshita 2019). At that time, I envisioned the world of reality placed on the lowest layer as something that existed only to be interpreted by the self (and others) who inhabited it. However, in fact, when someone performs various actions in the context of a 'story,' this will naturally have an impact on 'reality.' In other words, stories are not only spun unilaterally out of reality, but are also fed back into and intervene in the real world.

When an anthropogenic modification of the environment takes place, as long as the degree of the intervention is sufficiently small and the environment is sufficiently resilient, the environment can accept and absorb this human agency. In that case, the anthropogenic modification at issue will not leave its mark on the environment in the long run, and we could rightly describe reality as dominant in relation to stories. It is in this kind of reality-ascendant world that great apes live, and fossil humans and the earliest representatives of *Homo sapiens* lived.

The story-ascendant world

However, if an anthropogenic environmental modification takes place at a force and scale that surpasses the resiliency of the environment, then the environment will lose stability, such that this human agency will leave long-lasting, even permanent, traces. This could be described as a state in which the dominance of stories and reality are reversed. And the fact that human agency is imprinted on the environment can be expressed as reality being encroached upon by stories.

The legitimacy of stories that people originally spun and shared in the process of their survival, however, should in the final analysis have been secured with reference to that external reality. But if it is the case that this referential reality has been encroached upon by stories—if it has become a product of those stories, so to speak—then reality will no longer be able to retain legitimacy as a point of reference. Stories will lose their basis in reality, so that instead of stories being spun from reality, stories will themselves begin to spin subsequent stories, so that stories begin to be self-propelled.

Anxiety and optimism engendered by self-propelled stories

To inhabit stories that have lost their external referential basis in reality and become self-propelling is anxiety-inducing. This is because we cannot confirm whether or not the stories we inhabit are appropriate by making reference to a reality in the here-and-now. The only way to verify the legitimacy of these stories is for them to continue, and the only way we can do so is simply by propelling those stories earnestly toward the future.

However, self-propelled stories also evoke a kind of optimism in that they no longer have to be bound to reality. We are free to spin the stories we wish, and to decide for ourselves whether they are legitimate or not. Even when they falter due to inconsistencies with reality, the fact that this reality is merely a stage set means that we can modify it as we please to suit our stories. If this gives rise to further inconveniences, then we can modify reality again.

Such anxiety and optimism over the legitimacy of stories will also hold true in a world where reality is dominant in principle. However, as stories become ascendant, so too will anxiety and optimism become more prevalent. And both anxiety and optimism will eventually drive us to propel those stories forward. Might we not think of this as the true nature of humanity's future orientation?

It seems to me that the phenomena whereby stories encroach on reality and start to become self-propelling is one that also took place frequently even in societies prior to the Industrial Revolution. The Anthropocene could be described as an epoch when the ascendancy of stories was established and became self-propelling on a global scale.

Two views of the Anthropocene as extreme

The Anthropocene as self-propelled story

If we regard the Anthropocene as such an epoch, then wouldn't the very concept of the Anthropocene, having been proposed at such a time, be one of the self-propelled stories that have no basis in reality? Earth scientists such as Crutzen and Steffen make use of a variety of environmental data to objectively show how humans 'have become a global geophysical force' (Steffen et al. 2007: 614). However, while it is a fact that human agency is exerting a powerful impact on our current global environment in many ways, they reach no conclusion as to whether the nineteenth century should truly be regarded as the boundary of this new geological epoch. Nevertheless, they continue to use the term 'Anthropocene.' Is this not itself precisely a self-propelled story no longer bound by points of reference in reality?

Steffen et al. (2007), after evaluating the environment of humanity since the Industrial Revolution (using twenty-four different indices of human activity), divide the Anthropocene into three phases. The first phase covers the period from the Industrial Revolution until 1945. They hold that with the end of the Second World War, humanity's impact on the environment increased rapidly after 1945, entering a second phase that they call the Great Acceleration. The third phase comes after 2010, when the sense of crisis about the destruction of the environment by human agency takes on wider social currency. This is a period when the growth rate of environmental impacts begins to slow. The solution for humanity to survive the Anthropocene, Steffen et al. (2007, 2011, 2015) say, is for scientists to marshal knowledge and technology to monitor environmental indicators and exert controls so that the Earth's environment does not cross a tipping point from which there is no possibility of return.

Christophe Bonneuil critiques the above discourse about the Anthropocene by referring to it as a *grand récit* (grand narrative) (Bonneuil 2015; Bonneuil and Fressoz 2016). He says that Steffen and his colleagues talk about the Anthropocene as reality, but notes that it is actually a story spun by combining multiple phenomena. And it is a story of anxiety and optimism. It exacerbates the anxiety that humans, through our ceaseless environmental modifications, are having an irreversible impact on the global environment, to the extent that we will no longer be able to inhabit the planet. At the same time, it professes optimism that the hard work and enlightenment of earth scientists will be able to open humanity's eyes—that if we can control the global environment that serves as the stage set of this story by harnessing the power of science and technology, then we will be able to overcome this crisis. And the attitude of these authors in seeking to use technology to overcome a crisis brought on by technology is nothing other than the future orientation that brought about the Anthropocene, in terms of attempting earnestly to move the story forward.[4]

Meanwhile, thinkers such as Bonneuil, who critique 'grand narratives,' are also trying to ride out the Anthropocene by way of stories. Their critique is targeted solely at the framing of this grand narrative as a single story. And they point out the need to build a 'new narrative' about the Anthropocene (Bonneuil and Fressoz 2016; Shinohara 2018). However, this does not mean that even such a new narrative would be one spun from a point of reference in reality. So, ultimately, no matter what story we build, we are no longer able to break away from the current, unstable story, self-propelled on the twin wheels of anxiety and optimism.

A standoff with bare reality

And yet it is also possible to think in a way that completely inverts this proposition. In other words, rather than a story completely encroaching upon reality and beginning to propel itself, we can think of this story as one that has self-imploded as a result of having lost its reference point in reality.

When we think about how stories evolved, good stories (that is, adaptive stories) should be those that are in line with reality and as self-consistent as possible. Should there be too many contradictions or discrepancies between these stories and reality, this would jeopardize the survival and flourishing of those who inhabit them, as though climbing an unknown mountain armed only with a haphazard map.

However, since the advent of modernity, people have come to think that where there are discrepancies between stories and reality, the solution is to modify the latter to fit the former, rather than vice versa. However, the modifications of reality thus realized can only ever be makeshift, ad hoc measures, no matter how grand the scale. As a result, the contradictions, far from being resolved, have ended up becoming deeper. If we regard this as the arrival of the Anthropocene, then the Anthropocene could be said to be a state in which stories fail to maintain consistency with reality and then collapse, and an epoch in which humanity must confront a bare reality unmediated by stories.

Thinking thus, we gain a more positive understanding of the construction of a new narrative as proposed by Bonneuil and his colleagues. The Anthropocene signifies the end of the great stories that have persisted in the wake of the Industrial Revolution. This grand narrative is merely an attempt to prolong the life of a story by allowing it to propel itself when it should already have met its demise. Instead, if we can spin new stories from the bare reality we are now facing, we will once again be able to inhabit the interaction of stories with reality.

Acknowledgments

In the conception and drafting of this chapter, I received many valuable suggestions from Professor Takakazu Yumoto of the Primate Research Institute, Kyoto University. I would like to take this opportunity to express my gratitude.

Notes

1. Another problem is that this over-simplified view of human evolution ignores the fact that contemporary people in so-called natural societies (e.g. hunter-gatherers), apparently far more technologically advanced than the ancient *Homo erectus*, are leading sustainable lives.
2. Until quite recently, we were 'one of the underdogs of the savannah' (Harari 2011: 12).
3. This is exactly what is being advocated by authors like Crutzen (2002) and Steffen et al. (2007, 2015), who raise the Anthropocene as a crisis. They argue that even though it was technology that brought about the Anthropocene, it is by means of technology, nevertheless, that we will be able to overcome it.
4. 'On est en présence d'une fable modernisatrice annonçant la fin de la modernisation...[We stand in the presence of a modernizing fable announcing the end of modernization...]' (Bonneuil and Fressoz 2016: 98).

References

Bonneuil, C. (2015) 'The geological turn: Narratives of the Anthropocene'. In C. Bonneuil, C. Hamilton and F. Gemenne (eds), *The Anthropocene and the Global Environmental Crisis*. Routledge, 17–31.

Bonneuil, C. and J.-B. Fressoz (2016) *L'événent Anthropocène: La Terre, l'historire e nous*. Seuil.

Bourdieu, P. (1977) *Algérie 60: Structures économiques et structures temporelles*. Broché.

Caro, T. M. and M. D. Hauser (1992) 'Is there teaching in nonhuman animals?' *The Quarterly Review of Biology*, 67(2): 151–174.

Crutzen, P. J. (2002) 'Geology of mankind'. *Nature*, 415(6867): 23.

Crutzen, P. J. and E. F. Stoermer (2000) 'The "Anthropocene."' *IGBP Newsletter*, 41: 17–18.

Harari, Y. N. (2011) *Sapiens: A Brief History of Humankind*. Dvir Publishing House.

Kaifu, Y. (2016) *Nihonjin wa doko kara kita no ka?* (Where did we come from? New revelations of the great human journey to Japanese Islands). Bungeishunju.

Kawai, K. (ed.) (2019) *Others: The Evolution of Human Sociality*. Kyoto University Press and Trans Pacific Press.

Kawai, N. (2010) *Hito no shinka: 700 mannen-shi* (Human evolution: A seven-million-year history). Chikumashobo.

Klein, R. G. (1989) *The Human Career: Human Biological and Cultural Origins*. University of Chicago Press.

Lovejoy, C. O. (2009) 'Reexamining human origins in light of *Ardipithecus ramidus*'. *Science*, 326(5949): 74e1–e8.

Lovejoy, C. O. (2014) '*Ardipithecus* and early human evolution in light of twenty-first-century developmental biology'. *Journal of Anthropological Research*, 70(3): 337–363.

Mcbrearty, S. and A. S. Brooks (2000) 'The revolution that wasn't: A new interpretation of the origin of modern human behavior'. *Journal of Human Evolution*, 39(5):453–563.

Mesoudi, A. (2011) *Cultural Evolution; How Darwinian Theory Can Explain Human Culture and Synthesize the Social Sciences*. The University of Chicago Press.

Morgan, T. J. H., N. T. Uomini, L. E. Rendell, L. Chouinard-Thuly, S. E. Street, H. M. Lewis and K. N. Laland (2015) 'Experimental evidence for the co-evolution of hominin tool-making teaching and language'. *Nature Communications*, 6: 6029.

Musgrave, S., D. Morgan, E. Lonsdorf, R. Mundry and C. Sanz (2016) 'Tool transfers are a form of teaching among chimpanzees'. *Scientific Reports*, 6: 34783.

Nishida, T. (1994) *Chinpanjī omoshiro kansatsu-ki* (Thirty-four stories of chimpanzees). Kinokuniya.

Ōtsuka, R. (2015) *Hito wa Koushite Huetekita* (How humans have grown: A 200,000-year history of population change). Shinchosha.

Premack, D. (1986) *Gavagai! or the Future History of the Animal Language Controversy*. Bradford Books.

Saito, N., G. Suwa, Y. Satta, T. Yamamori, M. Hasegawa and K. Okanoya (2006) *Hito no shinka* (Human evolution). Iwanami Shoten.

Shinohara, M. (2018) *Jinshinsei no tetsugaku: Shiben-teki jitsuzai-ron-igo no 'ningen no jōken'* (Philosophy in the Anthropocene: The human condition after speculative realism). Jinbunshoin.

Steffen, W., W. Broadgate, L. Deutsch, O. Gaffney and C. Ludwig (2015) 'The trajectory of the Anthropocene: The Great Acceleration'. *The Anthropocene Review*, 2: 81–98.

Steffen, W., P. J. Crutzen and J. R. McNeill (2007) 'The Anthropocene: Are humans now overwhelming the great forces of nature?' *AMBIO: A Journal of the Human Environment*, 36(8): 614–621.

Steffen, W., J. Grinevald, P. Crutzen and J. McNeill (2011) 'The Anthropocene: Conceptual and historical perspectives'. *Philosophical Transactions. Series A, Mathematical, Physical, and Engineering Sciences*, 369(1938): 842–867.

Suddendorf, T. (2013) *The Gap: The Science of What Separates Us From Other Animals*. Basic Books.

Suddendorf, T. and M. C. Corballis (2007) 'The evolution of foresight: What is mental time travel, and is it unique to humans?' *The Behavioral and Brain Sciences*, 30(3): 299–313.

Takenoshita, Y. (2019) 'Society as a "story": Work sharing, cooperative breeding and the evolution of otherness'. In K. Kawai (ed.), *Others: The Evolution of Human Sociality*. Kyoto University Press, 387–406.

Wrangham, R. (2009) *Catching Fire: How Cooking Made Us Human*. Profile Books.

PART • TWO

Realizable Extremes I

Existence and Social Environments

CHAPTER 6

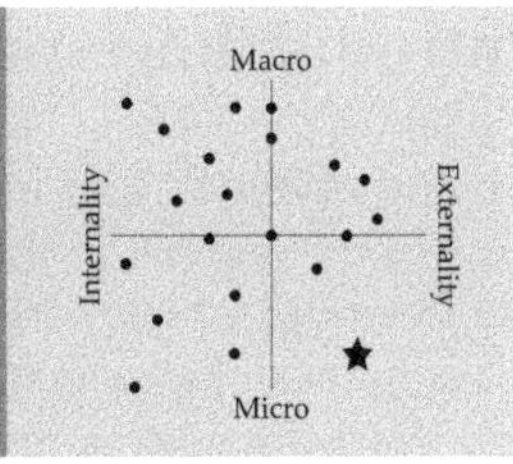

On the Survival of Orphaned Chimpanzees:

Does a Mother's Absence Constitute an Extreme Social Environment?

Michio NAKAMURA

Keywords: chimpanzees, orphan, mother's absence, allocare, life history

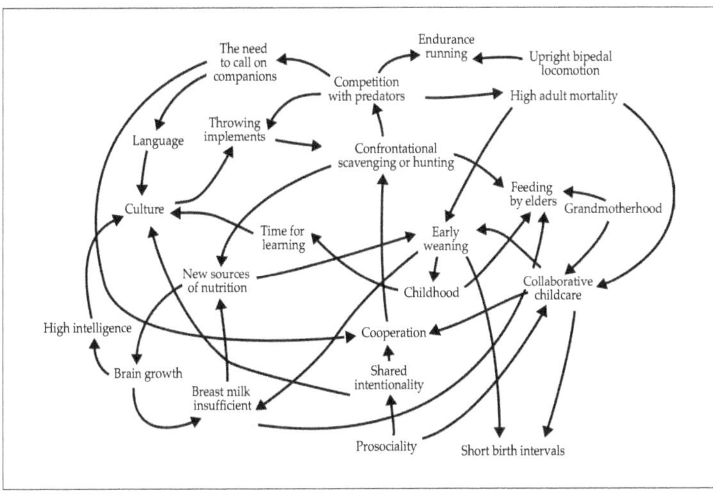

This diagram shows the various keywords and their mutual associations featured in recent discussions of human evolution to explain the 'special' character of humans. Arrows indicate the posited (sometimes speculative) causal relationships. The point here is that life history traits such as 'early weaning,' 'grandmotherhood,' and 'short inter-birth interval' are indirectly linked to various traits such as 'language,' 'high intelligence,' and 'culture' that have traditionally been perceived as demonstrating the uniqueness of *Homo sapiens*.

Introduction: special life history traits for humans?

Among mammals, species whose members are born in a very immature state are described as 'altricial' species, while those born at a relatively mature state are described as 'precocial' species. The former corresponds to animals like mice, for example, which are born with their eyes closed and little if any hair, and which are safeguarded in the nest by a parent for a length of time, while the latter are those like horses, which are able to stand up and keep pace with their mothers almost immediately after birth.

When we consider the case of primates, including humans, it seems they are often classified as precocial, since they do not make nests and in most species are born with hair. However, when compared to horses or the like, most primates are also clearly born in a more immature state. About the most that newborns can do is cling to the hair on their mothers' bellies (Photograph 6.1)[1] and suckle at their breasts. And while it varies depending on the species, for months and years after birth, the mother is an absolute supporter for the infant, as well as a means of transportation, the sole provider of nutrition, and a shelter from the rain and cold. Accordingly, for the immature individual, the absence of a mother—that is, becoming an orphan—should represent an extreme situation in the sense that this also jeopardizes its own survival.

In fact, in most mammals, the death of a nursing mother also spells death for her offspring. This being the case, it would be better for an offspring to become independent of the mother as early as possible, but this is not so simple. In principle, it is not impossible to ingest nutrients independently immediately after birth (think of fish fry as an example). But independence for an animal with a small body would make it more susceptible to predators, as well as

Photograph 6.1: A three-month-old infant chimpanzee clinging to its mother's belly

disadvantageous when competing with other individuals in its own species. Also, and particularly in the case of gregarious animals, the time spent under the mother's protection is also a period characterized by opportunities to socially interact with other individuals and learn a variety of new behaviors while under her wing, so to speak. The behavior acquired flexibly by such learning is also particularly advantageous for animals with relatively long lifespans.

Details of an individual's temporal change may, by and large, vary depending on species; such as the age at which animals reach sexual maturity, the speed at which they grow, how long they remain under their mothers' protection, the age to which they can expect to remain fertile or expect to live, etc. In biological terminology, this temporal flow of an individual's growth, maturation, and lifespan is referred to as life history (for a more detailed discussion of life history theory, see Sprague's discussion in Chapter 4). Certain aspects of life history can be understood from an animal's teeth or bones, and represent a biological trait prized by paleoanthropologists and others who study human evolution based on the fossil record.

Humans (*Homo sapiens*) are also a biological species with a unique life history. What follows is an overview of several life history traits that biological anthropologists have argued as being exceptional in the case of humans as compared to chimpanzees (*Pan troglodytes*), which, among extant species, are the most closely related to humans. Particularly central to this discussion is the period of 'childhood,' which has often been defined as a developmental stage unique to humans. It is said that humans in childhood are unable to live on their own, needing to depend on their parents or other older members of their species to obtain nourishment. Conversely, this means that apes are thought to be able to survive by finding food on their own as long as they have been weaned. In this chapter, I undertake a critical examination of these stereotypical images based on the existence of orphans (i.e., absence of mothers) in the context of a wild chimpanzee society that I have been observing for a few decades.

Early weaning

Because humans are also mammals, our newborn babies are also nurtured with breast milk. Reportedly, the timing of how long dependence on the mother's milk continues—and, from the mother's perspective, when breastfeeding will end—is considerably earlier among humans than chimpanzees. Based on the results of a survey of the literature pertaining to 113 non-industrial societies, humans cease breastfeeding at around the age of 2.4 years (Sellen 2001). Another study finds an almost identical value for this age, around 2.5 years (Kennedy 2005).

In chimpanzees, weaning is said to take place at around the age of four or five years. This is because they continue to suckle their mother's breasts until the age of 5.2 years on average (from fifty to eighty-six months) (Clark 1977). Even so, continuing to suckle does not necessarily equate to exclusive breastfeeding. According to Takuya Matsumoto (2017), who investigated the feeding behavior of infant chimpanzees, although suckling behavior continues until about the age of five (Photograph 6.2), feeding on solid foods undergoes a quantitative shift at around the age of three. Accordingly, Matsumoto argues that although nipple contact continues until around the age of five, the actual dependence on nutritional breast milk likely declines from around the age of three. In other words, while behavioral weaning takes place at around age five, nutritional weaning seems to take place around the age of three.

As different animal species have different lifespans and growth rates, we cannot compare their weaning schedules simply in terms of absolute age. Therefore, typically, interspecific comparisons are made based on events such as the eruption of the first molar (M1), which is the first permanent tooth to develop. Generally speaking, since permanent teeth are also called adult teeth, it is reasonable to think that around the time a permanent tooth erupts, an individual will move away from dependence on breast milk and start consuming the same diet as an adult. Since the eruption of M1 in chimpanzees takes place around age three (Smith et al. 2013), this is generally consistent with what Matsumoto (2017) calls the period of nutritional weaning. However, given that the eruption of M1 for humans takes place around age six, the weaning period takes place considerably earlier. The explanation for this discrepancy between the timing of weaning in humans and the eruption of M1 is the existence of childhood, as described below.

Photograph 6.2: Agano, soon to turn five years old, but still suckling his mother's nipple

The existence of childhood

According to the American biological anthropologist Barry Bogin, childhood is defined as 'the period following infancy, when the youngster is weaned from nursing but still depends on older people for feeding and protection' (Bogin 1997: 64; 2009), that is followed by juvenility. Here, infancy corresponds to the period from birth until weaning, and juvenility from the establishment of independence in feeding until sexual maturity.[2] Accordingly, this means that childhood refers to the period during which the youngster has started to wean but is not yet feeding independently. In evolutionary historical terms, Bogin believes that childhood likely appeared around the time of *Homo habilis*.

Childhood for humans generally corresponds to the period from the completion of the eruption of the full set of deciduous teeth to around 6.9 years old. Owing to their deciduous teeth and small digestive tract, humans in childhood need to eat nutrient-dense food that is easy to chew. Such foods are gathered, prepared, and then provided by the older members of the social group (see Sugiyama's discussion in Chapter 19). In other words, this means that it is considered that childhood-aged individuals cannot obtain food on their own (and, by extension, cannot survive on their own). It is thought that this period is required by humans due to the extra time needed for the development of their large brains and cultural learning. What makes these things possible is that individuals other than the mother (e.g., the father or grandmother) also play a cooperative role in bringing up the children.

On the other hand, among extant nonhuman primates, weaning (i.e., the end of infancy) equals the entry into full-fledged juvenility, without having passed through any intermediate state of childhood. This period of juvenility is common to most social mammals. What is important here is that juvenile mammals, although sexually immature, are nevertheless *considered to be* physically and mentally capable of caring for themselves. Among humans, juvenility corresponds to the period from around age seven (the eruption of permanent teeth) to just before the growth spurts of adolescence (around age twelve for boys and age ten for girls).

Placing emphasis on the fact that childhood is unique to *Homo sapiens* leads to the idea that, among apes, as long as weaning has taken place, the death of an individual's mother does not inevitably result in the death of the individual itself (e.g., Gibbons 2008), unlike the case with humans.

The existence of adolescence

Bogin (2009) also considers adolescence as a uniquely human stage of development. Adolescence in this case is characterized by a rapid growth spurt. According to the physical development data presented by Bogin, it is certainly the case that a tendency is apparent among humans to experience a rapid increase in the growth rate in girls around age twelve and in boys around age fourteen. These peaks are not apparent among chimpanzees. However, since such growth spurts can be seen among Japanese macaques (*Macaca fuscata*) (Hamada 1999), there is room for doubt as to whether this trait is truly exclusive to humans.

In the field of primatology, it is generally regarded that nonhuman primates also experience adolescence (e.g., Caine 1986). In this case, it is defined as the period from sexual maturity to the attainment of full adult body size (or the start of reproduction). In other words, during adolescence, the body is still small, despite having attained sexual maturity. While this definition does not necessarily entail a growth spurt, unlike Bogin's (2009) definition, the existence of a time lag between sexual maturity and the attainment of complete adulthood is now recognized by many researchers who observe extant primate societies.

In the case of the Mahale[3] chimpanzees that I have been studying, although males are able to ejaculate semen by around eight or nine years old, their body size at this age is still small, and it is not until they are around fourteen or fifteen years old that they reach their full size and are treated as fully grown adult males. Accordingly, the period of adolescence lasts between six and seven years. In the case of females, they begin to show sexual swellings at around eight or nine years old.[4] Swelling of adolescent females is not as pronounced as in adult females, nor is the swelling cycle as regular. Nevertheless, mating with young males in the natal group will also sometimes be observed. Some of these copulations are accompanied by ejaculations, but pregnancy is rare during this period (adolescent infertility). Later, around the age of ten or eleven years, females will begin to show full-blown sexual swelling in both size and cycle, and it is around this time that they will leave their natal group to immigrate to another. Females that have immigrated from other groups (see Hanamura's discussion in Chapter 7) are often estimated to be around ten or eleven years old, based on their physical characteristics. And although such females will mate frequently after their immigration from another group, they will not immediately become pregnant, and will typically be thirteen or fourteen years old when they give birth for the first time (Nishida et al. 2003; Nakamura 2015). Generally, a female will be treated as an adult from around the time she experiences her first birth.

While the phenomenon of adolescence is also an interesting topic for thinking about primate society in its own right, due to space limitations I will not delve into much detail in this chapter.

The existence of grandmotherhood

Another life history stage that may be cited as being unique to humans is that of grandmotherhood.[5] This period is associated with the long human lifespan and the lives of women after menopause (Hawkes et al. 1998). Sometimes also known as the 'grandmother effect' (Hawkes 2004), the idea is that it became adaptive that postmenopausal women, who are no longer giving birth themselves, survive even after menopause by supporting the reproduction and parenting efforts of their own offspring as grandmothers (i.e., taking care of their grandchildren). Hence this period, from postmenopause (after the end of fertility) to death, is known as grandmotherhood.

If we consider the mission of biological organisms purely as being to increase the number of their own offspring, we will no longer be able to explain why humans experience menopause or live for decades after menopause. It would seem more reasonable to continue reproducing until death, or for the life of the individual to end once it no longer has the capacity to reproduce.

Incidentally, it has been concluded that chimpanzees, *on average*, continue giving birth until death (i.e., without grandmotherhood) (Emery Thompson et al. 2007). I use the term 'on average' advisedly here because it is observed that some individuals stop giving birth quite a long time before their eventual deaths. For example, a Mahale female named Caliope, after giving birth to her final daughter at the (estimated) age of thirty-seven years, went on living for another fifteen years without any subsequent births before dying at age fifty-two in 2012 (Nakamura and Nishie 2012). This means that she would have lived without breeding for at least ten years, even after deducting the five years until her final daughter was behaviorally weaned. Caliope had given birth to at least five young (three daughters and two sons), four of which had survived to sexual maturity. Thus she could be said to have been a relatively successful female in terms of reproduction. Moreover, there are other species, like cetaceans, that usually go through a postmenopausal period (Johnstone and Cant 2010), and it remains necessary to verify whether the existence of grandmotherhood is truly a trait that is exclusive to humans.

The existence of human grandmotherhood is, in fact, highly consistent with the existence of childhood. In other words, the reason a mother can stop breastfeeding (i.e., wean) her immature offspring and quickly give birth to a next one is that individuals other than the mother are able to feed the older offspring in childhood, and it is the grandmother—who is related to both mother and offspring—who would be considered suitable for this child-rearing role.[6]

Discussions on human evolution in relation to life history traits unique to humans

Various life history traits thought to be unique to humans are discussed in relation to various 'humanly' traits that *Homo sapiens* has acquired in the course of its evolution. Below, I show a few such examples. It should be noted here that the kinds of behavioral changes that have actually taken place over the course of human evolution are necessarily speculative. And while there is some degree of evidence to suggest these changes, we must always keep in mind that such evidence will be interpreted differently by different researchers.

The transition to meat-eating (scavenging)

Chimpanzees and bonobos (*Pan paniscus*), which are closely related to humans, are basically frugivores. And while it is possible that the common ancestor of both humans and chimpanzees also enjoyed a frugivorous diet, at some stage in the human lineage this apparently shifted to a diet that includes meat. This is evident from the fact that the proportion of animal-based food consumed by extant humans is clearly higher than among other ape species, as well as the fact that human intestines are shorter than those of apes (as is typically the case for carnivores relative to herbivores).

It was once thought that active hunting began at the stage of early hominins such as *Australopithecus*. This would have been around the time that they acquired erect bipedalism and began to set out into the savanna. Although these images caught the public imagination in connection with catchphrases such as 'the killer ape' (Ardrey 1961) and 'man the hunter' (Lee and DeVore 1968), they have now fallen out of fashion. Even though they walked upright on two legs, the brain capacity of *Australopithecus* was not significantly different from that of extant ape species. This fact has led to questions as to whether they would have been able to hunt the large savanna-dwelling herbivores. Rather, the idea that these 'bipedal apes' had become the prey of large carnivores is gradually becoming more mainstream (Hart and Sussman 2005).

Nevertheless, at some stage in human evolution, there was doubtlessly a transition to a diet that included much more animal flesh than was consumed by extant apes, and the prevailing theory at the moment holds that this was initially accomplished by scavenging. Particularly in the context of the increase of brain capacity that began with the genus *Homo*, it is thought that the deficiency of protein in breast milk may have been supplemented with meat (Kennedy 2005). Human breast milk has less protein in comparison with cow's milk, and a diet consisting solely of breast milk will reportedly lead to protein deficiency by the age of four. For this reason, G. E. Kennedy (2005) argues that early weaning and the introduction of the same food eaten by adults is essential, and that animal carcasses brought home by adults would have represented an important source of protein.

In addition, if our ancestors had practiced confrontational scavenging[7] in which they actively chased away carnivores and deprived them of their meat, we can imagine that the mortality rate of young adults was quite high. Obviously, confronting a carnivore is a dangerous proposition. Thus it would have been advantageous for an immature individual to be weaned at an early stage so that the immature individual could survive even if a young adult in the parental generation were to die. This is also linked with the concept of allocare, as described in the next section.

Hyper-cooperation

Early weaning and the existence of a distinct stage of childhood serve to promote cooperative breeding. Since youngsters in childhood could be given foods other than breast milk, individuals other than the mother were now able to care for them. Mothers, conversely, were able to stop breastfeeding earlier, making it easier for them to become pregnant again (lactation typically inhibits estrus and fertility). Then, with the birth of subsequent offspring, it would become even more difficult for a mother to raise her offspring without assistance from other individuals.

The care of a young individual by an individual *other than* the mother is known as allocare. In the case of humans, it is taken for granted that biological fathers will also care for their own offspring, but this 'common sense' is by no means shared with other primate species. Particularly in species where there are multiple males in a group, paternity will not be clear, with the result that males will not often cooperate in rearing the young. Thus care by males (even biological fathers) can also be regarded as a kind of allocare.

Among types of allocare, some that are comparatively easy to understand are those provided by biological kin such as the offspring's father, siblings, or postmenopausal grandmother. However, in the case of humans, the existence of 'hyper-cooperation' (defined as cooperation that transcends kinship) has been discussed in relation to allocare (Burkart et al. 2014). Chimpanzees are positioned as a species with a relatively low degree of allocare, even among primates (Mitani and Watts 1997; Ross and MacLarnon 2000).[8] Thus it follows that the association between hyper-cooperation and allocare is a trait acquired by humans after their divergence from chimpanzees.

Everyday cooperation that transcends biological ties also affords advantages when confronting carnivorous animals. Thus the idea that traits such as the existence of childhood, the transition to meat-eating, hyper-cooperation, allocare, and a large brain evolved in close relation to one another (see the frontispiece to this chapter) has become a popular hypothesis, one assisted by its easy-to-follow story.

Other traits

But this discussion does not end there. Other traits, including acquisition of language (Bickerton and Szathmáry 2011), advanced cognitive ability (Isler and van Schaik 2012), shared intentionality (Tomasello and Carpenter 2007), and bipedal endurance running (Ruxton and Wilkinson 2013), are all also involved in the story in a complicated manner (see the frontispiece).

Hence, although childhood (the period when an offspring has been weaned but is not yet able to secure food for itself on its own) hardly seems at first glance to be that significant in its own right, the idea that it is closely related to various other traits that distinguish humans as a species has become a mainstream idea in contemporary biological anthropology.

The assumption here is that because childhood is exclusive to humans, nonhuman primates, which do not have a childhood, will be able to survive on their own as long as they are weaned, without having to depend on other individuals. But is this really the case?

Orphaned chimpanzees in the wild

Orphans in Mahale

To determine whether or not young individuals can live independently without having to depend on other individuals so long as they are weaned, it is useful to look at the case of individuals who have lost their mothers—that is, have been orphaned. Lacking any particular difference in the behavior or survival rate of orphans relative to identically aged individuals whose mothers are present, we would be able to conclude that the mother has no (or only limited) influence on her offspring after the weaning stage.

The number of orphans in a group of wild chimpanzees is by no means uniform. For example, whereas the M group, the current study group at Mahale, contained a comparatively large number of orphans between the 1980s and 1990s (e.g., Hayaki 1987; Matsumoto 1993), in the early 2000s, there were very few orphans. At the time of writing, there are once again several individuals who have recently been orphaned. In 1994, nine of thirty-one immature individuals were orphans. In 2005, this ratio had dropped to one of thirty, then rose again to three of twenty-eight in 2016 (see Table 6.1). Hence it needs to be kept in mind that the number of orphans is not constant. And although the behavior of a large number of orphans will be easily noticeable and thus attract the attentions of

	1994 Individual (age)	2005 Individual (age)	2016 Individual (age)
Males	**adolescent (14)**	adolescent (14)	adolescent (15)
	adolescent (13)	adolescent (14)	**adolescent (14)**
	adolescent (12)	adolescent (13)	adolescent (13)
	adolescent (12)	adolescent (10)	adolescent (12)
	adolescent (12)	adolescent (9)	**juvenile (6)**
	adolescent (11)	juvenile (6)	juvenile (5)
	adolescent (9)	juvenile (7)	infant (3)
	adolescent (9)	infant (4)	infant (2)
	adolescent (9)	infant (3)	infant (2)
	juvenile (6)	infant (2)	infant (2)
	juvenile (6)	infant (1)	infant (0)
	infant (3)	infant (0)	infant (0)
	infant (3)	infant (0)	infant (0)
	infant (2)		
	infant (2)		
Orphan ratio	5/15	0/13	2/13
Females	**adolescent (12)**	**adolescent (10)**	adolescent (9)
	adolescent (12)	adolescent (9)	adolescent (9)
	adolescent (8)	juvenile (8)	adolescent (9)
	juvenile (7)	juvenile (7)	**adolescent (8)**
	juvenile (7)	juvenile (7)	juvenile (7)
	juvenile (7)	juvenile (7)	juvenile (6)
	juvenile (7)	juvenile (6)	juvenile (6)
	juvenile (6)	juvenile (6)	juvenile (5)
	juvenile (6)	juvenile (6)	infant (4)
	infant (4)	juvenile (5)	infant (3)
	infant (4)	infant (4)	infant (3)
	infant (4)	infant (4)	infant (3)
	infant (4)	infant (3)	infant (1)
	infant (3)	infant (2)	infant (1)
	infant (1)	infant (1)	infant (0)
	infant (1)	infant (1)	
		infant (0)	
Orphan ratio	4/16	1/17	1/15
Total (males + females)	9/31	1/30	3/28

Table 6.1: Number of immature individuals and orphans in the Mahale M group for three different years

'Immature' here means individuals other than adults (thus 15 years or younger for males and 12 years or younger for females). However, as this table shows only natal individuals, immigrant adolescent females are not shown.

Ages (in years) represent those at September in each year.

Individuals whose mothers had died in each year are shown in bold face with underlines.

researchers who will attempt to record this as data, in their absence, no one will pay attention to orphans at all.

While I have had occasion to observe several orphans, as I show in the next section, orphans are clearly in dire straits (or at least seem to be). Conversely, the presence of mothers seems to be significant, even after the weaning stage.

The disadvantages of a mother's absence

To start with, there have been no observed cases in Mahale in which individuals survived as orphans before the age of three years. This is consistent with the idea of nutritional weaning mentioned above—in other words, should the mother die before her infant reaches the age of three (while it is still deriving nourishment from breast milk), it will perish at almost the same time as a result of the loss of this essential source of nutrients. Because an infant that has reached age three may be able somehow to get nourishment from solid foods, it will have a chance of survival (although not always, as we shall see). In what follows, I would like to focus on cases in which individuals aged three or older, who are conceivably nutritionally weaned, become orphans.

Several cases demonstrate how a mother's absence can become an extreme social environment. The most prominent of these is that of Flint, a Gombe male who was orphaned at the age of 8.5 years. Flint is said to have suffered from lethargy and a loss of appetite after his mother's death, ultimately succumbing himself three weeks later (Goodall 1983). Jane Goodall, in her report, speculates that Flint's psychological dependence on his mother must have been extraordinarily large.

In one Mahale case, a mother temporarily joined a neighboring group, leaving behind her five-year-old son, Katabi. In this case, although not strictly an orphan, Katabi was the victim of aggression more often during his mother's absence (Hiraiwa-Hasegawa and Hasegawa 1988), something the authors assumed would not have been the case had she been present.

Moreover, individuals orphaned even after they have been weaned can sometimes experience delayed physical development in adolescence (Goodall 1986). Although based only on a few observed cases, adolescent orphans often exhibited rough self-scratching,[9] which is an indicator of anxiety, and played less frequently (Botero et al. 2013). A Mahale female named Teto, whose mother died in 2014, also seemed to have reduced her frequency of play at first. Meanwhile, other individuals will play readily despite being orphans, so this seems to be an area that bears on individual personality and the offspring's relationship with the mother while she was alive. Also, looking at long-term data from Mahale, orphaned males, at least, have tended to die earlier than expected (Nakamura et al. 2014).

How do orphans survive?

Aside from extreme cases like that of Flint and those of individuals up to the age of about three who obviously still need breast milk, the death of the mother does not necessarily spell the immediate death of the offspring. So how do these orphans survive?

Individuals who are orphaned at a relatively young age (three to four years old) may in some cases be 'adopted' (Nishida 1994; Boesch et al. 2010; Nishida 2012; Nakamura and Hosaka 2015). A chimpanzee who becomes the orphan's adoptive mother will carry it when she is on the move, share her bed at night, and take care of almost every aspect of care aside from breastfeeding.[10] While adoptive individuals in Mahale were almost always unrelated females without any offspring of their own, it seems that in Côte d'Ivoire's Taï National Park there have been cases of orphans being adopted by adult males (Boesch et al. 2010). Also, siblings may provide care in cases where the orphan in question has an older brother or sister.

Since a young chimpanzee with a living mother will be carried or allowed to share a bed until the age of four or five, an orphan that loses its mother after this age will not need to be carried by another individual. Even so, the orphan will still need someone that it can follow around (Hayaki 1988; Nakamura and Hosaka 2015). For example, a female named Puffy who was orphaned at the age of six always followed a female named Gwekulo, who had been a close companion of her mother's. Gwekulo, who had no offspring of her own, would sit at the base of a tree waiting for Puffy when she would not come down from feeding in the branches, and frequently groomed Puffy, just as the mother usually does. This close relationship between Gwekulo and Puffy persisted until Puffy bore a baby of her own. Eventually, when Puffy attained adolescence, she did not transfer to another group, but instead, stayed in her natal group.

Allocare among wild chimpanzees

As noted earlier in the discussion about hyper-cooperation, based on a review that conducted an interspecies comparison, the practice of allocare among chimpanzees is quite limited. Few studies, however, have undertaken any detailed investigation of allocare among chimpanzees, and Toshisada Nishida's (1983) findings from his study of allocare for infants whose mothers were still alive, conducted with the Mahale K group (no longer extant) more than thirty years ago, remains the most detailed source of quantitative data on this topic. According to that study, the infant who received the most allocare accounted for 8% of observation time, while the median value for five individuals was only 2.6% (Nishida 1983). Despite the lack of quantitative studies of allocare since that time, even in Mahale, there have been a number of interesting anecdotal observations.

For example, while an infant in its first year of life will, under normal circumstances, rarely be separated from its mother, it is possible for it to be away from its mother for a considerable length of time if it has someone to provide allocare on a daily basis. In 2009, an eleven-month-old infant named XT08 spent more than half of the time during which he was under observation in the company of individuals other than his mother (Nakamura 2010). In this case, most of that allocare was provided by an unrelated adult female who was close with the mother, while for his part the infant beneficiary was riding around on her back and being held in her arms as though accustomed to such treatment. Nor did the mother appear to be particularly flustered, and after feeding for a while without the infant, she returned to him as though she had only just remembered, and gave him some milk.

In another case, the infant in question seemed to have some kind of handicap, and was lacking in strength in its lower body (Matsumoto et al. 2016). Usually, an infant will be able to cling to its mother's belly using the grip strength in its hands and feet. In this case, however, the mother had difficulty carrying her infant, since the infant could only grip with its hands. Especially when the mother was to climb a tree to feed, the infant seemed as though it were about to fall, so she once gave up eating. What was quite interesting in this case was that the mother left the infant in the care of her own adolescent daughter (i.e., the elder sister of the disabled infant). While the adolescent daughter was providing allocare, the mother went somewhere to finish feeding. In usual allocare, the 'babysitting' caregiver is more active in doing so (in other words,

Photograph 6.3: Caesar, an adolescent male, grooming his younger sister Cissy not yet a year old

he or she wants to be with the infant), but there are some cases, as here, in which care of the infant is entrusted to others by the mother.

Although none of the infants during the period when Nishida (1983) was carrying out his research had elder brothers, elder brothers up until adolescence apparently also sometimes provide allocare (Photograph 6.3). This means that the frequency of allocare will vary considerably depending on what and how many potential caregivers are available at any given time.

Thus, even among chimpanzees, a species in which the practice of allocare is reported in the literature to be almost non-existent, allocare can still be carried out for long periods of time depending on the circumstances and parties involved, and situations may also arise that could be interpreted as active recruitment for allocare on the part of the mother. Here, too, it is vitally important to develop a much more finely grained understanding of allocare, rather than simply in terms of average values.

Care still required after weaning

According to the theory of childhood, *Homo sapiens* is the only species in which care by mature individuals is still necessary after weaning. This way of thinking holds that chimpanzees and other apes have no need to receive care once they are able to feed on their own after weaning. However, this idea is open to the charge of unduly limiting care to the provision of nourishment (food). In fact, the care that chimpanzee mothers and other individuals provide to their young is not necessarily limited to the direct provision of food.

The problem of whom to follow

First, both before and after weaning, pre-adolescent chimpanzees, in principle, need another chimpanzee to accompany them to the feeding trees. During infancy, prior to weaning, this basically entails being carried on the mother's belly or back. As noted above, with orphans or in the case of allocare, such infants may be carried by individuals other than their mothers. Even an infant that has not been fully weaned, having arrived at a foraging tree, will move away from its mother into the treetops on its own, where it will eat the same fruits and leaves as its mother on a nearby branch. Foraging in the branches of a tree, the infant moves freely, and the mother allows it to do so. Looking at the distance between the mother and infant, it is often the case that although they are comparatively close when they are on the ground, they will be quite far apart in the tree (Photograph 6.4). Both mother and infant appear to feel more comfortable in the treetops than on the ground.

From slightly before weaning, and afterwards, the offspring will stop being carried as much by its mother and will begin to walk behind her. Basically, it is up to the mother to decide which tree to feed in, so the offspring will also feed in the same tree even after it stops being carried by its mother. Seen thus, the mother's presence is still important in terms of obtaining feeding opportunities, particularly when searching for food.

In most studies of chimpanzee ranging and party sizes,[11] individuals younger than juveniles tend to be excluded from total counts as dependents. As an example, 'Since infants and juveniles always accompany their mothers, they are not counted as independent animals' (Doran 1997: 188). In other words, even if chimpanzees are able to eat on their own before the age of eight, they may still be considered almost entirely dependent on their mothers until they arrive at the feeding site. Of course, once a chimpanzee becomes a juvenile, it will be able to eat on its own even when its mother is not eating—such as when, for example, an offspring may seek out and eat food on its own, staying close to its mother while she is engaged in grooming with another individual. However, even here it is not the case that the offspring is fully independent. This is because the mother will almost always wait for her young until it has finished eating. In very rare cases, the mother may lose patience and get moving first. In such cases, the now-stray juvenile will wander around searching for its mother, uttering the same whimper that an infant makes when feeling discontented (Photograph 6.5). Usually, the mother will be waiting somewhere not too far away.

To return to our discussion of orphans, it seems that the hardest problem experienced by most orphans is the question of whom to follow once their mothers, who had waited for them unconditionally, were no longer available (Photograph 6.6).

Photograph 6.4: Zamma (age one, right) sits separated from his mother, Zola (left), high up in a fig tree

Psychological dependence

By the latter half of an individual's adolescence, he/she finally becomes comfortable even without the mother's company at all times. An adolescent will, for the most part, no longer start whimpering and wandering around in search of its mother simply because she is out of sight. Even so, it is hard to reject the possibility that some degree of psychological dependence remains.

For example, Azam, a male in late adolescence, already spends little time ranging together with his mother, Abi. For her part, Abi tends to avoid raucous scenes where large numbers of individuals gather together, often preferring to situate herself slightly apart from adult males, even when moving as a member of a large nomadic party. On the other hand, because Azam often ends up being close to the adult males, they range separately most days. Still, Azam seems delighted whenever he first encounters his mother after they have spent some time apart. When he is near the adult males day to day, he appears to feel a certain sense of tension. Yet when he encounters his mother, the tension seems to visibly drain away from his face, at least temporarily, whereupon the mother and her adolescent son will continue grooming each other at length (Photograph 6.7). Relationships between grown sons and mothers are generally like this even when the sons are older—not that they usually stay close to one another, and may even sometimes give the impression that they are estranged. Once they do encounter each other, however, they will often spend a long time engaged

Photograph 6.5: Eight-year-old Liz screeches after losing sight of her mother—although she is already an adolescent, she still seems to feel uneasy when not in the company of her mother

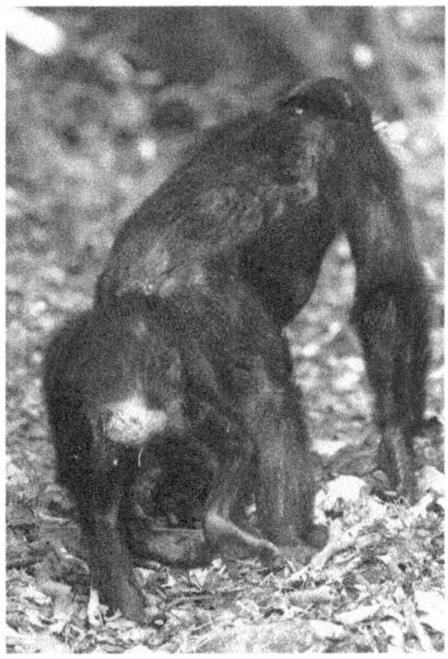

Photograph 6.6: Figaro, an orphan, following Darwin, an adult male

in grooming each other. Once, a son in his thirties and a mother in her fifties were even playing together and raising laughing voices. These brief yet intimate interactions with the mother are likely an important time for sons (especially for young adult males).

It was reported that a Gombe male named Figan, when he was a twenty-year-old fully grown adult male, hurried over to sit in contact with his mother during a strained encounter with another male. On another occasion, the mother, too, also ran over to the spot where he had had a fight with this other male. Having been injured in the struggle, Figan reportedly remained in his mother's company until he had recovered (Pusey 1983). Among chimpanzees, it is not uncommon for adult males, who stay with their natal groups, to be in the same group with their mothers (who, in most cases, will naturally be elderly).

In Gombe and Taï, cases have been observed of sons becoming alpha males with the support of their mothers (Goodall 1986; Boesch 2009). In Mahale, there have been cases in which mothers have still been alive when their sons became alpha males (Ntologi, Alofu, and Pimu), as well as cases where the mothers were already dead (Kalunde, Nsaba, Fanana, and Primus). Evidently, the mother's support is not always necessary (in the first place, females are not usually involved in struggles between males). However, it seems difficult to think that a mother would leave her son completely bereft of support in extreme situations—for example, where his life was at stake—but because it is rare even to be able to observe such cases, it is possible that their importance may have been underestimated thus far. It will be necessary to gather more information on this topic in the future.

Photograph 6.7: Azam, an adolescent male (right), encountering and grooming his mother, Abi (center), after a long separation; Azam has already grown to be as large as Abi

Conclusion

As we have seen from the whirlwind overview above, the binary oppositional view that 'humans require care after weaning while apes do not' is clearly too strong. Even among chimpanzees, the young individual's need for its mother does not end with weaning. This is precisely why becoming an orphan is such a disadvantage even after weaning has taken place. Yet anthropologists studying the life history of humans ultimately tend to place emphasis on the 'special' character of *Homo sapiens*.

As for the reason that chimpanzees have been summarized as not having allocare, it is simply that they practice it only infrequently. Certainly allocare among chimpanzees is by no means common when considered in quantitative terms. But this is partly because chimpanzee mothers are for the most part highly protective of their own offspring. As long as a mother is always holding and protecting her infant and unless she suffers some particular disadvantage as a result thereof, there will be no need for allocare. That said, the degree of allocare that does take place depends on the personalities of the mother and offspring, as well as the available caregivers. If the mother is still alive, any need for allocare will itself be low to begin with, but when it does arise (as in the case of the youngster becoming an orphan), allocare will become much more important.

Because the presence of orphans is far from uniform and is historically contingent, the extent of adoption and allocare will vary depending on when such data is recorded. Conceivably, group composition and social factors will also have a complicated impact in terms of the availability of individuals suitable as candidates to provide allocare (e.g., older siblings or close females without dependent offspring of their own).

In the case of humans, supporters bring food to youngsters after they have been weaned. This period is childhood. On the other hand, chimpanzee juveniles, once weaned, move to feeding sites in the company of their mothers (or foster mothers). In other words, whereas human youngsters are brought food, chimpanzee youngsters are brought *to* food. Thus, even among chimpanzees, weaning does not signal the complete withdrawal of care. We should consider, rather, that there are differences in the manner in which that care is expressed.

In the case of humans, who live a sedentary life (or at least maintain some kind of home base), food will end up being brought back to such places. Chimpanzees, on the other hand, will move to the places where food is found. Therefore, while 'care' for humans represents 'bringing food,' in the case of chimpanzees it takes the form of 'being brought/going together to the places where food is.' The definition of childhood should also be considered in this regard. Humans certainly differ from chimpanzees in that they 'bring food home.' However, they hold in common the fact that some sort of care is necessary with regard to offspring eating, even after they have been weaned. Before emphasizing the gap that separates humans from nonhuman primates by saying that only humans experience a 'childhood' during which they require care, it may be that we should think first of all about what 'care' entails according to the lives of each species. Among the prevailing orthodoxies about traits exclusive to humans, there are likely yet to be some characteristics that can also be found (if not in precisely the same form) in the societies of nonhuman primates.

Notes

1. Human newborns are unique in that they do not cling to their mothers' bellies (and their mothers also lack hair to cling to). However, even human newborns are known to possess sufficient grip strength to support their own weight.
2. Age classes typically used in primatology include infancy (up to the age of weaning), juvenility (from weaning to sexual maturity), adolescence (from sexual maturity to the attainment of full size), and adulthood.
3. My field site is Tanzania's Mahale Mountains National Park.
4. 'Sexual swelling' refers to the genital tumescence of the female. In adult females, it corresponds to the estrous cycle and shows maximum tumescence for roughly ten days prior to ovulation, and reduces in size at other times. Females will mate with males only when they show maximum tumescence.
5. A counterargument also exists to the effect that the life expectancy of early *Homo* was not long enough for grandmotherhood to have been adaptive (Kennedy 2003).
6. Other theories hold that the father assists with rearing offspring.
7. Researchers are divided as to whether our ancestors took part in such confrontational scavenging or were engaged in passive scavenging, which involves taking advantage of bone marrow and other scraps after the carnivorous animals had completely departed.
8. The extent of allocare is quantitatively judged according to involvement in aspects such as transportation, feeding, and protection by individuals other than the biological mother.
9. This is the behavior of scratching one's own body vigorously while making a loud sound. It is often carried out in times of stress, including those stemming from anxiety.
10. As a general rule, it suffices to say that breast milk is rarely given to orphans (in other words, foster mothers are rarely wet nurses). This is because lactating females will be nursing their own offspring, who they are much too busy raising in any case to become adoptive parents to an orphan.
11. Chimpanzees in the same unit group will routinely split up to range in smaller temporary groups called 'parties.' Many studies were carried out in the 1990s to investigate the factors that cause their size to vary.

References

Ardrey, R. (1961) *African Genesis: A Personal Investigation into the Animal Origins and Nature of Man*. Dell Publishing.

Bickerton, D. and E. Szathmáry (2011) 'Confrontational scavenging as a possible source for language and cooperation'. *BMC Evolutionary Biology*, 11: 261.

Boesch, C. (2009) *The Real Chimpanzee*. Cambridge University Press.

Boesch, C., C. Bolé, N. Eckhardt and H. Boesch (2010) 'Altruism in forest chimpanzees: The case of adoption'. *PLoS ONE*, 5: e8901.

Bogin, B. (1997) 'Evolutionary hypotheses for human childhood'. *Yearbook of Physical Anthropology*, 40: 63–89.

Bogin, B. (2009) 'Childhood, adolescence, and longevity: A multilevel model of the evolution of reserve capacity in human life history'. *American Journal of Human Biology*, 21: 567–577.

Botero, M., S. E. MacDonald and R. S. Miller (2013) 'Anxiety-related behavior of orphan chimpanzees (*Pan troglodytes schweinfurthii*) at Gombe National Park, Tanzania'. *Primates*, 54: 21–26.

Burkart, J. M., O. Allon, F. Amici, C. Fichtel, C. Finkenwirth, A. Heschl, J. Huber, K. Isler, Z. K. Kosonen, E. Martins, E. J. Meulman, R. Richiger, K. Rueth, B. Spillmann, S. Wiesendanger and C. P. van Schaik (2014) 'The evolutionary origin of human hyper-cooperation'. *Nature Communication*, 5: 4747.

Caine, N. G. (1986) 'Behavior during puberty and adolescence'. In G. Mitchell and J. Erwin (eds), *Comparative Primate Biology. Vol. 2A: Behavior, Conservation, and Ecology*. Alan R. Liss, Inc., 327–361.

Clark, C. B. (1977) 'A preliminary report on weaning among chimpanzees of the Gombe National Park, Tanzania'. In S. Chevalier-Skolnikoff and F. E. Poirier (eds), *Primate Bio-Social Development: Biological, Social, and Ecological Determinant*. Garland Publishing, 235–260.

Doran, D. (1997) 'Influence of seasonality on activity patterns, feeding behavior, ranging and grouping patterns in Taï chimpanzees'. *International Journal of Primatology*, 18: 183–206.

Emery Thompson, M., J. H. Jones, A. E. Pusey, S. Brewer-Marsden, J. Goodall, D. Marsden, T. Matsuzawa, T. Nishida, V. Reynolds, Y. Sugiyama and R. W. Wrangham (2007) 'Aging and fertility patterns in wild chimpanzees provide insights into the evolution of menopause'. *Current Biology*, 17: 2150–2156.

Gibbons, A. (2008) 'The birth of childhood'. *Science*, 322: 1040–1043.

Goodall, J. (1983) 'Population dynamics during a 15 year period in one community of free-living chimpanzees in the Gombe National Park, Tanzania'. *Zeitschrift für Tierpsychologie*, 61: 1–60.

Goodall, J. (1986) *The Chimpanzees of Gombe: Patterns of Behavior*. Harvard University Press.

Hamada, Y. (1999) 'Kodomo-ki ga nagai to iu hito no tokuchō' (Long childhood as a human trait). *Kagaku* (Science), 69: 350–358.

Hart, D. and R. W. Sussman (2005) *Man the Hunted: Primates, Predators, and Human Evolution*. Westview Press.

Hawkes, K. (2004) 'Human longevity: The grandmother effect'. *Nature*, 428: 128–129.

Hawkes, K., J. F. O'Connell, N. Blurton Jones, H. Alvarez and E. L. Charnov (1998) 'Grandmothering, menopause, and the evolution of human life histories'. *Proceedings of the National Academy of Sciences of the United States of America*, 95: 1336–1339.

Hayaki, H. (1987) 'Yasei chinpanjī no minashigo ni tsuite' (Orphan chimpanzees in the wild). *Kirisutokyō Hoiku Senmon Gakkō Hoiku Nenpō* (Christian Childcare Training College Childcare Yearbook), 7: 1–8.

Hayaki, H. (1988) 'Association partners of young chimpanzees in the Mahale Mountains National Park, Tanzania'. *Primates*, 29: 147–161.

Hiraiwa-Hasegawa, M. and T. Hasegawa (1988) 'A case of offspring desertion by a female chimpanzee and the behavioral changes of the abandoned offspring'. *Primates*, 29: 319–330.

Isler, K. and C. P. van Schaik (2012) 'Allomaternal care, life history and brain size evolution in mammals'. *Journal of Human Evolution*, 63: 52–63.

Johnstone, R. A. and M. A. Cant (2010) 'The evolution of menopause in cetaceans and humans: The role of demography'. *Proceedings of the Royal Society of London B*, rspb20100988.

Kennedy, G. E. (2003) 'Palaeolithic grandmothers? Life history theory and early *Homo*'. *Journal of the Royal Anthropological Institute*, 9: 549–572.

Kennedy, G. E. (2005) 'From the ape's dilemma to the weanling's dilemma: Early weaning and its evolutionary context'. *Journal of Human Evolution*, 48: 123–145.

Lee, R. B. and I. DeVore (eds) (1968) *Man the Hunter*. Aldine Transaction.

Matsumoto, A. (1993) 'Koji no shakousei: Mahale no chinpanjī no kansatsu kara' (The sociability of orphan chimpanzees in the Mahale Mountains). *Reichourui Kenkyu* (Primate Research), 9: 107–112.

Matsumoto, T. (2017) 'Developmental changes in feeding behaviors of infant chimpanzees at Mahale, Tanzania: Implications for nutritional independence long before cessation of nipple contact'. *American Journal of Physical Anthropology*, 163: 356–366.

Matsumoto, T., N. Itoh, S. Inoue and M. Nakamura (2016) 'An observation of a severely disabled infant chimpanzee in the wild and her interactions with her mother'. *Primates*, 57: 3–7.

Mitani, J. C. and D. Watts (1997) 'The evolution of non-maternal caretaking among anthropoid primates: Do helpers help?' *Behavioral Ecology and Sociobiology*, 40: 213–220.

Nakamura, M. (2010) 'Shakai ga fukuzatsu dearu to wa dou iu koto ka?' (What does it mean for a society to be complex?: Questions about relationship between society and individual). *Reichourui Kenkyu* (Primate Research), 26: 131–142.

Nakamura, M. (2015) 'Demography of the M group'. In M. Nakamura, K. Hosaka, N. Itoh and K. Zamma (eds), *Mahale Chimpanzees: 50 Years of Research*. Cambridge University Press, 82–93.

Nakamura, M. and K. Hosaka (2015) 'Orphans and allomothering'. In M. Nakamura, K. Hosaka, N. Itoh and K. Zamma (eds), *Mahale Chimpanzees: 50 Years of Research*. Cambridge University Press, 421–432.

Nakamura, M. and H. Nishie (2012) 'Death of the oldest female at Mahale and some notes about longevity of wild chimpanzees'. *Pan Africa News*, 19: 19–21.

Nakamura, M., H. Hayaki, K. Hosaka, N. Itoh and K. Zamma (2014) 'Orphaned male chimpanzees die young even after weaning'. *American Journal of Physical Anthropology*, 153: 139–143.

Nishida, T. (1983) 'Alloparental behavior in wild chimpanzees of the Mahale Mountains, Tanzania'. *Folia Primatologica*, 41: 1–33.

Nishida, T. (1994) *Chinpanjī Omoshiro Kansatsu-ki* (Thirty-four stories of chimpanzees). Kinokuniya.

Nishida, T. (2012) *Chimpanzees of the Lakeshore: Natural History and Culture at Mahale*. Cambridge University Press.

Nishida, T., N. Corp, M. Hamai, T. Hasegawa, M. Hiraiwa-Hasegawa, K. Hosaka, K. D. Hunt, N. Itoh, K. Kawanaka, A. Matsumoto-Oda, J. C. Mitani, M. Nakamura, K. Norikoshi, T. Sakamaki, L. Turner, S. Uehara and K. Zamma (2003) 'Demography, female life history, and reproductive profiles among the chimpanzees of Mahale'. *American Journal of Primatology*, 59: 99–121.

Pusey, A. E. (1983) 'Mother-offspring relationships in chimpanzees after weaning'. *Animal Behaviour*, 31: 363–377.

Ross, C. and A. MacLarnon (2000) 'The evolution of non-maternal care in anthropoid primates: A test of the hypotheses'. *Folia Primatologica*, 71: 93–113.

Ruxton, G. D. and D. M. Wilkinson (2013) 'Endurance running and its relevance to scavenging by early hominins'. *Evolution*, 67: 861–867.

Sellen, D. W. (2001) 'Comparison of infant feeding patterns reported for nonindustrial populations with current recommendations'. *The Journal of Nutrition*, 131: 2707–2715.

Smith, T. M., Z. Machanda, A. B. Bernard, R. M. Donovan, A. M. Papakyrikos, M. N. Muller and R. Wrangham (2013) 'First molar eruption, weaning, and life history in living wild chimpanzees'. *Proceedings of the National Academy of Sciences of the United States of America*, 110: 2787–2791.

Tomasello, M. and M. Carpenter (2007) 'Shared intentionality'. *Developmental Science*, 10: 121–125.

CHAPTER 7

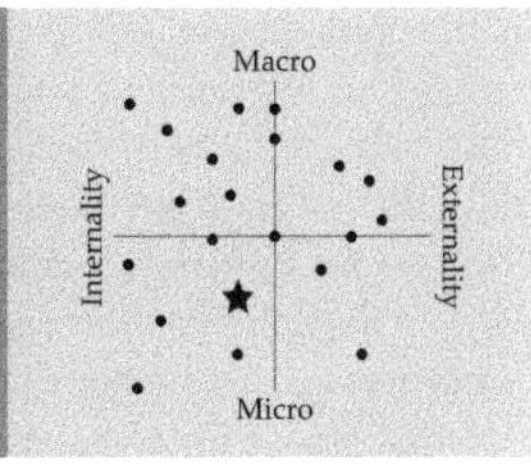

When Newcomer Females Become Separated:

Human Modes of Coexistence and Their Limits from the Perspective of Chimpanzee Partings and Reunions

Shunkichi HANAMURA

Keywords: ways of parting, parting greeting, reunion-enabling mechanisms, being a member of the same group, newcomer females / resident individuals, process-oriented—story-exploratory / goal-oriented—story-performative

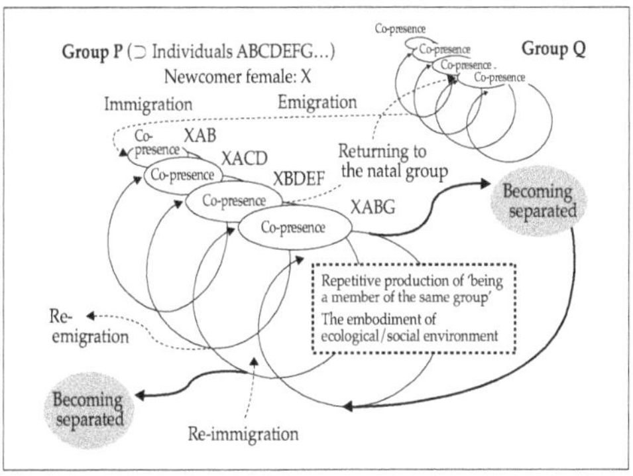

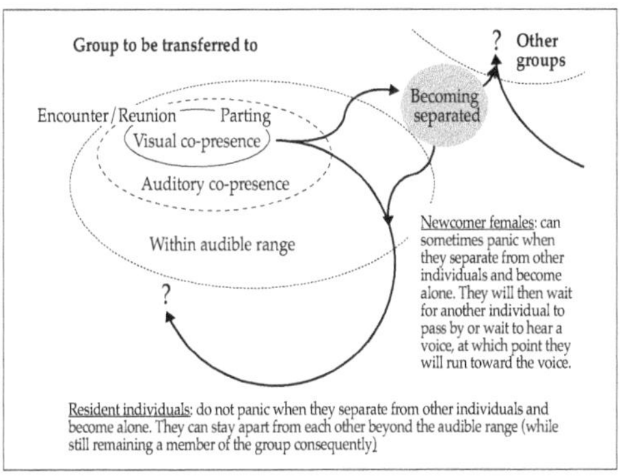

In previous studies (Hanamura 2017, 2019), I have argued that groups of wild chimpanzees are established solely on the basis of repeated co-presence. Adolescent females move into and out of such groups. Although newcomer females will follow various individuals, they sometimes get into a panic when they lose sight of others. This behavior, rarely seen among resident individuals, offers a glimpse of the possibility that newcomer females in such situations perceive they are in the 'extreme' situation (i.e., limits, edge, or border) as to whether or not they will remain a member of the group. By means of a comparison with such newcomer females, this chapter seeks to illuminate the ways in which resident individuals take leave of other individuals (ways of parting), as well as their behavior after such partings, thereby recapturing the mechanisms by which chimpanzees generate groups. Moreover, I also consider differences between modes of coexistence among humans and those among chimpanzees by focusing on ways of parting and reunion-enabling mechanisms—namely, the presence or absence of the following two matters: parting greetings that often take the form of a promise of reunion; and various 'stories' (i.e., conventions and institutions) that make reunions believable or normative by guaranteeing 'remaining a member of the same group (or social relationships)' in the future, as well as 'homes' that serve as a base for such stories.

Modes of coexistence seen in partings and reunions

Although the word 'extreme' is polysemous, in this chapter I use it in two particular senses. The first is in terms of the 'limit' or 'edge' as to whether or not an individual will be able to continue as a member of a group. I discuss this kind of extreme by focusing on partings and reunions among chimpanzees (*Pan troglodytes*) in order to explore modes of coexistence among humans (*Homo sapiens*) and what differentiates them from those of chimpanzees. The second sense, which I only touch on without necessarily analyzing in a concrete fashion, is that of conditions of physical or psychological extremity.

A society woven of partings and reunions

Usually, we engage in repeated partings and reunions without much conscious thought. We form multilayered and pluralistic groups (or social relationships)—from friends, lovers, and families to ethnic groups, nation-states, and religious groups—that are extremely diverse in scale, ranging from two into the billions. Yet no matter the level of such a group, the possibility of long-term separation is always present for its members. In other words, we could say that being a member of the same group is not something that is easily unstitched even when members stay apart from each other. While for now we acknowledge here that in the case of ethnic groups, nation-states, and religious groups, we may never even meet their 'imagined' members, in the case of friends, lovers, and families, one's membership will continue for as long as he or she can expect the possibility of reunion, even if such never actually takes place, just as cutting off the possibility of reunion means the end of the relationship or a definitive break from the group. In the context of the evolutionary history of human society, the establishment of families with a division of labor between spouses in subsistence activities such as hunting (or scavenging) and gathering—and of residential groups (bands) made up of multiple such families—is considered to have been one of the major events constituting the basis of human society (Imanishi 1961; for a recent review of Imanishi's idea, see Idani 2013). Partings and reunions would have needed to be conducted smoothly in order to maintain the family unit while some of its members were absent for prolonged periods due to the division of labor, or so that individuals from multiple families could repeatedly gather and disband in order to form subsistence groups.

However, looking at vertebrate societies in general, it seems that species that engage in the constant repetition of partings and reunions after somewhat long-term absences, such that specific individuals will be gone for a matter of days at least (hereinafter, 'species that live in societies interwoven with partings and reunions'), are not so common (Hanamura 2021b).[1] I have the sense that it may come down to animals living in the following two types of societies. The first corresponds to species such as chimpanzees or elephants, in which a certain number of individuals constituting a group repeat fission and fusion, whether at the level of individuals or of fixed subgroups of regular members (for a review, see Hanamura 2015a). The other corresponds to species, such as the pipefish (Sogabe 2011) or Adélie penguin (Davis 1988), that often form pairs with the same partners, interspersed with periods of separation. In this chapter, I would like to think about the features of human modes of coexistence and their basis in terms of evolutionary history from the perspective of ways of parting and reunion-enabling mechanisms. As a clue to consider this, I focus on instances of parting and reunion among wild chimpanzees, who, though they shared a common ancestor with *Homo sapiens* about seven million years ago, have today evolved a much different society than our own.

Chimpanzee modes of coexistence and female transfer

Chimpanzees in the wild form stable multi-male, multi-female groups whose members, while using the same home range (i.e., the habitual ranging area for the group as a whole—about 30 square kilometers in the case of the M group at Mahale, Tanzania, which is the focus of this chapter), engage in repeated visual encounters and partings (i.e., fission and fusion) at the individual level (Itoh 2017; for a review, see Hanamura 2015a). The time interval between when two individuals take their leave of one another until their next reunion may be minutes or months. The distance between individuals, likewise, could simply involve temporarily losing sight of one another, such as when obscured by thick bushes, or may stretch for several kilometers. In the process of this fission and fusion, in each instance chimpanzees form temporary and fluid gatherings of individuals, referred to as 'face-to-face parties,' that remain within visual range of one another. While these can sometimes grow quite large in terms of size (i.e., number of constituent individuals), it is unlikely that all the members of the group will meet at the same time. These gatherings of about four individuals on average will serve as a field for various face-to-face interactions, including the synchronization of activities such as movement, feeding, rest, and sleeping, as well as for grooming, play, and occasionally antagonistic interactions. In addition, in the M group, the size of gatherings, referred to as 'nomadic parties,' which consist of individuals who roughly synchronize the direction of the day's travel while repeating fission and fusion, fluctuates significantly with the seasons in accordance with the density of the fruits that represent their staple diet. For this reason, we can distinguish between a 'congregating season'—in which many individuals of the group loosely gather while undergoing frequent fission-fusion and such a large nomadic party moves across the group's entire home range on a wide scale—and a 'dispersing season'—in which they form multiple smaller and more cohesive nomadic parties that tend to move independently of one another (hereinafter, 'party' will refer to face-to-face parties unless otherwise specified).

When observing wild chimpanzees, in addition to the voices and noises of the individuals under observation, one can also hear various voices and noises made by other parties beyond the field of view. Of these, most frequently heard are long-distance vocalizations known as 'pant-hoots' that are audible over a range with a radius of 1 to 2 kilometers. While observing the M group chimpanzees, pant-hoots would be heard from beyond visual range once or twice an hour on average throughout the year. This does not mean they would be constantly heard, however. They were often heard during the congregating season, with apparent peaks during the period shortly after the animals woke in the morning (08:00 to 10:00) and during the transition from dusk to evening when they bedded down in the trees for the night (18:00 to 20:00)—that is, the times when they could coordinate the direction and timing of the day's travel as well as the site and timing to sleep for the night. During the dispersing season, however, they were not heard nearly so often and, when they were heard, the peaks were not so clearly distinguished (Hanamura, unpublished data). These vocalizations, accompanied with various other voices and noises they make, may also be deeply involved in the seasonal variation in the size of nomadic parties.

Thus wild chimpanzees experience the state of co-presence/non-co-presence with other individuals in the group both visually and audibly on a variety of temporally and spatially diverse scales. Here I would like to introduce interactions via pant-hoots among the M group chimpanzees, and some aspects of their experiences of others in such interactions that are linked to the issues raised in this chapter, by summarizing discussions in my contributions to the predecessors to this volume, *Institutions* and *Others* (Hanamura 2017, 2019).

While pant-hoots often take the form of chorus (synchronized or overlapped) among individuals of the same party, M group chimpanzees accomplish non-face-to-face encounters by exchanging such pant-hoots between parties within a span of approximately ten seconds, and continue a state of auditory co-presence between the parties concerned by repeating such exchanges.[2] However, the 'goal' of the interaction to be developed there is not predetermined, and chimpanzees will coordinate their actions with a 'process'-oriented attitude in which each coordinates his or her own behavior depending on how the other party behaves, all the while being influenced by their own respective activity rhythms and involvements with other individuals within the party. Thus, by means of this way of action connection, conceivably, it will always be possible for them to seek an encounter and attempt to maintain a state of co-presence as long as it is within the audible range, and at the same time it will also be possible for them to suspend such interactions to continue their own activities, let each other's voices pass while remaining apart, or go their separate ways even beyond the audible range.

Moreover, although M group chimpanzees basically adopt a process-oriented attitude also to the (rarely heard) voices of strangers (i.e., individuals of other groups), the actual reaction is very different from that given to voices by acquaintances (i.e., individuals in the same group). For most M group chimpanzees, strangers are individuals with whom they have never established a mutually non-hostile co-presence; therefore, for as long as the possibility of an encounter is manifested by hearing their voices or discovering their new tracks, such strangers are experienced as others whom 'they cannot stay apart from' in the sense that they will continue to listen and look for the subsequent voices and tracks in preparation for the possible encounter where 'they do not know what will happen.' In contrast, acquaintances are those with whom they have repeatedly established a mutually non-hostile co-presence; as a result, even if the potential for an encounter were to be manifested through voices or tracks, such acquaintances are experienced as others whom 'they can stay apart from' since little preparation is necessary for any encounter which will be 'likely to be the same as before.' However, because wild chimpanzees do not have language, they will have no a priori means of explicitly conceptualizing and sharing things (i.e., a means of confirming group affiliation using categories such as 'my group' and 'another group') prior to experience (i.e., actual interactions). For this reason, this difference between strangers and acquaintances must derive from the presence or absence of (or more or less of) co-presence with each of those two kinds of others. Although, in practice, this results in a large disconnection, it seems to just be a spectrum.

Wild chimpanzees lack the language and institutions that humans use as clues for confirming 'being a member of the same group.' Nor do they form the troops that many other primate species can use as such a clue. In the case of species that form troops, they are able to directly perceive that 'we are members of the same group' via the constant proximity of all the constituent members. As a result, groups of chimpanzees seem to be established solely on the basis of having repeated non-hostile co-presence (partings and reunions).[3] From our sensibilities as humans, who are heavily dependent on language and institutions for recognizing in-group and out-group affiliation, this would seem to be very unstable (Itoh 2003, 2013, 2019; Kitamura 2013; Hanamura 2019). What ingenuity in the way of co-presence / non-co-presence underlies the establishment of such groups that lack concentration? I am thinking here, in particular, of ways of parting and reunion-enabling mechanisms. In order to approach this issue, in this chapter I focus on behavioral differences observed between individuals who have recently come into a group and those whose residence is longstanding.

In chimpanzee society, adolescent females who have attained sexual maturity will transfer between groups whose constituent members and home range there is no way of knowing until they have actually spent some time therein. Although some females will remain in or return to their natal groups (Matsumoto et al. 2021), most females will stay for the rest of their lives in the non-natal group in which they first give birth. Therefore, in this chapter, I discuss the kinds of relationships fostered with other individuals in M group by 'newcomer females' (defined as females who have immigrated from other groups but have not yet given birth)[4] and the nature of their experiences of the state of co-presence / non-co-presence through a comparison with 'resident females' (namely, females who are not newcomers). In particular, I focus my analysis on scenes in which such females 'get separated from other individuals and panic.' In such scenes, we should find glimpses of the limit of whether or not chimpanzees who can stay apart from each other will continue being members of the same group. From there, it should be possible to recapture the mechanisms by which chimpanzees generate groups, as described above. In addition, through comparisons with chimpanzees' ways of parting and reunion-enabling mechanisms, it should also be possible to glean clues for considering how humans have dealt with such limits and created a multilayered pluralistic group in the evolutionary history of their society.

It should also be noted that the qualitative and quantitative data used in this chapter is based on field notes collected over a period of approximately one year spanning 2005–06 (the first field study period, hereinafter 'First Period') and four other occasions over a total of approximately five months between 2012 and 2014 (the second field study period, hereinafter 'Second Period'). During both periods, I employed primarily an observation method known as 'focal animal sampling' and recorded the details of my observations in a field notebook and the traveling routes of focal animals on blank maps in conjunction with GPS. The main individuals targeted by focal animal sampling were four newcomer females (accounting for approximately 250 hours of observation in total) and ten resident females (approximately 690 hours) during the First Period, and six newcomers (approximately 180 hours) and nine residents (approximately 140 hours) during the Second Period.

Newcomer females can sometimes be difficult to observe because they are leery of humans, especially immediately after their arrival in the group. Nonetheless, in M group, which is well habituated to human observers, the presence of humans is seen to be part of the environment, in the same way as other animals such as red colobuses and leopards. It could even be said that becoming acclimatized to humans is a basic prerequisite to becoming a member of M group. While the task of newly habituating a group of wild chimpanzees to humans will take several years even if all goes well, more than a few of the newcomer females who joined M group came to accept being subjected to focal animal sampling within a matter of a few days. Although there are significant individual differences—some individuals are not so tolerant no matter how many years have passed—newcomer females, including such cautious individuals, are more likely to fear humans when they are alone, and less so when they are in the company of other chimpanzees who have already acclimatized to humans. Taking these tendencies into account, it seems that many newcomer females would quickly become accustomed to humans through the attitudes toward humans evinced in each instance by other chimpanzees around them, in which attitudes, as a rule, do not engage as long as they do not experience anything untoward, such as being surprised or having their movements or interactions disturbed.

Of course, the practice of observation in general—and focal animal sampling in particular—constitutes an interaction with the target animal (Hanamura 2015c), and effort on the human side of the equation is also essential. Chimpanzees have different attitudes depending on whether the human observers are tourists or field researchers, as behaviors of these observers differ considerably (see also note 7). Sometimes the identity of the particular field researcher will matter, and attitudes may change even toward the same individual, depending on their manner of observation. Therefore, especially in the case of newcomer females immediately after immigrating, I suspended focal animal sampling on occasions when I judged that it was having too much of an impact (i.e., being felt unwelcome) on how these individuals were moving, as well as how they were interacting with others. On the other hand, in order for the newcomer females to accept me and my research assistants as quickly as possible, I also attempted focal animal sampling on an intermittent basis after hours or days, even if it seemed to prompt annoyance (in principle, however, the analysis below excludes observational data for such females immediately after the arrival in the group). While I learned a lot about how to observe chimpanzees without making them uncomfortable—from how to keep my distance to how to walk—from the methods of elderly local assistants who were deeply familiar with the forest and its denizens, more directly, I feel that I learned even more from the attitude to my observation practices shown by the newcomer females, as well as some older females who had remained wary of humans over the years.

Relations between newcomer females and other individuals: comparison with resident females

As a result of natural selection (sexual selection), female mammals are considered to have evolved behavioral tendencies that lead them to secure the food resources essential for ovulation, breastfeeding, and child-rearing. On the other hand, males, who are able to produce as many offspring as they have mating partners, are thought to have evolved behavioral tendencies that lead to an increased number of mating opportunities. Therefore, females are more likely to threaten or attack newcomer females who can be their potential competitors for foraging, while males are more likely to support or protect these newcomer females as potential new mating partners in their agonistic interactions with other individuals. Therefore, it is predicted that newcomer females will generally avoid other females and spend more time traveling with males. Findings that support these predictions were reported around forty years ago for the Mahale M group (Nishida 1989) as well as for several other local groups (for a review, see Hanamura 2015b).

However, my own recent research on M group yielded the results summarized below, which differ from the prevailing view. That being said, some of the results included here are only preliminary, as the volume of data remains insufficient for statistical processing. Below, the terms 'male' and 'female' are used to refer to sexually mature males and females aged nine years or older, regardless of whether or not they are adolescents or adults, while I omit gender distinctions for individuals aged four or under, whom I refer to as infants, and for those between the ages of five and eight, whom I refer to as juveniles.

(1) Focusing on those females on whom I had conducted focal animal sampling during the First Period and had obtained sufficient data without major seasonal bias, I analyzed time proportion by party composition ('party' is defined here as 'mature individuals within the observer's field of view at fifteen-minute intervals while conducting focal animal sampling'). I found that newcomer females tended to spend less time alone than resident females (although the majority of resident females were accompanied by their own children, usually one or two infants or juveniles), and spent most of their time in a 'mixed-sex party' that included both males and females. However, the longest newcomer female in years after immigration (Qanat) spent most of her time in a 'female-only party,' a tendency that was shared by the findings for more than half of the resident females. I then calculated the percentage of time spent by newcomer females in the same party as each individual on a dyadic basis (i.e., in terms of the two-way relationship) and compared this by age-sex categories. Doing so, I found that with the exception of one individual who tended to spend less time with other females, there did not appear to be any major discrepancy in the time that newcomer females spent in the company of males versus other females.

(2) While newcomer females followed a variety of individuals, including both males and females, immediately after their arrival in the group, at least two newcomer females (Qanat and Genie) began to follow the same

individuals consistently and repeatedly within a few years of arrival (usually older resident females, though not necessarily only one). Of these two, Qanat was confirmed as having maintained a similar relationship with the same resident females between the two field study periods separated by six years, even after giving birth herself. For each female (in the same sample as Result 1 above), if I defined 'co-ranging females' as those with whom the percentage of time spent in the same party on a dyadic basis was greater than or equal to the median percentage of all dyads and more than half of that time was accounted for time spent in female-only parties (hereafter, I refer to such relationships among females as 'co-ranging relationships'), most resident females—not only the relationship with two newcomer females above for specific resident females—also had co-ranging relationships with other resident females.

(3) Newcomer females were, in fact, more often threatened and attacked by males than by females, and even most of the rare instances of threats or attacks by resident females stemmed not from foraging conflicts but, rather, from high levels of interest shown by newcomer females to infants and juveniles, or the resident females themselves who are their mothers. One of the most common occurrences, for example, was for a scream to be uttered by a playing infant or juvenile over a trivial matter,

Photograph 7.1: March 2014, Enju (left, five months after arrival) and Mao (right, within one week of arrival). Although I am unable to touch on relationships between newcomer females in this chapter, it appears that females who arrive within a relatively close period of time tend to get on well together.

which in turn prompted a threat or an attack by its mother, who came rushing onto the scene. Also, during agonistic interactions between newcomer females and other individuals, other females (in most cases older resident females) sometimes even supported or protected newcomer females, and such behavior was not limited to co-ranging females of the newcomer females.

Accordingly, relationships with males and females seem to be equally important in the context of the process whereby a newcomer female joins a group. Theoretical predictions aside, in practice, a newcomer female—particularly immediately after her arrival—can be the focus of interest for resident individuals irrespective of age or sex. In one instance, shortly after some days of her first observation day, a newcomer female (Vera) was surrounded and being simultaneously groomed by four other individuals, including both males and females. They can also be popular among juveniles, especially in the case of young newcomer females—likely because of their high-spirited play (e.g., Yuri and Enju; see also Case 1 in the appendix to this chapter). Of course, I also observed some instances that might support the theory above; for example, a resident individual, Ikocha, adopted a threatening attitude by stopping in her tracks in surprise and then suddenly beating the ground at the sight of a just-arrived newcomer female, Vera, who would have had little previous

Photograph 7.2: A newcomer female, Genie (foreground) following a resident female, Ikocha (background); such interactions in traveling might be cultivating their 'co-ranging relationship' in the future. A scene from Case 3 in the appendix to this chapter (around 15:02, see Case 3 for information on the individuals involved).

involvement with her. In general, although resident individuals tend to take an ambivalent attitude toward newcomer females directly after their arrival, their presence seems to invite curiosity (see also Itoh 2013).

To examine why some of these findings differed from the prevailing theory, we need to compare a variety of factors against previous studies. Such factors include the characteristics of newcomer females at the time of the research (e.g., age, weight, personality, any previous experience of group transfer), the number of newcomer females, group size, and socionomic sex ratio (which refers not to birth sex ratio but to the intra-group sex ratio of individuals who have reached sexual maturity), and the type, amount, and distribution of foods, which all can affect the degree of feeding competition. Also, the earlier study of M group (Nishida 1989) was conducted in the early 1980s, at a time when a large number of females were joining the group in the process of the disappearance of the neighboring K group. Consequently, many of the newcomer females who immigrated at this time were parous females (i.e., females who had previously given birth) who had previously settled in K group and were now secondarily transferring into M group, some accompanied by dependent offspring. This difference from the usual case, in which newcomer females are nulliparous (i.e., have never given birth), will likely have influenced the survey findings to some degree.

That newcomer females sometimes attach themselves to and travel with specific resident females is a phenomenon also apparent among bonobos (Idani 1991), who belong to the same genus and share a quite similar social structure with chimpanzees, and something that features in case reports of Mahale chimpanzees (Nishida 2012). Also, as chimpanzee females are known to form long-term friendships with each other (for a review, see Itoh and Nakamura 2015), the 'co-ranging relationship' noted above can be considered to express one aspect of such a relationship.

Chimpanzee partings and reunions

Now that we have pinned down an overview of the relationships that newcomer females (and resident females) have with other individuals, I would like to introduce five specific cases (see the appendix to this chapter). Later, I will touch on the passages on which to focus for the analysis emphasized by some symbols (for details, see also 'Note to symbols' in the appendix), but for now I would like you to read through the cases. The overview of each case is spelled out in its title. Each case, excluding Case 2, revolves around a scene in which a newcomer female became separated from other individuals and was left alone but was then reunited with other individuals, though not always the same ones as those from whom she had been parted. The cases are listed in order of the length of time since the arrival of the female who represents the protagonist of the case, from shortest to longest.

Let us check just one point before entering into the main analyses. As outlined in the previous section, whereas newcomer females—especially just after their arrival—tend to attract the interest of various individuals, for their part these newcomer females also attempt to follow various other individuals in each instance. In this process, it is thought that newcomer females come to know where and what kinds of feeding trees are available (see the ★ enclosures in Cases 1 and 4) and what is edible,[5] as well as how many members there are in the group and how far their home range extends (Itoh 2013; Hanamura 2019). The newcomer females, who, like me, follow the other chimpanzees, should be in the midst of embodying the group's ecological and social environment through their association with other individuals. From the perspective of surviving an unfamiliar environment, we could argue that they are in a kind of 'extreme situation' that cannot be experienced by other individuals or in other situations. However, something that we can glimpse in the following analyses is the way in which they spend their time living carefree and comfortably in their own way, even under such circumstances. Though they can sometimes grow impatient or troubled, to that extent it can also be a life full of stimulation.

Partings: newcomer females who panic at becoming separated/resident females who calmly see others off

When females get separated from other individuals and left on their own, or—more accurately—when they become aware of this, they may exhibit a panicked behavior (as seen in Box A in Cases 1, 3, 4, and 5). This behavior entails rushing off in the direction thought to have been taken by the other individuals who had been there until moments earlier, and repeatedly pausing to scan the environs, sniff the ground nearby, and listen carefully while ranging back and forth. In the sense that these individuals are clearly seeking the other individuals who have disappeared and are troubled at not knowing where the other individuals went, the concerned individuals in such instances could be said to be panicking at becoming separated.

This panicked behavior is hardly ever seen when resident females part from others. While it is certainly not uncommon for chimpanzees, through their senses of sight, smell, and hearing, to seek the direction in which other individuals set off, in the entire period of my research (approximately one year and five months), I observed only two cases in which resident females (whom I observed in greater number of individuals and for a longer time than newcomer females), when parting from other individuals (excepting cases in which they were separated from their own children), clearly exhibited panicked behavior in the same way as newcomer females (i.e., behavior that not only involved searching for other individuals, but which was also accompanied by running and rushing around or moving around on the trees). The first case involved Ako, who, having spent several days traveling during the dispersing season in the sole company of her children, encountered Sally—one of her co-ranging females—and her children. After traveling together for a while, Ako suddenly noticed that Sally and her children had disappeared, whereupon she spent about a quarter of an hour frantically and quite persistently searching for them (they were subsequently reunited). The other case, which was also observed in the dispersing season, occurred whereupon Nkombo, a female, noticed the absence of Kalunde, a male with whom she had kept close company like a married couple for many years (see Hanamura 2015b), although several other individuals were still there (afterward, I lost sight of her briefly and by the time I found her again their reunion had already taken place).

I had observed instances of parting between Ako and Sally and between Nkombo and Kalunde on many occasions, and I do not understand why they became particularly upset on these occasions only. However, both cases share the common point that their behavior was observed when becoming temporarily separated from other individuals with whom they had a close relationship. While I observed five other cases of male individuals panicking at being separated from others, these cases also share this point, since they occurred when the mother (in one case) or a male with whom he had then kept close 'alliance' (in three cases) had suddenly disappeared in the midst of agonistic interactions with other males or during periods where dominant–subordinate relationships among males were unstable. In addition, except the one case that involved the disappearance of a female with whom the concerned male had kept 'consort' (i.e., one of the well-known mating strategies for chimpanzees where a male and a female in or around estrus keep exclusive company for anywhere from several hours to several weeks while avoiding encounters with others) and he was consequently left alone, these cases all occurred under the situation where other individuals (other than the disappearing particular individual) were still there, as in Nkombo's case.

In contrast, among the ten newcomer females I tracked with focal animal sampling (including those I did not observe for very long or on very many occasions), seven individuals panicked upon becoming separated from other individuals on at least one occasion (thirteen cases in total); and each instance was not one that occurred in a situation where particular individuals disappeared while other individuals remained there but, rather, occurred when an individual was left alone as a result. Also, while I observed multiple instances of such behavior in four newcomer females, the partners whose disappearance triggered their panicked behavior (or last individuals who had remained there until the females were left all alone) were different individuals in each case. Further, whereas I observed nine cases for individuals within one year of joining the group, for those who had been in the group for at least a year, I saw only four cases despite a greater amount of time or a greater number of occasions on which I observed them, which suggests that newcomers are more likely to panic just after they join the group.

For chimpanzees who are continuously engaging in fission and fusion, with the exception of mothers and their immature children, parting from other individuals is the common event; accordingly, they not infrequently travel alone and, basically, do not panic when they are parted from others or left alone. In turn, the fact that newcomer females do panic at becoming separated from other individuals and being left alone means that they are unable to engage in fission and fusion with others in the same way as resident individuals. Also, the fact that the partners whose disappearance triggers the panicked behavior vary on each occasion is a corollary of the attempts on the part of the newcomers to follow the other individuals they were accompanying on those occasions (i.e., they want to follow someone, and anyone will do). It could be said that newcomer females are highly motivated to maintain visual co-presence with the others they encounter on each occasion (i.e., they endeavor not to become separated from their partners, whoever they might be). As a result, as stated in the previous section, it seems that newcomer females tended to spend less time on their own than resident females.

For newcomer females, the triggering event that results in becoming separated from other individuals and then panicking in some cases was that of 'being left behind by the ones they were following during traveling' (the ▼ enclosure in Case 4). In the majority of cases, however, including in cases not presented here, the triggering event was that of 'being preoccupied with feeding (and thus not noticing when the other individuals moved off)' (the ▲ enclosures in Cases 1, 3, and 5). Although newcomer females pay attention to the movements of the other individuals around them (the ◆ enclosures in Cases 1, 2, 3, and 5) and such behaviors can be observed even when they are stretched out in day beds (Case 2) or when separated from others at distances of up to 100 meters (Case 5), perhaps owing to their youth, it seems that they can be easily left behind when they become preoccupied with feeding.

Since resident females also occasionally panicked, but did so when they noticed the disappearance of certain other individuals with whom they were close, rather than when left alone, we cannot perform a simple comparison with newcomer females. By and large, however, it was very rare for them to panic at parting from other individuals. In the background here, first of all, would be the fact that the resident females were already accustomed to the ecological and social environment of the M group. So, if the parting had just happened, they would have some idea of where the other individuals had gone, such as a nearby foraging tree. Also, perhaps because of experience commensurate with their age, resident females more nonchalantly but more assuredly pay attention to the movements of other individuals within their field of view than newcomers, so that they are less likely to be oblivious to the departure of other individuals. While I lack the space to give an example here, it is certain that resident females are as cognizant of the movements of other individuals within their field of view as newcomer females, if not more so. This is easily understood, particularly in the dispersing season, under circumstances in which parties are smaller in size with a relatively constant membership. Generally, these females cast glances at the departing individuals, and even when they may not watch them in particular, they usually prick up their ears to listen to their departing footsteps as they continue the activities they are currently engaged in, such as feeding, breastfeeding children, or nursing. That is, just because they do not panic at parting from other individuals as a rule, this is by no means to say that resident females are disinterested in partings from other individuals (i.e., in whether or not their state of co-presence continues). In terms of contrast with the panicked behavior of newcomer females, it might be said that they calmly see others off. Thus, in such situations, a resident female would find it an easy matter to interrupt the activity in which she was then engaging to follow the departing individuals, or, after a little while, may move toward where the individuals left off to meet up with them again there (in some cases by taking another route). Conversely, as against the choice by other individuals to leave the area, when a female chooses to remain in the area or to begin moving in the opposite direction from the one taken by the others, even I can understand that she has made the choice to part ways with the other individuals for the time being. However, this does not necessarily mean that she will remain apart from them, and it is still possible for those other individuals to return after a little while or for a reunion to take place at another location. Also, even after a female leaves other individuals behind and begins moving on her own, a frequently observed scene would see her slowing down, pausing, looking back, or sitting down in order to wait for a little while for other individuals to come along with her (though they did not always do so), in the attitude where she leaves it to the choice of the other party as to whether or not the parting will take place.[6]

Resident individuals who tolerate being followed, but will not considerately wait

However, when resident females are followed by newcomers, they will generally remain aloof to such behavior, though they accept being followed in the sense that they do nothing to discourage it through intimidation. Since a resident female will not spend much time waiting, even in situations where it seems clear that she is being followed by a newcomer (particularly in instances where she is being followed when it is just the two of them), the newcomer female may not even be able to stop to rest and eat grasses at the verge of the path (the first

▼ enclosure in Case 3). When a newcomer female panics at being left alone, it may not necessarily be due to her own carelessness, but rather due to the fact that the departing resident individual—who could be either female or male—without any special consideration, did not wait for the individual tagging along just because she was a newcomer (the ▼ enclosure in Case 4).

Furthermore, it seems that resident individuals are not only aloof, but in some cases may actually attempt to slip away without being noticed so that they will no longer be followed by the newcomer female. For example, in Case 3, when Genie, a newcomer female, becomes separated from a resident female, Ikocha, who until that time has been consistently aloof to the pursuing Genie (see the ▼ enclosures in Box C: she does not wait for her to feed on the grassy verge of the path and does not return the favor of grooming at all), Ikocha quietly slips away without pausing or looking back while Genie is busy feeding near a tree not far away (see the beginning of Box A). Given the aloof attitude exhibited by Ikocha up to that point, her timing does not seem coincidental.[7]

However, as outlined in the previous section, newcomer females also form co-ranging relationships with specific resident females within a few years of joining the group, and can end up repeatedly traveling in the company of these females for days on end (for a good example, see the case of the newcomer female Qanat and the resident females Gwekulo and Fatuma in Case 5). Without going into details here, in these cases the newcomer may often spend time frolicking with juveniles of the resident females, or may even carry an infant on their behalf. In such cases, the question of who follows who will be somewhat bidirectional, and the resident female will drop her attitude of aloofness toward the newcomer. Such close relationships may also arise as a result of following a resident female by a newcomer female, as happens to Genie in Case 3, or such newcomers may try to follow resident females to find out whether or not they would be likely to get along. In fact, although Case 3 took place only about six months after Genie joined the group, about a year later I was able to confirm that Genie had at least temporarily established such a relationship not with Ikocha, but with another older female called Fatuma (who may have been more welcoming toward newcomers, as she was also in a co-ranging relationship with Qanat). Although my records are intermittent, dating as they do from the Second Period of my study and I was unable to continue the investigation thereafter, during each of the two roughly month-long research periods that were separated by an interval of half a year, Genie was frequently seen in Fatuma's company. She would carry Fatuma's infant many times during the daytime, and when night fell would make her bed and sleep in a tree next to the one in which Fatuma made her bed.

While it is impossible to generalize, given the variability of attitudes on the side of newcomer females (i.e., the ways in which they associate with resident females), I have the impression that, over time, resident individuals, not limited to these particular close females, gradually become less standoffish about being followed by newcomer females. For their side, it also seems that newcomer females gradually stop following others so persistently, and to some extent settle on particular partners to follow. While the number of cases is small and not without reservations, such changes of relations with residents is probably the underlying basis for why newcomers are less likely to panic at being separated from other individuals after they have been in the group for at least a year.

Reunions: newcomer females who rely on voices/resident females who let voices pass

At the end of the analysis through our cases, let us consider the behavior of newcomer females after becoming separated and left on their own up to the point of their reunion with other individuals, though not necessarily the same ones from whom they were separated (Box B in Cases 1, 3, 4, and 5). As we have seen, it is certain that newcomer females panic at becoming separated from other individuals. However, even if they search around for the individuals who have disappeared but do not see them, and there are no tracks, signs of feeding, or scents left on the ground, this does not mean that they will then become so troubled as to lose all hope, and after a little while they will regain their composure. To begin with, they arrived from other groups most probably alone, and even if they have become separated from other individuals and left alone, they will still be able to seek out a suitable tree or site for feeding and sleeping. Even so, this is not to say that they will go just anywhere.

For example, in Cases 1 and 4, though they occasionally took time to feed or doze, the individuals in question remained still and quiet, staying on or near the observation trail or else in a nearby tree, as though waiting for someone to pass by or to hear a vocalization. As they did so, when they actually did hear a distant vocalization (pant-hoot), if the distance was short enough (as in Case 1), they decided to make their way toward it, or, if the distance was too far (as in Case 4), they made their way toward it, little by little, while waiting for another vocalization. When one was eventually heard nearby, they naturally rushed toward it. Then they rejoined a party consisting of the individuals from whom they had been separated and many others (Case 1), or else a party of other individuals entirely (Case 4). Enju, the newcomer female in Case 4, even let out a cry of excitement, sounding like 'faoo,' at seeing the others after breaking into a run in the final scene, likely because she was so exuberant at being reunited with the others.

From this, it could be said that, after having regained their composure, these females were seeking an opportunity simply to be reunited with any other individuals, and not necessarily those whose disappearance had triggered their panic. In Case 3, just before Genie, the newcomer female, started panicking after noticing that Ikocha, the resident female, had disappeared, she heard the pant-hoots of another party in the distance. This may be why she abandoned any attempt to rejoin Ikocha without making much effort to find her. Afterward, she stayed put, as though waiting for the next vocalization, just as Enju did in Case 4. It seems that she resumed her search for Ikocha while she waited, probably because she could not hear any sign of a subsequent vocalization. In the end, however, when she did hear another vocalization and began moving toward it, she eventually joined the party of individuals who seemed to be the vocalizing individuals.

With resident females, on the other hand—though, again, I do not have the space here to give detailed examples (see Hanamura 2010, 2017)—once they become separated from others and are left on their own (though individuals accompanied by offspring, who account for the majority of resident females, will still be together with their young), they will sometimes let any voices they hear pass and continue feeding, breastfeeding, or grooming even without looking back in the direction of the call, or may even move in a different direction than the one from which the voice came. During the dispersing season, in particular, resident females will often do this, such as when spending several days with only their own children. Of course, even so, they will also occasionally meet up with other individuals somewhere during those days, or meet up with the individuals presumed to have uttered the vocalization by moving in the direction from which it was most recently heard; in some cases, they establish states of auditory co-presence with other individuals by responding to the pant-hoots that they hear or by receiving responses to pant-hoots that they initiate themselves (for the definition of this call-response pattern, see note 2). However, unlike newcomer females, even when they are alone, they do not appear to seek opportunities simply to be reunited with any other individuals. It is also not uncommon that they will carry out such relaxed traveling in the company of a number of co-ranging females and their children. As a result, as outlined in the previous section, it seems that resident females tend to spend more time alone and in female-only parties than newcomer females.

The difference in attitudes between resident and newcomer females with respect to the voices they hear can also be verified to some degree based on quantitative trends. When we compare the frequency of pant-hoots heard by resident and newcomer females that came from beyond visual range during focal animal sampling, we find no significant difference during the congregating season, when newcomer and resident females alike take part in large nomadic parties and when pant-hoots are being uttered all around, whereas in the dispersing season the frequency is higher for newcomer females (Figure 7.1). This happens because newcomer females, when they hear a pant-hoot, will be more likely than resident females to coordinate the movement and timing of their traveling to stay within the audible range of subsequent pant-hoots. Alternatively, it could also be the consequence of a tendency to follow individuals or parties that utter pant-hoots frequently. Also, when we compare the proportion for the behavior of running toward the direction of a

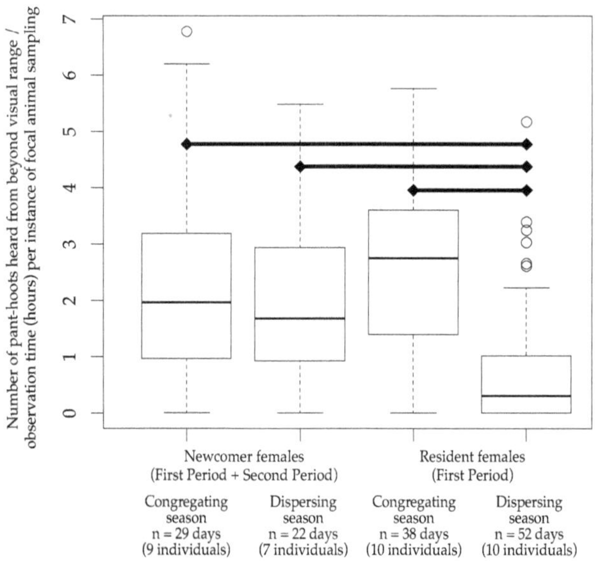

The thick horizontal line of the box indicates the median value (the second quartile), the upper and lower horizontal lines of the box indicate the upper and lower hinges (the first and third quartile), the whiskers indicate the maximum and minimum values, and data outside the whiskers indicate outliers.

Kruskal-Wallis test: df = 3, $\chi^2 = 37.81$, $p < 0.01$

◆—◆ indicates combinations with significant differences at $p < 0.05$ in the Nemenyi-Dunn test (multiple comparisons).

Figure 7.1: The frequency of pant-hoots heard from beyond visual range during focal animal sampling for newcomer or resident females and the seasonal differences (limited to sample days with observation duration of more than three hours)

Note: Since these frequencies are not determined solely by the behavior of the female who was the subject of focal animal sampling, but also by the number of surrounding parties and the frequency of pant-hoots they utter, the behavior of the females will only be partially reflective of these results. Moreover, even during the dispersing season, especially when the congregating season is not far away, relatively large nomadic parties are sometimes formed, so that pant-hoots may be heard more frequently. With regard to the finding for newcomer females that the frequency of pant-hoots heard from beyond visual range during the dispersing season was as high as that during the congregating season, I cannot rule out the possibility that the samples for newcomer females were recorded with a bias to such days. Although I strove to conduct focal animal sampling without any time bias, it will be necessary to conduct further study, either by increasing the number of samples or changing the method of analysis. Note also that the congregating and dispersing seasons are defined for each month according to whether the size of the monthly 'nomadic party' (see 'Chimpanzee modes of coexistence and female transfer' early in this chapter) for the First Period, for which there is a year-round record, is greater than or less than the median (see Figure 9.2 in Hanamura 2015a).

pant-hoot within ten seconds of hearing it when on their own, the median value for resident females is 0%, whereas for newcomer females it approaches 10%, regardless of season (Figure 7.2). This proportion is even higher when we limit the consideration to pant-hoots heard from nearby (i.e., within approximately 500 meters), though the data is not shown here. In other words, even in the congregating season when there is not much difference in the frequency with which pant-hoots are heard, we find that newcomer females will be more likely to begin running toward the voices they hear.

Accordingly, newcomer females may be said to rely on voices when they become separated. However, this is only in the situation where they become separated and are left alone. In cases where other individuals are visibly present, it seems they will more often privilege following those individuals over the distant voices (Box C in Cases 2 and 3). Voices are not as reliable, since one might wait but not hear the next voice, and one might arrive at the location from which a voice was heard only to find that the individual who made it

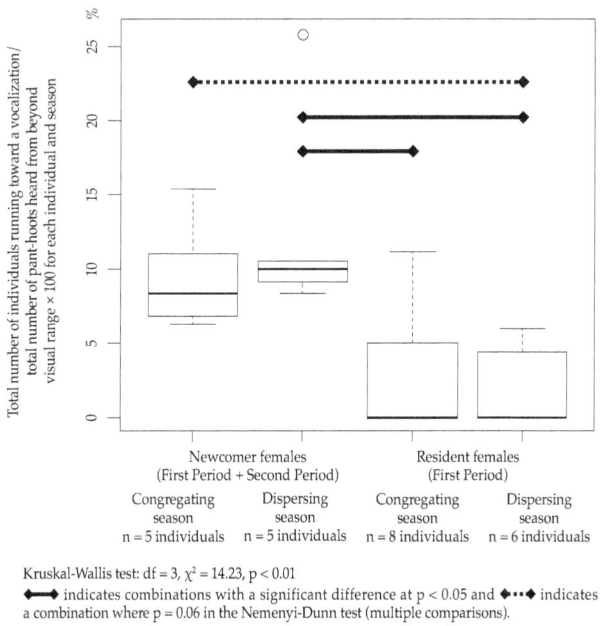

Figure 7.2: The proportion of the behavior of running toward the direction of pant-hoots heard from beyond visual range within ten seconds of hearing it when they are on their own for newcomer or resident females and the seasonal differences (limited to sample individuals with more than ten instances of hearing pant-hoots in a season)

Note: For definitions of the congregating season and dispersing season, and how to read the figure, please refer to Figure 7.1 and the accompanying note.

has already moved on (see Case 3 in Hanamura 2017). This is precisely why such individuals endeavor to keep others within visual range (i.e., a distance at which, even when another party temporarily disappears from view in the trees or bush, it remains possible to visually confirm their presence by moving slightly) and can end up panicking when left alone.

If they are relying on voices when they become separated, one would think that they should attempt interactions via pant-hoots, such as by initiating a pant-hoot after becoming separated in order to receive a response. Simply responding to pant-hoots they have just heard should be enough to at least let one's own location be known to the vocalizers and the other parties within earshot. However, in the cases presented in this chapter, the newcomer females do not respond to any of the pant-hoots that they hear, aside from Enju in Case 4, who breaks into a run while responding to the pant-hoots of other individuals just before joining them. Even given this situation, the following circumstances may underlie the fact that newcomers neither initiate nor respond to pant-hoots, although, first of all, the frequency with which females initiate pant-hoots is lower than that of males (though there are large individual variations) and their rate of response (i.e., the proportion of the number of times a given focal animal uttered a pant-hoot within ten seconds of hearing pant-hoots from outside the field of view in the total number of times pant-hoots were heard from outside the field of view during focal animal sampling) is only 10% to 20% (but this rate is basically the same for males). That is, especially for newcomer females who have only just arrived, even if they were to initiate calls, it is still not sure where and who will be able to hear their calls, and since their voices are not yet familiar to the other members of the group they can hardly expect a response, and even if a response does come, they will still not know from whom it has come. As described at the beginning of this section, for newcomer females who have recently joined the group, neither the identity of the other members of the group nor the extent of the group's home range will be obvious. In a situation such as this, it would be difficult to try to create a state of auditory co-presence with other individuals solely through non-face-to-face vocalizations.[8]

Interesting from this perspective is the behavior of Qanat in Case 5, who, at this time, had been in the group for almost four years and was already pregnant. To begin with, Qanat moved away herself from the other individuals she had been with up to that point, and until she began looking eagerly through the trees where the others had been, she spent about thirty minutes on her own feeding and resting without seeming overly perturbed (Box A). Also, after she realized that she had become separated from the others, she initiated a pant-hoot, though it ended as only an attempt (Box B). Since Qanat began this initial call while desperately searching for the others, who had disappeared, and then broke it off when she immediately afterward found some of the individuals she had been looking for, it seems that she had been hoping for a response from these particular individuals, rather than from just anyone. Qanat's slightly leeway behavior (i.e., moving away from others on her own and continuing to feed and rest) and her initiation of a pant-hoot differ from behaviors of the other newcomer females after becoming separated from other individuals analyzed above. Given that the individuals from whom she was separated were all her co-ranging females with whom she had repeatedly traveled, it seems likely that this behavior stemmed to some degree from Qanat's anticipation, based on her experience of associations with the co-ranging females up to that time, as follows: they could probably continue feeding there for a little while and thus she could rejoin them if she returned, or, if not, she could more or less guess where they were headed (though these anticipations were disappointed, which was why she panicked a little), and if she initiated a pant-hoot in this situation she would probably receive a response from them (though this anticipation was left hanging in the air as she reunited with some of them during the vocalization). In light of the fact that, in the two exceptional cases in which a resident female panicked at becoming separated from other individuals, the other party whose disappearance triggered her panicked behavior was her particular close individual (in one case, a co-ranging female, similar to Case 5, or in another case, a male with whom she had been close for many years), Case 5 may suggest that Qanat was by this time already beginning to behave as a resident female.

The limits of whether to continue being a member of the group

As we have seen, newcomer females are far more likely than resident females to follow and maintain visual co-presence with the various other individuals they encounter. As a corollary, they will sometimes panic at becoming separated from these others, no matter who they are. This is because, when these other individuals disappear without their noticing, they are left alone without knowing where the others have gone. Once they regain composure, in the attitude that these others do not have to be the individuals from whom they became separated but could be anyone, they will seek an opportunity to be reunited with others by waiting for another chimpanzee to pass by, or else to hear a vocalization from nearby. If they hear a distant vocalization, they will gradually make their way toward it; when they hear one from nearby, they will start running toward it. Accordingly, although such panicked behavior of newcomers directly resulted from becoming separated from various individuals in each instance, it may be that this also represented a separation from the collectivity that included all other individuals with whom they would have been able to be reunited by following the individuals who disappeared; in other words, *the collectivity of individuals that they had been following up to that point*—a collectivity that approximates to our own image of M group. One aspect of this may be that newcomers have trouble finding food or a place to sleep without following someone, but when we recall the intensely panicked behavior they occasionally exhibited, at that moment, it seems as though they have sensed, if vaguely, that they were placed at the 'limit' of whether or not they were able to continue to participate in this collectivity.

While it was hard for me to follow the newcomer females who panicked at becoming separated from others, as I often temporarily lost track of them, and although I felt sorry for those who were desperate, I sometimes found myself laughing at such panicked behavior. The reason I felt this behavior was so funny was that now we were in the M group's home range, it was almost self-evident to me that they would be able to rejoin other members of the group sooner or later, even if they did not panic, if they stayed around here. To me, who had such leeway of perspective (or, on the flip side, a kind of stereotype about the probability of the future), the panicked behavior of the newcomer females seemed excessive. Resident females, unlike newcomer females, basically do not panic at becoming separated from other individuals and calmly watch other individuals who are about to leave, even when it means they will be left on their own. That they can do so likely stems from a leeway of perspective similar to mine through their past experience of repeatedly reuniting with the other individuals after parting with them, as well as their past experience of having engaged in such fission and fusion with them in this same area. Thus, in exceptional cases where a resident female panics, this has nothing to do with whether or not she will be left alone as a result, but seems to be limited to times when she loses track of the whereabouts of a specific close individual with whom she has made a particular effort to maintain a state of co-presence on that occasion.

As it has been said that chimpanzees do not become concerned about parting from other individuals and, after a parting, will frequently 'carry on with the next thing, as though nothing had happened' (Osawa 2016: 176), their partings can seem surprisingly indifferent to human sensibilities. However, as I described above, this does not mean that they were not interested in the movements of other individuals who were about to depart or whom they were about to leave behind. It is not easily apparent from ways of parting among resident individuals who usually part as a matter of course and it may not even be something of which they are conscious so long as reunions are repeated, but as we have discussed through the behavior of newcomer females, since it is possible that one might never be reunited with those individuals and by extension with other individuals who could be encountered through those individuals, partings are in principle always the limit of whether or not an individual will continue being a member of the group.[9]

Itoh (2003), from the perspective of the lags of movements among individuals (e.g., in terms of their timing, speed, or direction) and their iterations or prolongations, has illustrated how the phenomenon of a chimpanzee 'group' that stays coherent while undergoing fission and fusion—and is thus difficult to perceive directly as a whole—is established solely on the basis of what (in the terminology of this chapter) we would call 'repeated co-presence' (see Itoh 2013 for her English literature that develops a similar discussion from different angles). The discussion above in this chapter has reconsidered this kind of mechanism of group generation from the perspective of partings and reunions. In other words, this is to say that from the standpoint of the females who transfer between groups, even if they become separated from various other individuals in their destination group, as long as they are reunited with someone in each instance, and as long as this is repeated, then they will remain members of the same group as those other individuals. In principle, this point should also hold true from the standpoint of males who are born and raised within the group. For example, while some males occasionally appear to lead a solitary life without encountering other individuals for long stretches of time, their social relationships with other individuals, including dominant–subordinate relationships with other males, and by extension, even their membership in the group, could be uncertain (see Nishie's discussion in Chapter 8), a fact that could also be (inversely) proof of the above.

Incidentally, the fact that partings always constitute a limit, as described above, also means that encounters with individuals from other groups may be waiting in store. In fact, adolescent females are thought to move in and out of multiple groups until they give birth. While it is difficult to confirm this process, the newcomer female in Case 1 (Yuri), for example, after having been in M group for several months, was then absent for about a year before reappearing, after which she was not observed again. If newcomer females are left on their own after becoming separated from other individuals in their destination groups and then head toward the direction from which they hear a voice, it may be that they will encounter individuals from another group. However, until they are able to understand the home range and the vocal character of the individuals in their destination group, it may not be easy to determine whether or not the vocalization that is heard belongs to that group. These conditions may lead them to transfer between groups and, in some cases, they may head toward such an uncertain vocalization in anticipation of that possibility (Hanamura 2019). In the cases presented in this chapter, as well, there are instances in which newcomer females appear to have an ambivalent attitude toward being reunited with resident individuals. Even as they tend to panic when left alone by standoffish resident individuals, they will soon regain their composure, after which they continue feeding on their own and sometimes even fall asleep. Of course, they will have become accustomed to this to some extent if some time has passed since their arrival, and even if they have only recently arrived, once they become separated there is nothing else they can do. At such times, however, it seemed that they were not so much obsessed with 'staying in this group' as simply leaving the matter to chance, waiting to hear a vocalization or else for someone to pass by from across the group's boundaries. Conversely, for resident individuals for whom 'being in this group' was self-evident, the option to transfer may have become latent. Not only that, but for resident individuals, individuals from other groups who 'should not be in this group' because they have not established repeated co-presence—especially if they are not the adolescent females unaccompanied by children (i.e., newcomer females) that show up each year (see note 4)—appear as 'strangers' who arouse a powerful sense of tension and anxiety (Hanamura 2019). Even so, just as shown in cases where resident females accompanied by children (see the section 'Relations between newcomer females and other individuals: comparison with resident females' above and Nishie's discussion in Chapter 8) or even males—though this was a case of a temporary stay (and merely a potential one, at that) in a region other than Mahale (Sugiyama and Fujita 2011)—transferred to another group, moving between groups is a contingent phenomenon that may occur for any individual, depending on the situation.

Ways of parting and reunion-enabling mechanisms: comparison with humans

Finally, I would like to shift my focus to resident females' ways of parting and behavior after parting, which I illuminate through a comparison with newcomer females, in order to think about differences in modes of coexistence between humans and chimpanzees.

Parting greetings peculiar to humans

Humans usually discuss plans for meeting again after meeting in person or talking on the phone with friends or lovers, particularly if those relationships have just been formed (and therefore, conversely, broaching this topic can also mean one's intent to leave). Also, in various languages and cultures, parting greetings often involve the use of words expressing an expectation of reunion (i.e., the continuation of 'our' relationship), as in examples such as 'See you again' (English), 'Matane/またね' (Japanese), 'Zàijiàn/再见' (Chinese), 'Nos vemos' (Spanish), and 'Tutaonana' (Kiswahili). (This is why words that do not explicitly state expectations of reunion, such as 'Sayonara' in Japanese (meaning, literally, 'if it is to be thus') or 'Goodbye' in English ('God be with you') are used for instances of final farewells, and sometimes ironically as a declaration of one's intention to sever a relationship.)

Because family members who live in the same household, employees who belong to the same division of a company, and members of the same university laboratory can all be considered to have a 'home' base for their activities, the venue for their reunion is generally the same. Thus, unless they follow an irregular schedule, there is no need for them to plan a reunion. Nonetheless, even if it is just an ordinary parting, they usually greet each other, and when leaving those homes to go to school, take a business trip, or head into the field, people will offer a confirmation about the time and date of their return before and after a parting greeting. And, for example, the Japanese parting greeting 'Ittekimasu/行ってきます' (which means 'I am going (and will return)'), as well as its companion reunion greeting 'Tadaima/ただいま' ('(I have come home) just now'), represents a confirmation that the site of utterance is a home to be returned to (i.e., 'our' home, at which to perform 'our' reunion). In this way, in parting we make a promise of reunion. We could say that the mutual communication of the date and location of the reunion, or of the fact that we are anticipating a reunion (even if such has not yet been planned), or else consider a reunion to be self-evident, has itself been formalized as the parting greeting.

The word 'greeting' resists easy definition. However, if we consider 'greetings exchanged upon meeting' as an interaction that affirms the fact of an encounter—whether this results in a subsequent interaction like grooming, mating, conversation, or sports competition, whether it is followed by engaging in different activities while maintaining co-presence, or whether it is followed immediately by another parting—then 'greetings exchanged at partings' could be characterized as an interaction affirming that a separation is about to take place (see also each of the definitions given in note 2). Not only that, but it also seems to constitute a 'reunion-oriented interaction' that leads to the stable reproduction of a larger framework (or 'story')—namely, the group or social relationship constituted by the individuals in question—that transcends the interaction being ended (see Kitamura's discussion in Chapter 14). Correspondingly, at a reunion, especially when the period of separation has been long, the greetings exchanged will evoke the time length until that reunion, such as in the examples of 'Long time, no see' (English), 'Hisashiburi/久しぶり' (Japanese), 'Hǎojiǔ bùjiàn/好久不见' (Chinese), 'Cuánto tiempo' (Spanish), and 'Siku nyingi' (Kiswahili). We are able to consider this as an assurance of the fact that 'our' group or relationship persists unchanged, despite the interruption by the separation.

Formulaic greeting-like interactions performed at the time of an encounter are not limited to humans, and are not so rare in species that live in societies interwoven with partings and reunions (see 'Modes of coexistence seen in partings and reunions' at the beginning of this chapter). This is because when an encounter constitutes a reunion after a long-term separation, it is necessary to confirm and update the basis of the relationship in terms of it as an affinity with the other party, a dominant–subordinate relationship, or as a confirmation (individual recognition) that the other party is still the same individual. A typical example from the behavioral repertoire of chimpanzees is the pant-grunt, though this is not always performed immediately after an encounter (de Waal 1982; Nishie 2017; Chapter 8 of this volume). As an example in a species that forms long-term pairs, we may cite the displays referred to as 'Cross' and 'Both arch' performed when pipefish pairs, which act separately during the day, are reunited every morning (Sogabe 2011). Even for one-time encounters, many species perform greeting-like interactions using behaviors typically known as 'conventional fighting' or 'ritualized displays' that serve to advertise fighting ability to other individuals of the same sex and attractiveness as a breeding partner to individuals of the other sex. Given that the interaction is inevitable in encounters as long as the presence of each other is perceivable, we can easily imagine that greeting-like interactions at the time of encounters evolved in many species as a way to ascertain how the other party behaves, and what kind of existence the other party could be for oneself (for one scenario, see Yabuta 2015).

However, it is probably only humans that exchange parting greetings. Even among species that live in societies interwoven with partings and reunions, reports of greeting-like interactions at partings, at least in the wild, appear to be unknown.[10] As we have seen in this chapter, among chimpanzees in the wild, group members frequently part with one another, and it is not uncommon for these periods of separation to exceed several weeks. Yet they do not perform any interaction that would seem to jointly confirm that they hope for reunion and regret the separation, or that a parting is about to take place (see also McGrew and Baehren 2016). The fact that they do not exchange greetings at partings, unlike humans, would be a major factor in why their partings seem very indifferent to humans.

Process-oriented/goal-oriented parings and story-exploring/story-performative reunion mechanisms

In the context of the evolutionary history of human society, we cannot know when our ancestors began to exchange parting greetings. However, it seems likely that, in parallel with the acquisition of language, alongside the mutual exchange of parting greetings expressing an expectation of reunion, humans also created various customs and institutions—'our' 'stories' and the 'homes' that served as the bases of these activities—which ensured that they would remain 'members of the same group (or social relationship)' even in the future, and which gave assurances about and normalized reunions with fellow group members.

Stretching our mind's eye, we can imagine how even small family groups of early *Homo sapiens* would have engaged repeatedly in partings and reunions as a result of the division of roles in the context of subsistence activities and interactions with other groups (see Terashima's discussion in Chapter 21). Such groups would have developed ways of partings that would reproduce groups in a stable manner even when bracketed by extended periods of separation. At the same time, they would also have a norm that its members are in a relationship that should be reunited by establishing stories about 'marriage' and 'family,' as well as about the 'domiciles (houses)' that provided the base for such activities. When individuals from multiple such families repeatedly came together to form subsistence groups and ritual groups, again there would have been a demand for the stories, such as 'relatives' and 'bands' (accompanied with the homes, such as 'settlements (villages),' 'meeting places,' and 'sacred sites'), that made up the units in which group activities were stably reproduced, as well as for parting greetings to support these stories by expressing expectations of reunion.

It is considered that as human began to live the multiple stories accompanied with their respective homes, the content of the varied activities that were carried out on the basis of those stories, the members who took part, and their location and timing came to be clearly defined. In line with this, the practice of parting by humans is also thought to have come to occur in order to conduct another activity (i.e., purposes or goal) after finishing one activity. For example, in the case of a family, members would take their leave of one another to engage in activities such as hunting and gathering. Conversely, members of a subsistence group would disperse to engage in activities in their respective homes (e.g., to rest, eat, reproduce, and raise children). This kind of human parting, like those we also perform in our day-to-day lives, could be described as the goal-oriented way of parting. Moreover, in terms of a mechanism enabling future reunions with those from whom one is taking leave, at each parting the expectation is expressed of such a future reunion (i.e., a continuation of 'our' story); where a home exists, the matter of where this reunion is to take place will also be preordained. In that sense, we can also say that humans engage in repeated partings and reunions in a story-performative manner. However, because a mechanism like this—i.e., one that enables reunions—could also act as a fetter impeding the imagination of alternative futures (and the stories by which they might be realized), it is also vulnerable to the fundamental contingency of the future. And in some cases, it may act as a shackle (or constrain) that forces reunions even when they are not desired.

Humans, in the same way as chimpanzees, form societies where members can stay apart in that each member of a group, like a family, engages in repeated partings and reunions, and all but immature individuals are able to act alone. However, because human groups have a strong aspect of being reproduced in a stable manner using 'stories' (in many cases accompanied with a 'home' that comprises a base of activity) to pretend that reunions are assured—i.e., that being a member of the same group is guaranteed even when an individual exists apart—it seems that groups are able to continue without repetitive co-presence on the part of their members, so that it is possible to retain membership in a group no matter how many years one spends apart. Of course, there are some—lovers, for example—who, when a relationship immediately becomes unclear or for some reason unstable immediately after it is formed, will repeatedly co-present in the attempt to maintain or stabilize that relationship (or see it through to its demise). However, as long as they share the same story, they will not necessarily need such repetition of co-presence (conversely, even with repeated co-presence, the existence of a story alone, with no substance, can sometimes leave the relationship as merely an empty shell). Even with the separation of death, a notional co-presence is still possible in imagined stories. In other words, although akin to the mechanism by which chimpanzees continue to be members of a group, with stories it would seem that humans have taken this 'limit' to an extreme.

In this way, humans, with reference to stories of relationships that keep the members of a group together, on the occasion of a parting will express their expectations for reunion through parting greetings, or sublimate their dead into ancestors and spirits through parting rituals (i.e., funerals). In doing so, these stories are repeatedly lived and historicized, and the souls of the departed are also mythologized and bound to the living. As a result, an inversion has taken place in which groups are felt to have an a priori existence as fixed entities, leading humans to always live those stories. Therefore, humans may be thought to lead social lives in which 'being co-present'—including notional co-presence—is the criterion as an ideal even though it is more common for the members of groups not to co-present in actual practice (i.e., in circumstances where interactions are not physically possible). And thus categorized groups must have been used as units to realize stories of superordinate relationships (i.e., social stratification) (see also Hanamura 2019). Furthermore, as with ethnic groups, nation states, and religious groups, it may be that even with an indefinite number of individuals who have never met in reality, identifying as members of the same group in the context of imaginary stories and choosing actions based on this sense of group belonging may have enabled people to live those stories in a concrete way, or, conversely, to conveniently change their membership in groups by regarding those groups ultimately as merely stories, and to select from among multiple memberships to make use of the one best suited to the situation at hand.

In contrast, with chimpanzees, a certain number of individuals will form a group by continual fission and fusion. Chimpanzees appear to lead social lives in which the criterion is this very act of alternating between co-presence and non-co-presence, as though being together and being apart were of equivalent value. Their partings and reunions are not so much carried out with an aim at some specific purpose (i.e., a goal) as they are autotelic interactions in which the *process* of repeated fission and fusion is carried out as an end in itself, where partings occur in each instance as the result of interactions over the timing, speed, and orientation of their movements with other individuals in their immediate presence. Accordingly, it is not known beforehand how long or with whose company various activities such as moving, feeding, or resting will continue. While close individuals will frequently range together, unlike humans, even individuals like these—although they can sometimes more or less guess their partner's next destination immediately after separating again—will have no home that prescribes when and where to reunite, and they separate without explicitly sharing a story that might guarantee their next reunion through a parting greeting (this is why, although rare, a member of the group will sometimes panic when separated from another chimpanzee with whom it is close). While it is still possible, after becoming visually separated, to create a state of auditory co-presence with the pant-hoot (call-response form) as long as the chimpanzee remains within audible range, even here the continuation of a state of co-presence is not the primary aim; depending on the behaviors of the two parties, it may continue to some extent or soon disappear, with no way of knowing when or where they will create it again (Hanamura 2017). They will remain members of the same group only as long as they actually continue separating and reuniting in this way.

In contrast to humans, chimpanzee partings can be characterized as process-oriented partings in the sense that whether they separate or not depends on the process of interactions in each instance. However, this does not necessarily mean that chimpanzees, unlike humans, 'live only in the present'—a claim that has been debated in recent years (e.g., Suddendorf 2013; see also note 10, as well as Takenoshita's discussion in Chapter 5). Given that new females who arrive in the group will no longer panic at separations through the experience of eventually being reunited after being separated from other members of the group, chimpanzees must also harbor an image ('story'), if vague, of a future reunion as a repetition of a past and, moreover, of 'the group' as something that continues to exist. Thus, in the sense that they build up this image by actually reuniting while there is the ever-present possibility that they will not be reunited, we could say that chimpanzees engage in these repeated partings and reunions in a 'story-exploring' (or 'story-generating' in the sense that a story accumulates as a result) manner.

Also, just because they do not make it appear as if their future reunion is assured through parting greetings does not mean that this manner of parting does not lead to a stable reproduction of the group. Rather, even given the ever-present possibility of not being reunited, it would seem that the story-exploring repetition of partings and reunions, all the while adopting a tough and flexible attitude toward an uncertain future, enables a set number of individuals to perform the stable reproduction of a decentralized group in which members continually engage in fission and fusion on an individual basis while forming temporary parties made up of different individuals each time.

Although this chapter has barely touched on extremes in the physical or mental sense, let us conclude by touching on these extremes in light of the discussion here, as well as of the topics and issues raised in the other chapters in this volume. At the same time as humans take leave from and reunite with one another in the story-performative manner, they are also powerfully constrained by the stories referred to therein (see Takenoshita's discussion in Chapter 5). In the case of broken hearts or bereavements, for example, people may have come to experience severe physical and psychological pain that could never be experienced by other animal species. Of course, chimpanzees and other (nonhuman) animals can doubtlessly feel great anguish, such as, for example, upon the death of a mother (see the discussion by Nakamura in Chapter 6). However, humans, as well as acquiring stress tolerance with which to withstand such extreme situations, will also sometimes even pursue such situations (see Kuroda's discussion in Chapter 18). And they have indeed created them, from rites of passage (or initiation ritual) that foster identity through pain suffered in extreme conditions (see Kasuga's discussion in Chapter 17 and Funabiki's discussion in Chapter 13) to magic and religion that assuage such pain or position it as fate, and, more recently, science that seeks to predict and relieve pain (on the shared roots of magic and science, see Nishii's discussion in Chapter 10), and to the status of the slave, which could never have been realized if a captive human could not tolerate pain.

Humans have also prayed for the miracle of being united with friends, lovers, and family members from whom they have been separated by circumstance or death, and have delighted or despaired at the fulfillment or failure of such reunions. Not limited to partings and reunions, no matter how much they may be prescribed by languages or institutions (i.e., 'stories' in the parlance of this chapter), future events such as natural and artificial disasters (a distinction that has grown increasingly ambiguous in recent years; see Tokoro's discussion in Chapter 11) are fundamentally indeterminate. Though usually rendered invisible by these stories, this indeterminacy occasionally reveals glimpses of itself; and while we may dread it, but use it as a resource nonetheless, we will continue to weave even more stories in future.

Appendix

Note to symbols

Alphabetic boxes: passages mainly referenced in the text (A: scenes of individuals becoming separated; B: scenes taking place after becoming separated and before reunions with other individuals; C: scenes showing a preference for following nearby others rather than seeking distant voices)

◆ ◆: examples illustrating individuals taking notice of movements on the part of other individuals who are (or were) nearby

▲ ▲: examples illustrating how easily an individual can become separated while preoccupied with feeding

▼ ▼: examples illustrating aloofness on the part of other individuals being followed

★ ★: examples illustrating the overlap between the ecological environment and the social environment

Case 1: Becoming separated and panicking, then waiting for someone to pass by or to hear a vocalization

Date: November 12, 2005 (congregating season)
Focal individual: Yuri (a newcomer female in her first month since arrival. She disappeared (transferred) after December 12, 2005. She reappeared (re-immigrated) one year later on December 22, 2006, only to disappear again soon afterward (transferred again).)

On this day, Yuri repeatedly moved, rested, and fed while engaging in fission and fusion with many other individuals from the early morning onward. During that time, she was grooming (mainly being groomed by) a total of seven other chimpanzees, consisting of two adult males, two young males, one resident female, and two juveniles, and played with five juveniles. She was quite popular. Over the morning, I heard pant-hoots as well as other boisterous voices, such as screams and barks, from here and there, but more or less ceased to hear such voices from just before noon.

★ **14:07** Yuri parts ways with several other individuals with whom up to that point she had been traveling for some time, and then heads into the bush, where Cadmus (a young male) had gone immediately before. In a little clearing stands an isakama fruit tree, with Cadmus sitting close by. After looking up at the fruit tree, Yuri rushes past Cadmus to climb up and start feeding on the fruit ★. Soon afterward, Tarnie (a newcomer female in her second year since arrival) arrives and lies down near the fruit tree. **14:18** Cadmus and Tarnie come to Yuri's side (in the tree) to peer at the fruit and her face as she feeds. When Christmas (a young male) arrives at **14:22**, Cadmus moves down to his side (on the ground). Tarnie begins grooming Yuri, who continues to feed, but soon descends from the tree and lies down on the ground.

Box A

◆ At **14:25** Tarnie departs into the bush to the south. Yuri turns to look back at the rustling of the leaves as Tarnie departs, then resumes feeding after ascertaining her whereabouts ◆. Thirty seconds later, Cadmus and Christmas also leave, heading south, so that no one remains. ▲ Yuri had been continuing to feed ▲, but at **14:26**, she suddenly leans down from the branch she has been sitting on, looks around, and after dropping the half-eaten fruit she has been holding in her hand, she descends from the tree at full speed, then heads south into the bush after Tarnie, Cadmus, and Christmas. Yuri rushes into the bush, then stops to look beyond through gaps in the undergrowth and sniff the ground, and then starts to run again at speed. She repeats this process, running in a different direction each time, and I lose sight of Yuri along the way.

Box B

Ten minutes after I emerge onto a nearby observation trail to try to ascertain Yuri's whereabouts, at **14:40**, I rediscover Yuri, who is peeping out at the trail from the nearby bushes. Yuri, who shortly emerges onto the trail, begins slowly making her way south while casting glances around her, then climbs up a fruit tree along the trail, where she begins feeding. After continuing to feed for about thirty minutes, she descends to the ground to resume resting in the bushes just off the trail (**15:20**). Just after Yuri emerges onto the trail and begins moving south again at **15:32**, pant-hoots from multiple individuals are heard from the southwest, not far away. Yuri turns toward the voice and begins sprinting south along the trail and, at **15:36**, comes upon a party consisting of one adult male, two young males, and two resident females with their young, all resting on the trail. Of the individuals who had parted ways with her earlier, only Christmas is here. The other two are absent.

Case 2: Preference for following nearby others over distant voices

Date: March 6, 2013 (dispersing season but well gathering, with pant-hoots frequently heard)
Focal individual: Devota (a newcomer female in her fifth month since arrival)

◆ At **12:10** Devota, who appeared while many adult males, young males, and resident females were feeding in the trees, climbs a separate tree all of her own, where she promptly prepares a day bed and dozes off. At **13:11**, most of the remaining individuals start to move, while exchanging pant-hoots with a few individuals who have just departed. At **13:25**, when some of the last remaining males start to move, Devota emerges from her day bed to follow the males ◆. The red colobus hunt begins at **14:10**, and I lose sight of Devota in the commotion.

At **17:20**, I find Devota again, moving along after some other males than the ones she was following when I lost sight of her. Afterward, Devota meets up with and takes her leave of several males while occasionally taking part in a chorus of those males' pant-hoot exchanges with another party.

Box C

At **18:14**, while feeding on the fruit of an ikubhila tree, the males depart, leaving Devota on her own. At **18:23**, when pant-hoots are heard from the north, not far away, presumably from some of the departed males, Devota stops feeding and begins heading in their direction. Soon, however, she encounters Darwin (an adult male), whom she begins following as he move east. At **18:56**, an exchange of pant-hoots is heard from the north and west, whereupon both Darwin and Devota stop and turn toward the voices. Darwin slowly resumes his movement east, however, and Devota follows him.

Case 3: Aloof resident individuals/relying on voices when becoming separated (I)

Date: March 23, 2013 (dispersing season but relatively well gathering, with pant-hoots occasionally heard)
Focal individual: Genie (a newcomer female in her sixth month since arrival)

In the morning, many individuals, including both males and females, traveled around strenuously while undergoing fusion and fission. In the afternoon, the movements calmed somewhat after brief rain. At **13:10**, Genie appears near where several other individuals are feeding. Some head northwest, and when pant-hoots are heard—presumably vocalizations from those who set out at **13:26**—of the three individuals who are still feeding, Michio (an adult male) is the first to head in that direction, soon to be followed by Devota (a newcomer female in her sixth month since arrival), and then Genie. They have headed northwest at high speed, so that I lose sight of the trio at **13:40**.

I follow the feces and footprint of a large number of individuals and, at **14:25**, I re-encounter Michio and Genie heading north on observation trail R3. At **14:29**, they climb a mukibhu fruit tree where several others are feeding, including Hadija (a newcomer female in her second year since arrival) in estrus, Primus (an adult male) who is following the estrous female on this day, Ikocha (the oldest resident female) who is accompanied by Iris (an infant), and her son Ichiro (a young male). Michio and Genie also begin to feed. At **14:43**, when a pant-hood is heard from the northwest, Hadija slowly descends from the tree to set out north on R3. Noticing her departure, Primus and Ichiro also set off after Hadija. At **14:47**, Michio descends from the tree, and he, too, sets off to the north on R3. ◆ Genie glances at the departure of these individuals, but continues feeding along with Ikocha until **14:57**, when Ikocha descends from the tree and sets off east from R3 into the bush. At this point, Genie also looks back in the direction of the sound, then descends close to the ground and stares at the direction taken by Ikocha and alternately north along the observation trail in the direction taken by the other individuals ◆.

Box C

As Ikocha heads further east, Genie rushes to follow. When a pant-hoot is heard again from the northwest at **14:59**, Genie stops to glance in the direction of the voice, but when Ikocha continues moving without changing course, Genie follows.

At **15:02**, when Ikocha eventually returns to the R3 trail and slowly begins to head north, Genie also heads north after her, staying close behind. ▼ Although Genie darts into the bush several times to pick itungulu fruit, Ikocha goes ahead, not paying her any mind, so that Genie rushes to catch up with Ikocha; and this pattern repeats ▼. At **15:15**, when Ikocha once again climbs another mukibhu fruit tree, Genie also climbs up behind her, and the two chimpanzees resume feeding. From **15:32**, Genie stops feeding and, while lying on a branch or staring at Iris, gradually approaches Ikocha, who is still feeding. At **15:35**, when Ikocha begins grooming Iris, Genie leaves to go and eat ilombo fruit. However, she returns soon at **15:49** to begin grooming Ikocha, who is still grooming Iris, and continues doing so for forty minutes until **16:29**. ▼ Halfway through, Genie pauses grooming several times and acts as though she has nothing to do, such as by scratching herself (which gives the impression that she would like to be groomed). Nevertheless, Ikocha does not groom Genie at all (at the end of the day, Genie has not been groomed even once) ▼.

At **16:30**, the two resume feeding on mukibhu a little further away, but ◆ at **16:32** Ikocha moves to the next tree to the east, then descends close to the ground. At the sound of this, Genie looks back and watches Ikocha for a while, but after seeing that Ikocha has sat down, she resumes feeding. At **16:39**, when Ikocha begins moving further east, Genie follows her, brachiating east through the trees. When Ikocha starts feeding in an ikubhila fruit tree, Genie also moves to a nearby tree from which she can keep an eye on Ikocha, then begins to feed on ilombo while casting glances at Ikocha ◆.

Box A

At **16:48**, ▲ when Genie is preoccupied with feeding on ilombo in a tree a short distance away from Ikocha ▲, Ikocha quietly descends from the tree and, without stopping or looking back, sets off to the east along observation trail N. Immediately thereafter, a pant-hoot is heard from somewhere far to the southwest. Genie turns in that direction while eating her ilombo, then continues feeding while looking southwest. At **16:58**, Genie suddenly turns to the east (seeming to have noticed Ikocha's absence). After staring at the ikubhila fruit tree where Ikocha had been sitting not long ago, she heads over, and then moves around through the branches, looking around the area while hurriedly cramming the ilombo she is holding into her mouth. But Ikocha can no longer be seen. Genie returns to the tree that she has just left and turns southwest, toward where she had heard the vocalization, then resumes feeding on ilombo.

Box B

Although Genie continues to feed on ilombo until **17:15**, looking to the southwest several times as though waiting for the next vocalization, there are no voices at all. At **17:18**, she once again moves closer to the ikubhila fruit tree where Ikocha had been feeding until just before she disappeared (Genie has confirmed things up to this point). At **17:19**, she descends close to trail N, then squints to the east (the direction in which Ikocha actually headed off, which Genie probably had not seen), then into the bush to the northeast. Then, at **17:20**, after brachiating northeast while paying close attention to the bush on the ground, she sits down in a tree. At **17:28**, another pant-hoot is heard from the southwest (only by my research assistant), and about two minutes later, Genie begins moving west along the trees. Afterward, as pant-hoots are heard again several times from the west and northwest, Genie moves further west while repeatedly feeding. While it was growing dark and I could not see very well, just before **19:00**, she met up with two resident females (not Ikocha), along with their children, as well as those whom she had been with earlier before beginning to follow Ikocha, such as Hadija, Devota, and several males, including Primus.

Case 4: Aloof resident individuals/relying on voices when becoming separated (II)

Date: October 13, 2014 (congregating season)
Focal individual: Enju (a newcomer female in her first year since arrival)

In the morning, many individuals were heading north while engaging in fission and fusion. During the movements, Enju followed various individuals, including other newcomer females, resident females, adult males, and young males, in such a way that she followed someone for now, and when encountering another individual beyond that, she would attempt to move with that individual next time. At **12:58**, she takes part in a grooming party of multiple males and females, where she grooms and is groomed by several others. But when a pant-hoot is heard from the northeast at **13:18**, many of the individuals present begin moving northeast again. Enju also moves, following various individuals. ★ From **13:30**, she begins to follow Devota (a newcomer female in her second year since arrival), which brings her to an mpila fruit tree ★. While Devota promptly moves off to the east, Enju does not give chase, but stays near Carter (an adult male), who is eating ilombo fruit in a nearby tree, and the two continue feeding together.

Box A

▼ When Carter heads off to the east at **13:51**, Enju follows, but Carter soon disappears from sight ▼. Although Enju rushes eastward for a while, stopping occasionally to look around the vicinity and sniff the ground, she eventually lays down on the spot.

Box B

At **14:03**, a pant-hoot is heard from the east (only by the research assistant), which is the direction that many individuals, including Devota and Carter, have headed. Enju gets up and heads slowly eastward, stopping several times to listen carefully. When she emerges onto the observation trail at **14:10**, she climbs a nearby tree, lays down on her stomach on a thick branch, and rests while occasionally casting her eye around her surroundings. At **14:19**, just as she is about to drift off to sleep, a pant-hoot is heard from the distant northeast. Enju looks in the direction of the voice, sits down again, and then stays still for a little while. At **14:24**, she lies down again, but then at **14:32**, another pant-hoot and a bark are heard, this time from the northeast, not far away. Enju gets up, looks in the direction of the voice for a while, then descends to the ground and begins heading northeast through the bush at speed. After moving for a while, Enju begins to sniff the ground, to stop and listen carefully, and look around. At **14:41**, she discovers the remains of a freshly eaten isakama fruit on the ground, and begins to head east while sniffing the ground in the vicinity. At **14:47**, when a pant-hoot is heard from the north, close by, Enju rushes in that direction while vocalizing a pant-hoot reply, and when she sees several males and females ahead of her in the trees, she lets out a cry of excitement ('faoo'). In the end, she joins Devota, Carter, and seven other individuals in the tree.

Case 5: (Attempted) initiation of a pant-hoot when becoming separated and the co-ranging relationship that underpins it

Date: September 8, 2006 (congregating season, but the focal individual on this day took part in only a small nomadic party, and we were not able to hear pant-hoots from any other nomadic parties)
Focal individual: Qanat (a pregnant newcomer female almost four years after her arrival)

The previous evening, Qanat had bedded down in the immediate vicinity of Fatuma (an older resident female, a co-ranging female for Qanat) and her children, as well as Gwekulo (a older resident female, likewise a co-ranging female) and Puffy (a juvenile 'adopted' by Gwekulo; see Nakamura's discussion in Chapter 6 in this volume). From around **7:00**, they wake up and all slowly begin feeding in a mukamba tree.

Box A

At **07:46**, ▲ <u>Qanat begins moving away from everyone while searching for ilombo fruit, and then continues northward along the branches and goes out onto the Kasiha River (**07:59**). The tall mukamba tree in which she has been with Fatuma, Gwekulo, and the others until a little while ago is visible about 100 meters to the south (it does not look to me like anyone is there anymore) ▲</u>. After resting in the tree for a while, Qanat resumes feeding at a nearby bhuhono tree at **08:15**, but soon descends to the Kasiha River. ◆ <u>After staring in the direction (south) of the mukamba tree where she had been with Fatuma, Gwekulo, and the others, she begins moving east along the river, and then looking intently to the south again, and so on repeatedly ◆</u>. At **08:19**, Qanat comes out onto observation trail R3 and then returns south at a brisk pace while sniffing the ground. At **08:21**, Qanat returns to the trail intersection (R3 × L) near the tall tree where she had been together with the others, but when she looks around, no one is there. She sits down in the intersection, looking alternately toward the east and west ends of observation trail L. At **08:27**, Qanat slowly begins heading east on the trail, but once again stops several times to repeatedly sniff the ground. Perhaps because she smells something along the way, Qanat goes off the trail into the bush on several occasions, but finding no one there, returns to the trail each time.

Box B

At **08:40**, Qanat is heading east along trail L when she begins to initiate a pant-hoot call. Partway through, however, at the sound of rustling, she looks up into a tree to find Gwekulo and Puffy fishing for ants. Qanat breaks off her pant-hoot, and then, after resting for about three minutes under the tree that Gwekulo and Puffy are in, joins them in fishing for ants.

After that, Qanat continued to travel along with Gwekulo and Puffy, and at **11:14**, they met up with Fatuma and her children. She stayed with these individuals for the rest of the day and at night bedded down in the vicinity of them that day as well.

Notes

1. One-time encounters and partings can easily take place between individuals of any species, and even individuals of different species, as long as they are able to perceive each other's presence. For a reunion after a parting, however, the other party must to some extent be identified and to some extent memorized. In the case of the innumerable individuals that make up 'anonymous gatherings' (*sensu* 'anonyme Schar'; Lorenz 1963) with no stable membership, as in the case of migratory fish or migratory birds, an animal may be physically reunited with another, but it is highly unlikely to be cognitively reunited (Hanamura 2021b).
2. By 'encounter,' I mean 'mutually visually perceiving that "the other party is paying attention to me"' or 'auditorily attaining the state of being able to expect that "the other party is paying attention to me" through exchanges of vocalizations.' While a state of 'co-presence' encompasses the state of an encounter, in the case of each party being engaged in different activities without directly interacting with each other, it will continue for as long as both parties can perceive and anticipate the presence and actions of the other, even if neither pays particular attention to the other. For this state to no longer apply is a 'parting,' and the state that follows a parting is 'non-co-presence' (on more approvable definitions, see Hanamura 2021a). However, while a visual parting is relatively easy to understand, an auditory parting (i.e., the juncture between co-presence and non-co-presence) is unclear. While these senses of co-presence and non-co-presence can be observed through the behavior of the individuals concerned (on hearing,

see Hanamura 2010, 2017), in this chapter, which does not go into such details, I refer to 'appearing within visual range' and 'exchanging pant-hoots within ten seconds' as visual and auditory encounters, respectively, and I consider cases in which this state continues or in which the exchanges of vocalizations are repeated as being co-presence. With pant-hoots, I consider vocalizations when there has been no vocalizations from another party beyond visual range for at least ten seconds to be 'calls' or 'initial calls' for accuracy, and those that take place within ten seconds of hearing the end of a vocalization by another party to be 'responses' (on these definitions, see also Hanamura 2010, 2017).

3. As discussed in Hanamura (2019), especially from the standpoint of males who do not emigrate from their natal groups, we cannot ignore the point that 'groups' are generated not only through the accumulation of non-hostile co-presence with 'acquaintances,' but also through hostile encounters with 'strangers.' Nevertheless, in this chapter I focus solely on endogenous mechanisms of group generation from the standpoint of the females who transfer between groups, rather than the promotion of grouping and cultivation of a sense of companionship or solidarity through contact with such 'external' factors.
4. The term 'newcomer female' here is an expedient definition based on the empirical fact that, in principle, females stay in the group where they give birth for the rest of their lives, which is also to say that they can leave the group at any time until they give birth. As also suggested by the qualitative and quantitative data presented in this chapter, the transition from 'newcomer female' to 'resident female' is considered to take place gradually, but the presence or absence of children has a major impact on female ranging patterns and their social relationships with other individuals. It is also worth noting that while there are large individual differences in terms of the time it takes for females who immigrate in M group to give birth, there are few cases of childbirth within a year of immigration. This usually takes place after a few years, and most individuals will give birth by the fourth year after their immigration. While the occasional individual will be infertile (or be presumed as such retroactively), this was not the case for any of the newcomer females dealt with in this chapter. Also, while the number of females who come to M group varies considerably each year, averaging numbers over the past twenty-five years or so, even if we restrict this to females who have given birth without emigrating again, gives us a figure of about one per year (Mahale Mountains Chimpanzee Research Project, unpublished data).
5. A good example of this is the lemon, a fruit eaten by all members of M group. Because it is not a native plant (would be sown by people who once lived in the area that overlaps the home range of the M group), it is possible that it may not even grow in the home range of natal groups of the immigrant females to M group. One newcomer, Vera, after closely watching other individuals eat lemons on the date she was first sighted, apprehensively reached out her hand to pluck a fruit of her own and, without peeling it, nibbled at it—skin and all—then promptly tossed it to the ground. Picking it up again, she gave it another sniff and a nibble, then dropped it again. Given that all the resident individuals in M group normally eat the fruit after peeling it with their mouths or hands, and that Vera began to eat lemons that way a few days later, it seems likely that Vera was not yet accustomed to lemons in her diet.

6. Chimpanzee movement constitutes a series of action choice made by each individual, such as 'to follow/not follow' others or 'to wait for/not wait for' others. The event of parting is therefore not solely determined by the action choice by one party (see Itoh 2003, 2019). Thus partings can sometimes happen quickly and completely, and at other times endless interactions—such that they seem to, or seem not to, part—can drag on all day.
7. Chimpanzees are acutely aware of whether or not other individuals can see them or hear the sounds they make, and it is quite possible for them to behave in such a way as to escape the notice of other individuals. One resident female (Ako) was about to emerge from the bush onto the trail when she became aware of a group of tourists approaching along the observation trail. Staying in the bush, she drew her two nearby children close and held her breath while staying completely still until the group had passed before coming out onto the trail (Nakamura (2009) also describes a similar case).
8. While I do not have space here to introduce a case study, maybe because of such uncertainty of their position as members of the group, it seems that newcomer females, especially those after the first year of immigration who thus to some extent are acclimatized to the social environment, are occasionally 'trying out' their pant-hoots to see if they can elicit responses from other individuals.
9. In other words, to borrow Nishie's terminology (Chapter 8), we could say that parting is always a movement laden with the potential to 'destructuralize' the stability (ideal structure) of the group's members, which is normally self-evident due to the repetition of partings and reunions, with the potential to become a singular point that causes a breakdown in the everyday order.
10. De Waal (1982: 185) reports the case of a female chimpanzee named Gorilla, one of a group of chimpanzees at the Arnhem zoo that are kept during the day in a large outdoor facility:

    > Gorilla has to bottle-feed Roosje [her foster child] every afternoon, and the keeper calls her indoors at the appointed time, while the rest of the colony stay outside. Before she goes inside with her foster child, however, she walks over to Yeroen and Mama and touches them briefly or kisses them. Sometimes this involves her making quite a detour.

    De Waal (1982: 185) interprets the explanation for this behavior 'to be that she is saying goodbye, because she can see ahead to the separation,' thereby positioning the behavior as an example of the kind of future-oriented actions that chimpanzees can take. There is also the case of Washoe and her companions, chimpanzees originally kept for military and medical experiments in the United States that became famous after learning sign language from the primatologist Roger Fouts. Fouts describes in his book how Washoe and her adopted child, Louris, who learned sign language from Washoe, would exchange parting greetings with him and his human colleagues (Fouts and Mills 1997). In other words, chimpanzees can also say goodbye as long as they are under human-style modes of coexistence—that is, when being raised in captivity where co-presence and non-co-presence are clearly distinguished and the timing of partings can be predicted because there are some 'homes', such as 'rooms' for each individual and 'outdoor' and 'indoor' facilities, that regulate times and places of co-presence and non-co-presence, as well as the content of activity to be done in each place, such as social interaction, eating, and sleeping—which in turn would indicate that a parting greeting ('saying goodbye') is something that occurs in conjunction with the human mode of coexistence. However, the human mode of coexistence discussed in this chapter may be biased by the mode of coexistence of settled agriculturalists—more specifically, the mode of coexistence and family and social values of city-dwellers since the advent of modernity, which are most familiar to me, personally. What are parting greetings like for hunter-gatherers and highly mobile nomads who undergo repeated fission and fusion in a manner slightly similar to chimpanzees, if only in terms of family units? There is no end to the questions that I hope to explore as a future challenge.

References

Davis, L. S. (1988) 'Coordination of incubation routines and mate choice in Adélie penguins (*Pygoscelis adeliae*)'. *The Auk*, 105: 428–432.

de Waal, F. B. M. (1982) *Chimpanzee Politics: Sex and Power among Apes.* Harper and Row.

Fouts, R. and S. T. Mills (1997) *Next of Kin: What Chimpanzees Have Taught Me about Who We Are*. William Morrow & Co.

Hanamura, S. (2010) 'Long-distance calls by chimpanzees: How they connect actions and organize their social field beyond visual contact'. *Primate Research*, 26: 159–176. [in Japanese with English summary]

Hanamura, S. (2015a) 'Fission-fusion grouping'. In M. Nakamura, K. Hosaka, N. Itoh and K. Zamma (eds), *Mahale Chimpanzees: 50 Years of Research*. Cambridge University Press, 106–118.

Hanamura, S. (2015b) 'Male–female relationships: Affiliative, interventional, and dominant-subordinate interactions'. In M. Nakamura, K. Hosaka, N. Itoh and K. Zamma (eds), *Mahale Chimpanzees: 50 Years of Research*. Cambridge University Press, 410–420.

Hanamura, S. (2015c) 'Saru to deai sono shakai ni makikomareru: Kansatsu toiu itonami nitsuite no ichikōsatsu' (Encountering monkeys and getting involved in their society: A study on the practice of observing animals). In D. Kimura (ed.), *Dōbutsu to deau I: Deai no sōgokōi* (Encountering animals I: Analyzing encounters). Nakanishiya Shuppan, 87–104. [in Japanese]

Hanamura, S. (2017) 'When keeping one's ears open for the distance voices of others: The process-oriented convention in chimpanzee and institution'. In K. Kawai (ed.), *Institutions: The Evolution of Human Sociality*. Trans Pacific Press and Kyoto University Press, 165–195.

Hanamura, S. (2019) 'When pricking up one's ears for the voices of strangers: Others in chimpanzee society'. In K. Kawai (ed.), *Others: The Evolution of Human Sociality*. Trans Pacific Press and Kyoto University Press, 177–216.

Hanamura, S. (2021a) 'Deai wo tassei suru "yobikake-ōtō," zyoutai no ikou to kankei wo kakunin suru "aisatsu"' ('Summons-answer' for accomplishing an encounter, 'greeting' for confirming a state transition and relationship). In D. Kimura and S. Hanamura (eds), *Deai to wakare: 'Aisatsu' wo meguru sōgokōikenkyū* (Encounters and partings: Interaction studies on 'greetings'). Nakanishiya Shuppan, 23–43. [in Japanese]

Hanamura, S. (2021b) 'Dōbutsushakai no kyōzon no yōtai to aisatsu' (Modes of coexistence and greetings in animal societies: Chimpanzees who can part without greetings, humans who cannot part without greetings). In D. Kimura and S. Hanamura (eds), *Deai to wakare: 'Aisatsu' wo meguru sōgokōikenkyū* (Encounters and partings: Interaction studies on 'greetings'). Nakanishiya Shuppan, 235–257. [in Japanese]

Idani, G. (1991) 'Social relationships between immigrant and resident bonobo (*Pan paniscus*) females at Wamba'. *Folia Primatologica*, 57: 83–95.

Idani, G. (2013) 'Origin of the human family: Comparison with ape societies'. *Japanese Journal of Adolescent Psychiatry*, 22(2): 81–93. [in Japanese with English summary]

Imanishi, K. (1961) 'Ningenkazoku no kigen: Puraimatorojī no tachiba kara' (The origin of the human family: A primatological approach). *Minzokugaku Kenkyū* (Ethnological study), 25: 119–135. [in Japanese]

Itoh, N. (2003) 'Matomaru koto no mekanizumu' (Dynamics of gathering). In M. Nishida, K. Kitamura and J. Yamagiwa (eds), *Ningensei no kigen to shinka* (Origin and evolution of humanity). Shōwadō, 233–262. [in Japanese]

Itoh, N. (2013) 'A group of chimpanzees: The world viewed from females' perspectives'. In K. Kawai (ed.), *Groups: The Evolution of Human Sociality*. Trans Pacific Press and Kyoto University Press, 111–119.

Itoh, N. (2017) 'Duality of the mode of coexistence and action selection: Groups and the emergence of "institutions" in chimpanzees'. In K. Kawai (ed.), *Institutions: The Evolution of Human Sociality*. Trans Pacific Press and Kyoto University Press, 141–163.

Itoh, N. (2019) 'Encountering the "others": How chimpanzees face *indeterminacy*'. In K. Kawai (ed.), *Others: The Evolution of Human Sociality*. Trans Pacific Press and Kyoto University Press, 149–176.

Itoh, N. and M. Nakamura (2015) 'Female–female relationships'. In M. Nakamura, K. Hosaka, N. Itoh and K. Zamma (eds), *Mahale Chimpanzees: 50 Years of Research*. Cambridge University Press, 399–409.

Kitamura, K. (2013) 'From whence comes human sociality? Recursive decision-making processes in the group phenomenon and classification of others through representation'. In K. Kawai (ed.), *Groups: The Evolution of Human Sociality*. Trans Pacific Press and Kyoto University Press, 59–77.

Lorenz, K. (1963) *Das Sogenannte Böse, Zur Naturgeschichte der Aggression*. Verlag Dr Borotha-Schoeler.

Matsumoto, T., S. Hanamura, T. Kooriyama, T. Hayakawa and E. Inoue (2021) 'Female chimpanzees giving first birth in their natal group in Mahale: Attention to incest between brothers and sisters'. *Primates*, 62(2): 279–287.

McGrew, W. C. and L. Baehren (2016) '"Parting is such sweet sorrow", but only for humans?' *Human Ethology Bulletin*, 31(4): 5–14.

Nakamura, M. (2009) *Chinpanjī: Kotoba no nai karera ga kataru koto* (Chimpanzees: What they have told me without language). Chūkō Shinsho. [in Japanese]

Nishida, T. (1989) 'Social interactions between resident and immigrant female chimpanzees'. In P. G. Heltne and L. A. Marquardt (eds), *Understanding Chimpanzees*. Harvard University Press, 68–89.

Nishida, T. (2012) *Chimpanzees of the Lakeshore: Natural History and Culture at Mahale*. Cambridge University Press.

Nishie, H. (2017) 'Who is the alpha male? The institutionality of dominance rank in chimpanzee society'. In K. Kawai (ed.), *Institutions: The Evolution of Human Sociality*. Trans Pacific Press and Kyoto University Press, 121–140.

Osawa, M. (2016) '"Shakaisei" eno hukanō na shinka' (Impossible evolution toward 'sociality'). *Gendai Shisō* (Contemporary thought), 44(10): 165–177.

Sogabe, A. (2011) 'Partner recognition in a perennially monogamous pipefish, *Corythoichthys haematopterus*'. *Journal of Ethology*, 29: 191–196.

Suddendorf, T. (2013) *The Gap: The Science of What Separates Us from Other Animals*. Basic Books.

Sugiyama, Y. and S. Fujita (2011) 'The demography and reproductive parameters of Bossou chimpanzees'. In T. Matsuzawa, T. Humle and Y. Sugiyama (eds), *The Chimpanzees of Bossou and Nimba*. Springer, 23–34.

Yabuta, S. (2015) 'Tekitaiteki sōgokōi kara "aisatsu" e: Dōkizuke no kattou ga hatasu yakuwari' (From agonistic interaction to 'greetings': The role of motivational conflict). In D. Kimura (ed.), *Dōbutsu to deau I: Deai no sōgokōi* (Encountering animals I: Analyzing encounters). Nakanishiya Shuppan, 151–184. [in Japanese]

CHAPTER 8

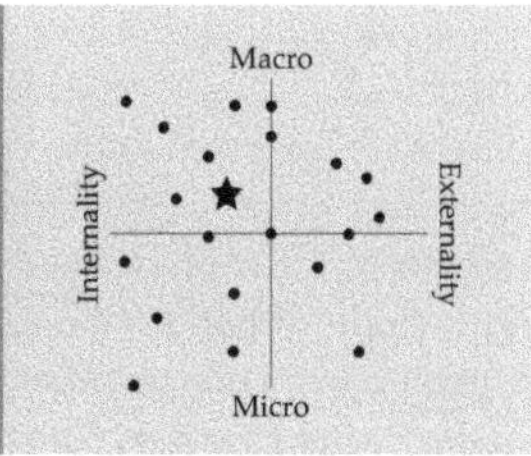

Isolation as a Singular Point of Society:

When Wild Chimpanzees Become Solitary

Hitonaru NISHIE

Keywords: structure / destructuralization, regularity / singular point, potentiality / realization

Exercise question

What does this figure represent?

Solution

Any number of things: the numeral eight, the letter S written horizontally, the symbol for infinity, part of a disentanglement puzzle, a twisted coil of rope, and so on and so forth. Even so, there are differences between these innumerable answers and the figure itself. The central intersection for the numeral eight and the symbol for infinity is not open, and a disentanglement puzzle or coiled rope are, first of all, three-dimensional shapes; thus these myriad solutions are only arrived at by ignoring part of the figure itself or, conversely, augmenting it by giving it a concrete form. That is, the figure becomes an abstract symbol by having a certain specificity removed, and it is when its abstract character is imbued with a certain specificity that its aspect as a certain object emerges. The relationship between ideal social *structures* and real *groups* can be likened to the relationship between a figure thus open to innumerable solutions, and the various solutions that are actually produced. The point found between 'the figure itself' and its actual solution is a 'singular point' that separates these differences. When we find one particular solution in this figure, a singular point appears in its difference from the pattern of regularity therein (i.e., in the figure itself). When we look at the way living things form groups in reality, we perceive a difference from the potential ideal structures that are latent therein. This vector/potential for destructuralization, which goes unnoticed when the same pattern is realized over and over again, deviates from the regularity of the structure when its realization is activated in a certain way. At the same time, in doing so, it also brings the regularity of the underlying structure to our attention. When an individual isolates themselves by withdrawing from a group that had previously appeared stable, the path of its deviation (the vector of destructuralization) at the same time illuminates an ideal structure that until that point had only been perceived in terms of a potential.

Living things and extremes

The goal of this chapter is to portray the aspect of living things that inhabit extremes. Generally speaking, the term 'extreme' calls to mind life-threatening or otherwise difficult situations like sudden disasters or abnormal weather conditions. In this chapter, however, I want to depict livable extremes in a broader sense—one that includes, but is not necessarily limited to, life-and-death situations like these.

The economist and political philosopher F. A. Hayek once discussed the ordering of human behavior and the evolution of institutions in terms of the concept of spontaneous order. The basic premise of this argument was that, as a general feature of organisms with sensory and cognitive systems, individuals selectively perceive information within an environment already delimited by the frame of their sensory apparatus and must therefore have a 'necessary and irremediable ignorance' of information about this environment (Hayek 1967a, 1967b, 1969, 1973: 12; Nishie 2017). This suggests that living things glean very limited information about their own a priori finitude from among the infinite information available in the environment, which they in turn use to lend certainty to their own survival. Extremes for living things, as I attempt to depict them in this chapter, belong to the class of problems that juxtapose finitude with the infinite. This class of problems, which is concerned with the finitude of living things with respect to their environment as a sea of infinite information, also includes 'extreme' states that constitute 'certain marginal situations' for living things, in the sense that these are realized in the form of actions chosen in each instance amid the pervasive uncertainty that haunts their normal lives, thus leaving the door open to more difficult, even life-threatening, situations. From the perspective of evolution, this constitutes the problem of 'livable adaptation' in terms of how living things perceive, act upon, and construct their respective living environments to be amenable to adaptation.

When stable daily repetition is possible, living things will automatically—in a manner of speaking—maintain stable interactions with their environments, more or less without any perception of such extremes. This corresponds to how living things form a spontaneous order with their environments, one that can then be repeatedly used for the selection of subsequent actions.[1] However, interactions between living things and their environments are always open to the possibility of unexpected disturbances and fluctuations, and an order shaped and maintained by past (phylogenetic or ontogenetic) history will not always remain stable or available.[2] When what had seemed to be a stable order becomes chaotic uncertainty, living things are forced to come up with new ways of coping at the point of their interface with the environment. At such times, it becomes clear that the everyday order, which had until then seemed stable, is fraught with potential singularities in terms of all manner of deviations and variations.

In this chapter, I consider the problem of 'livable extremes' (which in this chapter I also describe as 'singular points,' as I describe later)[3] in the sense of the selection of actions in what, for living things, are contexts of fundamental unknowability and uncertainty. This represents a problem that living things must deal with as the emergence of singularities (deviations and variations) that exist as potential within a smooth daily order (regularity) that seems otherwise stable. The phenomenon dealt with in this chapter concerns an ongoing series of uncertain situations wherein certain individuals one day suddenly disappear from groups—whose membership had up to that point seemed stable—to pursue a long period of solitude in some unknown location before (sometimes) returning to the group—though whether they will return, when they might return, and what will happen should they do so is in each case unknown. I want to think about how living things, which are bound by a certain finitude, deal with such peculiarities in instances where the future is difficult to predict, and the ways in which such behavior relates to social 'structure.' As a concrete example, I investigate how male individuals in wild chimpanzee (*Pan troglodytes*) society sometimes withdraw and isolate themselves from their groups to live a solitary life for long periods of time. By examining the relationship between the 'ostensibly stable order' of chimpanzee society and its 'unexpected disturbances and fluctuations,' and the range of how living things can deal with such instances, I sketch the relationship between isolation as a singular point in chimpanzee society and the 'structure' that it reveals.

Jun'ichirō Itani's 'structure/anti-structure'

To examine the 'structure' of chimpanzee society, I would like first of all to summarize the concepts of 'structure' and 'anti-structure' as formulated by Jun'ichirō Itani in his discussion of the evolution of primate social structures.

In his theory of primate social structures (Itani 1987c), a 'structure' appears in the form of a Basic Social Unit (BSU) that is unique to each species as prescribed by their respective sexual composition and patterns of membership transfer. Conversely, the idea of 'anti-structure,' as conceived by Itani (1993) in his later years, is defined by beings that do not fit into a structure; examples include transitory individuals who exist on the fringes of BSUs, groups of gorilla males, large flocks of birds, and polyspecific associations made up of multiple species. However, whereas Itani's concept of structure was carefully modeled and theorized as a tool to explain the evolution of primate society, the concept of anti-structure was never fully theorized by Itani himself, a fact that later led to some confusion (as we shall see). This confusion over Itani's idea of anti-structure is plainly illustrated by the fact that the simple question of 'what sort of individual (lifestyle) does anti-structure describe?' does not lend itself to any easy answers.

As a typical example of anti-structure, let us consider, for example, the solitary male Japanese macaques (*Macaca fuscata*) on which Itani (1993) focused. Japanese macaques generally form mixed groups of multiple males and females. But whereas females spend their entire lives in their natal group, males transfer out of their natal group upon reaching sexual maturity and spend a period of time living a solitary life before joining another group. In this way, groups of Japanese macaques are considered to have a matrilineal structure in which the BSU is reproduced by having females remain with their natal group. And in this case, males who transfer between BSUs (i.e., between structures), who are referred to as solitary males (*koen* in Itani's terminology), lead a highly solitary life apart from the group. However, Uchibori (2013: 45) argues that because the solitary male Japanese macaques that Itani regarded as model examples of anti-structure are a phenomenon that is limited to individuals of the sex (male) prescribed by the structure itself, they are actually 'part of a structured process.' On the other hand, Adachi (2013: 30), who has carried on and elaborated Itani's theory of anti-structure, positions these same solitary male Japanese macaques as 'anti-structured' solitaires. Thus even something that Itani, as the progenitor of the anti-structure concept, considered a model example of the phenomenon lacks consensus over whether it is structure or anti-structure. It seems to me that such discrepancies may have arisen because, as shown clearly in his own subsequent elucidation of structure and anti-structure, Itani conceived anti-structure in complementary terms as something 'outside structure'—that is, as a 'non-structure' on the same plane as structure. Moreover, it is also conceivable that Itani was not even himself aware of the ontological distinction between a 'real group' and 'ideal structure.'

So how is it possible to conceive of anti-structure 'as such' rather than complementarily? To answer this, we first need to clarify this difference of ontological level.

Itani considered that the social structure of organisms is formed and maintained by the behavior of individual organisms, in accordance with rules as a function that supports coexistence. If we focus on the expression, 'a function that supports coexistence,' we can see that Itani's structures and rules prescribe how individuals coexist, which is to say they prescribe how individuals arrange and connect themselves in relation to one another. In fact, Itani (1993: 198–9), likening a piece of paper to the entire *specia*,[4] describes how a pattern of dots on the page corresponds to the basic unit group/BSU/structure, while the many points found in the voids of the dotted pattern are likened to the solitary actors (i.e., anti-structure). Elsewhere, Itani (1987c) distinguishes among six categories of BSU (Figure 8.1) and illustrates the process of their evolution (Figure 8.2). However, it will be clear that the encircled BSU schema represents 'ideal arrangements of individuals' (i.e., structures), just like the dotted pattern on paper. That this structural schema is 'ideal' means that we do not actually observe these encircled groups or the individuals moving in and out of the encircled boundaries in reality. These are ideas that can only be captured by

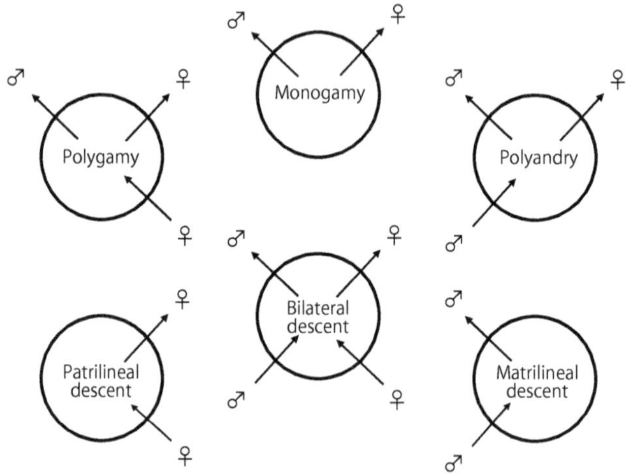

Figure 8.1: Six types of BSU (adapted from Itani 1987c: 310)

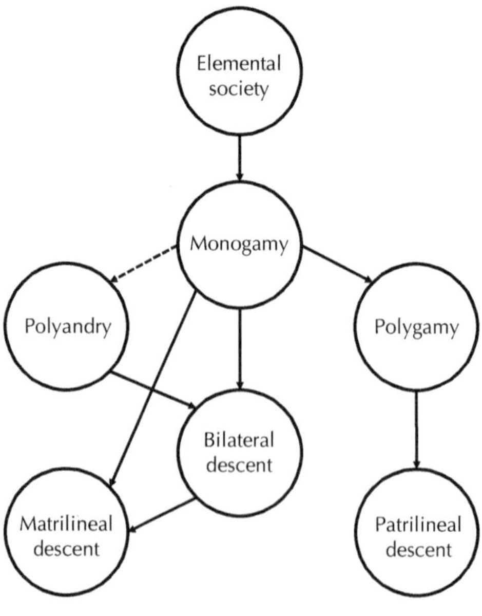

Figure 8.2: BSU evolution process (adapted from Itani 1987c: 318)

some sort of abstraction or schematic illustration—that is, by thinking. (What we *can* actually observe is the 'movement' of the association or dissociation of discrete individuals; we *cannot* observe the boundaries drawn by the lines that delimit the association or disassociation of those individuals.) In addition, this ideal structure is not merely a classification of a pattern in how individuals associate or dissociate; it can also point to how the way in which real individuals form groups (i.e., associate and dissociate) constitutes a 'generative principle' (i.e., potential). This is because, as well as classifying structures as described above, Itani also held that structures are supported by the actions of individual organisms in accordance with 'rules as a function' that makes the coexistence of individuals possible (Itani 1987a).

The difference of ontological level between the structures that Itani conceived as ideal and real groups that he fixated upon as phenomena is thus clearly demonstrated. Put simply, structures exist at the invisible level of the ideal and the potential, but are realized as the visible concrete and actual ways that groups are formed by the behavior of individuals acting in accordance with rules (i.e., association and dissociation). Incidentally, when thus distinguishing between real groups and ideal structures, individuals that do not associate (solitaires) are nonetheless practicing a mode of congregation that actually exists. For example, the structure of the *specia* of Japanese macaques, by means of rules as functions that support coexistence, may be thought of as realizing an association of individuals (Group A) and solitary males (X) in the process of transferring between groups. In other words, the structure of the *specia* of Japanese macaques is conceived of as an ideal schema (a BSU unique to the *specia* of Japanese macaques) and that structure realizes individual groups with varied characteristics (Group A, Group B, Group C...) and various solitary males (Individual X, Individual Y, Individual Z...) in accordance with their respective boundary conditions (environmental conditions, historical background, etc.). These Groups A, B, C... and Individuals X, Y, Z..., which for the sake of convenience are described separately here, differ merely in terms of their mode of congregation that actually exists. While they have the same ontological status (realized association or dissociation), the structure has a different ontological status in the sense that it remains within the realm of ideas. However, because Itani did not clarify the ontological distinction between 'ideal structure' and 'real group,' it seems that individuals who belong to the sex that transfers membership as prescribed by the structure (here, solitary males) must be positioned as what we might call a 'complement-structure' (i.e., complementary anti-structure). In this sense, solitary Japanese macaques are prescribed by structure, suggesting that there is some merit in Uchibori's (2013) indication that they are structured processes (in the sense that both groups and solitaires are realized in various forms by structures).[5]

We need to recall, however, that the schema of 'ideal structures and real groups produced by rules supporting those structures' was just a problem of 'the arrangement of individuals.' Itani actually explained this schema by likening it to the plane of a piece of paper, and it is static enough to convey a certain sense of beauty. Although Itani's explanation of anti-structure could hardly be called sufficient, solitary males are also contained in this static arrangement of individuals as a concrete embodiment of anti-structure. Contrary to Itani's intention, however, the solitary males contained in this static schema can be interpreted as a 'complementary structure' that 'fills the gaps between structures' as a part of a 'structured process.' But we must also pay attention to another aspect that seems to have been incorporated into Itani's concept of anti-structure. Namely, the aspect of 'destructuralization as a singular point' that is intrinsic to structure.

Here, let us return briefly to the description of the solitary male that Itani gave as a typical example of anti-structure. Itani (1993: 197) states that 'the departure of the solitary male is in some ways beyond the realm of our analogy,' and describes this behavior as 'having some aspects that are vagabond in that they stray from the rules of group society, transcending the secular world.' In this description, the ideas of 'straying from the rules,' 'vagabond,' and 'transcending the secular world' will no doubt catch the eye. These expressions go beyond merely considering something as outside the rules; they evoke an atmosphere that might almost be called a kind of sanctification. In fact, Itani himself likened anti-structure to Victor W. Turner's concept of 'communitas,' also positing the opposition of 'sacred' anti-structure to 'profane' structure (Itani 1993).

Although these expressions, which verge on the sanctification of anti-structure, suggest a whiff of indulgent romanticism on Itani's part, it seems that this may be because he felt that the proximate factor for the migration of males from the natal group—that is, the motivation for young males to leave the group—was their self-identification with the solitaires outside the group. This can also be seen in his statement that 'we may say that it is none other than these [the solitary males] who lure the young males of the group with the promise of a life of proud independence' (Itani 1993: 197).

Accepting for now the validity of Itani's explanation (that the migration of males from their natal groups is prompted by their identification with solitary males), we should note that the phenomenon to which Itani is paying attention here is not that males 'are solitary' (a state of being), but rather that males 'become solitary' (a vector of movement), and the fact that a 'vector of force' is in operation at such times. As discussed above, 'being solitary' (a state of being) is prescribed by the structure, and thus—as a complement-structure bridging other structures—can itself be interpreted as a static mode of solitary life. In that sense, it is also a 'structured process' (Uchibori 2013). On the other hand, 'becoming solitary' (a vector of movement) and the 'vector of force' at work therein indicate a dynamic mode that will not fit into the static arrangement of Itani's theory of social structure. Hence the conceptualization of anti-structure as the source of this movement and force.

However, if we are talking of anti-structure as a source of movement and force, perhaps concepts like 'destructuralization' or 'disorganization'—which indicate the nullification of structure or a movement away from structure—might be more suitable than Itani's 'anti-structure.' Itani's (1993) attempt to position anti-structure as a deviation from structure and, similarly, Adachi's (2013: 29) attempt to conceptualize solitary Japanese macaques in the process of transferring to a new group as 'individuals [that] comprise the non-structural portion of a *specia*' seem to be consistent with the positioning of destructuralization as a movement (vector) away from structure. In other words, Itani's concept of anti-structure confuses the two different aspects of a static mode (of being a 'complement-structure' bridging a gap between structures) and a dynamic mode (of 'destructuralization' as the source of a movement and force that deviates from and nullifies structure). This confusion is what seems to be at the root of the aforementioned inconsistency as to whether the solitary male in the process of transferring between groups of Japanese macaques constitutes structure or anti-structure (Adachi 2013; Uchibori 2013).

Let us recap so far. It seems that Itani's concept of structure does not distinguish clearly between ideal structures and real groups, and his concept of anti-structure does not distinguish between anti-structure as a state (complement-structure) and destructuralization as movement (disorganization). When these distinctions are made, Itani's theory of structure/anti-structure can be newly assessed by extending and clarifying it as follows. (1) Each *specia* has its own unique ideal structure. (2) Ideal structures are realized as various ways of congregating (e.g., groups A, B, C..., solitary individuals X, Y, Z...) by the action of rules that make coexistence possible. (3) As a state, anti-structure (complement-structure) corresponds to solitary individuals (X, Y, Z...) as a subset of these various ways of congregating, which can be interpreted as part of the structure (rather than in terms of anti-structure). (4) As movement, destructuralization refers to a structure that has remained at the level of ideal or potential by deviating from the framework of the structure (rules), thereby nullifying it.

As summarized here, steps (1) to (3) can be conceived of as a realization of structure, whereas (4) refers to a phenomenon that is on a different level than the ideal structure or its (normal) realization, in the sense that while being intrinsic to the structure, it is also accompanied by a vector of movement toward the 'outside.' The dynamic of destructuralization indicated in (4) can be conceived of as a 'singular point,' so to speak, that transcends the structure while being intrinsic to it. In so far as the structure is supported by the behavior of the individual according to the rules and repeatedly realized in a stable pattern, we will never become aware of its latent axiomatic character. However, when a movement that deviates from and nullifies the structure appears at 'a given point' intrinsic to that structure, we will perceive the path traced by that deviation and become aware of the potential structure—*which is what makes us see that movement as a deviation*. Taking the phenomena dealt with in this chapter as an example, when you see that an individual leaves the group, along with the movement of the withdrawal, you will also unconsciously become aware of the structure of the 'stable order' (i.e., the group) as an axiom. The point that separates structure and destructuralization in this way—the point that is intrinsic to the structure but transcends it—is referred to here as a 'singular point.' The singular point, as mentioned at the beginning of this chapter, is a 'livable extreme' for living things. That is, it is to be found in the path of the vector of the deviating movement (destructuralization) that brings about a modulation in the structure as a point separating the stable day-to-day order from chaos filled with uncertainty. This chapter seeks to clarify the nature and process of such deviation in chimpanzee society in the context of this withdrawal (isolation) as a singular point. As we shall see below, it is at this singular point that we should be able to see both the order brought about by the structure and the deviating movement brought about by destructuralization.

But why would Itani, who conceived of anti-structure in the first place, have focused on solitary males in the process of transferring between matrilineal groups? As has already been argued, males in matrilineal groups are the sex that transfers membership as prescribed by the structure, and therefore can be interpreted not as anti-structure, but as part of the structured process. So why did Itani not focus instead on isolation among females in matrilineal groups? For example, even in matrilineal groups of Japanese macaques, although isolation among females is a rare phenomenon, it was already known when Itani first announced his theory of social structure (in 1972) and, in fact, Itani does mention isolation among female Japanese macaques in his writings (1987b: 123–4). In addition, Itani first published his theory of anti-structure in 1991 (see Itani 1993), and as I will describe later, around that time several cases of the phenomenon of male isolation in patrilineal chimpanzee groups had already been reported. These instances of isolation—among females in matrilineal groups and among males in patrilineal groups—concern individuals that do *not* belong to the sex that transfers membership as prescribed by the structure, and would therefore also seem to be appropriate cases when conceiving of anti-structure. Nevertheless, Itani accorded these rare phenomena only the status of exceptions, seeming to take the negative view that they were not generalizable (Itani 1987b: 124). However, as discussed above, when we focus on destructuralization as a singular point, then being an exceptional event can, in fact, take on important meaning as a deviation from 'commonplace regularity.' Unique things hit an unpredictable and uncontrollable 'point of heterogeneity' as a deviation from the regularity of structure. However, they illuminate the nature of the 'ideal structure' that made it possible to act in accordance with the rules in the first place, not as a 'sacredness' that transcends structure, but simply by virtue of their being intrinsic to a structure while deviating from the rules that support it. By considering the withdrawal (isolation) of males in the context of chimpanzee society as a singular point of that society, this chapter gets to the bottom of the paradox that emerges in society at the singular point of isolation.

The limits of fission and fusion in chimpanzee society: withdrawal from the group and a prolonged solitary life

Wild chimpanzee society is characterized by the formation of unit groups with memberships that are, to a certain degree, fixed. Chimpanzee unit groups are made up of multiple males and females (multi-male multi-female groups), but whereas females usually move away from their natal groups to join other unit groups once they reach sexual maturity, males basically do not transfer their membership between groups (thus they are said to have a patrilineal society, since the succession of the group takes place along the male line).

Chimpanzee unit groups do not maintain a high degree of cohesion, but are instead characterized by their frequent fission and fusion (Itoh and Nishida 2007; Nishida 1968). It is almost never the case that all members of the group will be encompassed within the field of view, and they usually dissociate and wander in small groups called 'parties' made up of one to five individuals. For example, when tracking an individual, there will only ever be a few animals within visual range. Moreover, as an animal wanders, the members of its small party will switch out one after another, and it is sometimes the case that it will suddenly find itself all alone. Among the great apes, this fission-and-fusion character of groups is found only in the genus *Pan*, which includes chimpanzees, and in our own human society.

Fission and fusion typically have aspects that can be described in terms of spatial scale and aspects that can be described in terms of temporal scale. By distinguishing and making links between these two scales, we are able to describe spatiotemporal modes in the way that chimpanzees engage in fission and fusion.

In terms of the spatial scale of fission and fusion, a distinction has been made between 'face-to-face parties,' which are associations of individuals in close proximity that fall within each other's fields of view, and 'nomadic parties' that consist of loose clusters of individuals that include some who are not within visual range (Itoh and Nishida 2007). Face-to-face parties are associations of individuals within visual range, and the composition of such parties often changes with dizzying rapidity, especially when the nomadic party that spreads out around the face-to-face party is quite large. Conversely, when the nomadic party is smaller, the composition and size of face-to-face parties stays relatively stable.

Nomadic parties are loose clusters of individuals across a wider range than would be encompassed by a face-to-face party, and constitute associations that spread out over a range within which their vocalizations will be mutually audible, and which corresponds roughly to the direction of their movement. The size of nomadic parties is known to fluctuate considerably from season to season depending on the availability of fruits, which constitute the chimpanzees' main source of food. Large nomadic parties will form when fruit density is high. Conversely, this kind of party will be smaller in seasons when fruits are scarce (Itoh and Nishida 2007).

In terms of the temporal scale of fission and fusion, first, the scale of fission and fusion over one to several days is apparent primarily in terms of changes in the size and composition of face-to-face parties. Seasonal fission and fusion takes place on a scale of several months, and is apparent primarily in terms of changes in the size and composition of nomadic parties (a congregating season when large nomadic parties form and a dispersing season when small nomadic parties are formed). The phenomenon dealt with in this chapter, of the prolonged withdrawal from the group and subsequent reinstatement, takes place on an even longer time scale that might last from several months to several years. In this case, fission and fusion are apparent as the prolonged withdrawal of a specific individual from a group to lead a life of virtual solitude. In this way, when we extend the composition and size of nomadic parties and everyday face-to-face parties, we are able to situate the phenomenon of a solitary life associated with the prolonged withdrawal of a specific individual, as in the cases dealt with in this chapter.

The solitary nature of male chimpanzees

Chimpanzee societies form multi-male multi-female unit groups and are considered to have a patrilineal social structure in which males remain in their natal groups while females transfer membership between groups, and this social structure has been explained in structural and functional terms by specialists in the fields of socioecology and primate sociology.

Socioecology emphasizes the cohesive nature of males and the solitary nature of females in chimpanzee society (Wrangham 1987). The solitary nature of females and their use of a small core area have been explained as an adaptation to the distributional pattern of food resources (fruits), which are the most important resource for females. It is thought that loosely clustered (or moderately scattered) females secure relationships with males in groups to safeguard food resources and reduce the risk of infanticide. Meanwhile, closely related consanguineous males are explained as securing relationships with scattered females and jointly defending them from other neighboring groups by maintaining a high level of cohesion as an adaptation to the distributional pattern of breeding resources (females), which are the most important resource for males. Thus socioecology provides a consistently functionalist explanation (i.e., an adaptive relationship in the context of their ecological and social environment) of the structure of chimpanzee society in terms of male cohesiveness and female discreteness. From this point of view, however, the withdrawal and isolation of male chimpanzees dealt with in this chapter is not considered a problem in the first place, even at the phenomenal level, and the dynamic mode of destructuralization is not included in its theoretical scope.

In the context of primate sociology, likewise, discussions have been carried out with a focus on the cohesiveness of males in chimpanzee society. For example, Kuroda (2013) sees male chimpanzees as having an 'aspiration for dominance.' He contends that alliances with other males are important for improving their ranking, and also that males must maintain a high degree of cohesion out of the need to always be paying close attention to ascertaining that this alliance relationship is being maintained. For this reason, 'chimpanzee males fear isolation, inhibit hostility and repeat interactions to confirm their relationships' (Kuroda 2013: 285). Moreover, and in a closer context in connection with this chapter, he goes so far as to state that 'chimpanzee males, who tend to stay in a group except when they temporarily go solo after losing a fight over their rank,' enjoy something approximating 'the freedom of the loser' (Kuroda 2013: 279). In other words, this suggests a certain inevitability to the formation of groups by dominance-minded male chimpanzees; as a condition for 'dominance' to be established, male chimpanzees must congregate with each other and must gather females, whose presence is both the advantage and object of dominance (Kuroda 2013).

In each case, these are attempts to explain that 'male chimpanzees congregate' and, in contrast to the default cohesiveness of male chimpanzees, they regard solitude as an exceptional state of privation in which an individual is unable to congregate (due the cost incurred in doing so). However, this approach seems to be burdened with the same problem as Itani's complementary definition of anti-structure as 'beings that do not fit into a structure' and the difficulties this entailed for conceptualizing anti-structure. After all, does this mean that male chimpanzees are (or should be) intrinsically prone to congregate and that they become solitary when they cannot do so for some reason? These explanations underestimate solitary life as a variant mode of congregation that actually exists and that actually occurs, albeit in a small number of cases. By extension, they might also underestimate the 'structural potentiality' of chimpanzee society, as well as even 'withdrawal (isolation) as a singular point that amounts to a destructuralization' that nullifies the structure.

To get a richer grasp of this variant mode of congregation that actually exists in chimpanzee society, and to address the peculiarity of the depth of its structural potentiality/the peculiarity of its destructuralization, further individual case studies will be needed. For example, Goodall (1986) describes sexual differences in the sociability of chimpanzees in Gombe, Tanzania, by emphasizing the *solitary nature of males and the highly social nature of females*:

> When a male starts to leave his mother at the age of about eight years, he will for the rest of his life spend many hours by himself...A female, when first she starts to travel independently of her mother, will also spend time by herself. But from the time when, two or three years later, she gives birth to her first infant, she is unlikely (unless that infant dies) to spend *any* time completely alone for the rest of her life. (Goodall 1986: 159, original emphasis)

Further, with regard to the male aspiration for dominance, she notes that not all males seek to raise their status. Referring to males who work for only very short periods of time, males that quickly abandon the prospect of achieving higher rank, and males who seem to have little interest in their own social status, she discusses diverse ways of life among males that cannot be assimilated to a simple scheme that equates male chimpanzees with an aspiration for dominance (Goodall 1986: 415).

The cases of withdrawal and solitary life on the part of male chimpanzees dealt with in this chapter also exemplify this diversity in terms of male ways of life, as well as a variant mode of how chimpanzees actually congregate (i.e., how they associate and dissociate) in the real world. For male chimpanzees, the possibility of transferring between groups is considered to be virtually nil due to the high degree of hostility between groups. Accordingly, the extended period of solitary life that accompanies withdrawal from the group may be considered an extreme state from which there is 'no way out' in the sense that 'although the group is not easily left, even so it is difficult to get back in.' By sketching out the destructuralizing dynamic that 'solitude' holds as a mode of existence realized by male chimpanzees under these circumstances as a singular point of society, I will at the same time re-examine the structural potentiality that informs their society.

When male wild chimpanzees become solitary

As we saw in the previous section, it has been reported that male chimpanzees in the wild generally tend to congregate together, and spend very little time on their own. But this, on the contrary, makes their absence from the group even more pronounced on those rare occasion when males do leave the group.

In fact, at Mahale Mountains National Park in Tanzania, several cases have been reported to date of solitary male chimpanzees living for long periods on their own. In most of these cases, an individual who had been the alpha male isolates himself by withdrawing from his group. Given that alpha males are usually the focus of attention while they are in the group and because reversals of rank are occasionally observed in conjunction with their withdrawal (i.e., their replacement by a new alpha male), the phenomenon of an alpha male's withdrawal to spend a long period of time living a solitary life has been reported in detail, although with limited firsthand observation, including the process of withdrawal and reinstatement and their state of being during this period of withdrawal. Here, by taking an overview of these cases of male withdrawal, I would like to sketch out the process by which male chimpanzees attain solitude.

Kasonta's case (Nishida 1983; Uehara, Hiraiwa-Hasegawa et al. 1994; Uehara 1994; Nishida 1981)

Kasonta, a male born in 1940 (estimated), was the alpha male of Mahale K group from 1969. When he lost out to Sobongo, the second-ranked male in May 1976, he withdrew from the group to wander on his own, and remained more or less completely solitary until August 1977. Kasonta was briefly reinstated as K group's alpha male in September 1977, but disappeared in January 1978 (cause unknown).

As of May 1975, the Mahale K group counted four adult male members, with Kasonta at the top, followed in decreasing order of rank by Kajabala, Sobongo, and Kamemanfu. In October of that year, however, Kajabala disappeared (cause unknown). In March 1976, Kamemanfu and Sobongo had their first altercation, and then in April, Kasonta and Kamemanfu joined forces to attack Sobongo. In May, Kamemanfu began joining forces with Sobongo against Kasonta, while occasionally opportunistically working with Kasonta, so that despite being ranked third, he came

Photograph 8.1: Kasonta (Photo credit: Toshisada Nishida)

to hold the deciding vote in the three-way relationship. On May 14, 1976, Sobongo and Kamemanfu joined forces to attack Kasonta, after which Kasonta withdrew from the group to begin wandering on his own.

Although the details are unclear due to the fact that no researchers were present between May and December 1976, research assistants continued to sporadically observe Kasonta, who was reportedly also seen to join forces with Kamemanfu to attack Sobongo over this period. Over the nine months from December 1976 through August 1977, Kasonta was sighted on only eight occasions, four times on his own and four times in the company of a female and her young. Sobongo was observed on 171 days and Kamemanfu on 189 days during this same period, which clearly attests Kasonta's relative solitude.[6] Kasonta returned to the group in September 1977, and although he resumed his old position as the alpha male with the support of Kamemanfu, he disappeared in January 1978.

Kajugi's case (Uehara, Hiraiwa-Hasegawa et al. 1994; Uehara 1994; Nishida 1981)

Kajugi, a male born in 1955 (estimated), served as the alpha male of Mahale M group between 1976 and 1979. It seems that Kajugi lost the fight to maintain his rank in 1979, while researchers were absent, after which his observation frequency dropped off sharply, with only thirty-five sightings over the thirty-three months between August 1979 and April 1982. Of those, he was observed twenty times on his own, seven times in the company of a female and her young, and four times in a large party that included other males. Over the same period, Ntologi, who became the alpha male after Kajugi's withdrawal, was observed on 432 days. Kajugi is thought to have avoided Ntologi due to the fact that he lacked an ally with whom to stand against Ntologi, and on the occasions when Kajugi was sighted in parties made up of both sexes in August 1980 and April 1982, he reportedly fled as soon as Ntologi arrived. Kajugi was last sighted in April 1982, after which his whereabouts became unknown.

Photograph 8.2: Kajugi
(Photo credit: Shigeo Uehara)

Ntologi's case (Uehara 1994; Hosaka and Nishida 2002)

Ntologi, a male born in 1955 (estimated), served twice as the alpha male of Mahale M group, between 1979 and 1991 and between 1992 and 1995. He took up a solitary life in March 1991 after being attacked by an alliance made up of the second- and fourth-ranked males, and was observed on only twenty occasions over the five months from August 1991 to January 1992. Of these, he was observed nine times on his own, five times in the company of one other adult male, four times in parties made up of both sexes, and twice in the company of multiple adult and adolescent males. On the single occasion during this time that he was observed in an encounter with Kalunde, who succeeded him as the alpha male, Ntologi ran away, with Kalunde and several other males in pursuit. During his solitude, Ntologi was sighted several times wandering together with Bakali, an elderly male, and Jilba, a young male who had withdrawn as the result of a concerted group attack. Toward the end of 1991, Kalunde lost his alpha status, but since Nsaba—the male who defeated Kalunde—took up a solitary life of wandering from around the same time, Nsaba never became the alpha male, so that in early January 1992, the group was in the unusual state of not having an alpha male. Around this time, Ntologi began to rove in parties of more than ten individuals made up of both sexes, and on January 20, 1992, it was confirmed that he had been reinstated as the group's alpha male.

Fanana's case (Nishie 2017, 2019)

Fanana, a male born in 1978 (estimated), served as the alpha male of Mahale M group from 1997 to November 2003. He suddenly disappeared from view

Photograph 8.3: Ntologi
(Photo credit: Kazuhiko Hosaka)

Photograph 8.4: Fanana

after being observed on November 26, 2003, since which time he has been sighted sporadically on his own or in the company of a few other individuals. Two encounters with Alofu (born 1982), who rose to the position of alpha male after Fanana's withdrawal, were observed prior to September 2004. On the first occasion, Alofu acted as though he were subordinate to Fanana, but the second time Fanana expressed subordination to Alofu, and then other males attacked Fanana. After a prolonged stretch of an almost entirely solitary life that lasted forty-four months, Fanana was reinstated to the group in July 2007, and was still alive in 2019 as of the writing of this chapter. Over the forty-four months of Fanana's period of withdrawal, he was observed on 158 days, a frequency well below the average of 655 days for other adult males. Of these sightings, he was most frequently on his own, and including the number of times he was observed in the company of only one other animal accounts for fully half of the total. In other words, it seems that Fanana remained solitary, or close to it, during this period of his life, avoiding opportunities to join up with large parties.

Isolation as a singular point of society

Up to this point, I have outlined several cases of withdrawal and solitary life among males in the wild chimpanzee society at Mahale. Below, I examine chimpanzee society and what this process of destructuralization—from withdrawal to solitude—deviates from and nullifies, how it does so, and the kinds of structure it reveals.

That which is neutralized and reorganized with a separation

All the cases presented so far have involved an alpha male withdrawing from a group to spend a long period of time living in solitude. In the cases of Kasonta, Kajugi, and Ntologi, it was reported (sometimes speculatively) that the withdrawal was occasioned in each case by the male in question losing a challenge by the second-ranked male and others. In Fanana's case, no clear attack or exclusion by the others was observed prior to his withdrawal, so the direct cause of his withdrawal is unknown.

However, when the alpha male is driven out of his position, it is not always the case that he will withdraw from the group. In fact, even in Mahale, cases in which males that have fallen from alpha status remain with the group without withdrawing have been observed on several occasions. A wide array of options seem to be available to individuals once maintaining alpha status has proven too difficult. Which of this array they actually choose will differ depending on their individual circumstances (e.g., the ferocity of the attack; the presence or absence of allies; relationships with other members of the group, including females; seasonality in terms of food availability and congregating status; and the age, physical condition, and personality of the individual in question and of his challenger). However, when withdrawal does take place, this does not necessarily mean that all these variables will be considered, which is simply to say that the choice of behavior will not necessarily be based on an optimal solution after processing infinite information. As argued earlier, living things pick up very limited information delimited by their respective finitude, in each instance refining provisional and immediate solutions as they get on with their lives. If we accept this, then an individual who has withdrawn from the group will not have taken into account all the circumstances that led to that point or considered all the prospects that lay ahead. Conceivably, rather, he will have left the group for a trifling reason, out of a tiny fraying of routine stability—which is to say, as an immediate and particular solution.

The withdrawal of an alpha male as an immediate and particular solution could be said to be a destructuralizing 'move' that temporarily cuts off and nullifies the previous order (i.e., way of connecting). First, the alpha male's withdrawal from the group cuts off and reorganizes the range within which hierarchy functions as a rule that enables the coexistence of individuals within the group, causing fluctuations in rank. As Kuroda (2013) and Nishie (2019) point out, this is because the functioning of the hierarchy is necessarily contingent on mutual encounters in which interactions abide by (and thereby affirm) dominance relationships, which must necessarily be left hanging with individuals who have withdrawn from the group and are no longer regularly encountered. As I myself observed, as a result of the alpha male's withdrawal, 'it was, however, difficult for me to determine who the alpha male was during this extended period of time' (Nishie 2019: 133). Arguably, this was my expression of the anxiety resulting from the fact that the hierarchical order—which could only be confirmed by repeated mutual encounters—could no longer be confirmed due to withdrawal as a destructuralizing move.

In the cases of Kasonta, Kajugi, and Ntologi, it was reported (sometime speculatively) that the individuals in question withdrew to take up a solitary life after losing challenges by (coalitions of) other males. However, as in the case of Ntologi, who was reinstated as the group's alpha male after his period of solitary withdrawal, and of Kasonta, who resumed his position as alpha male by forming an alliance after his reinstatement to the group, withdrawal and solitary life are not in themselves necessarily directly linked with a decline in status. In this regard, Hosaka and Nishida (2002), reporting on Ntologi's 'expulsion (ostracism),' argue that in considering the options available to the alpha male when his dominance wavers, one strategy involves 'dropping to a lower rank in the hierarchy to remain in the group,' while another involves 'taking up a life of seclusion without admitting the upset in the ranking.' In order to remain in the group, a male will need to find a new way to establish relationships (connect) with other individuals, which in practice amounts to accepting a drop in ranking. However, because withdrawing from the group to lead a solitary life minimizes opportunities for encounters with the group's other individuals, dominance relationships that manifest only in the context of mutual encounters will necessarily be ambiguous. Accordingly, when an individual later rejoins the group, whether he will be reinstated to alpha status (as with Kasonta and Ntologi) or become, instead, a lower-ranking male (as with Fanana), or whether he will not rejoin at all (as with Kajugi), will be impossible to predict at that time, which is to say that it will depend on particular conditions. In this way, the newly formed ambiguity of each other's ranking in the hierarchy acts as an uncertain factor in the context of the chance (if there is one) of a subsequent encounter, and in some cases will be something that occasions considerable confusion when encountering or rejoining the group (as in the case of Fanana; see Nishie 2019).

In addition, the boundary (extension) of the unit group's stable membership also begins to waver in conjunction with a withdrawal from the group. Chimpanzee unit groups do not necessarily always act together. Rather, while undergoing constant fission and fusion, they maintain stable membership boundaries through occasional mutual encounters. However, when an individual withdraws from the group to begin leading a solitary life, these 'mutual encounters'—which are the only way to ensure that an individual is a member of the group—are minimized. Accordingly, the relationship between the individual who has withdrawn and the other members of the group will also undergo a qualitative shift away from the way of relating (connecting) that had existed with other members of the same unit group up to that point. In such a case, virtually (i.e., potentially), it is conceivable that the individual who has withdrawn may transfer his membership to another group at that point, as is the possibility that he will form a new group together with other individuals (resulting in a split of the group). Although it is not possible (at least in the structurally ideal sense) for males to transfer membership in chimpanzee society, when we consider that both patrilineal and matrilineal social structures are found in various phylogenetic branches, even just among primates, and that structural transformations must have occurred at some point in evolutionary history, it would be unnatural, rather, to think that deviations from the structure could not take place at all. In fact, moreover, in a few rare cases, transfers of male membership and the coexistence of males across different groups have been observed in reality (as described below). Only one case of a split of a group of wild chimpanzees has been observed in the past, in Gombe, Tanzania. However, unless we believe that the current distribution of wild chimpanzee groups has remained utterly unchanged since their appearance as a species, group splits must have taken place repeatedly in the past, and the possibility of a group splitting is conceivably embedded in its potential structure. Thus withdrawal as a singular point intrinsic to the structure reveals a structure (order) that until that point had not yet been made manifest and implies the sign of a transition to another structure.

The emergence of a new order due to prolonged withdrawal

Partly because males who have withdrawn from their groups are sighted so infrequently, observers have been able to obtain only fragmentary information as to their appearance during their withdrawal. Nevertheless, some common characteristics are evident. Looking closely at past reports, we find that males during their prolonged withdrawal are generally described as looking 'no different than usual.' For example, Toshisada Nishida (1981: 178–9), who observed Kasonta during his withdrawal, has left us with the following description:

> Surprisingly, it was Kasonta, looking hale and plump. When he grabs a piece of sugar cane, he deliberately eats it about seven or eight meters away from me. The same with the second one. Perhaps he is afraid that Sobongo might also have heard my voice and be approaching nearby. Kasonta has no scratches anywhere on his face, back, or arms. He is the picture of health itself. He hides away in the bush to eat the third piece.

In this example, even though Nishida speculates that Kasonta is endeavoring to avoid any encounter with Sobongo, the fact that Kasonta himself has not undergone any physical changes ('hale and plump,' 'no scratches anywhere,' 'health itself') seems to be acknowledged with some surprise. Kasonta had lost his alpha status and had been 'forced' into an extended period of solitary life, yet it was apparent that aside from the fact that he was endeavoring to avoid encounters with his old group, including Sobongo, he seemed to be perfectly fine.

In the same context, Nishida (1981: 179) also touches on Kajugi's experience of solitary life in the following terms:

> That Kasonta was not a lone exception was decisively shown when it was later observed that after Kajugi in M group lost his alpha status, he went on to lead a solitary life primarily centered in the southern fringes of M group's home range. Unlike Kasonta, Kajugi was still in his prime, and when the other M group males went to the south, Kajugi went further to Kansyana [an area in the northern part of M group's home range that served at the time as a feeding ground-cum-research camp], he even managed the rare feat of monopolizing several estrous females for several days.

Here, too, Kajugi is memorably described as having been able to take advantage of the absence of other males to achieve the 'rare feat' of monopolizing estrous females, despite the fact that he had been forced into a solitary life. Uehara (1994) wrote that 'the elusive Kajugi always seems to be in good health.' Based on my own observations of Fanana, I wrote that he 'looked healthy and no different than prior to his disappearance (except that he was alone) as far as I was able to observe' (Nishie 2019: 135). These descriptions of males living solitary lives hardly convey any negative impression (about not being able to be together) that might be conjured from the words 'withdrawal' or 'solitude.' Basically, we could say that aside from being solitary (not meeting up with other individuals), another common feature of these cases is that the males in question appear to be as healthy as they were prior to their withdrawal.

Other commonalities we can point out are that the individuals in question are engaged to some degree in 'collecting information' so that they can maintain their solitary lives, and that they seem to be avoiding unexpected encounters with their old groups on the basis of that information. Kasonta cautiously receives sugar cane from Nishida while listening for vocalizations from the group that Sobongo is in, while Kajugi gets wind of the movements of the group so that he can proceed to the northern part of their home range after the group leaves it empty. In the case of Ntologi, Hosaka and Nishida (2002), under the heading of 'information warfare,' summed up the solitary individual's acts of monitoring the trends of the group and selecting the appropriate timing for reinstatement (i.e., when male relationships become destabilized). Fanana, as well, was also observed to utter his own occasional long-distance vocalizations during his solitary existence, presumably to check for responses or to sound out the movements and compositions of the group from their distance vocalizations (Nishie 2019). From these descriptions, we can see that solitary males living on their own were not 'simply (somehow) separated from the group' but actually 'making a reasonable effort to sustain their withdrawal (and maintain their solitude).'

Over the longer term, a solitary life maintained in this way stabilizes into a new order (another way of connecting). In the group, a new male moves into the alpha status and relationships are resumed based on the same rankings as before. The now-isolated individual becomes accustomed to the routine of living alone while avoiding encounters with other males. In fact, we might say that it is only with this formation of the new order that it finally becomes possible for us to observe the phenomenon in question as a prolonged solitary life after withdrawal. In that moment, the withdrawal—which was seen to rend apart the previous order as a vector of force (a destructuralizing movement)—now shifts to a new order (another way of connecting) by producing a separate regularity as it stretches out over the longer term. And when the solitary life has lasted long enough that it flattens out (becomes latent), then the opportunity for the next encounter will engender chaos in the new order as a new singular point (destructuralization).

Reinstatement to the group

For an individual who has led a solitary life to be reinstated into the group, he will first need to meet up with the other members of the group in some fashion. However, once avoidance (i.e., *not* meeting) has become normalized, the longer the period of avoidance persists, the more difficult it is for a new meeting to take place with reference to the order of their previous mutual relationship. Hence forming appropriate relationships in the context of a new encounter seems to be fraught with difficulty.

Encounters between individuals who have withdrawn from the group and other members of the group have been observed only a handful of times. Even so, considering the case where Fanana fled after being approached by members of the group, prompting the others to give chase (Nishie 2019), for example, or the case in which the solitary Ntologi ran away from the scene of an encounter with Kalunde, who had succeeded him as the alpha male (Hosaka and Nishida 2002), it seems that a confusion emerges that is engendered by the destructuralizing character of the encounter after a long period of withdrawal—in other words, by *the deviation from the 'order of mutual avoidance.'* That is, although the structure of the unit group, which is (only) secured through mutual encounters, is temporarily nullified by the state of withdrawal, isolation, and mutual avoidance, once the period of withdrawal has lasted long enough for mutual avoidance to become normalized, this 'mutual avoidance' will have been newly rendered a latent structure, at which point a mutual encounter will act to destructuralize (nullify) the restructured order of the state of withdrawal and mutual avoidance.

The most detailed account of the ultimate reinstatement into the group of an individual who had previously withdrawn is that of Ntologi. Ntologi achieved reinstatement in the group as the alpha male almost as if he had gauged the moment when relationships between males in the group became unstable, casting the group into the unusual situation of being without an alpha male

(Hosaka and Nishida 2002). With the reinstatements of Kasonta and Fanana to their respective groups, as well, although they share with Ntologi's case the fact that they were observed to return to their groups 'one day out of the blue,' it took several months for Kasonta to then be reinstated as the alpha male, while Fanana was never reinstated to alpha status at all. In addition, although the details are unclear, Fanana's reinstatement to the group coincided with the downfall of the alpha male, Alofu, and in this regard may be said to have followed a process very similar to the reinstatement of Ntologi.

Thus, with regard to their ultimate reinstatement to the group, it has been observed that these males return to their respective groups relatively smoothly and unobtrusively. While this may be thought to some degree to reflect the limitations of observation (given the multitude of times and places that cannot be observed in the wild), it may also indicate that the power of restructuring/reordering to resolve the confusion that accompanies the reinstatement is acting particularly strongly. However, although only one example, we also have a report of 'isolation accompanying the disappearance of a group,' in which such reinstatement to the group was simply impossible; finally, I would like to consider this case.

The surviving male after the disappearance of the group: the case of Limongo (Uehara, Nishida et al. 1994; Uehara 1994, 2003)

Having looked over the features of the process of withdrawing from a group to lead a solitary life and then being subsequently reinstated, as an even more extreme example let us now consider the case of a male who was left on his own as a result of the disappearance of his group. Although this case should be characterized not as 'an individual who became isolated after withdrawing from the group' but as 'an individual who became isolated after he was left behind,' the fact that it is an

Photograph 8.5: Limongo (Photo credit: Shigeo Uehara)

extremely rare case means that it is still a particularly important case for looking at the range of variation within chimpanzee societies.

Limongo is a male that was born into Mahale K group in 1972. The number of adult males in the K group, in which Limongo was born and raised, numbered six in 1965, and decreased each year thereafter until 1979, when Kamemanfu became the last adult male in the group. At this point, the female members of K group began transferring in droves to the neighboring M group, after which Kamemanfu disappeared in 1982. At the beginning of 1983 only seven individuals remained who could be said to be members of K group, including Limongo (who, at the age of ten, was a juvenile on the verge of adolescence) and several females who were already in contact with M group. By 1987, the six individuals other than Limongo had either left to join M group or disappeared, so that Limongo, by now fifteen years old (an adolescent on the cusp of adulthood), was left as the sole remaining member.

Limongo was sighted nine times over the sixty-four months between September 1987 and December 1992. The sightings all took place inside what had previously been K group's home range. He was sighted five times on his own and four times in the company of another individual (aside from the one time when the identity of his companion could not be confirmed, these were all individuals who had once belonged to K group). Of these, in an observation from 1992, a male named Masudi (who was born into K group in 1977 and who transferred to M group, together with his mother, in 1983) was sighted feeding with Limongo in a tree. Typically, male chimpanzees exhibit very hostile behavior toward males from other groups, so this peaceful encounter between two males from different groups suggests that it was possible for Limongo and Masudi to interact as 'old acquaintances,' so to speak, even in 1992, nine years after Masudi's migration.

Limongo was sighted sporadically thereafter, including a solitary sighting in 1994 and a sighting of an individual that was thought to be Limongo in the company of two females in 1998 (Hosaka 2015; Sakamaki and Nakamura 2015). In any case, it is thought that Limongo likely lived a mostly solitary life for at least ten years after the rest of the K group males disappeared, and for at least five years after the last of the females left, leaving him completely alone. In a note accompanying the only photograph ever taken of the mature Limongo, Shigeo Uehara (2003), who sighted Limongo with Masudi in 1992, wrote, 'For my own part, I would not be able to stand such a life of solitude.'

This example concerned an extremely rare case in which a male who, with the disappearance of his group, was left behind in what had been until that point the group's home range, but who nevertheless continued to live an almost solitary life for at least (and likely more than) ten years, growing from adolescence into adulthood. Moreover, this case shows that a male who was transferred to a neighboring group in his childhood (Masudi) and the male who was left behind (Limongo) could still coexist in peace even about a decade after the former's transfer. That is, while it is generally held that neighboring groups of chimpanzees are hostile to each other, and that this tendency is particularly intense between fellow males, even if they had been 'old acquaintances' as juveniles, it was still possible for males to enjoy peaceful relations across group boundaries even ten years later. This means that the stability of chimpanzee unit group boundaries and the stability of the affiliation of individual members to their respective unit groups cannot be captured with a structural/ideal schema such as one that demarcates each individual association with a line; rather, they should be conceived of as the range within which flexibly fluctuating coexistence and communication are feasible. In the coexistence scenario presented by this case, at least, conceiving of Masudi and Limongo as exclusive members of different groups would seem to broadly disregard actual variation in an example of a rather forced structural explanation (i.e., the general proposition that each individual belongs to a patrilineal unit group whose individuals are mutually exclusive).

As we see in this example, the singularly long solitary life of a male and relations between fellow males across group boundaries amount to a clear deviation from the typical explanation of chimpanzee social structure (i.e., patrilineal multi-male multi-female groups that are strongly hostile toward other groups and in which males have a high degree of cohesiveness). First, a point in common with the cases we saw in the previous sections, it is indeed possible for a chimpanzee male to live a virtually solitary existence for an extremely long period of time, and this is a deviation from the structural explanation that holds that congregating is itself the basic mode of existence for chimpanzee males. In addition, the fact that it was possible for fellow males to enjoy peaceful relations in this way across group boundaries—even in the context of chimpanzee society, which is normally regarded as being characterized by hostility between patrilineal groups—and the fact that (although Masudi and Limongo had been born into the same group) Masudi (admittedly as an infant) had transferred between groups, together with his mother, are in themselves deviations from the structure of the patrilineal unit group. So the fact that Masudi, as a mature adult ten years after his transfer, was spending time peacefully with Limongo, who was still an individual from a neighboring group, could be said to be a deviation from the structure—from the idea—of mutually hostile patrilineal unit groups.

It should be noted here that this kind of deviance (destructuralization) indicates not the absence of structure but, rather, its potentiality. The reason that structural explanations in terms of mutually hostile patrilineal unit groups generally seem to work is that patterns like this are indeed realized in actual practice over and over again, and are thus observable to us. However, the mutually hostile patrilineal unit group is itself a potential and ideal structure.

In the context of chimpanzee society, which is constantly undergoing fission and fusion—so that it is, first of all, never the case that the entire group comes together at the same time—we are never able to observe even a unit group in its entirety. In this sense, even the ostensible substance of the unit group itself is actually a potential and ideal structure. What we actually see are only the vectors of congregation and dispersal that accompany the movements of individuals at any given moment, and the phenomena of the unit group—and withdrawal and isolation therefrom—can only be traced retrospectively as trajectories that can be obtained once those selfsame vectors are integrated into time and space.

Destructuralization as the inception of a transition to other potentialities/structures

Itani's theory of the evolution of social structure drew a schema in which each organism's *specia* had its own unique structure, and in which this social structure was itself subject to change (i.e., evolution). In this chapter, I have organized and extended Itani's theory of social structure by distinguishing between ideal structures and actually existing ways of congregating, as well as by repositioning 'complement-structure' as a static mode that bridges structures and 'destructuralization' as a dynamic mode (vector) that deviates from and nullifies structure. For individuals, structure exists a priori in the sense that it is potential and ideal. It is not absolute, however, in the sense that it can change (evolve). Structure supports the society of living things as a horizon (potentiality), but deviation (destructuralization) punctures each and every aspect of daily life, so that living things inhabit situations in which other ways of connecting (shifts to a different order) can take place at any time. This seems to be consistent with Itani's (1993: 200) statement that 'In some cases, the structure itself can destructuralize, which may in turn give rise to a new structure.'

In this chapter, I have outlined the process of the alpha males' withdrawal from groups of wild chimpanzees into prolonged periods of solitary life, possibly followed by their reintegration to the group at some later point, as a process of destructuralization at a singular point intrinsic to the structure with a subsequent reordering. For an individual to be the alpha male, it is necessary that he repeatedly encounters other individuals and that he accrues a track record of dealings in those encounters that is in keeping with his rank. Given these requirements, when the alpha male leaves the group to lead a (near) solitary life, his 'social dominance' will have no practical meaning. In other words, with the withdrawal of the alpha male from the group, the boundaries of the order of social interaction, whose iteration in keeping with the hierarchy had until that point been a matter of course, will be temporarily nullified. Furthermore, as the solitary life of the individual is prolonged—that is, as a result of a certain 'normalization of absence'—the social boundaries of the hierarchy and the members of the unit group will be reorganized. Over time, the longer that solitary life lasts, the more that state of affairs will form a new order, so that a subsequent encounter will bring about a deviation of this new order of mutual avoidance as a singular point of destructuralization.

I have already touched on how the daily life of chimpanzees is characterized by repeated encounters and withdrawals (fission and fusion). I have also argued how the 'isolation that accompanies prolonged withdrawal' as it is described in this chapter can be positioned as something that appears as an extension of the temporal and spatial scales of this everyday fission and fusion. This suggests that what we are actually able to observe in reality are always certain 'movements,' certain subtle 'vectors,' and also that it is only when these vectors are 'integrated' in time and space that the 'pattern' of everyday fission and fusion (e.g., seasonal fluctuations in the size of nomadic parties) and the 'prolonged withdrawal and solitary life' described in this chapter can be conceived of as certain trajectories, or individualized events. However, isolation as a singular point—that is, a point of heterogeneity that is discovered as a deviation from regularity at the moment of withdrawal—cannot be differentiated from its surrounding context (i.e., from the function of which it is a part), which means that a trajectory that includes this point cannot be integrated. This reveals the impossibility of perceiving the moment of withdrawal as an individualized event, and suggests that conceiving of withdrawal as a moment of isolation will require an ensuing 'flow of regular time' and retrospective awareness (integration) commensurate with the thickness of that time. This is conceivably a problem not only for human observers like us, but also for the chimpanzees themselves. Even for the members of the group, the withdrawal of an individual will be observable only as an accumulation of absence. Even for the individual who has withdrawn, this withdrawal from the group and prolonged solitary life can only be achieved through a sustained avoidance. This is because to meet again after only a few days would be no different from ordinary fission and fusion. In this sense, when a certain individual withdraws, there is no way—either for an observer or for the parties concerned—to determine whether the withdrawal is part of ordinary fission and fusion or the first step toward a period of prolonged isolation. It is only when the absence becomes prolonged and is conceived retrospectively as a certain event that the initial withdrawal is imbued with meaning as the catalyst for a prolonged isolation. If this is correct, then dispersal as part of ordinary fission and fusion is fraught with the possibility of turning into prolonged withdrawal at any point. In fact, with most disappearances of individuals (including presumed deaths and out-migrations), we have no way of witnessing the moment of withdrawal, and in truth we barely know even retrospectively that a given sighting was the final observation (and thus the moment of withdrawal).

In the context of chimpanzee society, where this sort of destructuralizing movement as a singular point is considered intrinsic to the everyday process of fission and fusion, the behavior of individuals in accordance with rules that support the structure realizes fission and fusion as an ordinary way of congregating, thereby bringing stability to the order of daily life. At the same time, it also has the potential to nullify and reorganize these rules simply by prolonging this 'ordinary' fission and fusion. Conversely, the ever-present potential for destructuralization to manifest also signifies that the structure that provides the basis for the destructuralization and the action of the rules that support the structure must be repeatedly made apparent and confirmed. This seems to render comprehensible, for example, why the greeting behavior from subordinate to dominant individual, known as a 'pant-grunt'—and which is thought to function in the context of chimpanzees as an affirmation of relative dominance—must be repeated so frequently, persistently, and in some cases more exaggeratedly than would seem to be necessary (Nishie 2017). While it is known that pant-grunts are most frequently addressed to alpha males, this makes no sense from the general explanation that pant-grunts function to affirm relative dominance. This is because alpha males are normally conspicuously dominant relative to other individuals, and thus there would no longer seem to be any need to affirm their relative dominance. So why, then, are pant-grunts repeated so frequently and persistently to the alpha male? The reason seems to be that, in the context of a chimpanzee society that allows for fission and fusion, the repeated mutual confirmation as potentiality of the structure and of the rules that support it will mitigate the risk that their mutual avoidance will aggravate uncertainty in their relationships by nullifying the rule of dominance hierarchy and the stable boundaries of group membership prescribed by the structure.

If this kind of inference holds true, this would suggest that chimpanzees are capable of referring to (or at least attempting to refer to) ideal concepts. This means that chimpanzees would no longer be limited to recognizing and using constantly repeated patterns, but would also be able to recognize and use structures and rules as a certain kind of representation underlying those patterns. We cannot know whether the social process of daily fission and fusion itself is what makes it possible to grasp potential (i.e., that which is not here-and-now) as an ideal, or vice versa, or whether they exert some mutual influence on each other, but this will serve to move society—as well as the creatures that inhabit it even as they build it—away from the world of the visible (the manifest) toward the world of the 'not here-and-now' (the potential). For our own species, *Homo sapiens*, who had freely manipulated language and representations and had extended the scope of society to 'distant groups' beyond those with whom we had daily interactions (Terashima 2013), this may have been a tiny but decisive first step in the direction of our own sociality.

In this way, my venture in this chapter—to conceptualize destructuralization as a singular point intrinsic to society—is an 'ideal' proposal to the problems of 'conversion from one structure to another' and 'the process of transitioning from one structure to another,' which Itani himself was conscious of as persistent issues (Itani 1987c: 307). In future, it will be necessary to continue working to confirm the validity of this proposal with the further observation of 'actually existing ways of congregating,' to provide detailed description to the movement of destructuralization as deviation (as a singular point), and to deepen our insights into structure as a latent and underlying idea.

Notes

1. Elsewhere (Nishie 2017), I have argued that the formation and use of such spontaneous order constitutes the evolutionary historical basis of human social institutions.
2. Elsewhere (Nishie 2019), I discussed the 'bare other' as the ultimate factor causing unexpected disturbances and fluctuations for ordered systems that would otherwise appear to be stable and available.
3. Some of the concepts introduced in this chapter, such as the singular point or ideal potential structures and the realization thereof, are borrowed from the philosophy of Gilles Deleuze. However, as the discussion in this chapter is neither intended as a discussion of Deleuze nor limited to the application of Deleuzian philosophy to reality, I shall refrain here from referring to them individually or discussing their minor differences.
4. According to Imanishi (in Adachi 2017: 278), the environment in which an organism lives and its activities are indivisible, and the environment and the organism/actor as a whole constitute a lifeform. Organisms that share the same lifeform constitute a species, which forms the *specia*. Different *specias* have different lifeforms—that is, they live in different fields and carry out different activities—hence habitat segregation is observed there. The occurrence of niche separation between different species is self-evident in Imanishi's argument on the *specia*.
5. To be more precise, in line with the discussion in this chapter, this means that groups and solitary individuals alike are better described not as 'structured processes' (Uchibori 2013) so much as 'things in which a structure has been realized.'
6. When seeking wild chimpanzees, we often rely first of all on the long-distance vocalizations that they make. However, because chimpanzees tend to make long-distance vocalizations more frequently when they form large nomadic parties, this strategy leads to more frequent encounters with large parties, and fewer with solitary individuals or those roving in small parties. Therefore, the frequency of encounters with a particular individual (i.e., the number of days on which an individual is sighted during a given period) generally correlates with the size of the party that includes that individual (in terms of the number of constituent individuals). In other words, individuals who are encountered less frequently are more likely to be in relatively small nomadic parties (i.e., roving in small parties, if not alone).

References

Adachi, K. (2013) 'The sociology of anti-structure: Toward a climax of groups'. In K. Kawai (ed.), *Groups: The Evolution of Human Sociality*. Kyoto University Press and Trans Pacific Press, 21–41.

Adachi, K. (2017) 'Living one's role under institution: Ecological niches and animal societies'. In K. Kawai (ed.), *Institutions: The Evolution of Human Sociality*. Kyoto University Press and Trans Pacific Press, 265–308.

Goodall, J. (1986) *The Chimpanzees of Gombe: Patterns of Behavior*. Belknap Press.

Hayek, F. A. (1967a) 'The result of human action but not of human design'. In F. A. Hayek, *Studies in Philosophy, Politics, and Economics*. Routledge and Kegan Paul Ltd, 96–105.

Hayek, F. A. (1967b) 'Dr. Bernard Mandeville'. *Proceedings of the British Academy*, 52: 125–141.

Hayek, F. A. (1969) 'The primacy of the abstract'. In A. Koestler and J. R. Smythies (eds), *Beyond Reductionism: New Perspectives in the Life Sciences*. Hutchinson, 309–333.

Hayek, F. A. (1973) *Law, Legislation and Liberty, Volume 1: Rules and Order*. University of Chicago Press.

Hosaka, K. (2015) 'Who's who'. In M. Nakamura, K. Hosaka, N. Itoh and K. Zamma (eds), *Mahale Chimpanzees: 50 Years of Research*. Cambridge University Press, 48–67.

Hosaka, K. and T. Nishida (2002) 'Osutorashizumu: Arufa o, murahachibu kara no fukken' (Alpha males and reinstatement from ostracism). In T. Nishida, K. Kawanaka and S. Uehara (eds), *Mahare no chinpanjī: 'Pansuroporojī' no 37-nen* (Mahale chimpanzees: 37 years of 'panthropology'). Kyoto University Press, 439–471.

Itani, J. (1987a) 'Shakai kōzō o tsukuru kōdō' (Actions that create social structure). In *Reichō-rui shakai no shinka* (The evolution of primate society). Heibonsha, 223–245.

Itani, J. (1987b) *Reichō-rui no shakai kōzō* (Primate social structure). Heibonsha, 35–208.

Itani, J. (1987c) *Reichō-rui shakai kōzō no shinka* (The evolution of primate social structure). Heibonsha, 297–325.

Itani, J. (1993) 'Shakai no kōzō to hi kōzō' (Social structure and anti-structure). In *Shizen ga hohoemu toki* (When nature smiles). Heibonsha, 187–217.

Itoh, N. and T. Nishida (2007) 'Chimpanzee grouping patterns and food availability in Mahale Mountains National Park, Tanzania'. *Primates*, 48(2): 87–96, doi.org/10.1007/s10329-006-0031-0

Kuroda, S. (2013) 'Collective excitement and primitive war: What is the equality principle?' In K. Kawai (ed.), *Groups: The Evolution of Human Sociality*. Kyoto University Press and Trans Pacific Press, 273–292.

Nishida, T. (1968) 'The social group of wild chimpanzees in the Mahali Mountains'. *Primates*, 9(3): 167–224.

Nishida, T. (1981) *Yasei chinpanjī kansatsu-ki* (Observing wild chimpanzees: Field notes). Chuokoron-Shinsha.

Nishida, T. (1983) 'Alpha status and agonistic alliance in wild chimpanzees (*Pan troglodytes schweinfurthii*)'. *Primates*, 24(3): 318–336.

Nishie, H. (2017) 'Who is the alpha male? The institutionality of dominance rank in chimpanzee society'. *Institutions: The Evolution of Human Sociality*. Kyoto University Press and Trans Pacific Press, 121–140.

Nishie, H. (2019) 'Who is the alpha male? The appearance of the "other" in chimpanzee society'. *Others: The Evolution of Human Sociality*. Kyoto University Press and Trans Pacific Press, 127–148.

Sakamaki, T. and M. Nakamura (2015) 'Intergroup relationships'. In M. Nakamura, K. Hosaka, N. Itoh and K. Zamma (eds), *Mahale Chimpanzees: 50 Years of Research*. Cambridge University Press, 128–140.

Terashima, H. (2013) 'From the "here and now group" to the "distant group": Hunter-gatherer bands'. In K. Kawai (ed), *Groups: The Evolution of Human Sociality*. Kyoto University Press and Trans Pacific Press, 199–218.

Uchibori, M. (2013) 'Assembly of solitary being: Between solitude and "invisible" groups'. In K. Kawai (ed.), *Groups: The Evolution of Human Sociality*. Kyoto University Press and Trans Pacific Press, 43–57.

Uehara, S. (1994) 'Lone male chimpanzees in the Mahale Mountains National Park, Tanzania'. *Primate Research*, 10: 281–288. [In Japanese with English summary]

Uehara, S. (2003) 'Mahare no chinpu (n?) Shōkai dai 2-kai: Rimongo' (Introduction to Mahale chimps: #2: Limongo). *Mahare chinbun* (*Mahare yasei dōbutsu hogo kyōkai nyūzuretā*) (Curious news from Mahale (Newsletter of the Mahale Wildlife Conservation Society)), 2: 6.

Uehara, S., M. Hiraiwa-Hasegawa, K. Hosaka and M. Hamai (1994) 'The fate of defeated alpha male chimpanzees in relation to their social networks'. *Primates*, 35(1): 49–55.

Uehara, S., T. Nishida, H. Takasaki, K. Norikoshi, T. Tsukahara, R. Nyundo and M. Hamai (1994) 'A lone male chimpanzee in the wild: The survivor of a disintegrated unit-group'. *Primates*, 35(3): 275–281.

Wrangham, R. W. (1987) 'Evolution of social structure'. In B. B. Smuts, R. M. Seyfarth and R. Wrangham (eds), *Primate Societies*. University of Chicago Press, 282–298.

CHAPTER 9

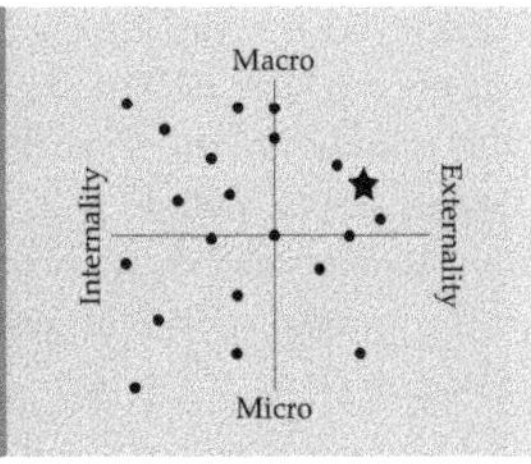

Rethinking Pastoralist Nomadism:

Dealing with the 'Extreme' among the Dodoth of East Africa

Kaori KAWAI

Keywords: East African pastoralists, animal camp migration, raiding, joint practice, memory and foresight

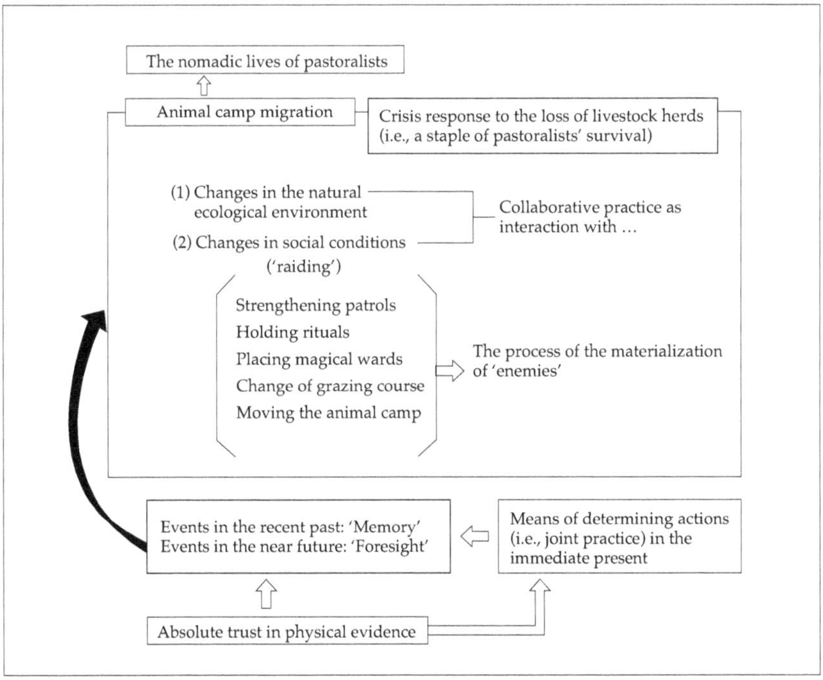

In effect, what makes up the nomadism that constitutes the basic way of life of the Dodoth, a pastoral people of Uganda, is the migration of their animal camps. As a means of dealing with critical situations, such as the loss of the livestock herds that are a staple of their own existence, people will, in various ways, perceive changes in the natural ecological environment and changes in social conditions—especially the danger of raids by other ethnic groups—and will develop a variety of joint practices. The ultimate resort in this regard is camp migration. This involves 'memory' as a grasp of events in the recent past and 'foresight' as a grasp of events in the near future, both of which cooperate to determine people's actions/activities (i.e., joint practices) in the immediate present. Both memory and foresight are grounded in an absolute trust in physical evidence.

Introduction: nomadic life and 'extreme' situations

This chapter focuses on nomadism, which represents the basic way of life for pastoralists living in arid regions with natural ecological conditions that can be extremely harsh. In doing so, I propose to think about the sort of things that constitute 'extremes' for human existence. In particular, I discuss the subsistence activity of animal camp migration as a practice by which nomadic life is effectively generated. As factors that give rise to this practice, as well as changes in the natural ecological environment, I also point to threats from 'others' (e.g., neighboring ethnic groups) as changes in the social context, and clarify the relationship of such threats to the practice of camp migration.

The arid land, with its poor vegetation and a scarcity of water, refuses human existence. As a form of subsistence, pastoralism is a pursuit that renders the harsh natural environment of arid land, which is not directly usable by humans, available through the medium of domestic animals. Pastoralists have developed nomadism as a way of life in which livestock are frequently driven over extensive territories in accordance with the temporal and spatial distribution of grasses and water, so as to make efficient use of the delicate natural environment, which is just barely enough to sustain human existence. However, it is widely known that nomadism among actual pastoralists is not necessarily always brought about solely by natural ecological conditions as an adaptation to the natural ecological environment. That is, the involvement of various religious, social, and political factors has also been pointed out, such as, for example, the holding of rituals, conflict in personal relationships, and 'fighting' against other groups.

Photograph 9.1: Dodothland (pasturage) in the dry season

In East Africa, among the Dodoth pastoralists of Uganda, existence is based on choosing activities to cope with changes in the natural ecological environment of the arid land as their physical/material environmental world, and consists of continuing to cope with ongoing changes in social situations engendered by the behavior that people elaborate on that basis.

An even more important change in the social situation to be 'coped' with is the relationship that the Dodoth maintain with neighboring pastoralist groups. Relationships between pastoralist ethnic groups in the area inhabited by the Dodoth have repeatedly alternated as hostile and non-hostile over time (e.g., Kawai 2004, 2019). When hostilities are in abeyance, amicable relationships can sometimes be seen, as in the case of joint grazing activities developed across ethnic boundaries. In times of hostility, however, frequent raiding takes place between ethnic groups, in which hostile groups are attacked and their herds plundered.

Falling victim to raids can entail the loss of the livestock herds that serve as the food source that literally supports the pastoralists' existence; in the sense that this threatens their (biological) survival, it can be said to be an event that plunges people into an extreme situation. The Dodoth are vigilant against raiding and take various actions to forestall it; by following this process, in this chapter I examine the ways in which extremes are perceived, understood, and inhabited among the Dodoth. Furthermore, by placing this situation in a more general context, I also examine the origins of extremes that are characteristic of humans by comparing them with those pertaining to other primates.

Nomadic life: distributed living and animal camp migration

The Dodoth are cattle pastoralists who inhabit the triple frontier of northeastern Uganda, Kenya, and South Sudan (Photograph 9.1). They raise four species of domestic animals; namely, cattle, goats, sheep, and donkeys. The nomadic life of the Dodoth is based on frequent movement with their livestock over a wide area. This is a way of life that typically enjoys broad recognition among East African pastoralists. The basic unit of production and consumption in Dodoth society is the patrilineal extended family, with family members and their livestock living in dispersed and autonomous *ere* (settlements) and *awi* (animal camps) in which they conduct their day-to-day affairs. Settlements are inhabited by the head of the family and his wives, elderly relatives, unmarried daughters, and young children, while youths and younger unmarried men stay in the animal camps, where they occupy themselves with grazing the livestock. In the basic form of livestock husbandry practiced by the Dodoth, up to a dozen or so domestic animals (lactating females and their young) are left in the settlements to provide milk (and blood) as a staple food for the inhabitants, and all other animals are kept at the animal camp for grazing.

The *ere* comprise robustly built structures and are semi-sedentary. They are moved only once every several years, and only over short distances of no more than a few kilometers at most. The *awi*, on the other hand, where life is essentially camp-based, generally do not involve the erection of buildings for human habitation. The camps move frequently, at intervals ranging from a few days to one to two months. For example, in one camp, over the nineteen months from August 1998 to February 2000, the camp was moved seventeen times (i.e., about once every 1.1 months), while over the nineteen months from February 2002 to July 2003, the camp was moved twenty-three times (about once every 0.8 months). Thus, in general, the camp seems to move every three to four weeks on average (Kawai 2002, 2006). The distances covered by these migrations range from short jaunts of only a few kilometers to longer stretches of more than 100 kilometers, which far surpasses the distances that settlements are moved. Accordingly, it seems safe to say that nomadic life for the Dodoth is shaped by the migration of their animal camps. In the following, I consider this animal camp migration in more concrete detail.

Impetus for camp migration

A case of animal camp migration

First, so that we can grasp a sense of the migration of Dodoth animal camps, I will introduce a case study of camp migrations that took place over a one-month period between June 2 and July 2, 2003. Table 9.1 shows an overview of these migrations. The camp was moved six times during this period, hence this

Migration date	Destination	Official 'reason' given for migration	Length of stay (days)
(1) June 2	Moru-angapion	Instructed by the *emuron*	1
(2) June 3	Moru-ese	To consume *engelei* (salty soil)	4
(3) June 7	Moru-angapion	Finished above	7
(4) June 14	Lokwakaramoe	Revelation from reading intestines	6
(5) June 20	Moru-angapion	Inhospitable topography at the previous camp site	12
(6) July 2	Loita	Revelation from reading intestines	–

Table 9.1: Chronology of the migration of an animal camp (June 2 to July 2, 2003)

case involves relatively frequent camp migration. I shall explain the sequence of events by which this process unfolded.

(1) All the livestock were in the village until June 1. Then, on June 2, all the animals except one lactating nanny goat and a few of her kids were moved to the animal camp. The site chosen for the camp was a steep slope facing a deep stream in Moru-angapion, a region characterized by sheer mountain ridges. This move was carried out in accordance with instructions given by the *emuron* (Dodoth religious specialists noted for their powers of divination and prediction).

(2) On the following day, June 3, the animal camp was moved to an area called Moru-ese. The Dodoth regard salinity as essential for the growth of their livestock. Grasses called *elet* (e.g., *Dactyloctenium aegyptium* [Poaceae], *Combretum aculeatum* [Combretaceae], and *Tribulus cistoides* [Zygophyllaceae]) and soil known as *engelei* are regarded as containing salt. Among the streams of Moru-ese, there is a place where the banks are deeply gouged to expose the *engelei.*

(3) The camp stayed in Moru-ese for four days, during which the livestock were fed *engelei* every day. Once the animals were deemed to have been given enough salt, the herds were returned to the camp site at Moru-angapion on June 7.

(4) A week later, on June 14, the camp moved to a site prepared on a steep slope in a mountainous area called Lokwakaramoe. The previous day (June 13), on the instructions of the *emuron,* a cow had been slaughtered as an *ajulot a arakala* (blessed sacrificial animal) at the camp and the move was carried out in accordance with the revelation obtained from a divination performed with its intestines.

(5) Six days later, on June 20, the herd returned again to the old camp site at Moru-angapion. This was due to the fact that the camp at Lokwakaramoe had been pitched on a particularly steep mountain ridge where the topography had been extremely poor.

(6) Twelve days later, on July 2, the camp moved to an area known as Loita, a move that had been discussed at meetings held every morning at the camp for several days previously. The move had become a pressing issue as the camp's inhabitants had become increasingly aware of deteriorating relations with the neighboring Turkana people. However, what finally decided the matter was the revelation obtained from

reading the intestines of three goats sacrificed on the day preceding the move. At this time, I overheard discussion along the lines of 'we cannot move until we've seen the entrails' (i.e., the camp migration cannot take place until the inhabitants learned the results of reading the intestines).

As described above, the de facto base of the animal camp during this period was at Moru-angapion, before eventually moving to Loita after moving back and forth to the two sites at Moru-ese and Lokwakaramoe. The direct impetus for this migration process, however, aside from the ecological reason that ingestion of *engelei* is directly involved in the husbandry of domestic animals, involved revelations such as those in the words of the *emuron* or the results of reading intestines. In light of these concrete examples, I would like to examine the manifestation of animal camp migration as an interaction of two factors—namely, changes in the natural ecological environment and changes in social conditions—in terms of the kinds of environmental and situational changes that provide the impetus for the migration of Dodoth animal camps.

Factor I: changes in the natural ecological environment

Although it did not feature in the case study above, among the Dodoth, much like with other East African pastoralists, camp migration is sometimes undertaken as a response (adaptation) to changes in the spatiotemporal arrangement of grass and water. Camp migration can take place for various reasons. In their own words, it might be that 'the distance between the camp site and the grazing pasture (foraging sites) has increased,' 'the distance from the camp site to the watering hold has increased,' 'the grasses around the camp have been exhausted,' or that 'the watering hole has dried up.' In each case, these migrations are occasioned by changes in the natural ecological environment, such as a decrease in the availability of grass and water. In fact, even when grass and water remain plentiful (or seem so to me, at least), in large camps with upwards of 100 animal fences/enclosures (i.e., with more than 100 livestock herds, numbering anywhere from 1000 to 10,000 domestic animals), the fact that so many herds are grazing in one place can itself cause camps to move or split up. Moreover, the Dodoth consider salinity to be essential for the growth of their domestic livestock, and as shown in the case study above, they will sometimes move their camps in order to allow their animals to forage on grass and soil containing salt.

The natural ecological environment in Dodothland consists of arid and semi-arid areas subject to erratic, unstable, unpredictable, and capricious climatic conditions, as well as frequent droughts. The region also occasionally suffers from 'great droughts' during which there is little if any rainfall throughout an entire year, and sometimes for periods of longer than a year. However, this does not necessarily mean that rainfall is always beneficial. Heavy rains and flooding are also an issue. This phenomenon has become more prominent since 2000, and may reflect the impact of global warming or global climate change or phenomena such as the El Niño climate pattern. Excessive rainfall is not desirable in dryland ecosystems, since livestock can be carried away or can drown in flash floods, and pasture land can end up being flooded for long periods of time with muddy water that will hinder the growth of grasses.[1] In this sense, arid areas are also vulnerable environments.

The natural ecological environment could be said to be a fundamental condition for the husbandry of domestic livestock, as well as constituting the basis of nomadic life for the Dodoth. Failure to adapt to the natural ecological environment (which amounts to securing water and grass) will result in the deaths of individual livestock, which by extension imperils the survival of humans. The Dodoth are well aware of this fact, and they accurately take note of signs of deterioration in the natural ecological environment, which they recognize as situations that might jeopardize the maintenance of their livestock herds. Urgent discussions about possible remedies are held among the members of the camp and, as a last resort, the decision will be made to move the camp to a more appropriate site as quickly as possible.

To start with, in regard to the establishment of a means of subsistence, the positioning of pastoralism as a dryland adaptation by which humans have come to inhabit land in otherwise unusable arid areas through the medium of domestic animals is universal. However, the image of nomadic pastoralists seeking grass and water is overly simplistic, and the perception and interpretation that pastoralism as a means of subsistence can only be a direct confrontation with the natural ecological environment is not realistic, or at least not in light of the specifics of Dodoth camp migration. This is because the Dodoth migrate even when there is no need to do so in view of natural ecological environmental conditions, sometimes intentionally moving to difficult sites even though conditions at the destination are inhospitable or overly severe. In the case study discussed above, three of the six migrations involved movements of the camp occasioned by the instructions of the *emuron* or the results obtained by divination in the form of reading intestines (Table 9.1). Even so, if the destination is a difficult site in terms of its natural ecological environment, camp life will not last long there. In terms of the case study, we can point to the example of how, as in migration (5), the camp was set up at an inferior site in order to avoid an enemy attack indicated as the result of reading intestines in (4). Because the terrain was extremely perilous, with cattle almost falling as they navigated a steep traverse along the route to and from the daily grazing site, the camp was abandoned in favor of another location after only six days—a reminder that success remained contingent on adaptation to the natural ecological environment. Despite this, in other words, it can arise that the Dodoth will willingly move their camps even when this means that they will immediately have to do so again. I would like to look at this situation next.

First, however, I would like to say a few words about extremes in the context of natural ecological environments. Fluctuations in the natural ecological environment undoubtedly have an impact on maintaining domestic animal herds. As noted above, the Dodoth are very aware of this fact. However, I do not personally have any field data about extreme conditions with regard to the natural ecological environment inhabited by the Dodoth. For a concrete example of a crisis threatening the very survival of pastoralists living in the arid expanses of East Africa, we need to turn to Jun'ichirō Itani's (1982) magisterial ethnography, *The Great Drought: A Turkana Diary*. The Turkana are a neighboring group of Kenyan pastoralists inhabiting the region east of the Dodothland. Itani's book is an extremely valuable record of the Turkana at the time of a once-in-a-century drought that occurred in 1980. While this chapter does not afford the space for a comprehensive recapitulation of the contents, I would like to excerpt a few passages (emphasis added).

> In contrast to most African famines, which have by and large been fraught with factors associated with anthropogenic disasters such as political turmoil and war, few if any such factors could be ascertained in the context of this famine, which stretched from northwest Kenya to Uganda. True, after the fall of the Amin regime in Uganda, scattered groups of demobbed soldiers repeatedly turned to looting, and their weapons did fall into the hands of the civilian population, thereby exacerbating the intertribal practice of raiding each others' livestock. Even so, this famine was basically a natural disaster. It revealed *the bare margins* of human existence in the arid zones of Africa, and truly presented a challenge for ecological anthropology (Itani 1982: 10–11).
>
> In 1978, I saw Turkana life stripped down of all excess and depicted in a simple geometric figure consisting solely of the sun, the earth, and livestock (Itani 1980). This time, however, that figure had been thrown completely off kilter. The sun shone mercilessly on the surface of a depleted earth, which was littered with the white bones of dead livestock as the Turkana struggled with starvation. Life among the Turkana, which had once appeared to me as a kind of *extreme* beauty, was clearly out of balance...In the face of this famine, I was reminded anew that Turkanaland represented *an extreme environment* for human life. This form of subsistence, in which the patrilineal extended family is dispersed to herd animals according to the respective ecological needs of five types of domestic animal was also...*a desperate adaptation* demanded of humans by this environment...This natural calamity led directly to *the collapse of their way of life*, and their culture was ill-equipped to resist it. (Itani 1982: 11)

> When conditions deteriorated, their only option was to move their herds to a better place...But...even so, they soon had no recourse but to eat away at their capital [i.e., individual heads of livestock]. (Itani 1982: 210)

East African pastoralists have routinely come up against unstable, capricious, and unpredictable weather conditions. Fluctuations in the natural ecological environment are a matter of course for the local inhabitants, who have developed nomadism as a way of life that entails the dispersal and frequent movement of people and domestic animals across a wide area. Nevertheless, the fact that people's survival in the face of the great drought was determined only by the size of their herds is quite grave. The Turkana reputedly adopted a consistently fatalistic attitude toward the disastrous situation they faced in the great drought, holding the view that 'all is the doing of God (*Akuj*).' They did nothing, and could do nothing, in the face of the great drought. Itani writes that 'They were completely powerless'—the Turkana's culture, institutions, and social system could hold no power against that great drought.[2] This situation was most likely the same in the Dodoth territory, given that the Turkana and Dodoth are closely related both linguistically and historically, and share mutually similar social institutions and cultural observances.

The factor of changing social conditions—the appearance of 'enemies'

The nomadic life of the Dodoth at first appears as an interaction with their natural ecological environment. People's lives—indeed, human survival—would not be possible without their interaction with the various elements of the environment. However, this is only part of the story. Recall how, even in the case study described earlier, camp migrations that could not be reduced to interactions with (i.e., adaptations to) the natural ecological environment took place with a frequency that cannot be overlooked. In the case study, the nature of the relationship with members of other ethnic groups was closely involved in the migrations of the camp. Adachi (2019: 381) writes that 'when we consider the evolution of human sociality, the most important environmental element for the actor is the existence of conspecifics [i.e., other members of the same species].' Taking my lead from this cue, in the following I would like to focus on the social context afforded by the specifics of interactions among groups of humans.

With regard to the use of land as the natural ecological environment and its attendant rights and obligations, Itsuhiro Hazama (2015:21), who has conducted fieldwork among the Dodoth as part of his ethnological research, points out that land use 'is based on an egalitarian style of social negotiations in which multiple ethnic groups that coexist over a larger area enjoy a loose connection to the land, mutually tolerating movements and the use of resources by each other's peoples.' This view holds that, in an environment marked by drastic fluctuations in climate, the availability of good pasturage and water will vary depending on the year and the season, so that the exclusive use of particular sites by particular groups will not be effective as a survival strategy. It emphasizes the overall survival of the 'meta'-ethnic group[3] (the collectivity comprising the multiple ethnic groups living in the area) and demonstrates the necessity of considering the social element of intergroup relations even in the context of the natural ecological environment.

The relationship between the Dodoth and the Turkana, their neighbors to the east, may be seen to be characterized by the following phenomena. First of all, Dodoth nomadism is geospatially complete within their own sphere of residence and activity. This contrasts with how the Turkana, who live in a more arid area, make use of Dodoth grazing lands and watering holes during the dry season. The relationship between the Dodoth and Turkana could be said to be commensal. The Dodoth have no objection to the Turkana using their lands (i.e., grazing lands and water). This is in spite of the fact that, over time, relations between the Dodoth and Turkana occasionally turn hostile, with repeated exchanges of raids and retribution. Here, I would simply like to point out that such 'allowances' and 'tolerance' are recognized by the Dodoth.

Within their own ethnic group, the Dodoth make adjustments by changing their day trip routes and moving their animal camps in order to avoid overtaxing grazing lands or watering holes within their own sphere of activity. Although these adjustments are always undertaken consciously, relationships with neighboring ethnic groups have a strong influence on both changes to day trip routes and camp migrations. This is well illustrated in the case study discussed above. Although Dodoth relationships with neighboring ethnic groups alternate between non-hostile and hostile, it is these hostile relationships that constitute an impetus for camp migration. Put simply, moving a camp to avoid raiding by other groups is quite commonplace. I often heard comments to the effect that 'animal camps do not remain in one place very long,' but the reason—rather than being due to changes in the natural ecological environment, such as 'a lack of grass and water [due to intensive use]'—that I more commonly heard was that 'if we stay longer, then our enemies will learn where our livestock are and discover our day trip routes, which will heighten the risk of raiding, such as ambushes while grazing, or attacks that target the camp.' Thus, even as a destination satisfied conditions in terms of being a suitable camping and grazing location (e.g., topography, vegetation, and the state of grazing and watering sites as elements of the natural ecological environment), camp migration was also inseparable from the exercise of caution against raids.

Camp migration, including the 'impetus' that people cite as the 'reasons' for doing so, thus emerges as part of the process of events and actions (practices) that take place in the camp (Kawai 2006). In the following, I would like to look at this process as a mutual activity premised on relationships with neighboring groups. In doing so, from the case study introduced earlier, I will trace the thirteen-day period from (5) to (6) in Table 9.1 in more detail. I should note that I was not permitted to stay in the animal camp during this time, owing to the risk of raids by neighboring groups. Accordingly, the data presented below is based on discursive information brought to the settlement from the animal camps. Although this amounts to only a tiny fraction of the events that took place in the camps, to the extent that it can be acknowledged to represent what was deemed important enough to convey to the settlement, it is not without meaning.

Events from June 20 to July 3, 2003 (from the arrival at Moru-angapion until the arrival at Loita)

June 20: the livestock were moved to a new camp set up at Moru-angapion. In the afternoon, an eyewitness spotted four figures at a stream about an hour's walk from the camp site. Believing the figures to be enemy spies planning an attack, the members of the camp armed themselves and hastened to the site where they had been spotted, by which time the four had already left the scene.

June 21: a patrol was sent to reconnoiter before the animals were pastured. The patrol found human footprints just off the trail that was to be the day trip route. The footprints were very fresh, and it was confirmed that they were linked with footprints at the site where the eyewitness had spotted the enemy spies the previous day. The patrol pushed on, only to see human figures in the stream. There were between twenty and thirty people, but when they caught sight of the patrol, they started running for the border. The patrol fired off shots and, intending to cut off the other party, took another route but was unable to close the distance in time. Once their quarry had escaped by running along the cliff route leading across the border into Kenya, they gave up their pursuit. On the way to camp, magical materials were put in place at two locations in order to repel the enemy. Three days earlier, on the instructions of an *emuron*, a red-haired goat had been sacrificed, and the 'placing of materials' had been indicated in a divination conducted using its entrails.

This two-day sequence of events was enough to call people's attention to the real threat of a possible 'attack.'

June 22: early in the morning, a patrol set out for the border. Whereas the previous day's patrol had focused on the grazing land, today's patrol targeted the route between Turkanaland and Dodothland, which included the escape route taken by the enemy party the day before. On this day, news arrived that an animal camp belonging to the Toposa, pastoralists from South Sudan, had been attacked by the Turkana. The Turkana had apparently organized a large armed force and, after a fierce struggle, had made off with eight herds of animals, leaving casualties on both sides. This news provoked a keen sense of crisis among the members of the camp, with some saying that the Turkana might have decided to attack the Toposa due to the fact that the Dodoth camp had been set up in a mountainous area that was difficult to attack.

June 28: the previous evening, the news had come that a herd of smaller livestock (goats and sheep) had been taken from a settlement in an area located across a stream from the campsite. Form the evidence of footprints leading to the border, it was determined that this had been the work of the Turkana. Although the settlement in question had given up hope of recovering their domestic animals, a tracking party was quickly dispatched from the camp. After finding the footprints, they made it partway down the cliff route to the border before becoming thirsty and giving up the chase. Later that evening, the Turkana invaded another camp in the same area and absconded with around thirty head of cattle. In the resulting firefight, one Turkana was killed, and one Dodoth was wounded. The following morning, a large-scale tracking party was organized at the camp, and a large number of people joined the pursuit. The fugitives' footprints led to the border. The tracking party followed the footprints down the cliff route at the border to near the Kenyan police post before turning back. The fact that a camp in the area had been attacked was considered to be a very serious matter. That night, the people at the camp discussed their next move.

June 30: in addition to the patrol sent out before grazing, armed herders were stationed on watch through the night along the cattle trail leading to the camp site.

July 1: despite hopes that the migration of the camp would take place as soon as possible, it was decided that the move would be put off until another day. A message instructing that a sacrifice should be given arrived from two *emuron*, and it was decided that the question of where to move would be determined by looking at (i.e., 'reading') the entrails of these sacrificial animals. A gray goat and a black goat were sacrificed in the camp's central clearing, following the instructions of the *emuron*. From the divination using the entrails of the latter, in addition to the revelation of an 'attack,' a further revelation was discerned suggesting 'that a domestic animal with red and black hair should be sacrificed on the road leading to the frontier.' It was also revealed that the enemy would take the cliff route leading down to Kenya, and that the path on that route leading down from the top of the cliff was poorly defended. This meant that the livestock would be stolen from the camp and then led down into Turkanaland along this route. Based on the view that 'we cannot sleep here for another two nights,' it was resolved that the camp would be moved the following morning. In the afternoon, a goat with reddish-black hair was sacrificed, as per the revelation. While Loita had already been mooted as a possible destination, from the goat's entrails it was read that 'there are no problems east of the Napoloto stream.' Loita is situated just east of this stream. That night, an armed party posted on guard outside the camp spotted human figures and immediately fired off warning shots, whereupon several people fled. Because alcohol had been brought into the camp that day, the guards were not posted until late. It was regretted that the guards had not been able to ambush the skulkers.

July 2: the camp is moved from Moru-angapion to Loita.

July 3: early in the morning, a large party of Turkana moved on the site of the vacated Moru-angapion camp. From the camp site at Loita, members watched a large number of figures converging on what until the previous day had been their camp site. After confirming these invaders to be Turkana, the camp members began to give pursuit, which deteriorated into a large-scale firefight. Although no one was killed, it goes without saying that this attempted raid, occurring so soon after the move, affirmed the accuracy of the revelation that had been obtained by reading the intestines of the black goat sacrificed two days previously, imbuing the revelation with an enormous amount of trust.

As described above, we can see how the various events and actions (practices) that take place in the camp occur as a sequence leading up to migration of the camp. Raiding is a matter of grave concern that relates directly to people's survival, since it involves the wholesale theft of the herds of livestock on which Dodoth subsistence is based. People 'knew' about the dangers of raiding and engaged collectively in various practices in order to avoid these dangers. Discursive information in the form of news of raids that took place elsewhere in Dodothland, together with eyewitness accounts of sightings of strangers and the discovery of footprints, served as reminders by which people 'knew' of the dangers of raiding. Moreover, the words of the *emuron* and revelations obtained by reading intestines also constituted information that prompted people to consider migration in the 'knowledge' of the dangers represented by raiding. In the camp, various actions were chosen on the basis of this information in order to protect the livestock. And the impetus for this joint practice was in turn provided by the enemies busily preparing these raids. While these enemies appeared in various forms, people attempted to deal with them collectively, such as by strengthening patrols, holding rituals, placing magical materials, changing grazing courses, and, as a last resort, undertaking the migration of the animal camp itself.

The 'enemy' mode of existence

The appearance of enemies as the result of divinatory revelation

As we saw in the previous section, relationships with enemy others (i.e., people from neighboring ethnic groups) represent a major impetus for animal camp migration and, as such, cannot be overlooked. Camp migration was also part of the process of interaction with these others.[4] The process of this interaction represents a linked chain of actions that include approaches on the part of others and responses by the Dodoth. Specifically, along with the dangerous acts (such as reconnaissance (spying) and raiding by the Turkana) that lead to the loss of the domestic animals that form the basis of Dodoth subsistence, and acts in which these others are either confronted directly (e.g., their pursuit or interception by the Dodoth in retribution for such action) or engaged with indirectly (e.g., by setting magical materials, reading intestines, or holding rituals), other acts are set in store in the form of subsistence activities like changing day trip routes or animal camp migration.

The enemy planning these raids does not always manifest to people as a (substantive) reality. While such enemies may actually appear in reality to Dodoth people as actors who invade Dodothland and engage in espionage or raiding, the 'enemy' that serves as the impetus for camp migration is frequently something created (or cultivated) via information. Here, 'information' should be understood to include 'discursive information' in the form of news conveyed about actual espionage or raiding by the enemy (including attempted raids in which attacks took place, but failed in their objective of taking livestock) and 'eyewitness information' in the form of the traces of such events (i.e., footprints or other evidence of human activity, such as traces of fires or meals), as well as what we could call 'symbolic information' in the form of the dreams of the *emuron* or revelations obtained by people's collective reading of intestines. As with discursive and eyewitness information, symbolic information acted substantially as a factor shaping the nomadic life of the Dodoth; for example, by encouraging changes in day trip routes or the migration of the animal camps. Next, I would like to consider revelations as symbolic information that leads to camp migration.

What I refer to here as 'revelations' include (1) the results of divination using sandals or dreaming by the *emuron*, Dodoth religious specialists who predict the future; (2) results obtained by reading intestines; and (3) the results of sandal divination or dreaming on the part of non-specialist individuals. Of these, (2) reading intestines, while sometimes performed by the *emuron*, is more often performed collectively by a group of people who have slaughtered a domestic animal at the direction of an *emuron*. The individual dreaming in (3) in most cases concerns dreams that are explicit, direct, and realistic, rather than suggestive, such as dreaming that an animal fence had become empty. Ritualistic acts such as simple charms or the placement of minor magical materials are sometimes performed in response to the results of sandal divination or dreams by individuals, but these ultimately amount to no more than individual practices; they do not extend to collective or communal practices, and nor do they have any bearing on camp migration. Although there are several types of revelation, the most frequently observed—a type that the Dodoth encounter on a routine basis and in which they place enormous trust, and which represents a practice that can provide impetus for camp migration—is (2), the practice of divination by reading intestines. In the following I would like to consider a case study concerning this practice.

Reading intestines: the entrails of the black goat at Moru-angapion

From Table 9.1, I would like to discuss (6) the instance of reading intestines that served as the impetus for the migration that took place on July 2, 2003. This was the divination using the entrails of a sacrificial animal (a black goat) that was slaughtered on the instructions of an *emuron* amid the deterioration of relations with the Turkana. The divinatory practice of reading intestines involves the discovery and pointing out of various minute 'abnormalities' on the surface of the entrails of the dead animal, and the contents of these objects are 'read' to obtain revelations (Photographs 9.2 and 9.3). Rather than being carried out by a particular specialist such as an *emuron*, the occasion proceeds in such a way as to have the large group of men[5] present at the scene gather around the entrails to share their opinions. In some cases, more than two hours can be spent interpreting and discussing the entrails before a final consensus is reached.[6] Here, I would like to provide a very brief synopsis of the main points of the revelation that was obtained on this occasion (Figure 9.1).

Two abnormalities were pointed out. The first was a red dot (a), a small blood clot. As something that indicated fire (i.e., fighting), this was read as an imminent threat of attack. The second was a slightly differently hued reddish-

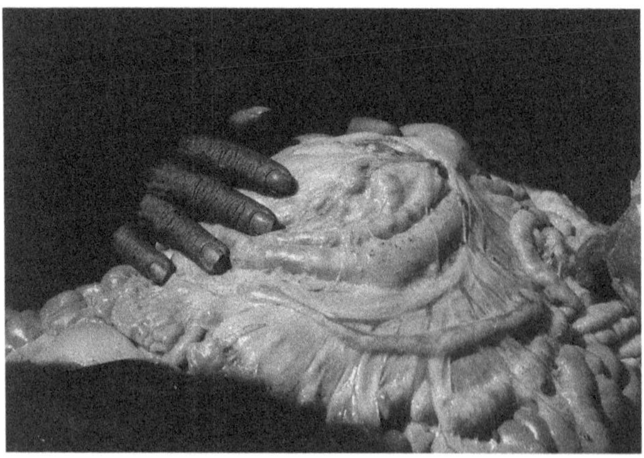

Photograph 9.2: Reading intestines (1)—'Reading' goat intestines

black dot (b) in the upper right of the intestine, which was read as the ritual sacrifice of an animal. Point (b) was located near the intersection of two thin sinews (c). The two sinews were deciphered as representing two cattle trails near the current camp site, and were read as a revelation calling for the offering of an animal with reddish-black hair as ritual sacrifice by which to repel an enemy attack at that location (i.e., the intersection of the two trails).

Aside from the two abnormalities, other problems pointed out included a small indentation in the surface of the mesentery (e), which indicated the camp site, and a sinew (f) that traversed the top (i.e., the north) of this spot. This sinew was judged to represent the road to Kenya, and the part of the small intestine at the end of this sinew (g) was empty of digestive materials (i.e., was filled with gas). The small intestine was read to represent the escarpment on the Kenyan border, but because the part (g) was empty, this was interpreted as a defensive gap, and it was read that the livestock would be taken by the Turkana into Kenya along this road (f) by passing through this point (g).

After this revelation, as directed by the intestines, a goat with reddish-black hair was sacrificed the same day at the intersection of the two trails near the camp site, and the camp was moved the following morning. The revelation obtained by reading the intestines served as an impetus for joint practices by the members of the camp in terms of the offering of a sacrifice and the migration of the camp. In this way, there is no shortage of cases in which reading intestines is strongly related to camp migration, which represents a basic activity in the context of pastoralist subsistence. When the arrival of an unfavorable event or phenomenon is read in the intestines, the steps that should be taken to avoid it are immediately taken up for discussion in the camp, and once a judgment is reached, all members of the camp work together to carry it out. Specific examples might include the stepping up of patrols, the performance of rituals,

Photograph 9.3: Reading intestines (2)—Crowding around a cow's intestines, looking, touching, and discussing ('reading' the intestines) all together

the placement of magical materials, or the changing of grazing courses. The most drastic and final recourse was the migration of the animal camp.

The materialization of enemies by means of joint practice

We can distinguish two modes of existence for enemies, each with a differing creation process. One is an enemy that appears in a form that alludes by means of discursive or eyewitness information to events or the traces of events that actually took place, while the other is an enemy that is shaped through symbolic manipulation by means of symbolic information. While the difference between these modes of existence stems from differences in the nature of information as the basis (both material and evidentiary) for their creation, it remains unclear whether or not they are regarded as distinct by the Dodoth themselves. It should be recalled here that symbolic information can lead not only to ritualistic and symbolic acts like the large-scale rituals that accompany magical charms and the sacrifice of domestic animals, but also, just as in the case of discursive and eyewitness information, to the fundamental subsistence activity of camp migration. I should point out here that with both these modes of existence, what is meaningful is how the enemy that is manifested (or has been foreseen to be) is preparing a raid and how, in order to deal with this, people step up patrols, hold rituals, place magical materials, change grazing courses, and ultimately

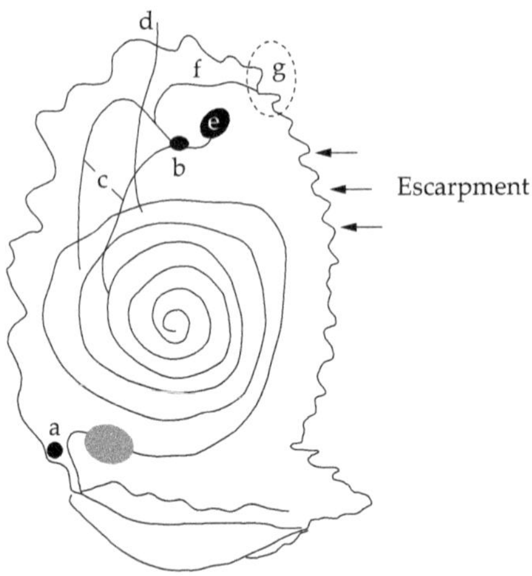

Figure 9.1: Schematic diagram for reading intestines

Note: This is a simplified sketch of the reading of the black goat's intestines as illustrated by a young Dodoth man. The central whorl is the large intestine (colon and rectum), and the surrounding wavy area represents the small intestine (jejunum and ileum). The figure is drawn with the top as north. Accordingly, the portion of the small intestine on the right represents the escarpment along the Kenyan border to the east of Dodothland.

undertake the migration of the entire animal camp. At the same time, in people's collective performance of these various activities, the enemy gradually becomes more of a real presence, materializing in the form of an 'other' who is preparing a raid. This could be described as the process by which the invisible enemy is made visible.

From individual dreaming and sandal divination to the performance of large-scale rituals accompanying the sacrifice of domestic animals based on the direction of the *emuron* and divination using their entrails, revelations pointed to (mainly dangerous) events in the near future—events that were recognized as something that people would have to somehow deal with. These revelations concern real situations, which is to say real problems, that people address in various ways. In addition to magical or ritualistic acts such as walking around and around an animal fence while balancing a clay pot on one's head, placing magical materials at particular locations, or holding large-scale rituals that involve the sacrificial offering of domestic animals, this might also involve altering daily grazing routes as a part of subsistence activities or, at the largest end of the scale, relocating the entire animal camp. The nomadic life of the Dodoth is substantially shaped by the movement of their animal camps, as described above. Revelations exert a direct impact on pastoralism as a subsistence activity, to the extent that we could say they represent (symbolic) information that shapes the nomadic way of life. Such information affects people's perceptions and practices in the same way as information about changes in the natural ecological environment, such as news that 'the grasses have disappeared' or 'the watering hole has run dry.' Together with 'signs' of raiding and 'awareness' of the same based on discursive information about where raids have taken place, eyewitness information concerning sightings of human figures or traces of activity-like footprints, it also influences the actions people choose to take.

The things that are indicated and shaped by revelation are the events themselves that will occur in the near future; these are perceived not as events occurring in some imaginary otherworldly location that is separate from the real world, but as real-world events, consistent with everyday experience. In other words, it was the situation of 'knowing' (based on the symbolic information of revelation) that 'the danger of raiding is now imminent' that led people to undertake the joint practice of having to 'somehow deal with this situation.'

Physical evidence that engenders 'signs' and 'awareness'

Symbolic information in terms of revelations is information that engendered 'signs' and an 'awareness' of raiding. It served as a device that convinced people that raiding would occur in the near future, that the danger of a raid was indeed already imminent, and that the social situation was changing. Changing day trip routes, in terms of shifting grazing sites or watering holes, and the migration of animal camps are fundamental practices at the core of Dodoth subsistence activities. While these are, first of all, actions and practices for dealing with changes in the natural ecological environment, these selfsame actions and practices can also be induced by symbolic information in terms of revelations. That is, whereas symbolic information leads to symbolic actions and practices such as the setting of magical materials and the performance of rituals, it also leads to subsistence-based actions and practices such as changing day trip routes and camp migrations.

However, people may not necessarily distinguish between symbolic actions and practices and subsistence-based actions and practices. For the Dodoth, the danger of raiding is a threat, and when this threat is imminent, they must immediately take some sort of action, since to neglect to do so is a serious matter that could threaten their very existence. In response to such situations, the 'correct' attitude for maintaining survival—from changing day trip routes to camp migration, and from trivial magical practices carried out at the individual level to large-scale collective rituals involving the slaughter of domestic animals—is simply 'to try everything' and 'to do what can be done.' To that end, these pastoralists were willing to sacrifice the livestock, which constituted their base of subsistence and directly guaranteed their own survival. Such sacrificial animals even included cattle, which were of paramount value as the principal domestic animals providing sufficient quantities of milk and blood to be able to substantially ensure Dodoth survival.

While symbolic information can be thought of as constructing the world by means of symbolic manipulation, for the Dodoth it is a stretch to say that this world departs from or is differentiated from the real world as a realm of the imagination. The Dodoth attach great importance to physical evidence such as footprints and other traces of human activity as definitive and creditable information. What is more, in their method of reading intestines, it is physical evidence that can be seen and touched (such as irregularities and physical abnormalities appearing on the surface of the entrails) that constitutes the decisive basis for meaningful clues as to how to read the intestines. It is because of their material characteristics that animal intestines were discovered and chosen as objects to be 'read' by the Dodoth, as 'realists' who place a paramount value on physical evidence. This is why, even today, the Dodoth remain so enthusiastic about reading intestines, believing firmly in their revelations, and unreservedly carrying out actions and practices in accordance with the signs that they read therein. On the other hand, the Dodoth manner of reading intestines lacks the characteristic mystique engendered and produced by symbols that is typically associated with divinatory practice. Beyond reading intestines, the world inhabited by the Dodoth is largely void of mystical or supernatural beings that cannot be seen or touched, and the function of mystical or supernatural forces seems to be limited. The power of symbols is weak, so to speak, almost as though symbols do not transcend the physical and material time and space of reality. This is not to say that there is no world in which symbols play a part. For the Dodoth, however, in that this imagined world often leads to actions that are not symbolic—indeed, which have no connection to symbols—it is merely the everyday world, consistent with reality.

Conclusion: memory, foresight, and joint practice

In this chapter, I have focused on the decisive impetus for animal camp migration (including with regard to its timing and destination), a practice that constitutes the substantive basis of nomadic life among the Dodoth. In the animal camp, a general meeting of all members of the camp was held for about an hour every morning. After exchanging information on pasturage and water conditions, day trip routes were decided by coordinating the grazing places and watering sites to be used by each herd that day. The morning meeting was also a time to share news (discursive information) about the details of raiding losses or attempts in Dodothland, and to confirm the movements of enemies by sharing news of tracks or other eyewitness information. Based on these reports, the camp members would step up patrols, alter day trip routes, or, based on what the *emuron* had to say, confirm plans to hold rituals, set magical materials, or slaughter an animal in order to read its intestines. The morning meeting was a setting in which to share various types of information, as well as to jointly confirm, coordinate, and decide upon the details of grazing activities for the day. Various aspects concerning the migration of the animal camp were also discussed at the meeting, and the timing and destination of any such move would be decided.

As well as the conditions of the grazing places, watering places, and camp sites, the decision to move the animal camp was heavily colored by the threat of enemies. Even when the natural ecological conditions posed no problems—in other words, where there were ample supplies of grass and water and the geographical conditions of the camp site were also suitable—when it was recognized that there was an imminent danger of raiding by another ethnic group, the animal camp moved immediately. If the danger of raiding were to be ignored and a raid did end up taking place, this would potentially result in the loss of entire herds of livestock. Families that owned the herds that might be stolen thus risked being cut off from their entire supply of food, which would place their very survival in crisis. Camp migration was frequently employed as an effective means to avoid this. The decision to move was strongly influenced by symbolic information obtained via revelation, such as the words of the *emuron* or results obtained by reading intestines, as well as by more direct and immediate discursive and eyewitness information conveying news about what was actually happening in Dodothland. In the Dodoth context, we can say that discursive and eyewitness information is information concerning the recent past that (as faithfully as possible) conveys what actually happened, while symbolic information is information about the near future that conveys (as accurately as possible) what will happen in the future. Based on this information, the Dodoth connected the recent past and the near future to the immediate present, allowing them to make an integrated judgment of their current situation and to use this information to decide the actions they should take in this immediate present. The 'recent past' is a grasp of events based on memory, while the 'near future' is a grasp of events based on foresight.

Although memory and foresight are diverse in the sense that they manifest in various ways, are positioned in various ways, and are used in various ways in different cultures and societies, they are broadly and generally recognized by humans and may therefore be considered to exert an influence on (i.e., act upon) their behavior. Even so, we cannot say for certain that the use of memory and foresight as something to rely on when deciding one's own behavior, or else as a basis for one's own behavior, is something that is unique or particular to humans. It would seem that we can also recognize them in the context of nonhuman primates, albeit in an incipient form; for example, they 'know' the fruiting season of edible plants and work quickly to engage in foraging behavior (e.g., by having all individuals in a group quickly move directly toward a foraging tree), and cases have been observed in which they seem to 'know' in advance what other individuals are going to do and choose their own actions on this basis (e.g., deceiving others). We could say that they also predict the (near) future in some way, based on certain things or events, or on the behavior of others. It is quite conceivable that these predictions also involve memories of the past. Just how far back this past goes for them will likely differ between apes and monkeys in terms of differences in their respective cognitive abilities, and variation will also occur between species and groups. But we can point out the possibility that past events and experiences will have an impact on (act upon) their behavior in the immediate present in the form of determining and guiding that behavior. While I cannot go into detail as to the mechanisms of memory and recollection in terms of how the past is retained as memory to be reborn in the present, what we need to recognize here is that this is not something that is unique to humans.

However, what I have called 'foresight' in the context of Dodoth society is subtly different from 'prediction' based on the memory and recollection of past events and experiences, which we also find in other primates. The Dodoth make use of discursive and eyewitness information to grasp their current circumstances, and by analyzing this information they can predict events that will occur in the (near) future, such as imminent raiding, and memories and recollections relating to past events and experiences may also play a strong role in this. On the other hand, the Dodoth have developed a way of 'knowing' things by acquiring information from the revelations obtained by reading intestines; in other words, from symbolic information. Although these revelations share commonalities in that they are based on some kind of (existing or formerly existing) thing or phenomenon, it is more appropriate to refer to this as a 'foresight' that intuits the future rather than as a 'prediction' that estimates the future. According to Michael Tomasello (2009: 59), 'human beings live not only in the physical and social worlds of other apes, but also in an institutional or cultural world of their own making…The specifics of this world vary greatly among different groups of people, but all groups of people live in some such world.' Foresight seems to be a uniquely human activity, one that goes beyond prediction, which is also possible in the physical and social world, and which is rather possible precisely by virtue of our inhabiting the institutional and cultural world.

However, as I have already pointed out, the Dodoth are not well versed in dealing with (complex) symbols, whose efficacy is relatively weak. They are people who are much more likely to live in the physical and social world and perhaps relatively less likely to actively inhabit the institutional and cultural world they have created. Even with reading intestines, which seemingly deals with symbols, rather than being a method of symbolic manipulation such as those commonly found in divinatory practices in the wider context of human cultures and societies, it would be more accurate to say that this is a method (technique) of situational awareness that relies on the 'big data' of collective knowledge that integrates past experience with the understanding of present circumstances. The important thing is that it elicits practice, and that this practice is a joint practice carried out together with compatriots. Reading intestines always presents people with concrete strategies that it calls on them to execute. Rather than a method or a means to decipher the future using symbols, it is a way to concretely elicit joint practices for dealing with the future.

The migration of an animal camp, for the Dodoth, is a communal practice on the part of the camp's members as a response to the critical situation of potentially losing their livestock, which are an extremely valuable resource that sustains them as a basis of subsistence. It is a practice undertaken as the result of people's perception of a critical situation—as a change in the natural ecological environment on the one hand, and as a change in the social situation on the other. At the same time, it might also emerge as a response to mental processes such as premonitions, signs, and awareness (although these are more like processes of collective psychological states than problems of individual minds). The 'extreme' in the Dodoth context is nothing other than a 'future' that is consistent with the immediate present, in terms of a collective dread that motivates people who have imposed upon themselves the proposition that 'we must deal with it.' This 'future' is engendered in the immediate present through people's specific joint practices, and the 'extreme' is always regarded as inseparable from practice as a form thereof. Is the origin of this situation that we could call an extreme condition—that is, the relationship between the way that extremes are recognized and the way that they are dealt with—an extreme way of being particular and unique to humans? Or is it something that is shared with nonhuman primates, albeit in an incipient and primitive form, together with the mental processes of premonitions, signs, and awareness? And when and how did it arise in the process of human evolution? These are future challenges.

Notes

1. This may mean that the grass in the pasture is rotting from the roots.
2. The Turkana are pastoralists who make use of both cattle and camels as large domestic animals, but who place a higher social and cultural value on cattle. As the drought became more serious, the Turkana preserved their cattle (retaining 70% of their pre-drought stock afterwards), choosing instead to slaughter their camels for food (retaining only 41% of their stock), despite the fact that their physiological characteristics were more resistant to drought. From an ecological perspective, Turkana should not have killed their camels, which would likely have been able to withstand the drought. The Turkana and their neighbors, the Gabra and the Rendille—camel pastoralists who live across Lake Turkana to the east—experienced similar droughts during the same period (with minimal differences in rainfall between the two areas), but these latter groups did not suffer as high a human cost as the Turkana. As camel pastoralists highly dependent on their camels, the Gabra and Rendille were arguably better skilled at exploiting the extremely arid environment via their highly drought-resistant camels, which allowed them to adapt their culture and society to the arid desert (Itani 1982).
3. Elsewhere, I refer to a higher-level group called a 'pastoral values-sharing sphere' (Kawai 2017: 234).
4. According to Adachi (2019: 381), 'when we consider the evolution of human sociality, the most important environmental element for the actor is the existence of conspecifics [i.e., other members of the same species].'
5. Men of various ages take part with equal standing; women do not participate.
6. For a detailed account of the specific progression of reading intestines as a divinatory practice, see Kawai (2007).

References

Adachi, K. (2019) 'Toward the environmental others: An ethnological essay on equilibrium and coexistence'. In K. Kawai (ed.), *Others: The Evolution of Human Sociality*. Kyoto University Press and Trans Pacific Press, 365–386.

Hazama, I. (2015) *Bokuchiku sekai no kyōsei ronri: Karimojon to do Dosu no minzoku-shi* (Commensal logic in the pastoral world: An ethnography of the Karimojong and Dodoth peoples). Kyoto University Press.

Itani, J. (1980) *Turukana no shizen-shi: Kashaku naki hitobito* (A natural history of Turkana: A people without remorse). Yuzankaku.

Itani, J. (1982) *Dai kanbatsu: Turukana nikki* (The great drought: A Turkana diary). Shinchosha Publishing.

Kawai, K. (2002) '"Chimei" to iu chishiki: Dodosu no kankyō ninshiki-ron josetsu' (Toponymic knowledge: Introduction to the environmental epistemology of the Dodoth). In S. Sato (ed.), *Yūbokumin no sekai (Kōza seitai jinrui-gaku 4)* (The nomadic world (Lectures in ecological anthropology 4)). Kyoto University Press, 17–85.

Kawai, K. (2004) 'Dodosu ni okeru kachiku no ryakudatsu to rinsetsu shūdan-kan no kankei' (Cattle raiding among the Dodoth and relationships with neighboring groups). In J. Tanaka, S. Sato, K. Sugawara and I. Ōta (eds), *Nomaddo: Afurika no gen'ya ni ikiru* (Nomads: Life in the African wilderness). Showado, 542–566.

Kawai, K. (2006) 'Kyanpu idō to chō uranai: Dodosu ni okeru rinsetsu shūdan to no kankei o meguru shakai kūkan no seisei kijo' (Camp migration and reading intestines: Generative mechanisms of social space in Dodoth relationships with neighboring groups). In R. Nishii and S. Tanabe (eds), *Shakai kūkan no jinrui-gaku: Materiariti shutai modaniti* (Anthropology of social space: Materiality, subjectivity, and modernity). Sekaishisosha, 175–202.

Kawai, K. (2007) 'Dodosu no chō uranai: Bokuchiku-min no yūdō ni kakawaru jōhō: Chishiki shigen no keisei o megutte' (Reading intestines among the Dodoth: Generating information about pastoralist migration as a knowledge resource). In C. Daniels (ed.), *Shigen jinrui-gaku 3 chishiki shigen no in to yō* (Resource anthropology 3: The yin and yang of knowledge resources). Koubundou, 29–71.

Kawai, K. (2017) 'Institutionalized cattle raiding: Its formalization and value creation among the Dodoth'. In K. Kawai (ed.), *Institutions: The Evolution of Human Sociality*. Kyoto University Press and Trans Pacific Press, 219–236.

Kawai, K. (2019) 'Teki to tomo no hazama de: Dodosu to rinsetsu minzoku Turukana to no kankei' (Between friends and enemies: Relations between Dodoth pastoralists and the neighboring Turkana). In I. Ōta and R. Soga (eds), *Yūboku no shisō* (Nomadic thought). Showado, 197–214.

Tomasello, M. (2009) *Why we cooperate*. MIT Press.

CHAPTER 10

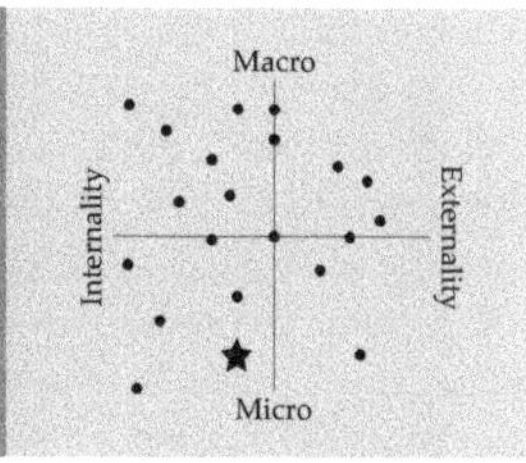

Approaching the World of the Living from the Extreme of 'the Invisible':

Thoughts on Spirits, the Dead, and Radioactivity

Ryoko NISHII

Keywords: the invisible, prediction and anticipation, spirits, the dead, radioactivity, magic and science

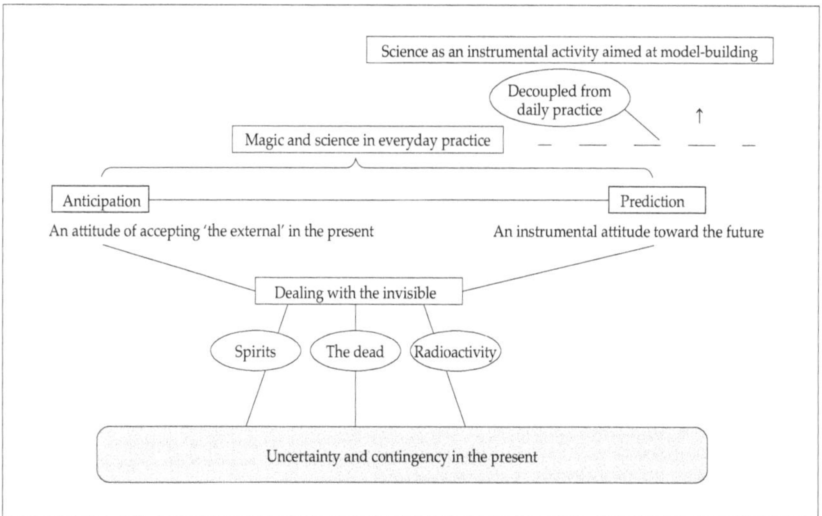

When dealing with 'invisible' phenomena such as spirits, the dead, and radioactivity, we can identify the attitude of 'prediction,' which deals with the future by rendering such phenomena visible, and the attitude of 'anticipation,' which accepts and addresses encounters with the invisible in the context of the present moment. Primates, including *Homo sapiens*, engage in their daily practices with varying degrees of future-orientation and present-orientation that cannot be neatly divided into either prediction or anticipation. Magic and science, as ways of dealing with the invisible, are indistinguishable in the context of everyday practice. The unique character of science only becomes discernable in instrumental activities that seek to build models that are decoupled from everyday experience. The extreme of 'the invisible' casts light on the art of living in a world of uncertainty and contingency.

The extreme of 'the invisible'

The world we live in is full of uncertainty and contingency. A well-known case of a people dealing with such uncertainty and contingency is Edward Evans-Pritchard's (1937) conception of magic in Azande society. When someone dies at the moment a granary on stilts collapses, and he is crushed underneath, why was that person in that place at that particular moment, so that he would be crushed to death? Why not someone else, like the person who had been with him until only moments earlier? Also, had he arrived just a few minutes or even seconds later, he would not have been caught in the collapse, so why then in particular? The Azande logic for explaining such things is magic and witchcraft.[1]

Masato Fukushima (1998) calls the logic of magic 'pseudo-causality.' By this he refers to the repeated connection between a certain action and its result, so that their coincidence is perceived as though it were causal. To illustrate this pseudo-causality, Fukushima (1998) gives the example of a high school baseball player who wins a game while coincidentally wearing a muddy uniform that he has not washed, and so decides to wear the same muddy uniform to his next match. It is also well known that the routines that sports athletes always perform before a game (e.g., the way the tennis player Rafael Nadal touches his nose, left ear, and right ear in the same order each time before hitting a serve) can contribute to their psychological stability by establishing in their minds and bodies a causal relationship whereby they will win the game by performing these actions. These are a way of somehow dealing with uncertainty and contingency by using everyday objects and events as props.

In this chapter, my goal is to rethink human activity in the world we inhabit through the lens of the invisible rather than what is visible to the eye. Here, 'the invisible' refers to the things that are generally not perceived by the human sense of vision, which is based in our biological organs. In our interactions with the world, the invisible is something that we position as the extreme of our dealing with reality. With regard to what I mean here by 'the invisible,' as a starting point, I am not considering 'the invisible' ontologically as the absence of 'the visible,' the way that Merleau-Ponty (1968) does.[2] That is, what I mean by 'extreme' in this chapter is that thinking from the perspective of the invisible in the context of our daily lives, conversely, sheds light on the aspect of how elements of existence beyond what we can perceive with our biological organs are still involved in our daily lives, and seems to allow us to rethink the nature of reality, which abounds with uncertainty and contingency, beyond a mind-body dualism that attributes that which we cannot grasp with our physical organs to the psyche.

In this chapter, as the invisible, I deal specifically with spirits, the dead, and radioactivity. To start, I would like to briefly explain my rationale for doing so. The first reason is that none of these are by any rights perceptible by our biological bodies. Another reason is that while spirits and the dead have been addressed in studies of magic and religion, radiation has been dealt with scientifically. In the context of anthropology, as well, separate lines of accumulated research have emerged, with the former being handled by the anthropology of magic and religion and the latter by the anthropology of science. There have even been attempts to forge a connection between these two lines of research (e.g., Nakagawa 2009; Kasuga 2016). Although this chapter is unconcerned with the genealogies of these research traditions as such, I hope that by focusing our attention on dealing with the invisible in the context of everyday practice, we also may be able to rethink the relationship between magic and science.

Prediction and anticipation

As ways of dealing with uncertainty and contingency in our contemporary context—that is, as ways of dealing with the invisible—we have the two diametrically opposed attitudes of prediction and anticipation. Yukio-Pegio Gunji (2013: 131) distinguishes these by defining prediction as the calculation of a single value or point for the future, and anticipation as a future-oriented calculation that cannot be determined as a single point or value, but only as an area or span. Here, by introducing the further perspective of the presence or absence of bodily sensation, I would like to think about prediction as the extreme of a future-oriented attitude that lacks bodily sensation, and which apprehends the future only by external indicators, and about anticipation as the extreme of a future-oriented attitude that has bodily sensation, and which projects a vague guideline toward the future. Here, I regard bodily sensation as something that is subjectively perceived, as opposed to an objective perception using biological organs.[3] Naturally, these attitudes relating to prediction and anticipation cannot completely be differentiated one from the other, and their relative degree, including the nature of any bodily sensation, will fall along a graded spectrum.

In the following sections, I would like to shed light on the characteristics of spirits, the dead, and radioactivity by dividing them in terms of these two attitudes and considering several case studies. In this chapter, I focus on instances of spirits and the dead drawn from the many years of fieldwork that I have conducted in Southern Thailand,[4] and in relation to radioactivity I rely mainly on an interview-based survey that I conducted in Fukushima, Japan.[5]

Prediction

In this section, I look at the attitude of prediction from a case of reading lottery numbers from events in Thailand and the case of measuring invisible radioactivity in Japan. In both cases, data is taken from the environment and linked to actions as an indicator for the future.

'Reading' winning lottery numbers in Thailand

In my field site in Southern Thailand, villagers often try to guess the winning numbers of Huai Tai din (underground lotteries) based on omens such as dreams and unusual events. In village parlance, buying lottery tickets is simply called 'buying numbers.' Up until around the 1990s, this involved guessing the final three digits of the first prize in the public lottery, which was announced every two weeks. Such lotteries are against the law, but since police officers themselves take part in buying the numbers, any crackdowns are merely perfunctory. Lottery promoters are rich men who live in the towns, and tickets in the villages are sold by villagers who subcontract as agents for these promoters. A villager will tell the agent the numbers that he wishes to buy, and the agent writes these down in a small, perforated notebook with a piece of blue carbon paper between the pages. The agent gives the written page to the buyer, and keeps a copy for himself. The most common way to buy a ticket is to buy the exact three-digit number itself, but as the number range is from 1 to 999, the probability of winning is 1 in 999. A winning ticket earns 600 baht, and the daily wage in the 1990s was about 50 baht (today it is closer to 300 baht). So spending 10 baht could earn a return of 6000 baht, which is a considerable amount for a typical villager. Many people spent 100 baht or 200 baht on tickets every two weeks. Even though the probability of winning was very small, people would lay out large sums when they thought they had obtained a 'sure' clue of their own.

As for how they obtain such clues, this is a matter of reading every sign that might lead to a winning number from the events around them. For example, when a car accident took place on a national highway and a car pitched into the grassy verge, I saw people try to wade through the long grass to approach the car. Although I presumed this was to check the safety of the people in the car, it was nothing of the kind. They were checking the numbers on the car's license plate. They were betting on an association between the 'car being hit' and 'winning the lottery' and buying those numbers. Since a car's license plate has four digits, they chose three of these. Also, Thai Buddhists are usually cremated at death, but until a crematorium could be built in a neighboring village, it was the practice to place coffins atop stacks of firewood and to cremate them in pyres. On one occasion, perhaps because of the way that the corpse's muscles were burning, one of its hands shot up from the casket with five fingers pointing skyward. People who witnessed this would buy numbers 854 or 458, interpreting the five fingers as five, the four corners of the coffin as four, and the corpse as eight (or sometimes nine, but no one could explain to me why a corpse might be interpreted using such numbers). Allegedly, many people had won the jackpot this way before. It is also possible to buy numbers left behind by people who died suddenly.

In addition to reading these numbers from events that take place in the context of their daily lives, people frequently also read them from their dreams.

For example, a dream about one's deceased father may prompt someone to buy a number corresponding to the date of his death, and people often buy numbers that are communicated to them by the dead in dreams. In another case, a dream about four monks bearing a coffin prompted one person to buy the number 446—four because of the four monks, four for the coffin, and six because the idea of four monks carrying a coffin is inconceivable, and thus a lie (*hok* in Thai), which is a homonym for six. There is no monopoly on dreams, and anyone who hears the story of a dream is free to glean what clues he can and then to buy any number he wants.

These lottery numbers are often read from events connected with death or supernatural beings. This is exemplified by the case of a monk in a neighboring village, who, in 1987, began working with a Muslim medium to divine lottery numbers. This monk attracted considerable attention with the sensational way he used a hand towel, allegedly used in a suicide, as a prop to invoke the spirit of the dead person and to have it possess the Muslim medium in order to ask it for numbers. When an increasing number of people donated money to the monk as they came to ask for the winning lottery numbers, or gave money to him as a gratuity to thank him after they had won, he quickly built a large concrete hall in a temple that had previously housed only a small wooden priests' quarters.

Some cases are matters of life and death, as in the following example. The father of an elementary school student made a vow to a sacrificial cow that if he won, he would spare the animal's life. Apparently, he promptly had a major windfall, whereupon, as promised, he brought the cow home and offered another cow for sacrifice in its stead.

In this way, the use of the dead, the spirits, the supernatural, and animals to obtain numbers is referred to as *tham boe* (making numbers). This is the act of reading signs that lead to winning numbers from things like slight anomalies, dreams, the dead, or spirits that are out of the ordinary routine, and then making bets on that basis. Guessing lottery numbers could thus be said to be an act in which people read various possibilities that would normally go unnoticed, things that are hidden in everyday life and would otherwise seem to be taken for granted, and then try to connect them to their own good fortune in a practical way.

Measuring radioactivity in Fukushima

In Fukushima, the nuclear accident caused by the Great East Japan Earthquake has engendered the need to deal with the invisible on a daily basis in terms of radioactivity. An effective way of dealing with this 'invisible' radioactivity is to render it visible. The principal method of doing so is to measure it with instruments that measure radioactivity or radiation.

On November 5, 2017, I was in Fukushima listening to a talk by Mr. Hiromi Abe, the director of the Fukushima 30-Year Project, an incorporated non-profit organization. A native of Fukushima Prefecture, Abe said that he used to work in anime production in Tokyo, but decided to return in the wake of the earthquake. The Fukushima 30-Year Project was established by citizens who came together with a mutual interest to minimize the damage caused by radioactive contamination in the aftermath of the March 2011 nuclear accident. Their main activities include measuring radioactivity, social functions, and study meetings, and in July 2011 they opened the Citizen's Radioactivity Measurement Center in Fukushima City. After starting with one simple measuring instrument, they began measuring radioactivity in food and crops, and then changed their official name to the Fukushima 30-Year Project in June 2014. Six years after 2011, they had measured a total of 4411 items, including towels dried outside in the air, raw rice, cooked rice, and food sold in supermarkets.

The project produces a large number of pamphlets. One of these is the *Kodomo hōshasen wākushoppu* (Children's radiation workshop) (Fukushima 30-Year Report 2015), which features the impressions of children who had learned about the importance of radiation measurement at a children's workshop that included hands-on measurement activities; for example, the following is from a girl in the first year of elementary school: 'I learned about radiation for the first time. I learned that you can't see it, but it's bad for your body' (Fukushima 30-Year Report 2015: 10). Another impression: 'I thought it was strange how you can't see, feel, or smell the radiation, so that you might wonder whether it is really there. But when I actually listened to the talk, I found out that there's a lot [of radiation] in apples and other fruits, which I thought was a little scary' (Fukushima 30-Year Report 2015: 13).

Another pamphlet, entitled *Mama-tachi no hottosupotto faindā taiken-ki* (Mothers' experiences as hotspot hunters), a manga illustrated by Kimidori Inoue (Fukushima 30-Year Project 2017), highlights an activity in which a mother and her children, before attending a parent-and-child tea party in a park, take measurements in various places at the public park—in the bushes, along the footpath, under trees, etc.—in order to mark these numerical values on a map to show where radiation levels are high. 'The important thing,' the pamphlet says, 'is not to live with a vague sense of anxiety, but to gain accurate knowledge about where we should be safe, and where we should be careful.' By taking measurements, people bring places of 'invisible' danger into view, thereby gaining the ability to plan their actions in order to deal with that danger. Conversely, it is by making sites of low radioactivity 'visible' that their safety can be confirmed. The pamphlet also features a comment by a high school student, who says, 'I thought by measuring radioactivity, we could prove the safety of Fukushima food products and that by communicating this to a large audience, the Fukushima brand would regain popularity. I felt that this was another reason that it was important to measure radioactivity.'

What impressed me about Abe's account was the way he described returning to his home town after working in Tokyo: 'Mount Azuma looked so beautiful, crisply outlined against the clear blue sky, and yet I knew that it was contaminated with radioactivity that I could not see and could not feel.' Here it seemed to me that the disconnect between the limpid sense of space that he obtained through his own body, through seeing with his own eyes, and his acknowledgment of the fact that it was a space with high levels of radiation, which could only 'be seen' through scientific technology, was being expressed as an acute sense of bitter frustration.[6]

This case from Fukushima is comparable to the act of reading signs for lottery numbers in that both provide guidance for selecting future actions by bringing the invisible into view. Reading signs from the presence of the 'invisible' dead or spirits and visualizing the existence of physical radiation are useful for prediction as a way for people to live in an uncertain world.[7]

Anticipation

As an attitude for dealing with the invisible, I have opposed prediction, which lacks bodily sensation, with anticipation, which has bodily sensation and which projects a vague guideline toward the future. In this section I would first like to consider the experiences of seeing, hearing, and detecting by smell what is normally invisible, or what remains invisible to others. In the Thailand case, experiences involving the 'invisible' dead and spirits are commonplace. And it has also been reported that some people can sense radioactivity.

Perceiving the dead, spirits, and radioactivity

The dead and spirits may not be invisible to those who have seen, smelled, or felt some hint of their presence. Even so, they are usually invisible presences according to what we sense with our biological organs. Here, what is perceived is an existence or world that is separate from the everyday, that is visible in ways that transcend physical perception.

Scent

Non, the third son of the family with whom I stayed during my first stint of fieldwork, was dying from the side effects of his AIDS medication. His flesh was rotting, to the extent that just before he died, it would slough away simply by grabbing part of his body. Though Non was being cared for at home, at length he was going to be taken to the hospital because he was in danger. However, his eldest brother was unable to stand the smell of his body, and did not visit him in the hospital in his final days. After Non's death, his eldest brother was tapping trees in the rubber plantation when he smelled a wafting odor, at which he sensed that Non had come. At the time, another man was tapping nearby with his child. When the child asked the father about the odor, the brother became certain that Non had come (from an interview conducted in August 2008).[8]

Seeing, hearing

Chai died of cerebral hemorrhage at the age of fifty-five. Before his death, his wife moved in with a widower in the house across the street, with whom she had been carrying on an affair. Perhaps because of the stress, Chai began drinking heavily and his health began to suffer. Her only daughter moved back and forth between Chai's house and her mother's house just across the street. Her niece, Pla, who had lived with Chai, Pla's son Kane (age five), and Chai's mother all felt Chai's presence several times after his death.

Pla alleged that she saw Chai on three occasions. The first time was on the morning of the second day after his cremation. She became aware of Chai sleeping next to her on the bed. She was sleeping on her usual mattress with Kane, and now Chai. She spoke to him, but he did not answer. The second time was four or five days later, also in the morning. She became aware of Chai at the entrance to the house. She asked where he had come from, but he gave no answer. The third time was on the morning of the ritual observance of the one hundredth day after his death, at around five o'clock. As Pla was preparing the meal that they would bring to the temple for Chai's ritual, she became aware of Chai sleeping next to Kane. At this, Pla told Chai that there was no need for him to menace her, and suggested he go to the temple.

A fourth time was in a dream. Chai was sitting under a photograph, fretting about one thing and another. Pla told him that there was no need to worry.

Chai's mother heard what sounded like a spoon scraping on a dish outside the room where she was sleeping. She said it had been Chai, eating his dinner. A woman who lived in the neighborhood also saw Chai. Riding her motorcycle home from work in the middle of the night, she passed Chai, who was heading for the neighboring village, and since someone from that same village had just died, she felt that Chai must have come to inform her of the death (interview, August 2017).

Sensing a presence

When Mrs. Pen, a teacher at a village elementary school, went traveling with her university friends in February 2017, she found that there was only one hotel available along the Mekong River in Chiang Saen, in Northern Thailand. It was called the Pink Inn, and it seemed to be in the middle of nowhere. Everything was pink, including the interior of the rooms, which were full of Hello Kitty characters. They were the only guests, and they split up into four rooms to spend the night. That night, they felt strangely ill at ease and had a hard time sleeping. When they rose in the morning, they discovered that they had all felt the same way (interview, August 2017).

Talking

In some cases people tell of how they have actually 'seen' and 'spoken with' the dead. These cases transpired in such a way that the person did not realize that the other party was dead. Bang Chaet dropped dead after suffering a sudden attack while playing mahjong with his friends. He was just slightly over forty years old and had become a Buddhist through his marriage, but he was called Bang Chaet after the Muslim fashion ('Bang' is a term attached to the names of Muslim men). He had been playing mahjong with his friends when he began to feel pangs in his chest, and then suddenly died. When I asked later, I was told that this happened around eight o'clock in the evening.

Pong (thirty years old) had been close with Bang Chaet, with whom he had worked in the same forest felling trees. That day, Pong was mending a fishing net behind his house when Bang Chaet came around from the front. They talked for about ten minutes about where he had come from, and then left by the front together. Pong offered to give him a lift on his motorcycle, but Bang Chaet said there was no need, as he would be walking back, then he left in the direction of

Photograph 10.1: Khao Tanaan, near the entrance to the village

the Thanaan Rock (Photograph 10.1), while Pong walked right to buy cigarettes. This took place at about ten o'clock at night. When he got to the store and heard that Bang Chaet had died, he did not believe it, saying, 'What are you talking about? I was talking to him just now!' But when the morning came, he had the shock of his life when he learned that, indeed, Bang Chaet had died prior to their conversation on the previous day (interview, August 2017).

The first three vignettes above are all cases of perceiving the 'invisible' dead in which the perceiver knew that the perceived person was dead. That is, the perceiving subject knows and perceives that the other is a being who should be invisible. In the fourth vignette, however, the person who did the seeing and talking perceived the other without knowing that the person in question was dead.

In any case, from these cases we can point out that the dead, who have relevance to the present of the living, are perceived as something that simply 'exists.' This is different from spirits or the dead who are found in everyday life to be disturbed in some way or to have some kind of purpose, or from tutelary deities or the anonymous 'dead.' In other words, rather than an instrumental attitude that seeks to achieve some goal, they point to an attitude characterized by a passivity that simply accepts what happens.

Sensing radioactivity

Radiation that should not be absorbed by the body's organs can also be sensed. Kate Brown (2017) introduces the words of Aleksandr Kupny, a man who is intrigued by the power of radiation and worked at Chernobyl in the aftermath of the nuclear accident.[9] Kupny sees nuclear energy as a natural force; he wants to understand this force, the immense power behind the accident. Holding the view that 'You can start to understand the power by seeing how the reactor fell apart. Unless you go there, you can't understand it,' Kupny sneaks into the sarcophagus that was built over the reactor after the accident (Brown 2017: 36). To Brown's question about how he can study radiation, which is invisible to humans, Kupny replies, 'With a dosimeter, you can hear radiation. With a camera, you can see it. If the levels are high enough, you can taste it on your tongue. It tastes metallic.' After forty years of experience in radioactive fields, Kupny claims that he can sense decaying atoms (Brown 2017: 40).

This is Kupny's individual experience as someone who is fascinated by the power of nuclear energy. The inability of others to perceive radioactivity, even in the same place, is similar to how perceiving the presence of spirits or the dead is likewise an individual experience. Here the invisible seems to be confined to individual experience. However, as we shall see in the next case, when we look at the aspect in which these sensitivities are linked to certain actions that implicate other people, we find that people do not necessarily take 'the invisible' to mean 'that which does not exist.'

Coping with the invisible

As an example of receiving a message from an 'invisible' being and, on that basis, making a connection with some action, we can cite messages received from the dead through dreams.

Bang Chaet, who conversed with Pong in our last vignette, appeared many times in the dreams of non-family friends, saying things like, 'I don't know which way to go,' and 'Let me stay with you, I don't know where to go. I have no place to go.' Bang Chaet had originally been a Muslim who converted to Buddhism through marriage. However, he was not an ardent Buddhist, and eventually converted to Christianity with his family in order to receive financial aid for his children's schooling at the invitation of a Christian organization. They were the first converts to Christianity in any of the nearby villages, whose residents were a mix of Muslims and Buddhists, and for some time they attended a newly established church in a nearby town every Sunday. In the village, when Muslims and Buddhists intermarry, one or the other spouse will convert. However, it is considered that the circuit for transferring merit from the living after death must still be one of these religions (see Nishii 2001). Since his religious affiliation was unclear, Bang Chaet's wife had him interred in a temple as a Buddhist. Later, however, when she heard from Bang Chaet's friends that he had appeared to them in dreams saying that he did not know what path to take, she went to the temple for advice. As a result, in order to lead Bang Chaet's lost and wandering soul to his final resting place, the plot where he had been buried was exhumed to hold a grave-opening ritual, after which the grave-sealing ritual was carried out a second time. According to the wife, 'the monk who was supposed to lead the cortege at the funeral to guide the soul of the departed did not do so correctly. That old monk [who led the cortege] was ordained ages ago, but he knows nothing.'

Here we can see that the appearance of the dead in dreams exerts an influence on the real actions of the living. Messages from the dead sometimes appear in the dreams of their descendants, telling them what they want to eat, for example, so that when someone has a dream, it is not uncommon for other family members to prepare these foods and hold rituals for the dead. As we saw earlier in the case of divining lottery numbers in dreams, dreams are a gateway to an external world that differs from our quotidian lives. Regarding what we see and hear in dreams as being messages from beyond is a shared human practice.

In connection with the relationship between dreams and spirits, Rane Willerslev (2007) says that whereas, for Freud, dreams are understood as fantasy, for the Yukaghir, the indigenous Siberian people among whom he studies, dreams are understood as 'accomplishing' and are regarded as something real. Here, dreams are believed not as manifestation of unconscious sexual repression, as in Freud, but through sexual engagements with spiritual beings in dreams, individuals can influence events to their own advantage (Willerslev 2007: 173–80). In Thailand, likewise, dreams are not fantasies; rather, dreams about the dead are received as messages from the dead, which the living reflect in their actual conduct in the real world.

Penetrating the invisible world

In some cases, rather than simply perceiving 'invisible beings' or receiving messages that encourage certain actions, people can become caught up in the other world, as in the following examples.

In the first of these cases, which concerns a mass possession at a school, we can see the emergence of a spirit world that differs from the everyday, as well as how the school's students become entangled in that world.

Mass possession at a school

This mass possession took place at a school near my village field site, which I have been visiting continuously over the three decades since my first trip in 1987. The area is populated by a mixture of Muslims and Buddhists, with both groups speaking the Southern Thai dialect on a day-to-day basis. I start with a brief overview of the incident.[10]

On December 18, 2004, an article with the headline 'Astonishing, possessed school: Convulsing students attempt to restrain each other' appeared in Thai-

Photograph 10.2: Spirit possession school

land's national newspaper *Thai Rath*. The incident in question took place over a period of about three months, from November 2004 to February 2005. Initially, only a few students were possessed, but in the course of events the possession spread to other students. The possession phenomenon was concentrated in the physical space of the school. It subsided when students returned home, only to reoccur when they returned to school. The school became a space of intractable possession, and various events began to unfold.

The first incident occurred on November 16, shortly after classes resumed after a month-long break that coincided with the arrival of a female principal, which was a rare appointment for this area.

The school in question is located in a predominantly Muslim area. Over the eleven years since its establishment, the school had observed an Islamic *tham nuri* ritual every two years, a memorial feast that involves offering goat curry to the local tutelary deity to ask for permission to use the land. Although it was the year of the ritual, the new principal, who, as a Buddhist, considered the killing of animals to be a sin, ordered that the ritual be canceled. The mass possession began when four second-year high school girls were stricken with fainting spells. They reported that their vision darkened, their hearts began to race, they felt a shortness of breath, and they felt a weight in their chests. A fifth girl began to faint frequently, sparking fears of contagion, at which point the principal asked all five to stay home from school to recuperate. However, no sooner had those five students stopped coming to school, another ten students became possessed. This time those affected included first- and second-year junior high school students, one of whom was Buddhist. This time the symptoms were much worse, beginning with gradually blurring vision, hands wandering in the air, fainting, and uncontrollable weeping. The students caused a major fuss, and tried to run up the hill behind the school and jump into the pond. They also said that they could see the figure of a person wearing *cong kraben* (traditional Thai trousers similar to Japanese *hakama*) standing next to the teacher.

By December, authority figures, including physicians, parents, and the police, were coming to the school to address the situation, and a television station sent camera crews early each morning to cover developments. The incident was also featured on the front page of the newspaper. Around this time, the principal had run out of ideas and had started to let the villagers take whatever steps they wanted, including *saiyasat* (magic). Would-be practitioners (*mor saiyasat*, or 'magic doctors') arrived from all over Southern Thailand, about thirty in all. Some of these practitioners themselves became possessed, even before arriving at the school, and turned back along the way. By now the commotion had grown so large that offerings of goat and beef curry had no effect. Although the disturbance subsided on weekends, when the students did not come to school, when Monday came it would begin again. The turmoil finally subsided when the principal left the school in February, by which time a total of thirty-nine students had been possessed.

The spirits that possessed the students were said to belong to Muslims, Thai Buddhists, and Chinese.[11] It was alleged that although the possession had begun when the local tutelary deity had been angered by the principal's decision to cancel the *tham nuri* ritual, as the number of possessed student had increased, spirits gathered at the school from a variety of places.

Here, a process took place in which spirits were 'visualized' in various forms in the course of the possessions, including as tutelary deities, the ghosts of the dead, and Muslim spirits. Something out of place had been sensed and somatically expressed, and this something had then been brought into the open by magical and religious practitioners. Characteristically, the world of the spirits that possessed the students revealed relationships that differed from those of everyday life. Apparently, fellow students who were normally on good terms with each other became antagonistic toward each other when possessed, while, conversely, students who were normally at odds got along well as fellow spirits. Here, a world of spirits had emerged that was outside of everyday experience.

Further, I would like to give an example of the dangerous, even life-threatening, nature of entering into this other world.

Taken into the To Naan cave

The cavern beneath Thanaan Rock, which towers over the approach to the village, houses a statue of To Naan, the village's guardian deity, to whom Muslims and Buddhists alike offer their most frequent prayers on day-to-day

Photograph 10.3: Statue of To Naan

matters. The villages say that both Muslims and Buddhists are descended from To Naan, though To Naan himself is believed to be a Muslim (Nishii 2001: 107–14). In November 1994, just after the busiest time of the shrimp farming season, a young Shan man from Northern Thailand who had been employed by a Central Thai aquaculture company went missing near Thanaan Rock, where the village deity is said to reside. Reportedly, he could not even speak the Thai language. He went missing when his company paid a sightseeing visit to Thanaan Rock. Since he did not reappear after three or even four days, it was widely rumored that he had likely been killed. But on the fifth day, the young man reappeared, appearing somewhat the worse for wear.

He said that he had been taken into the cave by an old man and had not been able to leave. In the cave, he had been handed Muslim prayer beads and made to recite 'Good deeds bring about good rewards' each time he counted a bead.[12] Afterward, he had been escorted by the old man from the cave back to his accommodations near the shrimp pond. The fact that he emerged on a Friday was interpreted by the villagers to signify that To Naan had returned him so that he could go to pray. For five days, he had been eating only tree roots passed to him by the old man, and since he was still alive and relatively healthy, the villagers whispered to each other that this was evidence that To Naan must really exist.

In Japan, being taken into the other world like this would be recognized as *kamikakushi* (divine concealment), and similar instances are often heard of in many places. For example, Willerslev (2007) reports a very similar case among the Yukaghir of Siberia. A hunter tells of his experience of living together with what he thought were humans but were actually reindeer. In his village, he was believed to be missing for more than a month. Willerslev (2007: 89–90) suggests that, as against the hunter's attempts to deceive the animals by assuming their bodily appearance and movements when approaching his quarry, he was in fact on the verge of undergoing an actual metamorphosis, having been tricked by the animals into seeing the world from the perspective of his prey. In this case, as in the case of To Naan, we could say that the person in question had entered into a world that was different from everyday experience.

Prediction and anticipation in everyday practice: future-orientation and present-orientation

Prediction and anticipation

In this chapter, we have considered cases by distinguishing between the attitudes of 'prediction' and 'anticipation' in dealing with uncertainty and contingency.

Prediction, as we saw early in this chapter, is defined by Gunji (2013) as the 'calculation of a single value or point for the future,' and does not necessarily involve perception. The cases presented above were as follows:

- predicting winning lottery numbers from various signs
- measuring levels of radioactivity with a radiation dosimeter.

Conversely, the attitude of 'anticipation' was defined as 'a future-oriented calculation that cannot be determined as a single point or value, but only as an area or span' (Gunji 2013). Anticipation, moreover, was based on sensory perception. The cases of anticipation that were presented were as follows:

- knowing the presence of the dead from their odor or the sounds they make
- sensing radiation
- receiving messages from the dead in dreams
- believing students to be possessed by spirits and to have incurred the wrath of a tutelary deity
- being taken and held in a cave by To Naan.

From these cases of prediction and anticipation, the following points have become clear. In the context of prediction, an instrumental attitude is evident that, while bringing the invisible into view, determines subsequent actions by reading (or measuring) certain information with a view to the future. Also, with regard to the encounters with spirits, the dead, or radioactivity that are described as 'anticipation,' their appearance is perceived with the body, and we see an attitude that accepts the invisible in its own right as an alterity that is outside our own experience.[13] In the context of everyday practice, the appearance of this 'outside' is always contingent, and we have no choice but to accept and live with it.

To recapitulate the above, we could say that prediction is an attitude that deals with the future by rendering the invisible visible, while anticipation is an attitude that accepts and copes with encounters with the invisible in the present. By focusing on the act of dealing with the invisible, we begin to see that the invisible may fundamentally support our attitudes toward the future and toward the present, even in our engagements with what is visible in our everyday experience. Here, rather, we open up a perspective that is consistent with Merleau-Ponty's (1968: 229) 'problematic of the visible and the invisible' as an ontology, which we distinguished at the outset.

Thus formulated, prediction and anticipation are not necessarily attitudes limited solely to humans. In Chapter 1 of this volume, Adachi also draws on Gunji's distinction between prediction and anticipation to discuss differences in modes of movement within polyspecific associations of Diana monkeys and red colobuses, with the former relying on prediction, calculated based on knowledge and memory, as distinguished from anticipation as a vague lattice of possibilities. This use of instrumental prediction would not seem to be inconsistent with the discussion presented in this chapter.

Hence, in an uncertain present, we could say that primates—including *Homo sapiens*—perform actions that fall variously along a graded spectrum between future-orientation and present-orientation that cannot be completely assigned to either prediction or anticipation.[14]

Invalidating the science-magic dichotomy

Here I would like to touch on prediction and anticipation in connection with the dichotomy of science versus magic. Since prediction is a scientific activity based on objective data, whereas anticipation is a non-scientific activity that assumes a transcendental existence, it might seem that the two could be interpreted in parallel with the conventional dichotomy between science and magic.[15] But if we look at the cases discussed here from the perspective of dealing with the real world in the context of everyday practice, we find that the existence or otherwise of a physical basis is inconsequential. From the perspective of everyday practice, the conventional view of whether or not we are dealing with the physical and natural world or with supernatural events is not quite so clear cut. What I have argued, rather, is that a process shared by both is a practice that renders the invisible visible. So what is the difference between science and magic? Simply put, we could say that the difference lies solely in their respective purposes, and how models relating to those purposes are built.

The purpose of science is regarded as being to build models that approximate real systems in their important points (Todayama 2005: 251–2)—in other words, that is, to build models that are as close as possible to the real physical and natural world.[16] Masato Fukushima states that 'scientific inquiry is an instrumental activity whose purpose is to acquire information that is considered to be meaningful in a specific field; it is a process that employs various resources in pursuit of that goal' (Fukushima 2017: 39). When instrumental activities aimed at model-building are cut off from everyday practice, science-like pursuits emerge as something that is now differentiated from magic.[17] However, the extreme situation of dealing with the invisible, which I have been discussing in this chapter, could be characterized as a situation in which the physical and natural environment, as well as the world of the spirits and the dead, both dovetail with our everyday experience as equally indeterminate externalities.

From the perspective of everyday practice

When we talk about seeing things from the perspective of everyday practice, we mean describing events from the point of view of the people who live them through their bodies, describing them as though they were intrinsic to those events.[18] Such a perspective can be found in Tim Ingold's 'dwelling perspective' and Willerslev's ethnographic descriptions.

Ingold (2000: 186) provides a contrast with the 'building perspective' of representationalist theory: 'What [the dwelling perspective] means is that the forms people build, whether in the imagination or on the ground, arise within the current of their involved activity, in the specific relational contexts of their practical engagement with their surroundings.' According to Willerslev (2007: 148), when using a computer or driving a car in daily life, for example, we do not necessarily need to understand how their technology works; rather, when things are going well, typing and driving simply continue in a 'transparent coping' mode. For hunters and the spirits they address in their rituals, something similar is taking place: 'They manifest themselves as "tools" that are used in a prosaic and matter-of-fact fashion in order to accomplish concrete objectives' (Willerslev 2007: 150). Willerslev shows just how much our fundamental understanding of things, whether spirits or technological instruments, is based on practical rather than theoretical concerns. From the standpoint of the 'dwelling perspective,' he makes no a priori assumptions about the separation of the subject and the world (i.e., he does not adopt the 'building perspective'), but rather sets out from 'a primordial condition of direct perceptual engagement, of dwelling, [to] argue that the subject has to then disengage himself from the current of his activity in order to reflect upon it in the abstract' (Willerslev 2007: 152). He says that it is only then that the subject will be able to ask, 'who are the spirits?' (Willerslev 2007: 148–53). To ask 'who are the spirits' in the context of practice means to start from the assumption of an objective reality that exists separately from the subject, and to develop an argument that follows the opposite path of describing spirits in the context of a given reality.

In other words, rather than positing in advance the external existence of spirits or the dead, I treat them instead as something that is precipitated from the midst of practice. The same could be said for everyday experiences that concern the invisible. It is often the case in our daily experience that the invisible is not synonymous with not existing in a physical sense. For example, when a family member needs to get around in a wheelchair, steps in the house that were never a concern before, which were not even noticed, suddenly become apparent. Also, things that were 'invisible' up until then suddenly become visible, and may even be incorporated into daily routines, as when you notice that a store near your home—one never given much attention since it was difficult to know what it carried just by passing by outside—is a store that sells nursing products.

In other words, this awareness means that your attention will be focused on some ongoing activity. Visualization is not only the measurement of radiation that is physically invisible, but also refers to the process by which things that are physically present but invisible (or unseen) are recognized and become visible.

Bodily sensations

When actions are positioned from the context of practice onto a graded spectrum in terms of their distance from bodily sensation and the degree to which they are future-oriented or present-oriented, we find that what is seen to be characteristic of prediction and anticipation alternate with each other in some ways, and progresses even as they become mutually entangled. We live our daily lives moving back and forth between prediction and anticipation.

When trying to connect and think about the seemingly outlandish 'invisible' phenomena of spirits, the dead, and radioactivity, an episode related by the anthropologist Tomoko Niwa illustrates the bodily sensations that make this connection.[19] I would like to conclude this chapter by describing her personal experience. Niwa conducts fieldwork in rural farming villages in Shaanxi Province, in the Chinese interior, as well as in Fukushima, where she is studying attempts on the part of high school students to visualize radioactivity through theater productions and the like. She describes how when she boarded a bus when she first came to Fukushima after the earthquake to conduct research, she had not known that the bus would pass through the village of Iitate, where radiation levels are considered high. When she found out, she was surprised and became flustered, afraid that the radiation would enter her even if she held her breath. She describes her experience as follows:

> As I passed through Iitate on the bus for the first time, I felt as though the radiation was being absorbed into my body through my breath and through my skin. I felt as though something like a force from the earth had entered into my body. I felt as though the air was materialized in my body and something like the vast dynamism of the earth was seeping into my body, as though my body were being undermined and soaking up [the radiation].

She added later that she heard that even when the radiation level was high, it would have been blocked by the chassis of the bus, and thus would have had little effect on the inside of the carriage.

If we replaced the word 'radiation' here with 'spirits,' we could say that in both cases the speaker would be talking about encounters with 'invisible' energy felt with the body. When I later asked her about this again, she said:

> For the first time, I understood the feelings of the people at my field site. I was able to feel with my body the sensation that people in the Chinese village describe, as though something like a vengeful ghost or wandering spirit has entered into their body to cause illness. I had the sensation that my body would change, that I would be changed, by this thing that came from outside of me.

The extreme of the invisible casts light on the art of living in a world of uncertainty and contingency. Perhaps it may also indicate that our foundation of living is our existence as a body, that the ways in which we exist transcend our perceptions of the physical environment as captured by our bodily organs, and that in acting as a body, we are engaged each day in the creation of a new world.

Acknowledgments

In connection with my fieldwork in Fukushima, I would like to acknowledge my profound debt to Prof. Yuichi Sekiya and Prof. Tadashi Yanai, both of Tokyo University, who have been conducting 'learning trips' in Fukushima and carrying out research related to the accident at the nuclear power plant. I would like to express my deepest gratitude to both. My research in Fukushima was conducted as part of a project on 'New Anthropological Approaches to Affective Studies through Fieldwork of Critical Situations' (17H00948). I have also been involved in discussions at meetings of the ILCAA Core Project on Anthropology, as well as the collaborative research project on 'Life as Dynamism: In Search of a Methodological Connection between Affect, Thinking, and Art.' I would like to thank all of those who took part.

Notes

1. While Evans-Pritchard distinguishes between witchcraft, which is unintentionally harmful to people, and sorcery, which is maliciously harmful, here I regard magic to include both practices.
2. Merleau-Ponty (1968: 227–8) holds that 'the invisible is not only non-visible… its absence counts in the world (it is "behind" the visible, imminent or eminent visibility, it is *Urpräsentiert* [fundamentally presentable] precisely as *Nichturpräsentierbar* [fundamentally non-presentable], as another dimension).' Moreover, he clearly considers the problematic of the visible and the invisible as an ontology, writing that 'the invisible is *there* without being an *object*, it is pure transcendence, without an ontic mask' (Merleau-Ponty 1968: 229, original emphasis).
3. Even for the act of perceiving by the same individual, I do not necessarily assume the subjective perception of the objective. What I have in mind here is a stance that is open to outside influence, such as Gunji's 'first-and-a-halfth-person intelligence.' Gunji refers to such intelligence as *ten'nen chinō* (natural born intelligence), which

he explains as follows. Natural born intelligence describes a being who somehow manages to live with 'a full-fledged exteriority that can never be seen, nor heard, whose presence must be felt, and which must be accepted when it appears, even though it is completely unpredictable'; this is contrasted with *shizen chinō* (natural intelligence), which is a third-person intelligence oriented toward objective knowledge of the world, and 'artificial intelligence,' which is a first-person intelligence (Gunji 2019: 9). The use of the sesquialteral term 'first-and-a-halfth-person' refers to the first-person 'I' that confronts the second-person 'you,' who is not included in the 'I' and is, in a manner of speaking, 'the other.' But while this other is not an objective exteriority that exists independently of the 'I,' neither does it belong to the closed world contained within the 'I.' In contrast to a third-person or first-person intelligence, which are not concerned with what cannot be perceived, Gunji (2019: 32) advocates an open way of being that 'accepts and awaits the existence of something that cannot be anticipated, regardless of the fact that it has not been perceived.' This is what he calls 'natural born intelligence.'

4. My fieldwork site has been in a village in Satun province, one of four border provinces in Southern Thailand with a majority Muslim population, in a country where 95% of citizens are Theravada Buddhists. The population of the village of about 600 residents is about half Muslim and half Buddhist.
5. Note that my selection of cases is not based on a dichotomous view that associates Thailand with magic and Japan with scientific research, but merely arises from my own personal experience.
6. Ingold (2015: 92) writes that 'To see the sky, Merleau-Ponty would say, is precisely to experience its luminosity from *within*...And it is precisely in this blending of the cosmic with the affective that the sky is constituted as a manifestation of atmosphere.' We could say that the sky that Abe looked up at also contained such affective elements.
7. Indeed, measuring the amount of radiation to 'contain' uncertainty is discussed by Sharon Stephens (2001) in her chapter on 'Bounding uncertainty: The post-Chernobyl culture of radiation protection experts.'
8. Concerning this case, see Nishii (2013: Chapter 6).
9. My attention was drawn to this case by Keiichi Omura, for which I would like to express my gratitude here.
10. Further details concerning this case are discussed in Nishii 2013: Chapter 2.
11. This is not to say that there is any systematic relationship to be found among the possessing spirits, but that they simply take on the aspect of the Muslims, Thai Buddhists, and ethnic Chinese Thai who live in the area. Magical practitioners, likewise, are both Muslim and Buddhist, and while the spells used in rituals are often adapted from the scriptures of their respective religions, the spirits that they deal with are not necessarily from the same religion.
12. While this expression represents the central doctrine of Buddhism, in the village it is regarded as an important religious teaching for Buddhists and Muslims alike (Nishii 2001: 88).
13. This seems to be consistent with the 'outside' for 'natural born intelligence,' which, as mentioned in note 3, Gunji (2019: 32) says 'accepts and awaits the existence of something that cannot be anticipated, regardless of the fact that it has not been perceived.'

14. See Kawai's discussion in Chapter 9 for a discussion of what is unique to human activity. She focuses on overlapping but slightly different attitudes with a distinction between 'prediction' and 'foresight.' Of these, she holds prediction to be an attitude toward near-future events that is based on past experiences, memories, and recollections—an attitude that is also shared with other primates. By contrast, she considers foresight as a way of surveying the future that is uniquely human, something that transcends the prediction that is possible in the physical and social world, and inhabiting, rather, the institutional and cultural world.

15. In anthropology, discussions of the duality of magic and science have taken various forms. Beginning with Tylor (1871), who viewed magic as pseudo-science, Malinowski (1948: 67) also contrasted science and magic, suggesting that the theory of scientific knowledge is governed by logic, while the theory of magic is governed by desire:

> Magic is based on specific experience of emotional states in which man observes not nature but himself, in which the truth is revealed not by reason but by the play of emotions upon the human organism. Science is founded on the conviction that experience, effort, and reason are valid; magic on the belief that hope cannot fail nor desire deceive. The theories of knowledge are dictated by logic, those of magic by the association of ideas under the influence of desire.

Tambiah (1990) identifies 'participation' as the basic attitude of magic, and 'causality' as the basic principle of science, as two modes of ordering the world. He proposes that 'it is possible to separate analytically at least two orientations to our cosmos, two orderings of reality,' which he labels as:

> *participation* versus *causality*. Causality is quintessentially represented by the categories, rules and methodology of positive science and discursive mathematico-logical reason. The scientific focus involves a particular kind of distancing, affective neutrality and abstraction to events in the world... The notion of causality is much out of place, and that of participation is very much in place, when describing aesthetic or religious orientations. [Earlier anthropologists] were speaking of a holistic and configurational grasping of totalities as integral to aesthetic enjoyment and mystic awareness. (Tambiah 1990: 105–6, original emphasis)

Moreover:

> while much of the discourse of causality and positive science is framed in terms of distancing, neutrality, experimentation and the language of analytic reason, much of the discourse of participation can be framed in terms of sympathetic immediacy, performative speech acts, and ritual action. If participation emphasizes sensory and affective communication and the language of emotions, causality stresses the rationality of instrumental action and the language of cognition. (Tambiah 1990:108)

Tosa (1997: 206) reformulates Tambiah by saying that the pole of the law of causality, which is based on an atomistic paradigm that represents science, and the pole of the law of participation, which is based on a holistic paradigm that represents ritual communication (including magic), respectively express two irreducible principles for the world.

However, against this dichotomous view, which looks to magic for sympathetic directness in contrast to the distancing of science from events as methodology, Satoshi Nakagawa points out that both magic and science share a reliance on the method of induction. Nakagawa (2009: 58–75) argues that there is no real difference, in the sense that science is also based on the groundless belief that 'the world is orderly.' However, the fact that science, like magic, presupposes perceptual faith was pointed out long ago by Merleau-Ponty (1968: 14–49), who was attempting to overcome the problem of philosophy and science. This way of thinking about scientific methodology is in line with the way of thinking about reality that constitutes Actor–Network Theory (ANT) as it has been advanced by Bruno Latour and his interlocutors, who have recently brought an anthropological perspective to the study of scientific activities in their own right (Latour 1987; Fukushima 2017: 16–17). In the context of ANT, rather than assuming a *tout faite* (ready-made) reality—i.e., a reality that is waiting to be discovered—it is acknowledged that existence is in the process of being constituted, that it is 'in the making' (Maniglier 2014: 38).

Moreover, recent studies of magic, rather than demonstrating the irrationality of the non-Western world (on the assumption of a dichotomy in relation to the West), instead treat magic as an aspect of apprehending humans that could exist in any world. Magic is understood as the way people make sense of reality. These studies show how magic exists as something that organizes the experience of people attempting to cope with a reality that is fraught with uncertainty and contingency as a part of their generation of reality, including the aspect of regarding contingent reality as something inevitable, as seen in Evans-Pritchard (in Shirakawa 2012: 27–9) and emergent acts that are in accordance with the irreducible by embracing possibilities that deviate from the irreducible (Ishii 2010: 163–6). See, for example, Abe et al. (2007), Hamamoto (2001, 2014), and Shirakawa and Kawada (2012).

16. Without getting into the controversy of whether or not to believe that any such thing exists, I will venture to say that, for scientists, in the context of scientific debates, a better constructed argument is said to be almost synonymous with existence (Fukushima 2017: 18).
17. This also parallels a distinction made by Naoki Kasuga (ed.) (2016: 17) in *Linking Science and Culture*, wherein he explains that from the relationship between the perception of will and the cognition of patterns, it could be said that perception of will creates a cultural field of meaning and value, while cognition of patterns serves as a pillar of science. In other words, the former focuses on dealing with reality in practice, while the latter focuses on model building.
18. While I am unable here to discuss the perspective of 'inward observation' raised by Gunji and Matsuno from inside scientific practice, I would like to point out that there are certain commonalities (Gunji et al. 1997).
19. I wish to express my heartfelt gratitude to Ms. Tomoko Niwa for sharing her valuable experiences in interviews I conducted with her on June 3 and July 8, 2018.

References

Abe, T., R. Oda and H. Kondō (2007) *Jujutsu-ka suru modaniti: Gendai Afurika no shūkyōteki jissen kara* (Enchanting modernity: Religious practice in contemporary Africa). Fukyosha.

Brown, K. (2017) 'Marie Curie's fingerprint: Nuclear spelunking in the Chernobyl zone'. In A. Tsing, H. Swanson, E. Gan and N. Bubandt (eds), *Arts of Living on a Damaged Planet: Ghosts of the Anthropocene*. University of Minnesota Press, G33–G50.

Evans-Pritchard, E. E. (1937) *Witchcraft, Oracles and Magic among the Azande*. Oxford at the Clarendon Press.

Fukushima, M. (1998) '"Gūzen" o kainarasu: Sekai no jujutsu-teki kōsei' (Taming 'chance': The magical structure of the world). In Y. Kobayashi and T. Funabiki (eds), *Shin chi no gihō* (New academic skills for the human sciences). University of Tokyo Press, 199–211.

Fukushima, M. (2017) *Shinri no kōjō: Kagaku gijutsu no shakai-teki kenkyū* (The factory of truth: The social study of science and technology). University of Tokyo Press.

Fukushima 30-Year Project (2015) *Kodomo hōshasen wākushoppu* (Children's radiation workshop): *Fukushima 30-Year Report, vol. 3*. Fukushima 30-Year Project.

Fukushima 30-Year Project (2017) *Mama-tachi no hottosupotto faindā taiken-ki* (Mothers' experiences as hotspot hunters). [Pamphlet; illustrated by Kimidori Inoue.] Fukushima 30-Year Project.

Gunji, Y. P. (2013) *Mure wa ishiki o motsu: kono jiyū to shūdan no chitsujo* (The swarm is conscious: This freedom and the orderliness of groups). PHP Institute.

Gunji, Y. P. (2019) *Ten'nen chino* (Natural born intelligence). Kodansha.

Gunji, Y. P., K. Matsuno and Otto E. Rössler (1997) *Naibu kansoku* (Inward observation). Seidosha.

Hamamoto, M. (2001) *Chitsujo no hōhō: Kenia kaigan chihō no nichijō seikatsu ni okeru girei-teki jissen to katari* (Order and method: Ritual practices and narratives in everyday life in Kenya's Coast Province). Koubundou.

Hamamoto, M. (2014) *Shin'nen no jubaku: Kenia kaigan chihō duruma shakai ni okeru yōjutsu no minzoku-shi* (The curse of belief: An ethnography of witchcraft in Duruma society in Kenya's Cost Province). Kyushu University Press.

Ingold, T. (2000) *The Perception of the Environment: Essays in Livelihood, Dwelling and Skill*. Routledge.

Ingold, T. (2015) *The Life of Lines*. Routledge.

Ishii, M. (2010) 'Jubutsu o tsukuru, <sekai> o tsukuru: Jujutsu no kōi suikō-sei to sōhatsu-sei' (Creating monsters, creating 'worlds': The performative and emergent character of magic). In M. Yoshida, M. Ishii and K. Hanabuchi (eds), *Shūkyō no jinruigaku* (The anthropology of religion). Shumpusha Publishing, 159–174.

Kasuga, N. (2016) 'Anarojī no hitaishō-sei kara kangaeru ito to patān: Kagaku to bunka o tsunagu shisaku' (Thinking about intentions and patterns from analogical asymmetry: Ideas linking science and culture). *UP*, 2016-8: 14–19.

Kasuga, N. (ed.) (2016) *Kagaku to bunka o tsunagu: Anarojī to iu shikō yōshiki* (Linking science and culture: Analogy as a mode of thought). University of Tokyo Press.

Latour, B. (1987). *Science in Action: How to Follow Scientists and Engineers through Society*. Harvard University Press.

Malinowski, B. (1948) *Magic, Science and Religion and Other Essays*. The Free Press.

Maniglier, P. (2014) 'A metaphysical turn? Bruno Latour's *An Inquiry into Modes of Existence*'. *Radical Philosophy*, 187: 37–44.

Merleau-Ponty, M. (1968) *The Visible and the Invisible: Followed by Working Notes*. Northwestern University Press.

Nakagawa, S. (2009) *Gengo gēmu ga sekai o tsukuru: Jinruigaku to kagaku* (Language games make the world: Anthropology and science). Sekaishisosha.

Nishii, R. (2001) *Shi o meguru jissen shūkyō: Minami Tai no musurimu bukkyōto kankei e no pāsupekutivu* (Death and practical religion: Perspectives on Muslim–Buddhist relations in Southern Thailand). Sekaishisosha.

Nishii, R. (2013) *Jōdō no esunogurafi: Minami Tai no mura de kanjiru tsunagaru ikiru* (The ethnography of affect: Feeling, connecting, and living in a Southern Thai village). Kyoto University Press.

Shirakawa, C. (2012) 'Kotoba, kōi, jujutsu' (Words, actions, and magic). In C. Shirakawa and M. Kawata (eds), *Jujutsu no jinruigaku* (The anthropology of magic). Jimbunshoin, 9–45.

Shirakawa, C. and M. Kawada (eds) (2012) *Jujutsu no jinruigaku* (The anthropology of magic). Jimbunshoin.

Stephens, S. (2001) 'Bounding uncertainty: The post-Chernobyl culture of radiation protection experts'. In Susanna M. Hoffman and Anthony Oliver-Smith (eds), *Catastrophe & Culture: The Anthropology of Disaster*. School of American Research Press, 91–111.

Tambiah, S. J. (1990) *Magic, Science, Religion, and the Scope of Rationality*. Cambridge University Press.

Todayama, K. (2005) *Kagaku tetsugaku no bōken*: *Saiensu no mokuteki to hōhō o saguru* (Adventures in the philosophy of science: Exploring the aims and methods of science). NHK Publishing.

Tosa, M. (1997) 'Jujutsu no gendai-sei: Jujutsu-ron ni miru gendai seiyō no tasha hyōshō' (Contemporary magic: Contemporary Western representations of alterity in theories of magic). In T. Aoki et al. (eds), *Shūkyō no gendai* (Contemporary religion). Iwanami Shoten.

Tylor, E. B. (1871) *Primitive Culture: Researches into the Development of Mythology, Philosophy, Religion, Language, Art, and Custom*, 2 vols. John Murray.

Willerslev, R. (2007) *Soul Hunters: Hunting, Animism, and Personhood among the Siberian Yukaghir*. University of California Press.

CHAPTER 11

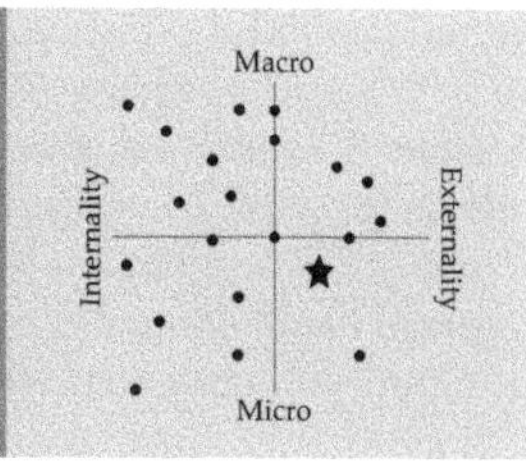

On the Environment and Existence in Hazard Situations:

Comparing Extreme Situations in the Mindanao Conflict and Radioactive Damage in Fukushima

Ikuya TOKORO

Keywords: hazard, Mindanao conflict, Fukushima Daiichi power plant, radioactive disaster, natural extremes and social extremes, state of exception, Anthropocene

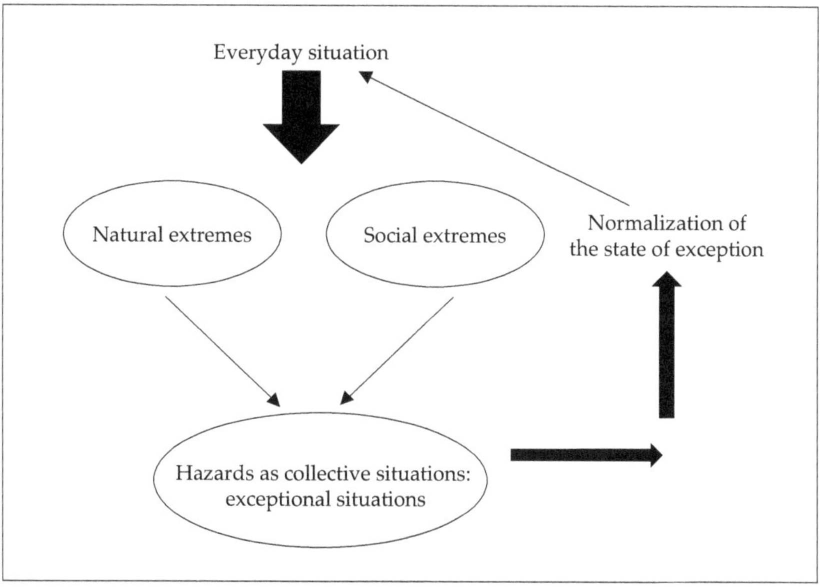

Extraordinary and extreme situations that pose a danger or threat to existence or survival for individuals or groups can be further distinguished as natural extremes and social extremes. Nevertheless, when we examine specific cases of hazards such as conflicts or nuclear accidents, it becomes evident that hazards are in fact engendered by the interplay of both social and natural (environmental) factors. In addition, although such hazards generally constitute unusual and exceptional situations, in some cases they can become the norm when observed at the microscopic level.

On existence and the environment in extreme situations

In this chapter, I conduct an exploratory study concerning the environment (in the broad sense) and existence under extreme conditions, as represented by so-called 'hazards.' More specifically, I take up two cases of extreme situations (hazards in the broad sense) to contrast and compare from the perspective of risk research; namely, conflict and violence surrounding the Muslim ethnic minority in the southern Philippines, collectively known as the Moro, and the radioactive disaster that resulted from the Fukushima nuclear accident that occurred in conjunction with the Great East Japan Earthquake. After this discussion, I examine some of the implications that these specific cases of extreme situations have for the more macroscopic level of the history of human evolution.

To start, I provide an overview of the case of what is known as the Mindanao conflict, which is a focus of this chapter. In the southern Philippines, conflict between the Moro and (mainly Christian) settlers and government forces began to intensify from the mid-twentieth century owing in part to land issues and demographic changes that were taking place in Mindanao. In the 1970s, this escalated into an armed conflict between the militant Moro separatist movement and the government's security forces (the so-called Mindanao conflict or Mindanao civil war). The conflict has progressed unpredictably and has yet to find a resolution. As recently as May 2017, an incident took place in which an armed group closely affiliated with the Middle Eastern Islamic State (IS) occupied Marawi, a city in Mindanao. Owing to this incident, fierce fighting continued for almost half a year between the government's security forces and the Maute group, which took hostages and holed up in the city. The conflict resulted in the creation of a huge number of internally displaced persons (IDPs)—as many as 400,000 by one local estimate. In the first half of this chapter I consider some of the extreme situations that this conflict has visited upon these displaced persons, including some that are literally critical to their survival.

In the second half of the chapter I discuss the Fukushima nuclear power plant accident that took place in March 2011 in conjunction with the Great East Japan Earthquake, as well as the resulting radioactive disaster. Within the area affected by the accident, a wide area located primarily within a 10-kilometer radius of the plant was designated as an evacuation zone. As a result, many residents of the municipalities around the plant were effectively forced out of their homes to live as evacuees. In addition, in zones within Fukushima Prefecture not designated as evacuation zones and, in some cases, even in prefectures other than Fukushima Prefecture, many residents opted to become so-called 'voluntary evacuees' due to concerns about the radiation risks in the wake of the nuclear accident.

Since 2017, as evacuation orders have begun to be lifted in some communities that had been designated as evacuation zones—and even where the evacuation order has been lifted—it has been pointed out that scattered pockets of relatively high levels of radiation (so-called hot spots) can still be found. Against the context of this background, more than a few residents of the areas where evacuation orders have been lifted continue to debate whether they should really return to their homes.

In this chapter, in discussing the two cases of the Mindanao conflict and the radioactive disaster in Fukushima, I attempt to examine each case from the perspective of their respective extreme situations, while at the same time comparing their mutual similarities and differences. As a reference framework for this undertaking, I develop the discussion from the tentative starting point of a conceptual distinction between natural extremes (e.g., situations of heightened risk or threats to the existence or survival of individuals and groups due to changes in the natural environment, etc.) and social extremes (e.g., situations of heightened risk or threats to the existence or survival of individuals and groups due to changes in the social environment). Furthermore, in the final part of this chapter, I raise some issues about the implications of this chapter's discussion of extremes in the context of human history at a more macroscopic level, particular in the context of debates over the Anthropocene.

What do we mean by extreme situations? Hazards, natural extremes, and social extremes

I would like to provisionally conceptualize the 'extreme situations' dealt with in this chapter as extraordinary events in which conventional daily life is disrupted by some change (or changes) in the surrounding environment that could directly or indirectly constitute an existential risk (danger) to the local community. Typically, these would include crisis situations associated with many forms of mass violence, such as war and conflict, as well as other so-called hazard situations.

The term 'hazard' typically refers to so-called natural disasters, such as earthquakes, tsunamis, or typhoons (think of 'hazard maps,' for example), but is not necessarily limited to these. 'Hazard' can also be used to refer to events that are caused by factors in human society, and which are deeply intertwined with human social conditions (e.g., vehicular 'hazard lights' or 'moral hazards'). The word 'hazard' can also simply mean 'chance' in Western languages (e.g., French), and may not necessarily carry the nuance of an existential threat. However, in this chapter, I would like to develop my discussion while keeping in mind hazards as typical events and situations that may directly or indirectly pose a danger or threat to the survival of individuals (or groups), particularly in the context of events and occurrences that take place accidentally.[1]

Moreover, as a tentative starting point for a discussion of extreme situations (including hazards), it would seem to be effective to distinguish broadly between two types of extreme situations—namely, natural extremes and social extremes. Here 'natural extremes' refer to situations of heightened risk or threats to the existence or survival of individuals and groups due to changes in the natural environment. In contrast, 'social extremes' may be thought of as situations of heightened risk or threats to the existence or survival of individuals and groups due to changes in the social environment. In this chapter, while using these two ideas of extremes as a provisional starting point for our discussion, I would eventually like to add a critical examination of how the two concepts intertwine.[2]

The Mindanao conflict in the southern Philippines

An overview of the conflict

In the Philippines, Muslims represent an ethnic minority estimated to comprise between approximately 5% and 7% of the total population. Generally, they are referred to collectively as the Moro people (or Bangsamoro, literally 'the Muslim nation'). The Moro population is estimated to be between five and seven million people, most of whom live in the southern Philippines, on the islands of Mindanao and the Sulu Archipelago. It should be noted here that 'Moro' is a collective term, and that it is possible to distinguish between about thirteen separate ethnic groups, such as the Maranao, Maguindanao, and Tausug, all of which have different languages and other characteristics.

Generally speaking, peace talks between armed groups on the part of the Moro and the Philippine government have been ongoing in conjunction with intermittent armed clashes between the two sides since the latter half of the twentieth century. The background of this situation is as follows.[3]

Until the end of the nineteenth century, the Moro people of the southern Philippines lived in their own Islamic provinces, such as the sultanates of Sulu and Maguindanao. By the twentieth century, however, as a result of the Spanish–American War, the United States took control of the Philippine archipelago from Spain and began the process of colonizing the Philippines, including Sulu and Mindanao. During the period of American colonial rule and subsequent Philippine independence, large numbers of Christian farmers were sent to Mindanao as internal migrants (i.e., settlers) as the direct result of the government's internal migration policies. As a result of these policies, Muslims in Mindanao were deprived of their lands in favor of the migrant Christian farmers, and were forcibly marginalized and impoverished, both socially and economically. Later, from the end of the 1960s into the early 1970s, clashes that had begun as land disputes between Muslim residents and the Christians who had arrived in Mindanao as settlers gradually escalated into armed conflict.

Against this backdrop, the Moro National Liberation Front (MNLF) emerged as a Muslim separatist organization in the early 1970s. At the time, the MNLF received financial support and bases for training from countries like Libya and Malaysia, and soon began an armed struggle with the goal of seceding from the Philippines. As a result, conflict accompanied by violent armed clashes took place between the MNLF and the Philippine National Army (the Mindanao conflict) in various parts of Mindanao and the Sulu Archipelago throughout the 1970s. In 1972, President Ferdinand Marcos imposed martial law, partly due to the political instability caused by the activities of Muslim separatist forces in the southern part of the islands, as well as those of communist forces, but the war continued to expand, notably resulting in the destruction of the city of Jolo, the urban center of the Sulu Archipelago, in a major battle that took place in 1974.

In 1996, after a tortuous process, the administration of President Fidel Ramos signed a peace accord with the MNLF in which the insurgents agreed to accept autonomy for Mindanao and parts of Sulu in exchange for renouncing their demand for secession. Since then, however, several other armed Muslim groups, including the Moro Islamic Liberation Front (MILF), a splinter group of the MNLF, have continued to stage terrorist attacks and engage in repeated clashes with the government's security forces, and the conflict has not yet reached a fundamental resolution. The armed conflict in Mindanao has caused the deaths of at least 150,000 people and the displacement of millions of refugees (including those forced to flee across the sea to Malaysia).

However, it can hardly be said that this explanation of the general situation based on the macro-level political and religious confrontation between the Muslim minority Moros and the mainly Christian-affiliated government necessarily adequately conveys the real situation as it is experienced at the micro-level of local individual actors. In order to provide a more concrete description, in the following sections I would like to examine the characteristics of the Mindanao conflict as an extreme situation, primarily with reference to the narratives of displaced persons that I heard in the field, where the conflict has been taking place.

Narratives of displaced persons in the southern Philippines

Here I would like to consider first of all the case of an individual who survived the battle that took place on the island of Jolo in the Sulu Archipelago in the early 1970s, followed by the case of a person who was displaced after a more recent battle in the city of Marawi in 2017.

Testimony of Aida, a survivor of the battle of Jolo

I would first like to showcase an excerpt from the testimony of Aida (a pseudonym), a Muslim woman who became embroiled in the fierce fighting that erupted on Jolo in 1974 as part of the wider Mindanao conflict.[4] Aida is a Sama woman from Tabawan Island in the southern part of the Sulu Archipelago, and in 1974 was working as a shop clerk in a Chinese store on Jolo Island. The following is her account of the battle of Jolo, which is generally regarded as the fiercest battle to have taken place in the Mindanao conflict.

> At the time, I was in my late teens and working in a Chinese man's shop in Jolo town. The shopkeeper and his wife were away, taking their child to the hospital in Manila, and I had been left to mind the shop alone with the shopkeeper's other child, who was seven years old. One evening, the sound of gunfire rang out all over town. At the time, Jolo town was already full of activists [i.e., MNLF guerrillas]. The shopkeeper's child and I were so scared that we hid behind a sack of rice in the storeroom. I was so scared that I couldn't even stand up. Eventually one of the activist *nakura* [leaders] came into the store. 'Are you with us?' he asked. 'Or are you with the government?' I replied that 'We're with you [MNLF guerillas].' At that, the *nakura* said, 'All right, then you should flee into the mountains. The government forces will be here soon.' So I fled with my employer's child toward the mountains. But there was another squad of government forces in the mountains, as well, and a battle broke out between them and the activists. That's why the Chinese child and I took refuge in the house of Mr. Abubakar, who was the mayor at the time. That mayor had 'two faces' in that he was on the government's side, while at the same time his nephew was one of the activist *nakura*. But around three o'clock in the morning, the house began to be fired upon by the government forces, and we were forced to flee again. We finally found shelter in the hospital in Jolo, but it was full of the bodies of the dead, and there was no food or even anything to drink. We had no choice but to leave the hospital. For the next three days, the seven year old and I wandered in each other's company with nothing to eat or drink. Every house in Jolo had been bombed, and the air was filled with the stench of corpses. We went back into the mountains to look for food. But all of the fruit and other food had been eaten up by the soldiers and guerrillas. And the fighting was still going on in the mountains, so we couldn't stay there. We finally reached the harbor and fled to Siasi Island by boat.

From Aida's testimony, we can point out some interesting features. First, Aida recalls that she was not only afraid of the government forces, but also of the MNLF guerrillas, or 'activists.' While there is not enough space here to examine every individual detail, a general characteristic of Aida's testimony is that its narrative is filled with indeterminacy and ambiguity that contrast with, and cannot necessarily be reduced to, something like the macro-level overview of the conflict described above, which is schematically based on the idea of opposition between Muslims and Christians. In this context, it was not only the government forces, of whom the majority were Christian, but also the MNLF guerrillas, who were ostensibly her co-religionists, who represented potential sources of fear and threat to her. In some cases, these characteristics also continue to be evident in the narrative of Muslim displaced persons in the more recent stages of the conflict, as we shall see.

Displaced persons and the Siege of Marawi

At the time of writing in July 2018, the peace process with the MILF had made some progress under the administration of President Rodrigo Duterte, which was inaugurated in 2016. In May 2017, however, an incident occurred in which the city of Marawi on Mindanao was occupied by the Maute group (MG), a cell of armed insurgents with links to IS, the Middle Eastern terrorist group mentioned earlier.

More specifically, on May 23, 2017, MG attacked the city of Marawi (population 200,000) together with Abu Sayyaf, another armed Islamic group affiliated with IS. The MG forces, numbering an estimated 700 people in total, occupied the city center, including Catholic churches and schools, took numerous hostages, and entrenched themselves before the situation eventually culminated in a large-scale battle with the government's security forces. Although the fighting

Photograph 11.1: A Muslim IDP (left) from Mindanao, Southern Philippines (outskirts of Manila)

finally drew to a close in October 2017, as a result of the fierce battles that had raged for almost six months, more than 1000 people died in and around Marawi, more than 400,000 residents of the city and the surrounding area became IDPs, and the city center was quite literally burned to ashes.

In this section, I introduce and discuss a selection of testimonial accounts of displacement based on multiple interviews I conducted in 2017 in the Philippine metropolitan capital region and in Davao City on Mindanao.

Narratives of internally displaced persons

To begin, I present the narratives of several IDPs living in S-town (pseudonym), a community in Metro Manila (based on interviews conducted in Metro Manila, June 18–26, 2017).

S-town is home to a community of Maranao (a Moro people) who migrated from Mindanao prior to the Siege of Marawi. Among other pursuits, these local residents are mostly engaged in the retail trade of various goods such as mobile phones and parts, luxury goods such as pearls, and timepieces. After the outbreak of large-scale fighting in Marawi on May 23, 2017, many citizens who could afford to do so fled the violence by escaping to Metro Manila, including S-town.

Among those concerned with these Maranao IDPs, Fatima (pseudonym), a Muslim woman I met at a market in S-town, described the fighting in her hometown by saying that 'Marawi is just like Syria…The people are being killed like cockroaches.' When I asked her about IS-affiliated groups like MG, she expressed her anger at MG—whose members, like her, were both Muslim and

Photograph 11.2: A security vehicle patrolling Mindanao under martial law (Davao City)

Maranao—saying that 'everyone is angry at IS' and 'I have no idea why they are doing such a thing during Ramadan.'

She felt no sympathy toward the actions of MG or other IS-affiliated armed groups, describing them, rather, as a bunch of outsiders who had ruined her relatively peaceful life. When asked her opinion of the government, she expressed mixed feelings, allowing that 'I support [President] Duterte, but Maranao locals are being killed indiscriminately by the government's airstrikes.' Reputed to have Maranao blood on his maternal grandmother's side, Duterte was, on the whole, a popular president among the Maranao and other Moro peoples from the beginning. However, the indiscriminate aerial bombing carried out by the government as part of the hostilities during the Siege of Marawi led to an increase in criticism toward the government and President Duterte among Maranao citizens. Her eyes glistening with tears, Fatima said, 'If Duterte really has Maranao blood, I would like him to listen more to the voices of the Maranao people, and to support those who have been displaced from their homes.'

Another displaced person, a Maranao woman named Lucida (pseudonym), said that many infants had died of malnutrition, diarrhea, and various other illnesses in shelters across Mindanao, including the one in which she was living. She said that the situation in war-torn Marawi was even more tragic, and that many residents who had been caught up in the fighting were still unable to escape the city. Many of these people had exhausted their supplies of food and were struggling with starvation. She spoke of how many in such circumstances were facing situations in which their lives were quite literally on the line, as some of those who remained in Marawi 'have no food left and are so hungry that they have no choice but to eat cardboard and the fibers of their blankets and bedding.'

In July 2017, when the fighting in Marawi was still intense, I visited Davao City on Mindanao, where I was able to hear the accounts of evacuees from Marawi and their families.[5] Salim, a Maranao IDP who has found refuge with a Maranao community in Davao, told me, 'I can't trust either the government [in Marawi] or IS [i.e., MG]. Some hostages were forced by IS to take part in the fighting. In Marawi, it's hard to tell friend from foe. You can't say that someone is an ally just because they're a Muslim. That's because IS regard Muslims who are unbelievers [in their eyes] as…infidels, and thus enemies.'

The IDP narratives convey several suggestive characteristics concerning the current conflict, including the events that took place in Marawi. One of these is the severity of the extreme situations in which local residents and IDPs find themselves. As clearly shown in the account of people who 'have no food left and are so hungry that they have no choice but to eat cardboard and the fibers of their blankets and bedding,' they are literally in a life-or-death situation that threatens the very survival of these individuals. Forced to resort to such unusual measures (such as eating things that are not food) to survive, we could say that their situation is literally extreme.

In the context of the conflict, this extreme can be understood as a social extreme—that is, a situation of heightened risk to survival due to changes in the social environment (including, in particular, the outbreak of violent clashes between armed groups and government security forces).

At the same time, we should pay attention to Salim's account that, in Davao, it was 'hard to tell friend from foe.' In other words, this suggests that in the context of a social extreme, there is an aspect in which it is not always clear which actors (i.e., which groups) represent the primary threat or risk.

In fact, as we can see from both Salim's and Aida's narratives, in the context of the fighting that has taken place against the background of the Mindanao conflict, the simple schematic of 'Muslims versus Christians' does not apply. For example, from the perspective of the (former) residents who became IDPs as a result of the fighting, we could say that both the government forces and the MG forces appear equally as fearsome 'others' (enemies) who threaten their survival. In other words, this could be said to be an extreme situation characterized by a certain incompleteness of information in the sense that the boundary between friend and foe is unclear and ambiguously defined.

Another characteristic of the situation of IDPs in recent years is that, as a result of the protracted conflict in the southern Philippines that has lasted since the 1970s, the inhabitants of conflict-affected areas have been forced to move back and forth between their villages and evacuation centers. This situation has resulted in a prolonged 'evacuee lifestyle' among IDPs, to the extent that, in some cases, evacuation—which is supposed to be a temporary and exceptional situation—has become the de facto 'routine' of everyday life. Maranao communities in the suburbs of Manila (including S-town) can also be said to have started out as groups of people who originally fled their native Mindanao in the 1970s to avoid the hostilities, and whose population has gradually been increasing. Now, in conjunction with the prolonged and consequent de facto 'routinization' of their life as evacuees (which was originally supposed to be an extraordinary situation), an influx of newcomer evacuees is arriving in such Muslim communities all over the Philippines, including in Metro Manila, following developments such as the fighting in Marawi City in 2017.

The Fukushima nuclear accident

'Refugees' in Japan: people displaced by the Fukushima nuclear accident

In the previous section, I discussed the extreme situation facing IDPs displaced by the conflict in the southern Philippines, but the problem of displaced persons cannot be said to be unique to developing countries like the Philippines.

In Japan, for example, there have been people who could be considered 'refugees' (IDPs) in the broad sense (even if only temporarily) as the result of past natural disasters. In the wake of the Great East Japan Earthquake of 2011, huge numbers of disaster victims, mainly in the Tohoku region, were forced to live as evacuees (including temporary cases), with some even continuing to live effectively as evacuees as of the time of writing in 2018. In this section, I would like to discuss the cases of evacuees (including people other than residents of Fukushima Prefecture) affected by the accident at the Fukushima Daiichi nuclear power plant that was triggered by the earthquake and tsunami in order to examine their situation from the perspective of extreme situations.

As of May 2012, the total number of evacuees from Fukushima Prefecture (including those displaced by the tsunami) was 164,865. As of March 2017, according to data released by Fukushima Prefecture, the approximate number of people still living as evacuees was estimated to be about 80,000 (including both in-prefecture and out-of-prefecture evacuees), of whom 12,381 were still living in temporary housing.

In the broader sense, those displaced by the nuclear accident include those who were evacuated from areas of Tohoku and eastern Japan other than Fukushima Prefecture. In terms of the places to which people were evacuated, the destinations range from other places in Fukushima Prefecture to other parts of Japan, and even to overseas locations such as Southeast Asia. There is also a great deal of diversity in the attributes of these evacuees that cannot be ignored, including whether they were forced to evacuate based on government or administrative evacuation orders and whether they evacuated voluntarily, temporarily, or semi-permanently (Institute of Disaster Area

Photograph 11.3: An entrance to a 'difficult-to-return' zone (in the vicinity of the Fukushima Daiichi nuclear power plant)

Revitalization, Regrowth and Governance et al. 2015; Satō and Taguchi 2016: 410–14). Among this diversity of persons displaced against the background of the nuclear accident, I would like to focus, first, on the narratives of those native to Fukushima Prefecture who were forced to evacuate as victims of the nuclear accident due to the fact that they were living in the vicinity of the nuclear power plant. Then I would like to turn to discuss the cases of those who were living in Tokyo at the time of the accident, but who voluntarily chose to evacuate due to concerns over the risks of radioactivity.

Narratives of victims evacuated due to the Fukushima nuclear accident

Extract from the narrative of K-san (male in his sixties, organizational employee)

Originally from the village of Iitate, K-san has for a long time worked in the coastal city of Minamisōma, which is where he was on the day of the Great East Japan Earthquake, working at his desk in the building of the organization he works for.

Even now, he vividly remembers the afternoon of the day of the earthquake. When the quake began, he says that he remained psychologically calm, so he was able to steady the shelves of documents in his office with his hands so that they would not topple over. As the trembling grew increasingly severe, the contents of the room began to fly past him through the air. Concerned about the possibility that the building itself might collapse, K-san left his office only to find that the nearby roads had liquefied. He called his wife first, and learned that she was safe, at least for the time being. However, he was unable to get in touch with his in-laws (his wife's parents), who lived in the coastal region of Hamadōri, which had been struck by the tsunami. Concerned that his in-laws might have been swept away by the tsunami, K-san called his relatives and acquaintances to inquire about their whereabouts, but no one knew where they were. At length, K-san returned to his home in Iitate after midnight without having found any leads.

At the time, between March 11 and March 14, he was unable to watch television or listen to the radio because of a power outage at his house due to the earthquake. During that time, he had no idea about the nuclear accident that was unfolding. He realizes in hindsight that he should have listened to the radio in his car, but this did not occur to him at the time. Therefore, unaware of the seriousness of the situation at the Fukushima Daiichi nuclear power plant, K-san prioritized searching for his wife's parents and checking school gymnasiums around Fukushima Prefecture (which at the time were being used as temporary morgues).

Then, on March 14, he returned to his office in Minamisōma to arrange some documents. Afterward, he went to the municipal office in Hamadōri at around eleven o'clock in the morning to inquire about his parents-in-law. As he arrived outside the municipal office, someone inside challenged him in a grave voice, 'What are you doing outdoors? Get inside!' At the municipal office, he learned for the first time of the seriousness of the situation at the nuclear power plant. Later, K-san sped back to Iitate in his car. Even so, he says that on his way home, he saw some people blithely working away in the fields.

At the time of the nuclear accident, K-san says that he did not hear any announcements by the municipal office or other authorities over the public address system loudspeakers. In any case, K-san returned to his house in Iitate, packed about three days' worth of clothing, then evacuated with his wife to Fukushima City. There they found shelter in one of the gymnasiums that had been designated as evacuation centers. Many of the evacuees there were from Futaba District and other areas in the vicinity of the power plant. However, aside from K-san, the residents of Iitate were not informed of the Speedi data (forecasts of radiation dispersal based on measurements of radioactive releases), so many evacuated precisely in the direction of the spread of the radiation.

At the time, K-san and his wife mooted the idea of fleeing to Yamagata or Aizu, since the Fukushima City center might have been dangerous if the situation were to continue unchecked. However, the snow in Aizu was deep and the tires on their car were not in good condition, so they abandoned the idea and registered as displaced persons at the gymnasium. At first, they simply lay down on the floor to sleep. After a while, they were provided with cardboard and a blanket. Food was distributed in the form of sweetened bread rolls. They stayed in the gymnasium-cum-shelter until April 1. They had left their cat in Iitate, so went back to feed her from time to time during their evacuation ordeal.

From April 1 to April 21, they temporarily returned home to Iitate. On April 22, however, they were suddenly told that the entire village had been designated as an evacuation zone and that they would have to leave immediately. The government had been saying that the village was safe until just beforehand, so this was a sudden shift in policy. At the time, many residents were bewildered at being told that the entire village would have to be evacuated.[6]

Extract from the narrative of H-san (male, mid-forties at the time of the accident, self-employed)

When the earthquake struck on March 11, 2011, H-san was living in Tokyo, where he experienced the quake in his car on his way to work in the city. At first, he thought the earthquake was just the car shaking as he drove, but he soon realized it was an earthquake when he saw the surrounding utility poles and the ground shaking violently. The shaking felt very strong and seemed to last a long time. Intuiting that this was no ordinary quake, instead of going to work, he returned home to his wife and three-year-old daughter.

For the next two days, he mostly remained home, scouring the television and Internet for information. He was shaken to learn that an accident had occurred at the nuclear power plant in Fukushima, leading to a state of emergency. He could not help but feel that the government was deliberately hiding a grave truth, as though information was being controlled or that the truth about the accident was being covered up. On the other hand, social media was rife with all sorts of untruthful information, including the possibility that a high-dose plume (an atmospheric flow of air containing radioactive materials) generated by the nuclear accident would soon reach the Kanto and Tokyo areas.

Given the circumstances, and fearing the worst, H-san decided to evacuate with his family. Unsure where the safest place to flee might be, he felt that, in any case, it would be best to leave Japan. On March 14, he hurriedly searched for tickets for himself and his family to destinations in Southeast Asia. Luckily, he was able to find tickets to Malaysia online. With these tickets in hand, H-san left his home in Tokyo with his wife and small daughter on the same day. Arriving at Haneda Airport, they found that a long line had formed at the ticket counter. It was late in the evening when they finally took off from Haneda. Looking out the window at Tokyo below, he found himself crying spontaneously. He says that he felt relieved to have escaped eastern Japan and that he would be safe for the time being, but he was also worried about his younger brother, as well as other family members and friends who were still in the Kanto area. Wondering when he might be able to return to Tokyo and what would happen to his life, he felt a complicated mix of anxious emotion.

On the morning of the next day, March 15, they checked into the hotel that he had booked online. When he turned on the television in their hotel room, they learned that the Fukushima Daiichi nuclear power plant was in a state of imminent danger. He says that watching the images of Self-Defense Force helicopters spraying water on the plant left him with a feeling of despair.

H-san subsequently traveled from Malaysia to the Philippines and, in April, decided that he would have to return to Japan to resume his work. Even so, rather than return to their home in Tokyo, he decided to evacuate his family to an apartment in Naha, Okinawa, which he rented on a weekly basis. H-san also took time away from his work, staying in Naha with his family from mid-April until after the May holidays. After that, once the situation seemed to have calmed down in May, the family decided to return to Tokyo. Primarily, this was due to the financial constraints predictably associated with their evacuee lifestyle, and the fact that they could not continue to take leave from their jobs, which they had put on hold due to the evacuation. Accordingly, after the May holidays in 2011, they resumed their lives in Tokyo.

Even so, for a while after returning to Tokyo, H-san wore a mask when he went out and subsisted primarily on foods produced in western Japan, and never completely returned to the life he had led prior to the earthquake disaster, when he had been utterly unconcerned about radiation. Also, he and his wife differed in terms of how concerned they should be about the risks of radioactivity, which prompted arguments over which foods they felt were acceptable to eat and occasionally strained their relationship after they returned to their lives in Tokyo.[7]

Characteristics of the radioactive disaster as a 'hazard'

From this point, in light of the narratives of K-san and H-san above, I would like to discuss the characteristics of the radioactive disaster in general terms as one kind of extreme situation. First, while it differs in many details from the case of the conflict in the southern Philippines, there would seem to be no objection to the claim that the nuclear power plant accident in Fukushima is also an 'extreme' situation in the terms used in this chapter—namely, that it is an extraordinary event in which conventional daily life is disrupted by some change (or changes) in the surrounding environment that could directly or indirectly constitute an existential risk (danger) to the local community. Nevertheless, the extreme situation surrounding the accident at the Fukushima Daiichi nuclear power plant has some unique aspects that are particularly characteristic of radioactive disasters, and which are quite different in nature from other types of hazards like the conflict in the Philippines or even the tsunami generated by the Great East Japan Earthquake.

One such aspect is that the effects of the radioactive materials released by a nuclear accident on the surrounding populace and environment are relatively uncertain, opaque, and invisible, unlike, for example, those that stem from the use of violence in conflicts or from the damage caused by a tsunami. In particular, the effects of long-term exposure to low doses of radiation on the human body are known to be probabilistic, which is to say that they can only be predicted or estimated in terms of probability and range. The flip side of this is that it is also almost impossible to determine whether a person who has been exposed to radiation is safe or unsafe in the long term simply from the fact that there is no apparent damage in the short term (no immediate effect). Hence, even among experts, there are considerable differences of opinion as to the long-term risks of radiation exposure, which can sometimes be the subject of heated debate. In this connection, some have indicated that the standard values for radiation doses and the risk assessments that form the basis for establishing evacuation zones are often determined not only from a purely medical perspective, but also from an economic and political perspective (Satō and Taguchi 2016: 110–11). Generally speaking, in the case of a radioactive disaster, the behavior of the victims and, consequently, the physiological impact on human bodies—and ultimately survival—will be affected not only by the actual 'objective' behavior of radioactive materials in the natural environment, but also by the ways in which risk is evaluated and interpreted on the part of human society.

At the time of the Chernobyl accident, Ulrich Beck (1987), a sociologist famous for his research on risk, had already pointed out that the dangers of radiation could not be detected by the five senses, and therefore that information equals reality. However, in the former Soviet Union at the time of the Chernobyl accident, information was centrally controlled by the state. By contrast, as Yoko Ikeda (2013) points out, the information environment in the case of the Fukushima nuclear accident was decentralized, which enabled the dispersal and transmission of a greater variety of risk interpretations.

In fact, as shown in the narrative of H-san, a voluntary evacuee, differences and discrepancies in the interpretation of radiation risks can occur not only between government officials and residents, but also, in some cases, even within the same family or between spouses. There have even been cases of divorce among evacuees due to family strife over such radiation risks, such as over what is acceptable to eat. Not only that, the assessment of the radiation risk in the surrounding environment can fluctuate and change with the passage of time, even within the mind of an individual victim. Some voluntary evacuees, like H-san, initially fled out of a sense of extreme danger, but after weighing up the economic factors ultimately decided to close one eye to the risks, so to speak, and to return to their original places of residence. Conversely, in other cases the constantly changing messaging on the part of the government caused confusion among the residents as to whether their homes were safe or dangerous. As noted by K-san in his narrative, in Iitate there was initially almost no official information from the government, or else there was a public information campaign by the government to assure people that the village was safe—a narrative that changed abruptly with the order to evacuate the entire village.

Radioactivity as an invisible actant

Compared to blatant terrorism and violence, or natural disasters such as tsunamis, volcanic eruptions, and floods, radioactive disasters are characterized by a greater degree of opaqueness or invisibility of their risks and effects. While the invisibility of radioactivity is also pointed out by Ryoko Nishii in Chapter 10 of this volume, this is in contrast with terrorism, violence, and natural disasters (narrowly defined), where the predominant background is the action and exercise of physical forces that are more visible (i.e., in which the risk is intuitively visible). In other words, in the case of a radioactive disaster, the main characteristic of the radioactive materials as the principal actants that cause harm to humans is that they are invisible, tasteless, and odorless to the senses—

that is, the difference stems from the fact that the actants are intangible and impalpable 'things.'

Of course, the presence of radioactive materials as impalpable 'things' can still be detected and visualized (and their intensity and distribution quantified) through the use of instruments such as dosimeters (Geiger counters). For example, at first glance, the numerical value of radioactivity measured by dosimeters in microsieverts per hour would appear to be an extremely 'objective' indicator. In reality, however, it is widely acknowledged that airborne doses can vary greatly even within the same municipality depending on contingent factors such as where measurements are taken—for example, how many meters from the ground the monitoring posts are or whether measurements are taken close to the ground, on road surfaces, or in places where radioactive materials are likely to remain, like gutters and rain spouts. Thus there are pitfalls in stating that a given town is safe in terms of the average value. This is sometimes also pointed out as the problem of so-called hot spots, where even if the average for the entire region is below the standard value that is considered to be safe, at a more granular level there may be particular places where the radiation dose is much higher than the average or standard value.

Thus, in general, in the case of radioactive disasters, we could say the behavioral and life patterns of victims (e.g., where to live, what to eat) and, as a consequence, effects on the human body, are governed not only by the actual 'objective' behavior of radioactive materials in the natural environment, but also by spatial patterns of information transmission and reception (i.e., the information environment in the broad sense) and interpretations on the part of human society as to how to evaluate the associated risks.

Summary and discussion: two extremes and a hybrid—on the matter of the Anthropocene

In this chapter, I have taken up for discussion two cases involving the conflict in the southern Philippines and the Fukushima nuclear accident that took place in conjunction with the Great East Japan Earthquake in order to consider their characteristics separately. Here, as a final step, by comparing and contrasting these two cases, I would like to attempt an exploratory study to summarize what we mean when we discuss extremes, which is the overall theme of this volume.

First, in the case of the southern Philippines, seen from a macroscopic perspective, we have a conflict of interests between different groups, based on land issues, that has escalated into a conflict characterized by mass violence. This is a typical example of a 'social extreme' consistent with the conceptual definition advanced at the beginning of this chapter. On the other hand, the accident at the Fukushima nuclear power plant can arguably be seen as a typical example of a 'natural extreme' (i.e., a natural disaster in the narrow sense) in that it was caused by the earthquake and accompanying tsunami that occurred on March 11, 2011, off the eastern coast of Japan as a result of tectonic behavior.

Next, in terms of conflict, the threat of the direct use of physical force (i.e., violence) exists, above all, as a direct risk. In situations that involve conflict as a social extreme, the risk will not only come from direct violence by IS (and MG) terrorists, but also from the risk of death by starvation that attends the conflict, which may be said to be perceptible to living people through physical senses like hunger (such as the need to stave off hunger by eating bedding fibers or cardboard). Also, actors who bring about danger (e.g., terrorists) are humans as visible presences.

In contrast, in the case of the Fukushima nuclear power plant accident, as discussed in the previous section, radioactive materials, as the actants that generate risk, are tasteless, odorless, and invisible 'things,' with the consequence that the risks entailed by radioactive disasters are more opaque and probabilistic than with other hazards. These materials are characterized by the fact that they are only perceptible to the general population through detection by dosimeters or other instruments, followed by the transmission of that information, its interpretation from various perspectives, and finally its acceptance. In addition, there is a great deal of diversity in the interpretation of risk information, and its interpretation and the responses it elicits leave considerable scope for the possible emergence of differences in subsequent rates of survival.

In this sense, even within the category of natural extremes, we may be able to highlight the specificity of nuclear power plant accidents as compared to other narrowly defined natural disasters such as tsunamis, earthquakes, and volcanic eruptions. Put another way, in the case of a nuclear accident, we could say that there is more scope for the intervention of the information environment, so to speak, in terms of risk detection, communication, and interpretation, whereas in the case of natural disasters more narrowly defined, it is the direct action of the natural (physical) environment that dominates.

Considered in this way, the case of the conflict in the southern Philippines and the case of the Fukushima nuclear accident may appear to be contrasting cases of extremes—the former being a social extreme and the latter a natural extreme—as well as in terms of the relative visibility of the actants posing risks and threats to survival. In fact, however, along with these marked differences, we cannot overlook the fact that the two cases share certain commonalities and structural similarities, and there is also some question as to the degree to which these differences need to be considered essential.

To start, I would like to revisit the issue of visibility and invisibility of risk. As I have repeatedly noted, a major characteristic of radioactive disasters is the fact that their actants are radioactive materials, which is to say 'things' that are imperceptible to people's physical senses.[8] However, as I pointed out above (in the section 'The Mindanao conflict in the southern Philippines'), when considering the testimonies of IDPs, even in the case of the conflict in the southern Philippines, we can catch glimpses of a characteristic feature of conflict situations—namely, that the boundary between who is a friend and who is an enemy can sometimes be vague and difficult to distinguish. In other words, even in cases of conflict, we can identify a characteristic that could be described as the 'invisibility of "the enemy,"' whereby the 'enemy' who is the source of threat and risk is sometimes difficult to see clearly.[9]

Considered in this way, the difference in terms of the invisibility of risk-causing actors (and actants) between the cases of the radioactive disaster in Fukushima and the conflict may actually be better understood as a difference that falls along a graded spectrum (a question of degree) rather than as any absolute difference.

With regard to the IDPs in the southern Philippines, I noted earlier that their situation had resulted in a prolonged 'evacuee lifestyle,' so that evacuation—which is supposed to be a temporary and exceptional situation, so to speak—has become the de facto 'routine' of everyday life. In fact, this characteristic is also recognizable among some people displaced by the Fukushima nuclear accident. In particular, areas that had been designated as evacuation zones until only recently, such as the village of Iitate and the towns of Namie and Tomioka, have seen relatively few evacuees return to their hometowns even after the lifting of the evacuation orders. In many cases, they have apparently decided to settle permanently in their new places of residence or, for various reasons, have been forced to continue living as evacuees in temporary housing or rented accommodation.

It should also be noted here that the routinization of something that was supposed to be an 'extraordinary' or 'emergency situation' is not limited to life as an evacuee. Another example of the 'routinization of the extraordinary' we can mention is the protracted and routinized use by governments and administrations of extrajudicial measures (that were only meant to be adopted exceptionally or provisionally). For example, in the case of the conflict in the southern Philippines, martial law was imposed on the entire island of Mindanao immediately after the occupation of Marawi City by the IS-affiliated armed groups and the outbreak of fighting in 2017, as discussed earlier. In the name of maintaining public order, the state of martial law enabled special measures that included nighttime curfews, frequent checkpoints, and warrantless detention by the security authorities. While the state of martial law was initially only meant to last six months as a temporary special measure, it was subsequently repeatedly extended; the state of martial law ended with the third extension lapsing on December 31, 2019. In this arguably insidious fashion, 'temporary' extrajudicial measures have become commonplace.

Likewise, in the wake of the Fukushima nuclear power plant accident, the Japanese government designated as evacuation zones areas where radiation exposure met or exceeded the level of 20 millisieverts per annum. According to Japanese laws and regulations concerning radiation exposure prior to the accident, however, the limit for annual radiation exposure on the part of the general public was 1 millisievert, the standard for designating a Radiation Controlled Area was 5.2 millisievert per year, and the standard for certifying leukemia among nuclear workers as a work-related injury was 5 millisieverts per year. By that measure, the measures taken after the Fukushima accident are extremely exceptional and could arguably be described as 'extrajudicial.' On this point, the philosopher Yoshiyuki Satō and others have pointed to the fact that, in the aftermath of the Fukushima Daiichi nuclear accident, the public has been cast into a situation that Walter Benjamin and others termed the 'normalization of the state of exception' (Satō and Taguchi 2016: 16).

Thus, in the contexts of both the southern Philippines and the Fukushima nuclear accident, we can point to a state of affairs that might be called the routinization of situations that are supposed to be extraordinary and exceptional ('the routinization of the state of exception'). (On the concept of the state of exception, see also Agamben 2005.)

Next I would like to revisit the distinction between social extremes and natural extremes in an attempt to consider it in the context of a long-term time span that could have a bearing on human evolution. Generally speaking, the dichotomy between social extremes and natural extremes proposed at the beginning of this chapter as a tentative starting point could also be said to have some validity as a tentative starting point for furthering the discussion. However, there would seem to be another pitfall in making a rigid distinction between the two and regarding it as absolute. Let me explain.

First, as we have seen thus far, we have understood the case of the conflict in the southern Philippines as a situation of increased existential risk that has arisen mainly due to the effects of the social environment (i.e., a social extreme). While not necessarily wrong as a précis of the situation, we also must not overlook the fact that the background of the conflict is fundamentally a land problem in which ecological conditions in the broad sense—for example, demographic changes in relation to resources such as arable land in Mindanao—are also deeply involved. In other words, even though the conflicts that have erupted are violent clashes between human groups, we could say that they are also deeply intertwined with ecological environmental conditions, especially the balance between the land area and its carrying capacity in a particular ecological environment.

Next, in the case of the Fukushima nuclear power plant accident, the proximate cause of the accident was a large earthquake caused by the tectonic behavior, which in turn generated the 'unanticipated' tsunami that followed, and which caused the emergency generators to lose power and prevented the delivery of cooling water. If we focus on these aspects, the accident seems emphatically to be a typical natural disaster (i.e., an 'act of God').

However, an important premise for any consideration of the Fukushima nuclear power plant accident is the fact that a radioactive disaster like this one simply could not have happened in the absence of the nuclear power plant as an artificial object based on the extremely special technology of generating power using a nuclear fission reaction. Moreover, in the process of reviewing the accident, questions have been raised as to whether there were any deficiencies in terms of seismic design or planned tsunami countermeasures. In the words of Yōtarō Hatamura (2011: 113), the nuclear accident and the radioactive disaster could also be said to be more akin to an 'organizational accident' (or a kind of 'human catastrophe').

Furthermore, to reframe this point from the perspective of risk research, in the case of the nuclear accident, we could point out that a factor behind this disaster was a relative lack of anticipation of the possibility of so-called 'black swan' events—that is, catastrophic events that have a very low probability of occurring, but which can have extremely serious consequences when they do happen (Taleb 2007, 2013).

Considered from an even more theoretical level, in many cases it would be more appropriate to consider not only nuclear accidents, but even hazards such as tsunamis, earthquakes, typhoons, and volcanic eruptions, which are regarded as natural disasters in the narrow sense of the word (i.e., disasters that are regarded as pure 'acts of God'), as the result or effect of a complex process implicating the natural environment as well as the social environment (e.g., population density, settlement patterns, infrastructure development, disaster prevention systems, and political and economic conditions).

For example, as an extreme case, imagine an uninhabited island, isolated from human society, where it rains extensively, causing the island's rivers to overflow and resulting in a flood of unprecedented scale. Even so, this would be difficult to characterize as a 'disaster' for humans (except in the case that humans had set up a facility or property of some sort on the island). However, were people to settle on the island and build houses near the river, so that it became a densely populated area, then even if the rainfall and flooding were physically and meteorologically of an identical magnitude to the previous case, this could still represent a major disaster for the human group now living there. Thus it would be more appropriate to understand hazards (including those that would be considered purely 'natural' disasters at first glance) as phenomena that occur as hybrid entanglements or as the effects of interactions between the natural environment and the social environment.

This idea is by no means original. For example, in an essay written prior to the Second World War based on his experience of the 1923 Great Kanto

Earthquake, the physicist and essayist Torahiko Terada (2011) noted the fact that 'the more advanced civilization becomes, the more violent natural disasters become.' In his essay, Terada pointed out that whereas shanty architecture was rather safe from earthquakes and easily rebuilt, the level of damage from disasters has tended to increase due to the recent proliferation of complex buildings based on modern technology.

I would like to conclude by suggesting that the idea of the hybrid intertwining of the natural and social environments that I alluded to above is useful for considering extremes on a long-term time scale, including that of human evolutionary history, and can be related to the recent debate over what has been called the 'Anthropocene,' a concept first proposed by the atmospheric chemist Paul J. Crutzen and biologist Eugene F. Stoermer.[10] Focusing on the human impact on the present-day global environment, these authors pointed out that phenomena such as the increase in the concentration of carbon dioxide and the mass extinction of plants and animals as a result of human activities since the Industrial Revolution suggest that the Earth has entered the Anthropocene as a new geological epoch that has superseded the preceding Holocene. According to their argument, the effects of human activities are now affecting the entire global environment, including the cycle of matter. As a result, living organisms, including humans, no longer inhabit a pristine natural environment but, rather, a special environment in which the results of human activities have a large-scale impact.

In an Anthropocene environment, extreme situations, including various types of hazards, can be thought of as appearing as the effects of complex interactions between the natural and social environments, as I have already pointed out. Moreover, if we also take into account the routinization (normalization) of extraordinary situations, then the hybrid situation described above—the mutual implication of the natural environment and the anthropogenic (social) environment—is not an exceptional or emergency situation in the model of a hazard; rather, it could be characterized as a fundamental precondition for any consideration of human existence and evolution.

Nevertheless, the proponents of the Anthropocene and their arguments invite certain questions. For example, even among proponents of the Anthropocene, there is a tendency (with certain exceptions) to emphasize the impact of human activities on the environment primarily from the industrial era onward. However, if we consider the scale of human evolution over a longer time scale, the question arises as to whether it might not only be in the modern era or from the industrial era onward that human activities began to exert an effect on the natural environment and nonhuman organisms. For example, surely it is possible to say that by the time they first began farming (and herding) at least 10,000 years ago, humans had already started actively intervening in and modifying the natural environment and wildlife species. Furthermore, the extinction of nonhuman animal species as a result of human activities is by no means an exclusively recent phenomenon (i.e., from the Industrial Revolution onward). And, in fact, it has been pointed out that this phenomenon had already occurred many times over, during the tens of thousands of years since early *Homo sapiens* began to spread throughout the world (Mazlish 1995; Diamond 1999)[11].

In light of these points, it would seem that there is still room for a re-examination of the debate over the Anthropocene from a more macroscopic perspective in terms of setting a time axis (time span) that encompasses the scale of human evolutionary history. However, since I have exhausted the space available for this present discussion, I will bring the exploratory discussion of extremes that I have undertaken here to a close, and this last point to a future endeavor.

Acknowledgments

Financial support for research activities concerning the conflict involving the Muslim ethnic minority (the Moro peoples) in the southern Philippines and their cultural background was provided in part by a MEXT Grant-in-Aid for Scientific Research on Innovative Areas (Research in a proposed research area) 'Construction of Face-Body studies in transcultural conditions' (17H06341). For my argument about the same conflict, I also relied on discussions that took place in a study group held in conjunction with the ILCAA Joint Research Project on the 'Multi-disciplinary Study on Islam and Cultural Diversity in Southeast Asia (The 3rd Term): Dynamics of Conflict and Co-existence.' With regard to my discussion of the Fukushima nuclear power plant accident, I drew inspiration from discussions held as part of the ILCAA Joint Research Project on the 'Anthropological Study of "Waza": Technology, Body, Risk/ Hazard (Anthropological Study of Things (3).' Also, for my June 2018 survey in Fukushima Prefecture concerning the victims and evacuees of the Fukushima nuclear power plant accident, I received support from a MEXT Grant-in-Aid for Scientific Research (A) on the theme of 'Child Migration in Asia: Emerging Transnational Class' (16H02737). I would like to express my sincerest gratitude to Prof. Kayoko Ishii and her colleagues for their support. I should also note that my interest in research on the Fukushima nuclear accident was greatly stimulated by a 'learning trip' to Fukushima Prefecture taken together with other members of the ILCAA Core Project (Anthropology) in November 2017, as well as by my participation in the related workshop under the auspices of the same project. I would like to take this opportunity to express my gratitude to Prof. Ryoko Nishii (ILCAA), who is the lead investigator of the Grant-in-Aid and the Core Project, and Prof. Yuichi Sekiya and Prof. Tadashi Yanai (both of the University of Tokyo), who organized the 'learning trip.' Nevertheless, it goes without saying that any omissions or errors that might be contained in this chapter are the sole responsibility of the author.

Notes

1. Even so, I would like to draw attention to the fact that the boundary between 'extraordinary and extreme situations' and '(non-extreme) normal everyday life' can in some cases be unclear and variable. For this reason, as I discuss in the latter half of this chapter in connection with the cases of the southern Philippines and Fukushima, I would like to caution against the possibility of what we might call the 'normalization of exceptional situations' (the routinization of the extraordinary).
2. With regard to these two types of extremes, Kawai's chapter in this volume (Chapter 9) also distinguishes between extreme situations premised on natural ecological environmental factors and socially extreme situations premised on relationships with 'others,' as represented in her discussion about the social phenomenon of raiding.
3. For a discussion of the historical background of Moro society in the southern Philippines and an overview of the background and current status of the Mindanao conflict, see Tokoro (2013, 2015, 2018).
4. For a more detailed presentation and discussion of Aida's narrative in the context of the anthropology of time and memory, see Tokoro (2011).
5. I conducted the interview in Davao City in July 2017.
6. This excerpted narrative is based on an interview with K-san conducted in Fukushima Prefecture in June 2018. The name is a pseudonym, as is that of H-san below.
7. This excerpted narrative is based on an interview with H-san conducted in Tokyo in June 2018.
8. While radioactivity (and radioactive material) as an invisible and unknowable presence may be said to be a richly interesting subject to examine from the perspective of the 'anthropology of things,' the discussion in this chapter does not afford scope for a more detailed examination. On this point, I would refer the interested reader to Nishii's discussion in Chapter 10 in this volume. As shorthand, I use the term 'actors' when the agents in question are humans, and 'actants' with reference to nonhuman presences such as radioactivity (or radioactive materials). My provisional adoption of this terminology here, in recognition of the power of nonhuman 'things' as agents, is based on recent anthropological research relating to 'objects' such as the framework of so-called Actor–Network Theory (ANT). For a more detailed discussion, see, for example, Tokoro and Kawai (2018).
9. Kawai's discussion in Chapter 9 of this book is also helpful in this regard. She points out that 'the enemy' in the context of the pastoralist Dodoth society of East Africa does not have an a priori existence in any clear-cut form. Rather, enemies become real only gradually, materializing as 'others who are preparing a raid' through various activities such as rituals and patrols intended to deal with raiding parties. This is described by Kawai as 'the process by which the invisible enemy is made visible.'

10. On the concept of the Anthropocene, see, for example, Crutzen and Stoermer (2000) and Christian (2014). The critical review of the literature on the Anthropocene undertaken earlier in this volume by Takenoshita (Chapter 5) develops a stimulating discussion that highlights the cognitive and social characteristics that have enabled humans to carry out such large-scale environmental modifications, including an orientation to the future that allows humans to break away from the here-and-now.
11. Furthermore, we should note that large-scale environmental modification is not limited to *Homo sapiens* (Nakamura 2019: 270). For example, it is well known that cyanobacteria released large amounts of oxygen and drastically changed the composition of the Earth's atmosphere about 3.5 billion years ago, long before the appearance of humans, and it has also been pointed out that global and drastic climate changes have occurred repeatedly in the past in response to the activity of vegetation (Beerling 2007). In this context, the historians of science Christophe Bonneuil and Jean-Baptiste Fressoz have developed a critical discussion of the latent 'anthropocentrism' underlying some of the debates over the Anthropocene (Bonneuil and Fressoz 2013).

References

Agamben, G. (2005) *State of Exception*, Kevin Attell (trans.). University of Chicago Press.

Beck, U. (1987) 'The anthropological shock: Chernobyl and the contours of the risk society'. *Berkeley Journal of Sociology*, 32: 153–166.

Beerling, D. (2007) *The Emerald Planet: How Plants Changed Earth's History*. Oxford University Press.

Bonneuil, C. and J.-B. Fressoz (2013) *L'Evénement Anthropocène*. Seuil.

Christian, S. (2014) *The Anthropocene: The Human Era and How It Shapes Our Planet*. Synergetic Press.

Crutzen, P. J. and E. F. Stoermer (2000) 'The "Anthropocene"'. *Global Change Newsletter*, 41: 17.

Diamond, J. (1999) *Guns, Germs, and Steel: The Fates of Human Societies*. W. W. Norton & Co. Inc.

Hatamura, Y. (2011) *Mizou to sōtei-gai—Higashinihon daishinsai ni manabu* (Beyond precedent/beyond belief: Learning from the Great East Japan Earthquake). Kodansha.

Ikeda, Y. (2013) 'The construction of risk and the resilience of Fukushima in the aftermath of the nuclear power plant accident'. In T. Gill, B. Steger and David H. Slater (eds), *Japan Copes with Calamity: Ethnographies of the Earthquake, Tsunami and Nuclear Disasters of March 2011*. Peter Lang: 165-200.

Institute of Disaster Area Revitalization, Regrowth and Governance, Japan Civil Network, Save Fukushima Children Lawyers' Network (2015) *Genpatsu hinan hakusho* (White Paper on evacuation in the wake of a nuclear power plant accident). Jimbunshoin.

Mazlish, B. (1995) *The Fourth Discontinuity: The Co-Evolution of Humans and Machines*. Yale University Press.

Nakamura, M. (2019) 'Hito to mono no Hazama de: Doubutsu kara jinruigaku eno shiten' (Between 'human beings' and 'things'—an 'animal' perspective on anthropology). In I. Tokoro and K. Kawai (eds), *Mono no jinruigaku 2* (Anthropology of things 2). Kyoto University Press, 259–278.

Satō, Y. and T. Taguchi (2016) *Datsu genpatsu no tetsugaku* (The philosophy of denuclearization). Jimbunshoin.

Taleb, N. (2007) *The Black Swan: The Impact of the Highly Improbable*. Random House.

Taleb, N. (2013) *Antifragile: Things that Gain from Disorder*. Penguin.

Terada, T. (2011) *Tensai to kokubō* (Natural disasters and national defense). Kodansha.

Tokoro, I. (2011)'Hukusu no jikan/jyuusou suru kioku- Sulu kaiiki sekai ni okeru souki to boukyaku ' (Multiple temporalities and overlapped memories–memory and oblivion in the maritime world of Sulu). In R. Nishii(ed.), *An Anthropology of Time*, Sekaishisosha-Kyogakusha Co., Ltd., 278-300.

Tokoro, I. (2013) 'Violence and the autopoiesis of groups: From the ethnography of pirates and feuds'. In K. Kawai (ed.), *Groups: The Evolution of Human Sociality*. Kyoto University Press and Trans Pacific Press, 143–166.

Tokoro, I. (2015) 'The re-emergence of Islam in the context of Muslim separatism in the Philippines'. In I. Tokoro (ed.), *Islam and Cultural Diversity in Southeast Asia*. ILCAA, 153–174.

Tokoro, I. (2018) 'The Mindanao conflict and peace process from the Aquino to the Duterte presidency: A perspective from the Sulu Moro community'. In I. Tokoro and H. Tomizawa (eds), *Islam and Cultural Diversity in Southeast Asia (Vol. 2): Perspectives from Indonesia, Malaysia, the Philippines, Thailand, and Cambodia*. ILCAA.

Tokoro, I. and K. Kawai (2018) 'Why the anthropology of mono (things)?' In I. Tokoro and K. Kawai (eds), *An Anthropology of Things*. Kyoto University Press and Trans Pacific Press, 18–34.

PART • THREE

Realizable Extremes II

The Power of Memory and Imagination

CHAPTER 12

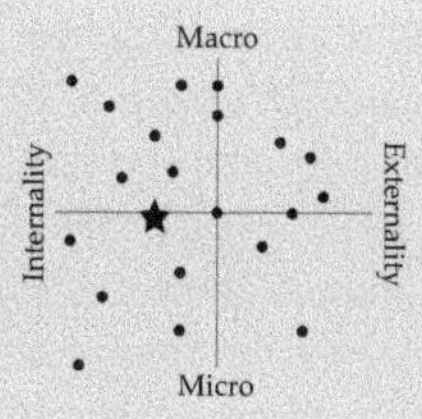

Between Ethnic Favoritism and Ethnic Hatred

Toru SOGA

Keywords: cultural evolutionary theory, altruistic behavior, conflict, pastoralists

Parochial altruism and tolerant non-altruism toward other ethnicities

How unreasonable it would be to be killed or persecuted simply for belonging to a certain group or category though you bore no responsibility yourself! The twentieth century was an era of ethnic conflicts and wars between states, an era when people attacked and harmed others simply for belonging to certain groups or categories. Even today, hate crimes remain a problem the world over. However, the twentieth century also saw the development of a legal system for judging those who foment wars, conflicts, and massacres. Those who instigate and direct wars are now liable to be punished for a variety of offences, including war crimes, crimes against peace, crimes against humanity, and genocide.

International military and criminal tribunals distinguish between those who order genocide, crimes against humanity, and war crimes and those who are ordered to commit them, holding the former responsible and the latter exempt from liability. Yet while it is certainly true that more than a few of those who are ordered to commit such crimes may be forced to participate in a war unwillingly, it is equally true that there must be many who willingly join the fight. These are the ones who answer the call to battle, who are stirred by the agitation to war, and who actively play a part, or those who support the fight by inciting and inspiring their comrades, though they do not engage directly in the battle. Legally, these people are exempt from prosecution under specific jurisdictions. And yet these actions still sanction politicians who stir up conflict between groups.

To prevent conflicts from occurring, we would need a deep understanding of the mechanisms that steer ordinary people toward such actions. In his book *In the Name of Identity: Violence and the Need to Belong*, Amin Maalouf (2001: 28) writes:

> When an otherwise normal man is transformed overnight into a killer, that is indeed insanity. But when there are thousands, millions of killers; when this phenomenon occurs in one country after another, in different cultures, among the faithful of all religions and among unbelievers alike, it's no longer enough to talk of madness. What we conveniently call 'murderous folly' is the propensity of our fellow-creatures to turn into butchers when they suspect that their 'tribe' is being threatened.

The purpose of this chapter is to examine this 'propensity'—that is, for normally peace-loving people to become killers when threatened—from the perspective of human evolutionary history. Recently, in the field of evolutionary psychology, a unique hypothesis has been proposed to explain how humans acquired altruistic behavior. In essence, human altruistic behavior has evolved through struggles with other groups (intergroup selection). In this chapter, then, I discuss the propensity Maalouf describes above by ethnographically describing conflicts experienced by East African pastoralists and considering them in light of the hypothesis advanced by evolutionary psychologists.

Living in the midst of conflict

Gabra pastoralists

This chapter considers conflict as it has been experienced by the Gabra, a group of East African pastoralists, as well as the people's behavior. There are two Gabra groups—namely, the Gabra Malbe, who live in northern Kenya, and the Gabra Miigo, who mostly live in southern Ethiopia. The focus of the chapter is on the latter group, and hereinafter my use of the term 'Gabra' should be understood to refer to the Gabra Miigo, unless otherwise specified. In addition to herding pastoral livestock such as camels, cattle, sheep, and goats, the Gabra also keep donkeys as work animals. Since the 1990s, some Gabra have also engaged in small-scale rainfed agriculture, but they remain highly dependent on their livestock.

Günther Schlee (1989), an ethnologist who has conducted extensive field research ranging from southern Ethiopia to northern Kenya, found that pastoralists who herd camels in this area share common traditions and rituals across ethnic groups, which he has designated as the Proto-Rendille-Somali culture. Schlee advanced the hypothesis that these people were influenced by the Somali and Oromo peoples and differentiated into the existing peoples. The Gabra also abide by this Proto-Rendille-Somali culture, sharing similarities with the Somali in terms of clothing, the shape of their houses, and Islamic beliefs, and with the Oromo in terms of their language and political systems.

When I began my field research in 2000, the Gabra people consumed most of their milk themselves, selling any surplus to local vendors with kiosks near their villages. Occasionally, they sold goats and sheep for ready money, and sold camels and cattle when they had major expenses, such as a wedding. By 2018, however, most of their milk was being sent for sale to the market town of Moyale, about 240 kilometers south. Also, since around 2007, the market for camels has become more active due to increased demand from the Arab countries, and the Gabra have started to earn cash by becoming small-scale traders, assisting in the buying and selling and loading and unloading of bull camels for transportation (Soga 2019).

A history of conflict

Historically, the Gabra have experienced four instances of ethnic conflict since the Second World War (Soga 2008, 2012). The first was a conflict that took place in 1963 in response to the Pan-Somalism movement between the Garre Somali, who lived in Ethiopia, and the Boran Oromo, who were supported by the Ethiopian government, at which time the Gabra fled after being attacked by both ethnic groups. The second was the conflict that broke out in 1974, when Haile Selassie I was deposed and arrested by a military coup, which created a political vacuum in southern Ethiopia. The Gabra were attacked by the Guji Oromo from the north side and the Boran Oromo from the south side, and fled into Kenya and Somalia, where they lived as refugees. Meanwhile, some Gabra received military training in Somalia in order to fight back against the Borana, becoming guerrilla fighters. The third instance was in 1992, after the Derg (the Provisional Military Government of Socialist Ethiopia) was overthrown by the Ethiopian People's Revolutionary Democratic Front, which created a new political vacuum that once again brought conflict to southern Ethiopia. At this time, the Gabra were attacked by the Borana but allied themselves with the Guji to fight back (see Case 5 below for a more detailed discussion).

The Ethiopian People's Revolutionary Democratic Front, which came to power in 1991, introduced a federal system along ethnic lines in 1994, which led to the creation of regions named after the country's major ethnic groups, including Tigray, Afar, Amhara, Oromo, and Somali. Each region was divided into several zones, which are also named after the groups that make up the larger ethnic identities. For example, the names of the Borana and Guji zones

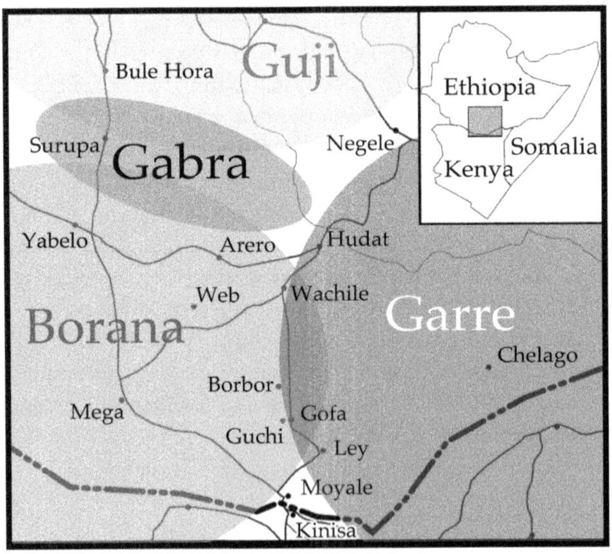

Figure 12.1: Ethnic distribution in southern Ethiopia

*This map shows a rough outline of the distribution of ethnic groups mentioned in this chapter. However, it is not intended to support the territorial claims of any particular group.

are based on the names of the Borana, Guji, and other groups that make up the Oromo people. Historically, the Oromo ethnic identity was created in the 1960s, so groups such as the Borana and Guji, while identifying themselves ethnically as Oromo, nevertheless maintain their own ethnic consciousness and political and cultural cohesion. Accordingly, for the purposes of this discussion, I regard all these collectivities as ethnic groups.

The fourth instance, which took place in 2005, was an internecine conflict within the Oromo (see Cases 2 and 3 for details). This conflict broke out between the Guji Oromo, who intended to establish a new West Guji Zone in addition to Guji Zone, and the Borana, who were wary of being deprived of land in Borana Zone. The Gabra, who inhabit the borderland between the two groups, were attacked by the Guji from the north side on two occasions, in 2005 and 2006, prompting them to flee south to Moyale and northern Kenya. According to a news report by the BBC (Henshaw 2006), the clashes in 2006 resulted in the deaths of upwards of 100 people and the displacement of close to 30,000 others as refugees.[1]

Currently, as the result of past ethnic conflicts, the Gabra are dispersed over a wide area, mainly in the Surupa area of southern Ethiopia (Figure 12.1), but also in places like Arero, Negele, Guchi, Hudat, and Moyale, as well as Kinisa in northern Kenya.

The period since the introduction of ethnic federalism has seen the frequent occurrence of conflicts over borders between Ethiopian regions (Abbink 2006). In southern Ethiopia, conflicts continue to take place on the border between the Somali Region and Oromia Region, as well as between Oromia Region and the Southern Nations, Nationalities, and Peoples' Region. Conflicts over the border between the Somali Region and Oromia Region, where the Gabra are concerned, have intensified in recent years, with a total of forty-seven armed clashes and killings recorded between September 2017 and July 2018 (see Cases 1 and 4 for details).

Changes in the nature of clashes

Anthropologists who have studied pastoral societies in East Africa have repeatedly reported how pastoralists have a strong attachment to their livestock, and how herds can sometimes be the focus of violent raiding (e.g., Evans-Pritchard 1940; Marshall Thomas 1965; Fukui and Turton 1979; Galaty and Bonte 1991; Sagawa 2011). As mentioned above, the Gabra have also been victims of conflict and the targets of attack, but the nature of the clashes in which they have been involved has changed significantly over time. The Gabra divide warfare into *gaada* (small-scale murder conducted by only a few), *duula* (livestock raiding and limited military expeditions), and *hadha* or *olki* (full-scale war). Let us turn now to consider how the nature of these has changed over time.

To start with, *gaada* has changed from being a cultural act associated with the Borana to being an act by the Somali and Oromo peoples seeking to achieve a political effect with respect to their shared regional border. Traditionally, *gaada* involved a conspiracy by several Borana to murder a member or members of another ethnic group. The Borana practiced a custom whereby they would target members of other ethnic groups who were grazing their cattle alone, and then kill them with a spear. Afterwards, the genitals would be cut from the corpse to be taken home as evidence of the kill. The killers would then be praised and receive *sarma*, or gifts of livestock (Baxter 1979). Some Gabra experienced near-death at the hands of the Borana as children. Even so, under the Derg government, the Borana culture of *gaada* apparently disappeared after public executions following a crackdown on such premeditated killing.

After the implementation of ethnic federalism in 1994, it appears that the practice of new *gaada* began to take place between the Garre Somali and the Boran and Guji Oromo with regard to the regional border between Somali Region and Oromia Region. In this form, however, *gaada* differed from the premeditated killings observed in Borana culture. Contemporary *gaada*, rather, is best understood as a political act aimed at driving other ethnic groups from the regional frontier in order to expand regional borders and has emerged as the immediate cause of interethnic wars.

Next, *duula* (livestock raiding and limited military expeditions) were carried out primarily for the purpose of taking the livestock of other ethnic groups. Although people were sometimes killed, doing so was not the purpose of a raid.[2] In the past, the Gabra and Borana had certain mutually understood 'rules' in the context of livestock raiding. For example, when they fought, they were not allowed to stab fleeing opponents from behind with their spears. It was only acceptable to kill someone who was coming at you head-on. Another rule held that once you took some livestock, you were to head back to your own village while attempting to evade pursuit by the rightful owner; once you were able to pen the livestock you had taken in your own enclosure, then the raiding would be considered complete, and the robbed party was not entitled to take the matter further.

In recent years, *duula* has taken on a stronger connotation of military expeditions, even while it continues to retain the aspect of livestock raiding. Moreover, *duula* has also come to be practiced between the Garre Somali and the Boran and Guji Oromo. In these cases, however, there are no rules, as in the past. Nowadays, *duula* refers to military expeditions that take place as part of interethnic warfare.

Third, the term *hadha* (full-scale war)[3] indicates armed conflict between ethnic groups to achieve their own goals, which may vary. In 1963, a major *hadha* took place in southern Ethiopia in which the Garre, a Somali people who live in Ethiopia, sought to answer the Pan-Somalism movement by separating from

Ethiopia and joining their territory to Somalia. In a 1974 *hadha*, Gabra youths who were attacked by Guji and Borana traveled from Ethiopia to Somalia, where they received military training and launched a guerrilla insurgency. While their nominal goal at that time was the realization of Pan-Somalism, a more tangible goal for these Gabra youth was attacking the Borana and Guji to secure a safe place for themselves to live. The 1992 *hadha* was a counterattack by the Gabra against the Borana after being attacked by Borana—in other words, a war of retribution.

While every ethnic group has its own purposes for raising a *hadha*, generally speaking, it seems that recent instances tend to be aimed at the territorial expansion of regions and zones in line with the framework of Ethiopia's ethnic federalism. For example, the Garre Somali and the Guji and Boran Oromo are trying to expand their respective boundaries in Somali Region and Oromia Region within Ethiopia. A large-scale *hadha* in 2005 seems to have arisen due to a clash of interests between the Guji Oromo, who (as mentioned earlier) are seeking to establish a new West Guji Zone, and the Borana, who are wary of losing territory on the frontier of the Borana Zone.

Taken as a whole, the culturally and socially lauded aspects of the Borana custom of *gaada* have disappeared, as has the ritualistic practice of *duula*. However, it could be said that *gaada* killings, *duula* raiding, and *hadha* wars have started being conducted in order to intimidate other ethnic groups and to force them to retreat from border areas with the intention of expanding regional and zonal boundaries.

Coping with war

Patterns at the outbreak of hostilities

Outbreaks of war tend to follow a definite pattern. Initially, they might manifest in the form of livestock theft, fighting among children and youths, or a suspicious accidental death (e.g., drowning at a watering hole). Livestock theft and fighting are usually triggered by dissatisfaction with other ethnic groups; for example, a given group would prefer to water their own livestock first, but cannot do so because of another ethnic group, and this dissatisfaction over the intrusion of another ethnic group's livestock into their own pastureland boils over into heated fighting. In recent years, political activities to expand the boundaries of zones and regions have also led to conflicts between ethnic groups that want to expand the boundaries and others who want to prevent the reduction of their own, which in turn has led to livestock theft and fighting. A drowning at a well might be the result of a careless accident, but people who read the death in the wider scheme of the conflict will rumor that the victim was murdered. Steps will be taken either to ease the conflict or to stir it up, with each ethnic group holding councils to analyze the situation and local government officials holding meetings to prevent further conflict.

Songs and poems insulting other ethnic groups will spread electronically via mobile phones. When people stop exchanging greetings when passing members of other ethnic groups on the street, they will know that war is imminent. Should a shooting or premeditated murder occur, then the tension will rise dramatically. Failure to catch the culprit will spawn a chain of conspiracies. And once looting begins in the regular markets, people flee en masse.

As people flee, uninhabited areas appear between different ethnic groups (Figure 12.2). Looking at the area around the town of Surupa in southern Ethiopia, we find three ethnic groups—the Guji, Gabra, and Borana—distributed along an axis from north to south. Near the boundaries between the ethnic groups, different peoples live in close proximity to each other, their respective settlements mixing together in a mosaic pattern. However, when the time comes to flee, people will flee in a direction that they consider to be safe. For example, when a *gaada* occurred between the Guji and Gabra in 2005, given the prevailing distribution of ethnic settlements, the Guji fled to the north, where many of their own people were living, while the Gabra fled south in the opposite direction. As a result, a no-man's land appeared between the two peoples where no one lived any longer.

Once an area becomes uninhabited, the real fighting finally begins. Men with guns will form groups and start attacking. The government and the military will call for people to return to their villages in an effort to prevent uninhabited areas from forming in the first place. However, unless government troops are deployed in sufficient numbers to restore security, a war will begin.

Below, I would like to look at the specific actions that people take during three phases—the beginning of the war, evacuation, and battle.

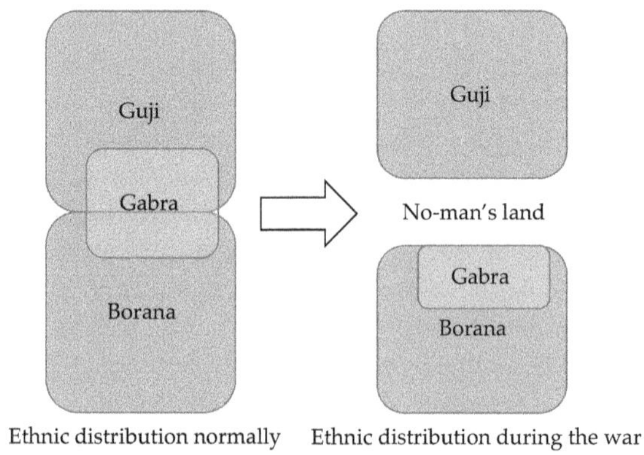

Figure 12.2: The no-man's land that appears at the beginning of a war

Efforts to halt the war

The first thing to discuss is what people do when they begin to see signs of war. Gabra living in the Surupa area become uneasy at news of fighting or the theft of livestock, and quickly undertake to understand what has happened. Nowadays the ubiquity of mobile phones means that people can quickly call to check on incidents that have taken place a long way away. Most rumors are false, but not all. People exchange information about the details of what has happened. Let us look at people's behavior using the example of an incident that actually took place. The following case illustrates the behavior of villagers in the Surupa area, where I was living during my fieldwork.

Case 1: Ethnic murder can be a precursor to war

On the night of September 6, 2013, Gabra living in the Surupa area received word that a Borana had been killed near Arero, about 80 kilometers to the southeast. Apparently, his head had been cut off and his face peeled away from the skull. However, according to information that arrived the following morning, on September 7, the Borana in question was alive and unharmed. Later, a Borana acquaintance living in a neighboring village brought news that there had been no murder, and people began to relax. But at noon, news came that the killer's footprints had been found leading from the Borana corpse toward the Gabra village, so people began thinking that a murder might have taken place after all. In the afternoon, word came that a Borana had, indeed, been killed and that his body had been buried that morning. Apparently, the footprints of two perpetrators had been found heading toward the Gabra village, and there were reports of a strong police presence in the Arero area to investigate the case.

That evening, a Guji youth from a neighboring village arrived in the village where I was living to share information with the villagers. By his account, the Borana had been shot to death, then decapitated, and then placed in the branch of a tree. He also said that only a few Borana had come to the regular cattle market held that day in Surupa (for fear of the Gabra). That night, news came that the culprit or culprits had been caught. It seemed they were Gabra. Around midnight, there was more news. Two Gabra men had been detained and were being held responsible for killing the Borana. The names of the Gabra who had been arrested were also disclosed.

By noon on September 8, people were saying that four Gabra had been arrested. One of these men was also suspected of smuggling guns, and a total of eight guns were seized.

> On September 10, we learned that one of the four was found guilty, while the other three were said to have been found *shake* (not guilty). Apparently, residue from the killer's gun suggested that it had been fired recently. Although the judgment was still to come, as a *dhiiga* (blood price), he would additionally have to pay the family of the slain Borana thirty head of cattle—of these, one would be paid by each of the twelve Gabra clans, and the remaining eighteen by his own clan.

The incident in Case 1 was not entirely unanticipated. Two days prior to the incident, another incident had taken place in the Alis area, east of Surupa, in which five Borana (two of whom were armed with guns and another with a spear) had stolen camels belonging to a Gabra herder. In this incident, the Gabra, upon learning that his livestock had been stolen, fired warning shots and succeeded in getting his camels back. The Borana who had committed the theft were caught by the police and the matter was supposed to have been settled, but the people in Surupa believed the second incident to have been a retaliation by Gabra who had been angered about the incident. In addition, since young men from the village near Surupa had taken their livestock into the Arero area where the incident had taken place, the villagers were very worried, for fear that the retaliation might escalate to embroil those young men.

From the way that the Gabra behave, it is evident that they exchange information with Borana and Guji acquaintances who live in neighboring villages. Like the Borana and Guji, the Gabra use their own ethnic network to obtain information. In addition, members of each group will visit acquaintances belonging to other ethnic groups to compare notes with one another to confirm their information.

Appadurai (1996) has proposed the concept of an 'ethnoscape' as a perspective for reading modernity. In the event of a murder, for example, people will not step back to consider the overall picture from a neutral standpoint. Rather, they will read its significance from the standpoint of their own ethnic group. This is not a 'map' that shows a bird's-eye view from above, but more like a 'landscape' that represents a way of seeing that is unique to each ethnic group. And just as the appearance of a landscape differs depending on whether one is leaving or returning, the same event will take on a different aspect depending on the perspective of the ethnic group from which it is seen.

This case shows that people are cognizant of the fact that members of each ethnic group see events differently. In this case, given the possibility that the Gabra were on the side of the perpetrators, Gabra people were especially nervous and keen to understand how the Borana, who were on the side of the victims, perceived the incident. People in the Surupa area, both Gabra and Borana alike (as well as the Guji who were not immediately considered to be party to the conflict), were all fearful of the escalation of the conflict, and so sought to understand the situation. No one simply tried to amplify ethnic hatreds. Of course, this may not necessarily have been the case for the Borana living near Arero, who were directly affected, but in and around far-off Surupa, rather than amplifying hatred based on rumor, people attempted to calm the situation by talking to each other about the importance of peace. Hence the relief that villagers felt when it was confirmed that the Gabra perpetrator had been arrested—this was actually relief that the tide of war seemed to have ebbed, at least for the time being.

Fleeing from warfare

How do people act when, despite their best efforts, signs of war do, in fact, amplify and lead to war? In the case of the war between the Guji and Gabra in 2005, the initial trigger was the disappearance of a Gabra child. People had been searching for the missing child, and one day they discovered the murder after a wild animal dug up the corpse from the place in which it had been buried. Although the police investigated, they were unable to apprehend the killer. After it was rumored that the Guji may have been responsible, more rumors began to spread that the Guji were going to attack the Gabra, which put people in a fearful and suspicious frame of mind. In response to these growing tensions, local government officials called an assembly involving the Gabra, Borana, and Guji. However, someone fired bullets into the assembly, killing several people. After people fled, the war began.

At this time, those who fled the hostilities faced severe hardship. I would like to examine, below, the ways in which they managed to secure their own survival.

Case 2: Difficult days

In 2005, Gabra people, suspecting the Guji to have been behind the killing of the child, fled en masse to the south where no Guji were present. Informant A, a man who lived 15 kilometers north of Surupa town, fled as soon as he heard that shots had been fired into the assembly. After sending his pregnant wife and parents on ahead, he took his camels and fled south, leaving their household property in the village. He sent his goats and sheep, as well as his infant daughter, in the company of his younger sisters. Since the government had ordered people not to flee, they were unable to take their household possessions. He said they acted as though they were preparing to put their livestock out to pasture, and then fled. Gabra men with guns were guarding the roads, along with Oromia's regional state police.

Over the next three days, Informant A's family traveled about 15 kilometers south and set up a camp near the northern limits of Surupa town. Each time they edged their way south, the men would return to their native village to gather information. On the third day, when they arrived at the regular mar-

> ket held in Surupa, they found people fighting over the goods. Sensing imminent danger, they decided to move further south. Informant A, his younger sisters, and his parents and wife split up into three groups. The national highway, Route C80, was teeming with Guji fleeing north and Gabra fleeing south
>
> .On the seventh day, the clashes began. Informant A was fleeing in a heavy downpour, leading the herds across the swollen river. Around this time, his aged father faltered, no longer able to walk, and told the others to leave him behind and go. They left him under a tree and moved on, eventually setting up camp near a Borana village. When word came that a Gabra elder had been killed, they feared that it was Informant A's father who had been slain, but a Gabra scouting party later found the father and took him safely into custody. The encampment lasted for almost a week, during which time they experienced severe hunger. A Borana woman brought them milk, saying that it was for the children. Since his wife was pregnant, Informant A decided that he would go into Surupa alone. Armed Gabra and Borana men were guarding the hilltop, and each time he was challenged he called out his name before passing. Finally, he managed to obtain some provisions.

Looking out for oneself

Just before the clashes broke out, people began to flee with their livestock. The family is the basic unit for managing domestic animals and other property, and in cases of flight, each family makes its own decisions about what to do and acts accordingly. Further, although I use the phrase 'each family,' it is not necessarily the case that all family members act together. A family may split up, some guiding camels, others goats and sheep, and others infant children and elders, with each group taking responsibility for its own evacuation. During the evacuation, they will each make decisions to secure their own safety while keeping an eye on the enemy's movements. Young children, pregnant women, and the elderly must all walk on their own, and those who cannot walk have no choice but to be left behind. Informant A's family, likewise, was forced to make the harsh decision to leave his father behind.

After the 2011 Great East Japan Earthquake, the term *tendenko* came to increased prominence in Japan. The word is a catchphrase which means that in the event of a tsunami, each individual should evacuate on their own, rather than trying to help others. The social psychologist Katsuya Yamori (2012) argues that *tendenko* has four meanings and functions: namely, (1) it emphasizes the principle of self-reliance (i.e., looking out for one's own safety); (2) it promotes the evacuation of others (i.e., the effect whereby someone who sees others flee is encouraged to flee themselves); (3) it fosters mutual trust in advance (i.e., as a prerequisite for *tendenko* to function effectively, it has the function of fostering advance trust in the fact that others who one considers important will also be sure to practice *tendenko*); and (4) it assuages the sense of survivors' remorse (i.e., it works to lessen the sense felt by survivors that they might have managed to avoid the deaths of their loved ones if only they had done more themselves).

Looking at the evacuation conduct of the Gabra in 2005, we find something similar to *tendenko*; namely, there is a sense of mutual trust in the fact that people will flee based on the principle of self-reliance. Gabra society does not have any term like *tendenko*; on the contrary, it actually discourages people from acting *kopa* (alone). However, at the time of the 2005 evacuation, families that enacted a division of labor already had a rough idea of where they would run. So, when the evacuation actually took place, they acted autonomously, each taking responsibility for their own livestock and family members. In addition, when Informant A's family took flight and left the father behind, there was a shared understanding that the father would naturally make decisions to try to ensure his own survival, such as by hiding from the enemy. Thus even as the Gabra view acting alone as undesirable, they have ways of avoiding danger when they are alone, as well as a mutual trust that they will each somehow manage to overcome crisis situations individually.

In their daily lives, Gabra repeatedly share stories of their experiences of fleeing after getting caught up in ethnic conflict, or of being attacked by Borana while grazing their herds and narrowly escaping with their lives. By sharing these experiences, they come to understand that everyone must use their own judgment to overcome crises, which in turn fosters autonomy and trust. Gabra people have also had the experience of herding livestock alone since they were boys and girls, which has helped them to develop a strong sense of autonomy from living with the constant danger of being attacked by enemies and wild animals.

The temporary dismantling of residential communities

Looking at people's evacuation behavior, we can see that Gabra residential communities were temporarily dismantled. People decided to evacuate as individual families, each heading for different destinations. Thus, for the time being, the residential communities that the Gabra had created were broken up and scattered.

When the Gabra evacuated their villages, they took their herds of livestock with them. This is essential for preventing the loss of their production system. Moreover, when they flee in a hurry—even when they do so separately at the level of individual family units—they will reorganize their community at their destination in an effort to rebuild a system of mutual support. In fact, during the evacuation in 2005, the Gabra fled to near Haro Bake, about 20 kilometers southwest of Surupa, where they organized their community anew. While they were evacuating, however, the community was temporarily dismantled. During this period, the Gabra were stricken by a host of difficulties.

For example, Informant A returned to Surupa *kopa* (alone). Since pastoralists typically travel with a *waheela* (partner) in case something happens, this solitary behavior was unusual. It was hunger that drove him to such behavior. Camping in the rain on an empty stomach is extremely cold and grueling. Seeing his pregnant wife and child growing weaker, he decided there was no option but for him to go alone. In doing so, he faced not only the danger of falling prey to the enemy, but also the risk of being accidentally shot by friendly fire. The reason he had to make such a dangerous journey alone was that his community was scattered, which meant that he was unable to find a *waheela* (which is yet another form of social support), even though he normally would have done so in times of peace.

It is noteworthy that the Borana woman provided the Gabra with milk. Gabra people who had fled their homes separately set up camp (took refuge) near Borana villages. The Borana were aware of the plight of the Gabra, and brought them milk from their village.

Toru Sagawa (2011), describing the Daasanach people of southern Ethiopia, notes that although they fight fiercely against their enemies, whom they kill without mercy, they nevertheless receive protection from their enemies, the Turkana, while also offering protection to their enemies, the Nyangatom. Sagawa (2011) explains this seemingly contradictory behavior on the part of the Daasanach in terms of Emmanuel Levinas's concept of 'vulnerability' as a characteristic condition of social life for humans; that is, the Daasanach maintain the 'attitude of co-existence with others' in full knowledge of the possibility that they may suffer physical or mental injuries (i.e., their vulnerability) in their interactions with different ethnic groups. In other words, they repeatedly interact with others who can be both 'enemies' and 'friends,' and when their vulnerability becomes apparent, they accept it and live with it.

The relationship between the Gabra and Borana is replete with such vulnerability. Tracing their history, we see how the Gabra were violently attacked by the Borana and deprived of their livestock three times in 1963, 1974, and 1992. Despite this past, the Gabra still chose to camp near Borana villages at the time of their evacuation in 2005. And they were given milk by a Borana woman who sympathized with the difficult plight that the Gabra now faced.

Engaging with the enemy

Finally, let us now turn to see how people behave during the phase of actual battle. The term 'battle' can refer both to cases of defense, as well as actively setting out on a campaign. Keeping this distinction in mind, let us look at how people behave in battle situations.

Case 3: Defense against an attack from the enemy
When the conflict began in 2005, Gabra armed with guns gathered together before moving to intercept the Guji attack. According to a Gabra, Informant B, just over ninety men apparently gathered near Surupa. They separated into bands of ten to ambush the enemy. They stole sugarcane from the fields to eat. When four Guji approached the site of the ambush, they killed one and took his gun. That night, they went back to camp, where they killed and ate one of their male camels. On the third day, they staged another ambush, and also sent an additional ten men out on a *salfa* (scouting party). These men encountered a Guji scouting party, which led to a fierce exchange of fire. In the battle, two Gabra men were killed, and one of their guns was taken. A week later, the Guji attacked again, killing a Gabra herder and taking ten camels. The Gabra retaliated, killing two Guji and taking their guns; they then gave chase to the others in order to recover their remaining camels, but they ran out of ammunition. At this point, Informant B returned to the village. He had not eaten for two days and had lost his shoes.

Case 4: On campaign
In 2018, the Gabra and Borana jointly launched a massive expeditionary attack against the Garre Somali near Hudat in Somali Region. Before the expedition was launched, an assembly was held at which it was decided that men who owned guns would be the ones to be sent on campaign. The number of men to be sent from each area was determined, with five men chosen for every fifty households. Since there were about twenty households in a certain area, two men were chosen, Informant C and D. However, Informant D, demurred to go on the campaign, saying that his gun did not actually belong to him. Other people tried to get him to change his mind, but he refused to go. Another three people from the neighboring area were also supposed to go, but did not.

Looking at these two cases, the first thing we notice is the difference in how people took part. In Case 3, people voluntarily assembled to take part in the battle, and these gatherings formed spontaneously in several places to take the battle to the enemy. On the other hand, in Case 4, an assembly was held to decide the assignment of personnel to be dispatched from each area. While some, like Informant C, put themselves forward willingly to participate, others, like Informant D, offered logical quibbles to try to justify their refusal. Not a single person from the neighboring area played any part at all.

How do such differences arise? Another man, Informant E, offered an explanation in the following terms: 'Between the man on the road and the man in his house, the man in his house is the more formidable.' He meant that people will fight more desperately when they are defending (in their homes) than when they are marching to war (on the road). In fact, when defending against enemy attack (Case 3), people came together voluntarily without any discussion. They understood that they were facing an attack by the Guji, and they took part in the defensive battle as an affair in which they had a personal stake. On the other hand, in the case of the campaign (Case 4), even after assignments had been determined by an assembly, there were still those who were unwilling to go into battle. There was no reason for them to deliberately head into danger at a time when they were not personally in any immediate peril. Besides, since the opposing side would be the one defending, they would be placing themselves in even more danger by attacking those likely to fight more desperately. It would thus seem only natural that they would not want to go into battle.

There was also another reason why people did not want to join the campaign. According to Informant D, who did not join the expedition, the Hudat area was also home to Gabra who had taken refuge with the Garre after the 2005 conflict; in his view, joining the expedition would put him at risk of fighting with Gabra living among the Garre. For the Gabra, the Garre are not an enemy to be hated as an enemy but, rather, a people with whom their brethren coexist.

Nevertheless, this is not to say that the Gabra did not actively attack their enemies. After the collapse of the Derg, the Borana took advantage of the political vacuum to surround and attack Gabra settlements and steal their livestock. At this time, Gabra who had escaped north from the Borana attacks gathered near the town of Melka Soda. Then they formed an alliance with the Guji, marched south, and violently attacked Borana settlements. At this time, the fighting dragged on for more than a year, until security was finally restored by the Ethiopian People's Revolutionary Democratic Front. The story of Informant F, who took part in several battles during this period, is truly horrific.

Case 5: Attacking the enemy

On one occasion, a Gabra *salfa* (scouting party) captured two Borana men and one woman near Arero. The scouts immediately killed the woman and one of the men, and threatened to kill the remaining man unless he led them to a Borana village. Near the village, they cut the throat of the Borana man who had guided them, and then attacked two villages, killing upwards of eighty people. They then proceeded to attack other Borana settlements in the Web area, without paying any attention to the livestock.

One reason the Gabra attacked the Borana so violently in this case was revenge. It is said that the 1992 war was sparked when the Borana surrounded and attacked a Gabra village, at which time the Gabra came under heavy fire for nine days.[4] This story was widely shared, prompting antipathy toward the Borana on the part of the Gabra. This is why the attack in this case differed from ordinary battle in that the Gabra killed women and children, massacring them without paying any heed to the livestock.

Breaking free of extremes

In the context of ethnic conflict, people are placed in extreme situations. If you were a member of a group that was party to such conflict, then in relation to a hostile group you would be in an extreme situation due to the fact that they are trying to kill you. Conversely, in order to defend your own group, you would be placed in the extreme situation of battle. What steps do the Gabra take to escape these extreme situations?

The first option is to abscond. When attacked by the enemy, people will flee. As we saw above in the section 'Fleeing from warfare,' the perception of an ominous atmosphere may prompt people to flee in various directions. And the distance and scale of their flight will depend on the scale of the suspected attack. In 1974, the Gabra fled in three different directions when they were attacked by the Guji and Borana. One group went to northern Kenya, where they lived tougher with Somali pastoralists, such as the Ajuran Somali and Degodia Somali. A second group fled eastward inside Ethiopia, and began living with Somali pastoralists. And a third destination was the Somali Democratic Republic (now the Federal Republic of Somalia). This group lived in refugee camps set up by the United Nations High Commissioner for Refugees. Old men, women, and children lived in the camps, while young and middle-aged men and women received military training in the Somali Democratic Republic in order to resist the Ethiopian government forces as guerrillas. That those who fled to the Somali Democratic Republic engaged in guerrilla activities could be said to indicate that they continued to maintain their identity as members of the Gabra people.

A second way to escape is to abandon one's ethnic group identity and become a member of another ethnic group. For example, the Gabra have deep historical ties with the Garre, a Somali ethnic group. At the time of the 1963 war between the Garre and Borana, the Gabra suffered attacks by both parties to the conflict. Even so, many Gabra still sought refuge with the Garre, apparently assimilating into the Garre people. Also, when tensions arose between the Gabra and Guji in 2012, one wealthy Gabra merchant who owned many fields and pasture lands in a primarily Guji area changed his identity by declaring himself to be Guji. Since the Guji and Gabra are culturally quite different, this came as a surprise to the Gabra, but was nevertheless accepted by the Guji. Later, the merchant in question married the daughter of a Guji merchant and has been living as a Guji.

Such changes in ethnic affiliation are historically common. Changing identity was commonly practiced among the pastoral societies inhabiting the areas of southern Ethiopia and northern Kenya, including the Gabra, and the presence of conflict was only one among several opportunities to do so.

Empathy and hatred for other groups

Thus far we have considered the various ways that people behave in the context of ethnic conflict. In this section, I attempt to interpret these behaviors in line with findings from the field of evolutionary psychology.

An evolutionary psychological hypothesis

At times of ethnic conflict, people have either viciously attacked other ethnic groups or allied with them to defend themselves. In other cases, people have avoided attacking other ethnic groups or else accepted charity from them. What explanations does evolutionary psychology offer for the evolution of human psychological mechanisms such as aggression toward other groups, in-group cooperation, tolerance toward other groups, and sympathy for others?

In their monograph *A Cooperative Species: Human Reciprocity and Its Evolution,* Samuel Bowles and Herbert Gintis (2011) explain that altruism in *Homo sapiens* has evolved through a combination of selection pressures both at the individual level and at the level of groups. This is because although there is a greater gain

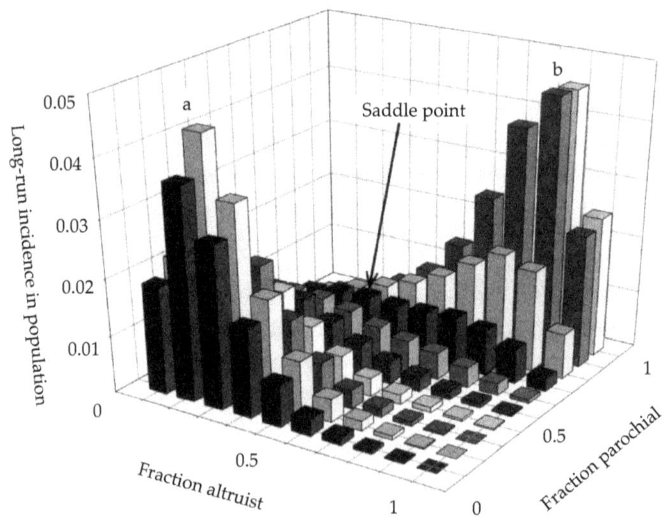

Figure 12.3: Frequency of parochial altruists versus tolerant non-altruists (adapted from Choi and Bowles 2007: 639)

to be had by behaving selfishly at the individual level, this is not the case in the presence of group conflict, where having more members acting altruistically in one's own group is associated with higher rewards.

According to another study that Bowles authored with Jung-Kyoo Choi (Choi and Bowles 2007), an examination of the process of evolution finds that altruistic behavior and parochialism toward other groups could not have evolved independently of each other, but must have coevolved under conditions of intergroup conflict. To better understand the relationship between individual selection and group selection, they combined potential human behavior in the form of parochialism and tolerance toward other groups and altruism and non-altruism within groups to develop a model consisting of four behavioral types: parochial altruism (PA), parochial non-altruism (PN), tolerant altruism (TA), and tolerant non-altruism (TN). They then simulated which of the four behavioral types would proliferate under the conditions of the late Pleistocene and early Holocene, which are regarded as having experienced major climate fluctuations (Figure 12.3). As a result, they found that two behavioral types occurred with higher frequency throughout the simulation period; namely, 'parochial altruists' (Figure 12.3a), who are correlated with the probability of war, and 'tolerant non-altruists' (Figure 12.3b), who are correlated with times of peace. The relative proportions of these two behavioral types were also found to shift over the course of generations. This is because although the frequency of warfare increases with the presence of more parochial altruists, having too many parochial altruists diminishes the probability of intergroup imbalances in the number of fighters, which are necessary for war to occur, even where there is hostility toward the other group (Bowles and Gintis 2011). Thus, in times of peace, tolerant non-altruism reaps more benefits.

Understanding the Gabra conflict

Finally, let us examine the case of the conflicts arising in southern Ethiopia in light of these hypotheses from evolutionary psychology. Given the frequency of ethnic conflict in southern Ethiopia, the results of the simulation in Choi and Bowles (2007) suggest that there should be a high proportion of individuals who are parochial toward other groups and altruistic toward members of their own group. In fact, in the case studies, as well, people willingly participated in defending against the enemy (Case 3) and violently attacked the enemy in retaliation (Case 5). In these cases, the Gabra people willingly cooperated in defensive measures to protect their own group (altruistic behavior), or attempted to eradicate the Borana community for revenge (parochialism toward other groups). They seem to be behaving consistently with the claims of evolutionary psychology—that is, parochially toward other groups and altruistically toward their own group.

On the other hand, Choi and Bowles (2007) state that the proportion of those who are tolerant of other groups but non-altruistic will increase in times of peace. In the section 'Breaking free of extremes' above, I described the presence of those who change their ethnic identity, and we could regard such people as being tolerant of other groups. However, we do not know whether they are non-altruistic. In the case of the Gabra merchant who assumed a Guji identity in 2012, we can assume from the fact that he was trying to protect his own fields and pastureland that he prioritized his own interests. Yet he changed his ethnic identity during the conflict, not in peacetime. Thus it is still possible that there is a mechanism at work which differs from the simulation results of Choi and Bowles.

To explore the possible mechanism, let us look at the case of the expeditionary campaign to attack another group (Case 4). People did not necessarily participate actively in this attack. One reason given for not participating was that some Gabra had sought refuge with the opposing Garre, and people were concerned that taking part in the campaign would mean they might end up fighting Gabra living among the Garre.

In cultural evolutionary theory, that humans possess a universal trait to behave altruistically is explained by the migration of populations to other groups, in addition to intergroup selection. Migration is considered to be an opportunity for the cultural diffusion of altruistic behavior to other groups. In fact, the model presented by Choi and Bowles (2007) assumes that 25% of the population migrates to other groups in each generation.

In reality, however, as we saw in Case 4, there were those who demurred from attacking another group when members of their own group were seeking refuge with that same other group. Choi and Bowles (2007) argue that in-group altruistic behavior (i.e., toward members of one's own group) evolved in tandem with parochial behavior toward other groups. Contrary to the claim advanced by Choi and Bowles, however, population migration has increased tolerance toward other groups while suppressing altruistic behavior toward members of one's own group.

Population migration, in other words, represents the embedding of one's own group into another group. This other group is now the 'outside inside' of one's own group, and so population migration has the function of weakening the alterity of the other group.

However, we must not overlook the fact that in Case 4, although there were many who avoided the attack, there were still many others who did take part. Those who took part in the campaign would thus have attacked even though Gabra who settled among the Garre were living there. In fact, about three months before the expedition, the Gabra called an assembly at which it was decided that Gabra staying in the Somali Region (i.e., Gabra who had migrated and settled among the Garre) would be called upon to return to Oromia Region. It was further decided that anyone who did not answer the call would be considered an enemy. Those who took part in the attack would thus have deliberately considered Gabra living among the Garre to be their enemy. This is due to the inherent human ability to manipulate and establish categories. As I argued in Soga (2013), humans apply categories to themselves, coordinating differences and working toward homogeneity therein. The primatologist Frans de Waal (2009: 213) writes that 'empathy needs both a filter that makes us select what we react to, and a turn-off switch.' The Gabra's assembly seems to have been oriented to 'turning off' empathy by manipulating categories.

The model developed in Choi and Bowles (2007) envisions the period of the late Pleistocene and early Holocene, and calculates the size of groups at about 150 people. To apply their small-group model to conflicts among ethnic groups with populations of tens of thousands, it will be necessary to gain further insights into the manipulation of categories, a task I intend to address in future.

Notes

1. The BBC article reports that the clashes were instigated by the movement of the Guji onto Borana land without permission, without mentioning the underlying conflict over the zone boundary.
2. Kawai (2013) also points out that the looting of livestock is an act that is different from war.
3. In addition to *hadha*, the Borana dictionary compiled by Father Leus (1995) also lists *olki*, a term of Somali or Islamic derivation. The Gabra use *hadha* as both a noun and a verb, while *olki* is used only as a noun. The first appearance of the term in Gabra history was in the Gabra calendar year *ahad olki guba* (Sunday year of the War of Guba (place name), August 9, 1931–August 7, 1932). This was a reference to a war between the Garre and the Boran (Soga 2006).
4. It is unclear whether the number nine in this context means exactly nine days or has the figurative meaning of 'many' days. In Gabra society, the number nine connotes a large quantity, as in *tula sagalan* (the Nine Great Wells).

References

Abbink, J. (2006) 'Ethnicity and conflict generation in Ethiopia: Some problems and prospects of ethno-regional federalism'. *Journal of Contemporary African Studies*, 24(3): 389–413.

Appadurai, A. (1996) *Modernity at Large: Cultural Dimensions of Globalization*. University of Minnesota Press.

Baxter, P. T. W. (1979) 'Boran age-sets and warfare'. In F. Katsuyoshi and D. Turton (eds), *Warfare among East African Herders*. Senri Ethnological Studies (SES) 3, 69–95.

Bowles, S. and H. Gintis (2011) *A Cooperative Species: Human Reciprocity and Its Evolution*. Princeton University Press.

Choi, J.-K. and S. Bowles (2007) 'The coevolution of parochial altruism and war'. *Science*, 318: 636–640.

de Waal, F. (2009) *The Age of Empathy: Nature's Lessons for a Kinder Society*. McClelland & Stewart.

Evans-Pritchard, E. E. (1940) *The Nuer: A Description of the Modes of Livelihood and Political Institutions of a Nilotic People*. Clarendon Press.

Fukui, K. and D. Turton (eds) (1979) *Warfare among East African Herders*. Senri Ethnological Studies (SES) 3.

Galaty, J. G. and P. Bonte (eds) (1991) *Herders, Warriors and Traders: Pastoralism in Africa*. Westview Press.

Henshaw, A. (2006, June 30) 'Ethiopian fears after land battle'. BBC News, http://news.bbc.co.uk/2/hi/africa/5133060.stm

Kawai, K. (2013) 'Forming a gang: Raiding among pastoralists and the "practice of cooperativity"'. In K. Kawai (ed.), *Groups: The Evolution of Human Sociality*. Kyoto University Press and Trans Pacific Press. 167–186.

Leus, T. (1995) *Borana Dictionary*. W. S. D. Grafisch Centrum Schijndel.

Maalouf, A. (2001) *In the Name of Identity: Violence and the Need to Belong*. Arcade Publishing.

Marshall Thomas, E. (1965) *Warrior Herdsmen*. Secker & Warburg.

Sagawa, T. (2011) *Bouryoku to kantai no minzoku-shi: Higashi Afurika bokuchiku shakai no sensou to heiwa* (The ethnography of violence and hospitality: War and peace in East African pastoralist society). Showado.

Schlee, G. (1989) *Identity on the Move*. Manchester University Press.

Soga, T. (2006) 'Changes in knowledge of time among Gabra Miigo pastoralists of southern Ethiopia'. *Nilo-Ethiopian Studies*, 10: 23–44.

Soga, T. (2008) 'Gabura: Nanmin to shite, gerira to shite ikita 20 seiki' (Gabra: Life as refugees and life as guerrillas in the twentieth century). In T. Ayabe, K. Fukui, S. Takezawa and Y. Miyawaki (eds), *Sahara-inan Afurika (kouza sekai no senjuumin-zoku 5)* (Sub-Saharan Africa (Indigenous peoples of the world 5)). Akashi Shoten, 161–182.

Soga, T. (2012) 'Kokka ni kousuru kyoten to shite no nariwai: Bokuchiku-min Gabura migo no nanmin senjutsu' (Livelihood as a base for resisting the state: Refugee tactics among Gabra Miigo pastoralists). In K. Matsui (ed.), *Nariwai to seisan no shakai-teki fuchi* (The social fabric of livelihood and production). Showado, 389–426.

Soga, T. (2013) 'Perceivable "unity": Between visible "group" and invisible "category"'. In K. Kawai (ed.), *Groups: The Evolution of Human Sociality*. Kyoto University Press and Trans Pacific Press, 219–238.

Soga, T. (2019) 'Nanmin o sasaeta rakuda koueki' (Camel trade to support refugees). In I. Ohta and T. Soga (eds), *Yuuboku no shisou* (Nomadic thought). Showado, 91–114.

Yamori, K. (2012) '"Tsunami tendenko" no yottsu imi' (Revisiting the concept of 'tsunami *tendenko*'). *Shizen saigai kagaku* (Journal of the Japan Society for Natural Disaster Science), 31(1): 35–46.

CHAPTER 13

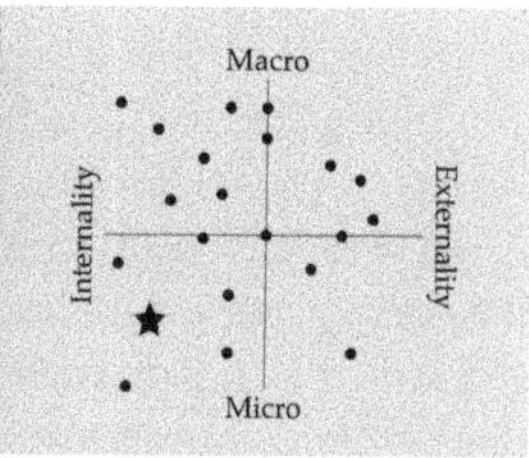

A Survival Strategy for Extreme Populations:
Mbotgote, 1974–76

Takeo FUNABIKI

Keywords: survival strategy, population decline, disappearance, ritual

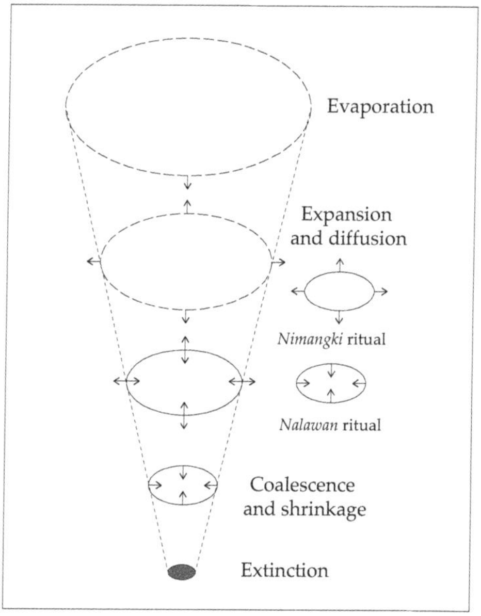

Two possible avenues for the disappearance of Mbotgote society:

- as defined by the *nimangki* ritual, the outer edges of Mbotgote society may broaden with the spread of the *nimangki*, so that the society diffuses and disappears
- as defined by the *nalawan* ritual, the scale of Mbotgote society may contract because of the inward coalescence of the *nalawan*, so that the society ceases to exist.

At the time of my fieldwork in 1974, the Mbotgote, a scattered population inhabiting the wooded hills of the southern part of Malekula Island in what is now Vanuatu (formerly the New Hebrides, a colony managed jointly as an Anglo-French condominium), consisted of only 146 individuals. Considering the society's sex ratio and the number of births in the preceding years (Tables 13.1 and 13.2), it was expected that the Mbotgote would either dwindle and disappear, in numerical terms, in the not-too-distant future, or else lose their identity and cohesiveness as a group in the very near future. In this sense, faced with a critical situation, how did they recognize and strategize with respect to their survival, both as individuals and as a group? This chapter describes the ways in which the Mbotgote thought and acted during a limited period from 1974 to 1976 and, from this data, considers the direction in which this population was headed, both individually and collectively. The purpose of this study is to infer how a particular group of people, the Mbotgote, strategized in the midst of crisis and—though it might seem a bit grandiose to do so on the basis of a single case study—to attempt to gain some inkling about how human groups assign meaning to, and choose their own actions in, the context of extreme situations.

Demographic overview

In 1974, when I began my fieldwork, dubious information about the existence of a 'still uncivilized' people living scattered in the wooded hills of Malekula (also spelled Malakula) had long been rumored both in the New Hebrides and beyond, but there was scant documentation in the form of written reports by local observers or the like as to the reality of the situation. What mention could be found—in travelers' accounts and in articles in magazines, newspapers, and research reports—customarily referred to these elusive people collectively as the 'Small Nambas' (meaning 'small penis sheaths'), in contrast to the people from the northern part of the island, known as the Big Nambas (i.e., 'big penis sheaths'). Among the Christianized *skul* ('civilized'; a concept derived from the English word 'school') islanders, who were also Malekula islanders but had moved to the surrounding coastal areas, these people were also called 'Mbotgote.'

To the *skul* living on the island's southern coast, the Mbotgote people were not as outlandish a people as they were described elsewhere, off the island. Rather, they were regarded as people who shared the right to use a large swath of land in the island's southern interior and had rights to certain ritual complexes (*nimangki* and *nalawan*).[1] At the same time, there was a nuance in the words of the Christians on the southern coast suggesting that they considered the Mbotgote to be a backwards people who kept to *kastom* (a word derived from the English word 'custom') and who lived in the traditional and customary manner of old, even in the latter half of the twentieth century.[2]

The first time I saw the island from an approaching ship, I was overwhelmed by the forests carpeting the sheer slopes that jutted skyward just in from the coast. What a thick mass of vegetation. Considering the people scattered among them, it was immediately apparent that my impending fieldwork would not be easy.[3] Naturally, there were no accurate population statistics for Malekula, and the situation of the people of the island's interior was unknown to the colonial administration. There were no clearly marked roads, of course, and although the map of the island showed the elevations of the land and the rivers, it was utterly devoid of any marked settlements.[4]

Before my own survey, descriptions of the people called Mbotgote—and of people calling themselves by that name—were sporadic at best. To date, my own monograph (Funabiki 2012) remains the only published work to have used the name Mbotgote to describe this 'linguistic, social, and cultural group' in any significant detail. In this chapter, I would like to consider the origins of this group by looking back on the process of how (and to what extent) I came to know the Mbotgote over the course of my fieldwork.

The people called the Mbotgote have never had a legal or political system defining themselves to the outside world. And while one could apprehend them as a language group named Mbotgote, in this region (which is characterized by a mixture of many languages and dialects) it would only be an a priori convenience for anthropologists to regard them as a 'social group.' If the question 'Who are the Mbotgote?' has any meaning for them at all, it is in the sense that each of them possesses rights to ritual involvement and land in the island's interior.

Seen from the outside, Malekula's entire coastline—aside from two spots in the northeastern and southwestern parts of the island—has only a few low points, like a mountain sunken into the sea. One has only to walk a few dozen meters inland from the coast to find the slopes thickly covered with trees and shrubs. The forested area stretching 35 kilometers east to west and 25 kilometers north to south in the island's southern interior is a massif that outsiders may not traverse on foot without being accompanied by an Mbotgote guide. Small settlements are scattered at intervals of about half a day's travel, and the people move back and forth between them, checking the routes already used and hacking at the grass and trees with machetes as they walk to open a space wide enough to pass through on foot. Table 13.2 lists the Mbotgote villages that existed as of 1975, along with their populations.

Penetrating the interior to survey the people who live in the forest, one finds that in addition to the Mbotgote, there are other people who call themselves Nabwol.[5] Thinking about the way that the Mbotgote and Nabwol are distinguished is a good example for thinking about what social groups were like during the era when the entire island adhered to *kastom*, so to speak. That is, it tells us about the interrelationship of social groups on Malekula Island before

	Island interior			South West Bay		
	Male	Women	Total	Male	Female	Total
Mbotgote	50	31	81	5	11	16
Nabwol	23	13	36	4	0	4
Netangalan	0	0	0	2	0	2
Other independent actors*	4+	3+	7+	–	–	–

Table 13.1: Mbotgote and other populations living in the island interior and on the coast (as of July 1975)

*I have included these independent others in this tabulation because they hold rituals jointly with the Mbotgote. Given the likelihood that there were more independent actors than the number listed, I have added a plus sign to the numbers (which thus represent a minimum value).

Settlement	Age category	Male	Female	**Total**
Lendemboi (including six small settlements without a ritual hut)	Adult	12	10	22
	Child	7	8	15
	Total	**19**	**18**	**37**
La'oitambus	Adult	3	2	5
	Child	1	1	2
	Total	**4**	**3**	**7**
Mabaruk	Adult	6 (4)*	1	7
	Child	1	0	1
	Total	**7**	**1**	**8**
Yabugatas (including four small settlements without a ritual hut)	Adult	12 (3)	6	18
	Child	8	3	11
	Total	**20**	**9**	**29**
Kamalwibru (including five small settlements without a ritual hut)	Adult	14 (8)	5	19
	Child	3	3	6
	Total	**17**	**8**	**25**
Yabukamabus	Adult	3	3	6
	Child	3	2	5
	Total	**6**	**5**	**11**
Mbotgote	Adult	33 (7)	19	52
	Child	17	12	29
	Total	**50**	**31**	**81**
Nabwol	Adult	17 (8)	8	25
	Child	6	5	11
	Total	**23**	**13**	**36**

Table 13.2: Population, sex, and relative age in villages with ritual huts (as of July 1975)

*Parentheses indicate the number of unattached young people who have completed initiation as adults.

the political power of the colonial government and the religious integration of Christianity. In other words, it teaches us about the societal strategies of these groups.

First, as a shorthand, we can think of there being two subpopulations within the higher-order population known as the Mbotgote—namely, the Nabwol and the Mbotgote in a narrower sense (hereinafter, I represent Mbotgote in this narrow sense in italics; otherwise, where I use Mbotgote without emphasis, I am referring to Mbotgote in a broader sense that includes the Nabwol). These two groups recognize each other as having the same language.

Next, these two groups also recognize that they are within the same ritual system. The two groups—the *Mbotgote* (in the narrow sense) and the Nabwol—are centered on two male leaders, each of whom is accorded paramount ritual rank and is trusted by others as a man of character. The *Mbotgote* inhabit the western part of the island interior, while the Nabwol occupy a smaller sphere of influence on the eastern part of the island, where they live while intermingling with their neighbors. In terms of the relative hierarchy between the two groups, based on the grading of their ritual huts, the Nabwol are subordinate to the *Mbotgote*. However, even this subordinate status is regarded as a qualification that can only be gained by being Mbotgote (in the broad sense), and the Nabwol could not have obtained this ritual authority had they been separated from the Mbotgote. This is a shared understanding on the part of the Mbotgote that derives from a broad notion of graded ritual[6] shared among the people of Malekula and of adjacent islands. And while this relative hierarchy is primarily a positive explanation given by the *Mbotgote*, the Nabwol also acknowledge it, albeit with some reservations. And the Nabwol not only acknowledge their inferiority, but also observe the ritual restrictions that it entails. This could be called a strategy of concession, in terms of securing one's (own) ritual rights even from a position of inferiority, as we shall see.

In concrete terms, this restriction is a hierarchical rule which holds that whereas a *Mbotgote* can gain the right to enter a Nabwol ritual hut by passing a *Mbotgote* initiation, the inverse is not true for a Nabwol. That is, when adult Nabwol men visit an *Mbotgote* village, they are not permitted to enter its ritual hut.[7]

The following vignette illustrates a special case of this. In 1975, among the Nabwol, there was a certain boy (I estimated his age at eight or nine years old) who could not get anyone to perform an initiation ritual for him because there were so few boys in the Nabwol group, even though the time had come for his initiation. Instead, he ended up undergoing the initiation ritual together with *Mbotgote* boys. Even so, when the ritual was finished, he could not get any assurances from the *Mbotgote* as to whether he would be able to enter the *Mbotgote* ritual hut. For my own part, I thought the Mbotgote were speaking in a 'pompous' manner, and that he would eventually be admitted. Since the boy would necessarily have opportunities to enter the *Mbotgote* ritual hut during the year-long period of initiation, I presumed that he would have the right to do so once his initiation was completed. Nevertheless, the *Mbotgote* men kept equivocating, saying that the matter had not yet been decided.[8]

My reasoning was that the Nabwol claim that they were ritually distinct from the *Mbotgote*, as I have already mentioned, carried the risk that they would lose their ritual rights, even if these were inferior to *Mbotgote*, and was thus likely deemed disadvantageous at a time when their population was already shrinking. Therefore, I inferred that the Nabwol would consider it advantageous to be recognized as members of one of two groups within the gradually shrinking Mbotgote population. This is the root of my understanding of the Mbotgote and their population in the broader sense. Thus, while the Nabwol admit their inferiority, this is not surprising in light of their own conception of ritual and (by extension) the wider conception of ritual that exists in the region, as well as, more broadly, on Malekula and the adjacent islands. For all men—whether Nabwol or *Mbotgote*, as we shall see—it is the case that even if they have traditionally possessed certain ritual rights, there are other ritual rights they have not possessed. And cannot acquire. This is thought to have been the perennial state of the world of *kastom* that once existed on Malekula. By 1974, however, it could only be seen among the Mbotgote, with the Nabwol or *Mbotgote*.

Under these circumstances, my own census of the population is shown in Table 13.1. First, among those living in the island's interior (i.e., the interior of the southern part of the island in the context of Malekula Island as a whole) who have not turned to *skul* and are roughly observing *kastom*, I counted eighty-one *Mbotgote* (in the narrow sense) and thirty-six Nabwol. An additional seven individuals still held rights to land in the island's interior by sharing in or jointly performing rituals with the Mbotgote. Accepting the possibility that my study was not exhaustive and that there were more Mbotgote living elsewhere in the forests, there may have been more than seven (indicated as '+' in the table).

Aside from these, there was an enclave of sixteen former *Mbotgote* and four former Nabwol living on the coast in the geographically named South West Bay, as well as two former Netangalan. (Though I will not go into the details of their origin here, see Funabiki (2012) for the background of the Netangalan branch of Mbotgote.) From the perspective of the Mbotgote who keep to *kastom*, these others are regarded as having turned to a life of *skul*. As such, they are considered to have lost their right to possess or perform rituals and to any land rights to the island's interior. However, from the point of view of these twenty-two people who have turned to *skul*, they have retained ritual rights to some degree (e.g., to take part and to perform rituals jointly with the Mbotgote who keep to *kastom*), and nor have they lost their land rights. And, indeed, some of these *skul* Mbotgote were among the participants at a *nimangki* ritual observed in 1975 in Fenambwass, a community near South West Bay. Also, some of these twenty-two people were continuing to use their own Mbotgote land rights near the coast, even while being censured by the *kastom* Mbotgote for doing so. Judging from these facts, I concluded that these twenty-two people could still be counted as Mbotgote, even though they now identified as *skul*. This accounts for the population of 146 individuals I mentioned at the outset. I should also note that while I was conducting fieldwork on the island between 1974 and 1976, I counted only one birth and one death.[9]

Land rights and ritual rights

Land rights

The relationship between the *Mbotgote* and Nabwol may be considered as a prototype for relations among social groups on Malekula during the era of *kastom*. In the main, the people of the island would have built their social relationships and intergroup arrangements on the basis of two conditions—land rights and ritual rights. Let us now turn to matters as they stood in 1974.

As alluded to above, in geological terms the island of Malekula has been thrust upward by volcanic forces, so that it looks like a mountain that slopes down to the sea. Food sources across the island as a whole consist mainly of yams and taro, as well as cassava, which is cultivated and believed to have been introduced relatively recently. In addition, cash obtained by the cultivation of coconut palms as a commodity crop allows the islanders to buy other foods (e.g., rice and canned goods) from the few stores operated by the government-organized cooperative or from Chinese Polynesian traders in South West Bay. In addition, pigs kept by individual households and cattle raised in common on the palm plantations are slaughtered for pork and beef. Only a limited amount of food is purchased, and there are not many opportunities when pigs or cattle are eaten as meat. Therefore, among both the people of the forest and the residents of South West Bay, time and energy for acquiring food is devoted to the horticultural cultivation of yams and taro in the forested hills. Even though they live on an island surrounded by the sea, they face the mountains, metaphorically speaking, and turn their backs on the sea, living as though the sea lay out of sight.[10]

The first thing that comes up in this discussion is land rights. No one knows anything about recent demographic changes with any accuracy. The posthumous monograph by anthropologist Bernard Deacon (1934) contains no such data. And while it is entirely hearsay, the following information comprises a list of what I have been able to infer from interviews with people on the island.

There were once many more people on the island, but the introduction of firearms led to an escalation in fighting; many people died, and many more died of disease. This is what we are told. However, a few reservations are necessary here.[11] In the past, it is certainly the case that disease and strife were factors pushing people out of the island's interior, and although we cannot offer accurate statistics, we can safely assume this fact in the context of the history of the Pacific. This is another way of saying that the spread of disease and fighting exacerbated by guns are already part of the conventional narrative, and that even if one doubts their veracity, the fear that people have of them is real.

On the other hand, that the factors pulling people to the coast were associated with the factors pushing them out of the interior can be illustrated as follows. In connection with guns and fighting, the 'peace' offered by the constraints set by the colonial administration conferred a certain stability to interpersonal relationships. With regard to disease, while I do not know the degree to which medical care or medicines were available, government-run health centers would have represented a bulwark against disease to some extent. At the very least, rumors of their efficacy would have had the power to attract people to the coast.

Other elements attracting people to the coast included *skul* goods pertaining to food, clothing, and shelter, including textiles, food products, and building tools and materials. And while arguing the degree to which the schools (from which the term *skul* derives) and Christian churches, as a religion, attracted people would not be easy, the merits the education they provided in terms of reading and writing English (or French in the areas governed by the French administration) and pidgin English (Bislama or Bichelamar, the English-based creole used in Vanuatu) must have helped to maintain peace by allowing people separated by many different languages and dialects to have a common language. Then there was the gradual arrival of accompanying goods (e.g., medicines and medical equipment, which would have provided increased safety in childbirth and delivery, for example) and information. And while goods were expensive and accordingly always in short supply, information brought with it the fascinations of the outside world. In short, this meant radio communication with other regions and coming to 'know' the outside world through wireless broadcasts in pidgin English. Taken together as a whole, these developments—that is, *skul*—attracted people to the coastal settlements.

If there was thus such merit in *skul*, then why had the Mbotgote in the island interior not come down from the hills as of 1975? Why had they stayed hidden in the highlands as a 'people of rumor' at a time when the entire population of Malekula was in excess of 10,000 people? The Mbotgote answer to this question is straightforward; namely, their land is in the middle of the island and to cultivate it from the coast would entail too much of a trek.

Let us consider a few points. First, what is life like for people in South West Bay? These people originally lived in the forested areas of the island, not far from the coast, subsisting by a type of slash-and-burn agriculture. However, when they came down to the coast, although it depends on which group they once belonged to, they continued to take the time to make the trek through the slopes and forests to the lands over which they had rights, and the sites that were suitable for cultivation therein. Given the significant population decline in recent years, determining which plots in these 'regions of rights' belong to whom is no easy matter. At this point, I would prefer not to stray too far into an anthropological discussion of descent. However, although the Mbotgote practice patrilineal descent, we must consider the fact that, as seen in other societies in Oceania, the patrilineal groups to which they belong are quite flexible (through discussions and disputes, and in combination with other conditions). Therefore, the few Mbotgote who converted to *skul* settled on the coast and negotiated and acquired the rights to cultivate land at distances that required as little trekking as possible while maintaining peaceful relations with the people of South West Bay in a web of descent, rituals, and various other relationships. As of 1974, apart from disputes in terms of suspicions of magic or entanglements with sexual relationships, there seems to be no overt conflict between the Mbotgote and other groups over land. The presumed decline in population in the region and in South West Bay in the recent past seems to have had the positive effect of suppressing the occurrence of land disputes, and it might be said that the region currently exists in a kind of equilibrium. As in the case of the discovery of oil in New Guinea, land disputes might still erupt suddenly in the event of the discovery of mineral resources, but this has not happened so far.

Mbotgote in the island's interior, naturally, should they descend to the coast, must either choose to commute via a difficult trek that takes them far from their cultivated land, or else build a shack on the cultivated land and spend extended periods of time away from their families on a regular basis. Either is a considerable burden. The Mbotgote who have already gone down to the coast are few in number, and while the fact that they were among the first to do so worked to their advantage in that they were able to negotiate for the right to cultivate their land, the same will not be possible for the dozens or hundreds of people who could potentially follow them.

More than this, however, those who went down to *skul* were not looked upon with envy by the majority of Mbotgote who remained in the highlands. Doubtless, there is some merit in taking refuge from the anxieties over strife and disease that I have already mentioned, and in enjoying the various privileges of *skul* that are available to those living on the coast. If the merits of *skul* and the security gained by taking refuge were perceived to be greater than the difficulties of accessing their cultivated lands, then they should do so. But these land-related advantages and disadvantages are not the only conditions to consider when moving to the coast. There is yet another factor in addition to these land rights, one to which the issue of distance is just as pivotal, and which is also deeply related to land rights. I am speaking here of ritual rights.

Ritual rights

Ritual once had great significance and value throughout the islands of the New Hebrides. As mentioned above, Malekula and its neighboring islands were once the site of considerable ritual and festival activity. Even if only for the Mbotgote and the residents of South West Bay, much of the value they created, aside from things that contributed directly to their survival, lay in ritual activities. When Queen Elizabeth visited the New Hebrides in 1974, two highlights of the hospitality shown to her were the 'bungee jump' in a ritual on Pentecost Island and the tusks of pigs raised for a ritual on Malekula Island. Both were rooted in ritual practices.[12] When the coastal *skul* people talk about the Mbotgote, underlying their scorn at how these people still follow an archaic lifestyle is also the sense of a certain prestige accorded to a people who still maintain and carry out their ritual rights. At the root of this sentiment, which could be likened to awe, is their awareness of the contrast between the Mbotgote, as adherents to traditional custom, and themselves, as a people who have broken with it.

In the southern part of Malekula, a clear separation between men and women was traditionally observed in that they ate their meals apart and did not eat foods cooked over the same fires. Even among men, distinctions were observed depending on whether or not they had completed their initiation. Furthermore, these kinds of divisions also existed at the higher and lower grades of the men's ritual hierarchy. Men's purpose in life was to rise through these grades. To become someone who is *skul* is to abandon these ritual practices through conversion to Christian ideology. From the point of view of traditional ideology, this means abandoning the distinctions made between men and women, and between men and boys, and breaking the rule of 'separate fires' to become *skul*, or, speaking metaphorically, 'fallen.'

A characteristic of this transition from *kastom* to *skul* is that, for men, it is an irreversible process. In short, once he has converted from *kastom* to *skul*, a man who has eaten a meal prepared on the same fire as a woman cannot return to *kastom*. Excluding the 'converts' of South West Bay, as of 1974 there were 124 purely *kastom* people in the southern part of the island, rather than 146. And if we count only the men who hold ritual rights, there are seventy-seven at most. Compared to these seventy-seven, there are likely closer to 10,000 people who live in South West Bay and other coastal areas around the southern part of the island. These people may fashion and display idols and artifacts from the old rituals and perform their associated dances at certain secular events (such as, for instance, on the occasion of a political campaign to bring an airfield to South West Bay in 1976), but they cannot do so in the context of a 'genuine' rite of ascension, as that would be a violation of the rules of the ritual. However, there would only be seventy-seven men directly responsible for enforcing control over such violations and, given the sheer number of people, this would be impossible. Adherence to the rules is grounded in the fact that the coastal *skul* islanders, who constitute an overwhelming majority, and the Mbotgote, who are so few, nevertheless share the same norms. The dances and the production of ritualistic objects performed by the people on the coast are permitted because both parties now regard these as a 'traditional performance' rather than a religious activity.

As of 1976, the people of Vanuatu were busy preparing for independence, which was scheduled for 1985, and the administration[13] was promoting modernization, including the 'invention of tradition,' while following the example of other Melanesian regions. As a result, 'tradition' began to be considered as an identity of the 'nation-state' to come—not as a barbaric custom or vice as it had previously been defined by the Christian church, but as something that should be preserved, which pertains to the topic addressed in this chapter.

A photograph of a Mbotgote ritual on the cover of *National Geographic* in January 1972 (Muller, 1972) provided an opportunity for the Mbotgote to become more widely known outside their own society, and the circulation of this attention likely became a turning point for the re-evaluation of people of *kastom* like the Mbotgote. Thus the Mbotgote, seen as a 'backwards' people still living in the hills, were gradually coming to 'represent' Malekula and, by extension, the New Hebrides (later Vanuatu) as a whole. Such national prestige, and the merit that would be derived from it, were not yet a concern for the Mbotgote as of 1976. However, their gaze from the coast into the forests was already changing. It may have been gradually becoming an element in the contingent determination of a social survival strategy in the microcosm of southern Malekula Island. Let us look into this further.

Transacting tradition in markets old and new

The *nimangki* ritual

How was this gradual change in the perspective toward *kastom* received by the Mbotgote? Rumors had reached them about how people on the coast were producing ritual idols and artifacts and selling them for cash. Such rumors provoked two thoughts. First, they were surprised that people would pay money for such things. Some Mbotgote were already earning cash by working as day laborers on the coast. Second, they were concerned that the *skul* might be infringing upon and selling the rights of *kastom*. Even the Mbotgote had already started paying for things like machetes, cigarettes, T-shirts, and various other products used as ritual paraphernalia.

Items sold as 'traditional' included the masks, headgear, wigs, and other costumes used in ritual dances, idols, dance accessories, wooden drums, and decorations used in ritual settings. I have elsewhere described two of the rituals practiced by the Mbotgote; namely, the *nimangki* and the secret *nalawan* (Funabiki 2019). For example, the Rhamburhambu, a life-sized mannequin embedded with the skull of a dead person that is made for funerary rites and which features in the *nalawan* ritual, not to put too fine a point on it, has become a staple exhibit in the world's major ethnological museums, much like impressionist paintings in art museums. The Rhamburhambu figure is specific to a ritual passed down by Mbotgote men, of whom there are now only seventy-seven. While the sheer number of specimens populating so many museums may seem strange, given this limited population, from the nineteenth century to the late twentieth century, visitors 'from away' likely brought them home, attracted by the uniqueness of the life-sized figures encapsulating the skulls of the dead. In the past, the Mbotgote of Malekula and others who performed similar rituals may have sold Rhamburhambu that were moldering away in the ritual huts in exchange for cash. Some may also have been made to order and sold by *skul* entrepreneurs. The market for 'tradition' was expanding into the Mbotgote world.

The Mbotgote are now aware of the value of their rituals as traditional performance to the outside world. However, their strategic thinking has not progressed to the point of producing ritual idols because they can command a high price. As always, they are concerned about the violations of *kastom* by those who are *skul*. That is, they wonder whether the people on the coast are performing *kastom* ritual, and whether those who do not trace their descent to the former ritual groups are arbitrarily claiming to be their inheritors.

It is only among themselves that the Mbotgote see any value in ritual. What is important is their own joy at ascending the ritual steps, their own sense of accomplishment in performing funerals and other services, and the respect that they gain from their fellow Mbotgote by doing so. In addition, the envy of those who have gone down to join the *skul* and the attention of the entire *skul* community that now surrounds the Mbotgote have become newly involved in the recognition of value.

A good touchstone of the relationship of the Mbotgote rituals to the outside and the *skul* was a 1975 *nimangki* ritual conducted by two men living in Fenambwass, the nearest Mbotgote settlement to the coast. The public nature of the *nimangki* is apparent from the fact that women also take part in the ritual, albeit in a minor capacity, and that neighboring groups other than the Mbotgote are invited to the festival. This was particularly true in 1975, when the festival was staged about an hour-and-a-half on foot from the coast. The two hosts allowed people from South West Bay to come to the festival as 'spectators.'

From the very beginning there had been constant rumors among the Mbotgote that the two men's true intention in establishing their own settlement (along with their two families) on the coast, near the *skul*, had always been to get closer to the *skul*. During the staging of the *nimangki*, a long-awaited event that had taken several years of preparation, the two organizers glowed with pride to see the many spectators during the festival. After considerable discussion, even the paramount leader of the Mbotgote, who served as the master of ceremonies during the ritual, did not object to the presence of *skul* people as spectators. However, among those Mbotgote who held a contrary view was a man who was highly regarded in society and who cited his wife's pregnancy and the long distance as reasons why he did not take part. Among the Mbotgote, it was said that this constituted a rebuke. From its own religious viewpoint, the Christian church was critical of such islander rituals, but with respect to the 1975 event, a pastor living in South West Bay visited the festival grounds with his family during the daytime, but his attitude suggested he was appreciating a local 'traditional performance' of the community to which he had been posted. For the coastal *skul* people, the pastor's actions made them realize that the Christian church would not deny 'tradition' as such.

However, something I felt to be more critical for the Mbotgote's future survival strategy was the participation of a former Mbotgote who was not in the *skul*. The man in question had grown up Mbotgote and completed his initiation, but owing to complications in a certain personal relationship, he had become involved in a feud with other Mbotgote and had some of his fingers blown off by gunfire and was now living in South West Bay. He had given up his ritual rights, but still claimed the right to cultivate land in the island's interior, for which he was still negotiating. He not only participated in the *nimangki* on this occasion, but also brought a small pig and joined in the dancing. This suggested that the Mbotgote recognized the bona fides of former Mbotgote to some degree, even those who had gone down to the *skul*. And although the complications of his personal relationships had led to the attack in which his fingers were blown off when he went up the mountain in the past, in the future he would not be ambushed again, even if he did go up to cultivate the land. However, it seemed that this acknowledgment did not extend as far as taking up the position of a ritual organizer.

One of the Mbotgote had dressed for the occasion in a matching red sports jersey and shorts; if we consider *skul* as modernity, then he was stretching the limits of its acceptance. Since he and others like him would eventually become central figures among the Mbotgote, one might expect him to treat *skul* with a certain degree of congeniality with respect to ritual. But this was only with respect to the *nimangki* ritual. When it came to *nalawan* ritual, someone like the red-shirted man, for example, would have a much different attitude.

The *nalawan* ritual

The case with respect to the *nimangki* seemed to be as follows. In 1976, when non-Mbotgote men in a different village on the southern part of the island performed the *nimangki*, I had my doubts about whether it was legitimate for them to do so, given that the men of this southern village had already been 'contaminated' by *skul*. To the Mbotgote men, however, it seemed they were willing to make this concession, up to a point. Of course, the *nimangki* and *nalawan* are two different and independent rituals. The *nimangki* attracts large crowds, which may also have some advantages for the Mbotgote. They get a certain amount of respect for performing the old rituals, and even as they engage in their longstanding tradition of joyful revelry—performing frenetic dances, and sacrificing pigs and consuming their flesh—in front of the Christian pastor, they receive no censure from the *skul*. This is not a bad situation. Of course, even with the *nimangki*, idols are created and can be sold. These command a high price because they are actually authentic, rather than fake. The public nature of the *nimangki* is advantageous to the social survival strategy of the Mbotgote, since it allows them to acquire cash despite their geographic disadvantage. However, the *secret* nalawan ritual is a different story altogether.

The *nalawan* is esteemed as a more rarefied ritual because of its secrecy. This is why the Mbotgote are respected, envied, and even feared as the hereditary stewards of *kastom*. As I have mentioned elsewhere (Funabiki 2019), even the matter of who has achieved what grade in the *nalawan* is a secret. As the Mbotgote tell it, the strength of Mbotgote 'rituality' has long surpassed that of any of the other groups near the coast, just as the *Mbotgote* surpass the Nabwol. That is, regardless of whether the people living on the coast were *skul* or not, the '*kastom*-quotient' of the *Mbotgote* was higher, even during the era of *kastom*. I gained a firsthand appreciation of the sheer magnificence and quantity of the rituals featuring the Rhamburhambu, mentioned above, on the occasion of a funeral held in 1976 in the *Mbotgote* village of Yabugatas, which at the time had a population of twenty-nine individuals. Even though their standard of living was not as affluent as it once would have been, either in terms of material or time, the ritual prowess of the Mbotgote was on display in the funeral square and the many ornaments arrayed there.

In 1976, South West Bay was visited by a high-ranking British administrator. The National Party, a political organization that by that time already covered the whole of the New Hebrides, organized a performing arts festival (in support of the construction of an airfield for the bay) that would serve as both a demonstration of local culture and entertainment for the official. The Mbotgote were asked to participate in the festival, and while they made use of the opportunity to show the coastal area to young people who had never had the chance to see the coast before, their dance performance was lackluster. However, former ritual groups now living in South West Bay staged their own ritual dances and displays, which in the past had been highly confidential. Some of these involved the presentation of phallic objects more than five meters in length, which greatly aroused my own ethnological interest. My Mbotgote companions, on the other hand, seemed to experience something like hesitation or unwillingness. Even so, they watched the proceedings from a slope well away from the soccer pitch that served as the staging ground for the festival, either because they were scandalized at seeing what had once been the mysteries of other groups ('hey, should we have seen that?') or as a way of avoiding censure for having seen something illicit. Of course, they did not present their own *nalawan*. They simply looked through their fingers at the former mysteries of the others. This was the current status of the Mbotgote in 1976.[14]

The problem is that the number of people setting any store by the *nalawan* ritual is gradually diminishing. This is not an 'invented tradition,' but an authentic tradition of the Mbotgote themselves, and investment in that tradition is dwindling. Even though the *nalawan*, due to its secrecy, allows the Mbotgote to obtain a certain kind of reverence from a wide audience, the reality is that it is only the Mbotgote who actually buy into its mysteries. The Rhamburhambu used in the rituals might command a high price when new ethnological museums are built, or from enthusiasts or curiosity seekers; even so, to make them simply for the purpose of selling them would be to degenerate into *skul*. In the end, they must be created in the ritual process by which the adepts convey the mysteries to the initiates. And a great deal of energy must be expended in doing so. Notwithstanding the fact that any statues left over after the festival can be sold to ethnologists or other interested parties, a ritual hut must be built, pigs with curved tusks suitable for the *nalawan* must be procured, and yams and other foods must be prepared for a large feast.

To derive some conclusion from the foregoing, I would like to look back on what we have seen thus far to discuss individual and collective strategies more holistically.

Conclusion: groups, family, and individual strategies

Keeping in mind the pointlessness of repetition, let us list the strategic issues that the future held for the Mbotgote in 1976.

First of all, they cannot move their land. If the Mbotgote live on the coast, their only option for using the land in the areas of the interior to which they hold rights is to move themselves. But even if the first few families were able to obtain land use rights near the coast through individual negotiations, it would not be feasible for the rest of the Mbotgote to go down to the coast to obtain similar land use conditions. Even looking not only at South West Bay, but also at the island's eastern and southern coasts, we find that the coastal areas no longer have the capacity to accept that many 'immigrants.' Even taking each individual Mbotgote in turn, at this point it would be hard to expect to inspire them to start searching for their own personal rights by comprehensively tracing their kinship back through the generations or extending their affinal relationships laterally.

It could happen that they move as close to the coast as possible, entering into social relationships with others on the coast while they cultivate the lands to which they have rights, but this would mean joining the *skul*. And joining the *skul* entails relinquishing their rights over *kastom*. More specifically, this means singing hymns at church and sending their children to school, but if they can overcome the mental objections to undertaking such a life, then perhaps they can take the plunge and start a life on the coast, even if this means taking on financial hardship, so that the balance of advantage and loss tips toward *skul*.

But I may have gotten ahead of myself. Just becoming a *skul* person is still a matter of doing well and easing into the lower tiers of the *skul* world. This is not just a matter of settling for economic inferiority. Men (to whom I should perhaps limit myself, as I do not have time to mention women in this chapter) will lose the meaning and values they had heretofore attached to their lives. For what they lose, they gain a new 'Christian worldview' and 'scientific understanding,' but this cannot be said to make up for the loss of their ritual worldview. They also lose the reverence that they had been accorded as 'the last remaining *kastom* men,' and their sense of satisfaction with their own lives.

Here, we need to revisit the distinction between the *nimangki* and the *nalawan*.

Even now, it may be possible for the Mbotgote to maintain a certain amount of ritual cachet by continuing to perform the *nimangki* in the time-honored fashion. Like the two men who performed the *nimangki* in 1975, they live in the *skul* as the 'ex-Mbotgote' listed in Table 13.1, carrying with them the standing of having until recently performed the *nimangki* as an authentic ritual as a people of *kastom*. When those two men performed the *nimangki*, we could say that they intentionally approached the *skul*, even as they kept their backs to it. Also, the instance of the *nimangki* performed in the southern part of the island in 1976 may have served as a demonstration that *nimangki* rituals can also be performed in the *skul*. Furthermore, in the case of the *nimangki*, there was also an aspect that the ritual, as it took place on Malekula, had the potential to be a 'modern form of self-expression' through the invention of traditions, such as the introduction to the ritual of a new idol that had appeared in the dream of the organizer.

A necessary and final consideration in this discussion, however, is the fact that the Mbotgote's prestige derives not from the public *nimangki* but from the secret *nalawan*, and that the very transmission of this ritual is endangered due not only to population decline, but also the attendant aging of those who have achieved the highest ritual ranks. If the Mbotgote *nalawan*, the characteristic ritual that has a mystical quality that allows the Mbotgote to regard other groups as subordinate, is not passed down within the community, they will be just another of the many groups that have come down to the coast. This would be the case even if they still had the *nimangki*, which in the southern part of the island was performed even by people in the *skul*.

The current state of the *nalawan* could perhaps be interpreted as follows. From the Mbotgote perspective, the *nalawan* and its appurtenances are not cultural phenomena in terms of the inheritance of a tradition in contemporary society. Rather, they involve the transaction of rights. From that point of view, the problem is that fewer people are buying into the secret of the *nalawan* ritual. Nowadays, the rights of the *nalawan* ritual have taken on the feel of wartime bonds in a country that is on the brink of losing a war, or paper currency during a time of economic turmoil. The source of their value may disappear, or else fall precipitously with hyperinflation. This time, specifically among the Mbotgote in 1974–76, curious tactics began to appear between potential 'buyers' and 'sellers.'

A seller, which is to say someone among the Mbotgote with higher-ranking rights in the *nalawan*, would think as follows. He has no desire to trade the *nalawan* for something of lower value (i.e., fewer pigs) than he gave when acquiring the *nalawan* himself. However, there are no longer as many buyers as there were, so he may not be able to sell it at a high price. A buyer would think the following. Since there are no longer many buyers in Mbotgote society, with its declining population, surely the price can be lowered. And when the seller is old and not long for this world, then he will have no reason not to sell.

However, the seller also has the following train of thought. He understands that there are not many buyers, and that if he does not lower the selling price, then the performance of the ritual will no longer be possible. In that case, he should sell it cheaply. But wait! Even if the price of the ritual rights decreases, their scarcity, and thus the ritual authority of those who hold them, increases the less they sell them. So even if his authority dies with him and the secrets are not passed on to future generations, then at least as long as he lives, he can gain personal advantage not by selling, but by virtue of the prestige of having it. The buyer also thinks as follows. The more he persists in this ritual transaction, which hides who holds which *nalawan*, the more he places himself at the mercy of the other party, so that it may end up being a complete waste of time. First of all, if he does buy it, to whom would he sell it? Would anyone among today's youth ever want to buy it? Rather than going through the mental anguish of being bested in any negotiation, better to simply ignore it and wait for an overture from the other party.

A synoptic view of all of these tactics would show the following. Even if the mystical *nalawan* were to be lost to the Mbotgote and the earth, there would still be a kind of benefit to the holders of high ritual rank. Their prestige would be preserved for as long as they live. For the buyers, as well, not being forced into the purchase allows them to conserve both their energy and economic resources. Nor will anyone blame them, since they can also say that they attempted to buy into the ritual, but some foolish men of higher rank damaged the legacy of the predecessors by spouting nonsense. But what does this diminishment and stalling mean for the Mbotgote as a whole? Even if it is a difficult life after coming down from the hills, when the reversal that we see in the current situation takes place—that is, when *kastom* starts to be appreciated and the Mbotgote are accorded a kind of dignity for having it—then who would lose out if the *nalawan* ritual, as the source of that dignity and awe, were to be lost? If the *nimangki*, which seems to be permitted no matter who practices it, should become a traditional performance, then the privilege held by the Mbotgote might disappear, leaving them only with the economic disadvantages of living in the hills.

In addition to predicting the fate of the Mbotgote after 1976, I would also like to reflect on the strategies available to groups in the extreme situation of potentially disappearing due to population decline.

In attempting to solve the problem of economic 'distance,' the Mbotgote are torn between overlapping individual and collective strategies. At the individual level, there is the option of going down to the coast early with family members and perhaps like-minded families to live in an economically inferior situation, which has the advantages of *skul*, but comes at the price of the loss of the prestige associated with *kastom*. Staying in the highlands would mean the opposite. However, even if they stay in the hills, the continuation of the *nalawan* ritual, which constitutes the core of the prestige accorded to *kastom*, is in danger due to a lack of action on the part of those with the highest ritual rank. The Mbotgote, taking the individual as a unit, can be divided into three groups:

- those who go down early to become part of the *skul* and slide into the lower echelons of coastal society

- people at the lower ranks of the *nalawan* ritual who abide by *kastom* (i.e., stay in the hills) and who, if they cannot afford to pay for rights in the *nalawan* ritual, resign themselves to staying in the lower ranks—but if they can save up the payment (i.e., the tusked pigs, goods, and energy involved in performing the ritual), this enables them to preserve their prestige as members of Mbotgote society

- people at the higher ranks of the *nalawan* ritual who abide by *kastom* (i.e., stay in the hills) and who, if they cannot sell their rights in the *nalawan* ritual, although they will not be able to recoup what they paid themselves, will be able to preserve their prestige as high-ranking adepts of Mbotgote *nalawan*.

Each of these three groups seems to be doing what is optimal for the individual. Curiously, however, they are alike in that they are all foreclosing on the future continuation of the present state of the Mbotgote—that is, they are doing nothing to prevent the disappearance of their society.

The penultimate question here is this. Why are individual Mbotgote unable to adopt a constructive strategy to ensure their collective survival? We can see this by thinking about the Nabwol. The *nalawan* intersects with rights to the ritual huts. Selling to the Nabwol would mean that the Nabwol could also enter the ritual huts of the *Mbotgote*. But the Nabwol are precluded from buying. The Nabwol do not represent a new market because, as we saw, Nabwol are considered to have a lower rank, even one who undergoes an initiation ritual together with *Mbotgote*. And, of course, one cannot sell to the *skul*. In other words, the prestige of the *Mbotgote* is considered relative to the Nabwol and the *skul*, but since the Nabwol and *skul* are defined in the first place as lower than the *Mbotgote*, they have no 'outside,' and must therefore live on their own. Although I am saying this as a metaphor rather than trying to suggest something in this case, there is a similarity here to the way that the human species inhabiting this planet Earth, when the time comes that they are about to perish because they likewise have no outside, will have no final escape or successor.

In terms of the alpha (α) schema I set have set out elsewhere (Funabiki 2017), when the α of ritual creates a semantic space, the *nimangki* is inadequate for unifying that semantic space as the Mbotgote identity. If the *nalawan* disappears, the Mbotgote will cease to exist. The Mbotgote in 1976 were about to lose the constitution of their semantic space due to the loss of a shared α in the hierarchical opposition between higher-ranking and lower-ranking *nalawan* initiates. Beyond the Mbotgote, the *nimangki* might have the potential to be the focal α for the wider area of South West Bay, or of the southern part of Malekula Island, or even the entire New Hebrides, which in 1976 was on the verge of independence. But then, the Mbotgote would be no longer.

As defined by the *nimangki* ritual, the outer edges of Mbotgote society may broaden with the spread of the *nimangki*. But as it expands, the outer edges will gradually fray, so that the Mbotgote social identity as defined by the *nimangki* ritual diffuses and disappears. As defined by the *nalawan* ritual, the scale of Mbotgote society may become rigid and fixed due to the inward coalescence of the *nalawan*. However, with the decreasing number of *nalawan* initiates, it is also possible that we may see an inward convergence and, in fact, the scale of the society will also contract until finally Mbotgote society as defined by the *nalawan* ritual may cease to exist. The conceptual diagram at the start of this chapter is an expression of this fact.

Even in the face of a group's disappearance, when doubts arise about the α of the group's very reason for being, as, for example, in the case of the Mbotgote *nalawan*, the survival of the entire group is jeopardized in the commonplace phenomenon of being unable to negotiate prices. However, it does not appear that individual strategies are accurately calculated according to relative gains or losses in the face of such disappearance. The Mbotgote can be seen as one example of such a situation.

And yet, as a kind of postscript, I would like to add that for a long time now, I have been seeking, researching, and discussing Mbotgote society and all its rituals in the 'primary' discursive space of the Mbotgote 'mainstream.' But what comes to mind in hindsight is not that I have not written enough, but how my research was lacking. That is, what was the involvement in these strategies of the women who make up half of the Mbotgote society? There are also graded rituals for women in Mbotgote society, or at least on Malekula. In the sense that I arrived at the determination that these rituals were subordinate, not major, I focused on the male rituals. When Ortner (1974) pointed out the existence of such female rituals, I avoided the idea of interpreting Mbotgote rituals from that perspective as something to be used 'for the sake of' a discussion about gender (Funabiki 1983b). I still feel that there was a point to be made in terms of a discussion about ritual.

However, the first article I wrote about the Mbotgote (Funabiki 1977), entitled 'An old woman's journey,' was a preliminary critique—a kind of postlude—to my own doctoral dissertation, which I was then in the process of completing. The article told the story of a woman in an *Mbotgote* village who had taken it upon herself to end her marriage (i.e., what we would call 'divorce') and return to make a life in the village where she had been born and raised. Although it is now only a dim memory, like something that happened to a stranger long ago, it may be that at the time I had picked up on something else. That is, I may have noticed that the semantic space defined by α—the semantic space whose highest value I accord in this chapter to the *nalawan* ritual—was recognized and interpreted by each and every individual in turn. But had I said that, it would have meant the absence of anything like a society, and so I did not take that route. I suppose I was afraid that if the public principle—that men maintained a single semantic space, even while overlapping with and being torn between the individual and the group as a whole, and that the entirety of this space corresponded to society—could be disrupted by the deviation of a single woman, then I would not be able to bring myself to finish writing my monograph

Though I can no longer do so myself, it could be that each individual person can act within a certain range of freedom, even within the semantic space defined by α, which is to say the Mbotgote semantic space sustained by the *nalawan* ritual. A discussion could be developed around such an assumption. An even if this results in the destruction of the basis of the Mbotgote, that will not matter so long as a meaningful space remained for something new. Only then will the individual be able to develop a strategy for living in the world in the face of the extreme being experienced by his or her own group. I feel that we have been able to develop our thinking in this direction and should be able to continue doing so.

Notes

1. As used in this chapter, the term 'ritual' should be understood to mean exclusively *male* rituals. Although women in this region also have their own ritual organizations, I will make a special note of this as necessary.
2. For more detailed discussions of *kastom* and *skul* (in Japanese), see Funabiki (1983a) and Yoshioka (2005).
3. Although this has little to do with the discussion here, I think the reason I did not lose heart at the time was that I took it for granted that I had already become an anthropologist. Thinking back now, however, I wonder at what a strange thing it is to decide to become a field anthropologist.
4. I was never able to pinpoint exactly where I was during my fieldwork. However, on one occasion, on a day when I was staying in the center of the island, I caught sight of a smaller island to the south and, from the angle, was able to estimate my current location.
5. I only ever visited the area where the Nabwol live, which extends roughly over the eastern third of the southern part of the island, on two occasions, in 1974 and 1976. Both visits were brief, and my data on the demographic makeup of the area are based entirely on hearsay.
6. This ritual organization, which has many variations in terms of both names and contents, is detailed in the work of Michael Allen (1967).
7. Both Nabwol and *Mbotgote* claim that although Nabwol are barred from the higher-ranking *Mbotgote* ritual huts, they can still enter the lower-ranking *Mbotgote* huts. For example, Nabwol men stayed in a ritual hut during the *nimangki* ritual in Yabugatas. However, there is a certain practicality to this, since the ritual hut in Yabugatas was built to accommodate visitors for the *nimangki* ritual. As there is no need to go into this further, I will leave the matter there.
8. In fact, this outcome was further complicated by the fact that the boy was not mentally prepared to withstand the year-long period of isolation, and thus unable to complete the year of initiation, but there is no need to go into the details of the matter for the purposes of the present discussion.
9. As both took place after July 1975, they are not reflected in Tables 13.1 or 13.2.
10. This contrasts with other Pacific islands that are made up coral atolls and have a maximum altitude of only a few meters. Although the people of Malekula fish for mullet and shrimp in the lower reaches of the rivers, they only rarely take food from the sea. Even in the villages of South West Bay, there are hardly any canoes, and those that do exist are only used as a means of transportation to other nearby coasts.
11. Of course, we cannot know what 'many' amounts to in this context. Even so, the sense that the population decreased as a result may reflect some basis in fact. While it is said that many were killed by guns, the actual number of guns seems to be limited, and purchasing ammunition for the guns that do exist is neither cheap nor easy, so their use would have been limited. Nor do we know what any 'diseases' actually were. However, it can be inferred from other cases

in Oceania that deaths would have increased due to the introduction of certain infectious diseases to which the islanders would have had a low resistance. The anthropologist Bernard Deacon died of blackwater fever contracted while conducting field research in South West Bay on his way to take up a research position in Australia (his monograph on Malekula was published posthumously as Deacon 1934). However, blackwater fever is not an epidemic disease as such, but a combination of the common cold with the complication of malaria that causes the vomit to turn black, and it would be a misunderstanding to suggest that a special epidemic of this particular disease had spread in this region.

12. The pig tusks used in the rituals mentioned here are special tusks that curl up into a circular shape, the artificial result of long-term selective breeding. Pigs with such tusks are used as a 'currency' with which to buy ritual rights.
13. At the time, the New Hebrides were jointly governed by Great Britain and France under a special political system known as a condominium. However, as this fact is not relevant to the present discussion, I will leave off any further explanation.
14. Once the festival was over, when we paused for a rest in a copse on the bank of a river a couple of hours into our two-day journey on foot back to the village at the center of the island, a young man imitated the dancing of another group we had seen at the festival, convulsing with laughter at how ridiculous they had seemed. Though it may stretch the point to say so, the only thing that they took home from their encounter with the *skul* was their experience of seeing children play soccer, so that after they left, they played their first game of soccer in the ritual square, using a citrus fruit as a ball.

References

Allen, M. R. (1967) *Male Cults and Secret Initiations in Melanesia*. Melbourne University Press.

Deacon, A. B. (1934) *Malekula: A Vanishing People in the New Hebrides*, edited by C. H. Wedgwood. Routledge.

Funabiki, T. (1977) 'Rōjo hitori no michi' (An old woman's journey). *Yasei jidai* (Wild age), 4: 158–165.

Funabiki, T. (1983a) '*Kastom* to *skul*: Vanuatsu, Marekura-jima no shakai henka ni kan suru bishiteki kentō to riromteki kōsatsu' (*Kastom* and *skul*: A microscopic examination and theoretical discussion of social changes in Vanuatu and Malekula Island). *Tōyō bunka kenkyūjo kiyō* (Memoirs of the Institute for Oriental Culture), 93: 31–66.

Funabiki, T. (1983b) 'Women's position to men in the social universe of the Mbotgote in Malekula Island, Vanuatu'. *East Asia Cultural Studies*, 22(1–4): 169–185.

Funabiki, T. (2012) *Living Field: Monograph and Models Concerning Human's Social Design, Based on the Mbotgote in Malakula Island, Vanuatu*. The University Museum, The University of Tokyo.

Funabiki, T. (2017) 'Basic components of institution: Understanding institution according to triangular and tetrahedral models'. In K. Kawai (ed.), *Institutions: The Evolution of Human Sociality*. Kyoto University Press and Trans Pacific Press, 309–323.

Funabiki, T. (2019) 'Epilogue—future agenda, others as an affliction: Tripartite relationships and the tetrahedral model'. In K. Kawai (ed.), *Others: The Evolution of Human Sociality*. Kyoto University Press and Trans Pacific Press, 425–442.

Muller K. (1972) (photograph of Mbotgote man during a ritual). *National Geographic*, 141 (1): front cover.

Ortner, S. (1974) 'Is female to male as nature is to culture?' In M. Z. Rosaldo and L. Lamphere (eds), *Woman, Culture and Society*. Stanford University Press, 67–87.

Yoshioka, M. (2005) *Han-posuto koroniaru jinrui gaku: Posuto koroniaru wo ikiru meraneshia* (Anti-postcolonial anthropology: Melanesia in the postcolonial world). Fukyosha.

CHAPTER 14

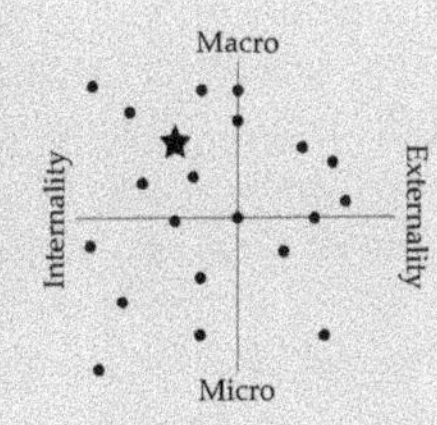

Interaction Communication Systems and the Context of 'Society':

Communication as a Touchstone for Approaching the Extreme of the Disappearance of Local Community

Kōji KITAMURA

Keywords: interactions, society, social systems, collaborative communication, double selection

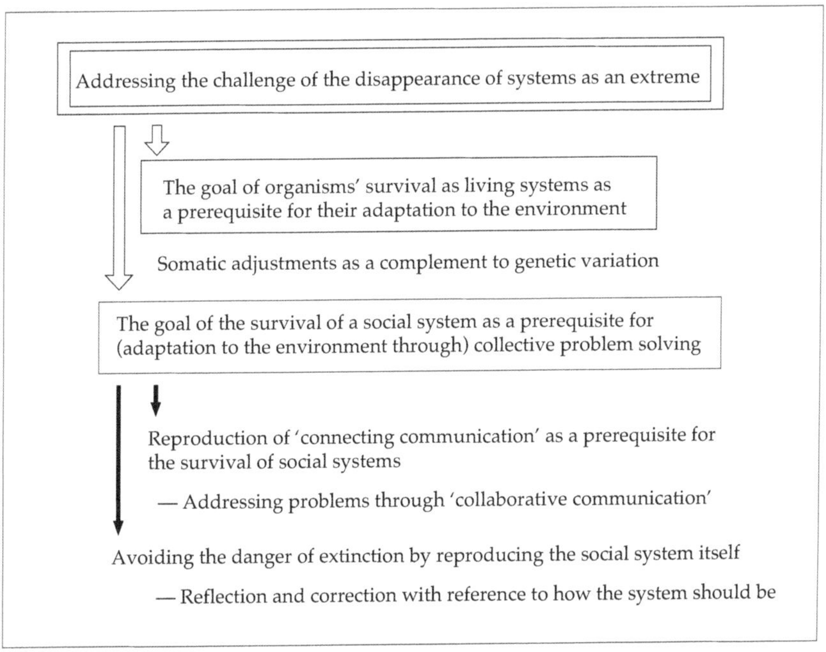

In discussions of biological evolution, only the presence or absence of fluke genetic mutations tend to be regarded as problematic as a response on the part of organisms to major changes in their environments. From the standpoint of an organism as a system, however, the question is whether or not somatic adjustments to ensure the organism's own survival are even possible under the impact that this has on its system as a whole. In this chapter, I apply this way of thinking to the field of how social systems adapt to their environments. Drawing on concrete examples, I consider the kinds of countermeasures that social systems have prepared to avoid the possibility that their choices for adaptation to the environment may endanger their own survival—in other words, what kind of 'double selection' do they perform in order to ensure adaptation to the environment simultaneously with their own survival?

Human sociocultural evolution

In this chapter, I undertake to consider human sociocultural evolution on the same theoretical basis as that of biological evolution. Specifically, I consider the social and cultural transformations that individual human societies (which serve as the focus of our study) have selected in order to adapt to changes in their environments as being equivalent to the adaptations of individual organisms. Here the subjects adapting to the environment are not individual people but individual social clusters (i.e., social systems).

As with biological evolution, even in the context of human sociocultural evolution, it is thought to be extremely difficult to adaptively transform social and cultural practices in the face of major environmental changes. In the case of biological evolution, not only is it extremely rare that seemingly random genetic mutations produce characteristics coincidentally suited to a changing environment, but even when they do, other difficulties are expected. Here, in addition to the difficulties associated with the level of manipulating changes in response to changes in the environment, I would also like to focus on a distinctly different problem of equal importance.

Individual organisms are complex wholes, intricately intertwined systems with many interdependent parts. When a change is engendered in one part of the system as the result of a genetic mutation, this change inevitably spreads to many other parts of the organism. While the organism's continued survival entails the need for complex coordination of these various changes, it is unlikely that a single mutation would be able to carry out such coordination.

Bateson (1972: 348–9) illustrates this point as follows:

> A hypothetical pregiraffe, which had the luck to carry a mutant gene 'long neck,' would have to adjust to this change by complex modifications of the heart and circulatory system. These collateral adjustments [i.e., those necessary for continuing to survive with a long neck] would have to be achieved at the somatic level. Only those pregiraffes which are (genotypically) capable of these somatic modifications would survive.

The reason that environmental adaptations are also difficult in the context of human sociocultural evolution is presumably because it would be incredibly difficult to sustainably maintain the structure of a society or culture that had changed in conjunction with a major transformation of its sociocultural practices to adapt to a changing environment. In the case of biological evolution, even if a mutation arises that should produce a change that represents a valuable environmental adaptation, this will be for naught unless the survival of the organism (i.e., the survival of the biological system) can be ensured by means of somatic changes that will address new situations internal to the organism itself. Applying this logic to human sociocultural evolution, when we undertake to make major changes to the structure of our own society and culture in response to environmental changes, we must at all costs avoid the situation whereby the social systems that encompass the practices in question become unsustainable because of our haphazard manipulation. Accordingly, it seems inevitable that we should attempt to transform ourselves to adapt to the environment while somehow maintaining the general framework of a society that enables such manipulation and ensuring our survival by repeatedly making flexible adjustments. Absolutely vital to this approach is, first, to ensure *appropriate choices in dealing with collective problems* and, second, to coordinate such choices so that they support the survival of the social systems that enable the reproduction of interpersonal communication. Trying to achieve these two things at the same time entails making a *double selection over two levels* of social activity: first, attempting to achieve the desired outcome so that those involved in each interaction can continue living, and, second, attempting to ensure the continued survival of the social systems that enable the stable reproduction of the activities involved in doing so.

However, in discussing activities supporting the survival of social systems, I am not referring to activities that maintain organizations such as those that require shared values and loyalty from each member, as in the case of 'organizational systems' in which membership is clear and people are tied together by their sense of belonging. Rather, the social systems in question here are spontaneous and extremely ambiguous, to the extent that they are not usually consciously noticed. They are the kind of thing that comes to our attention only when situations arise that threaten their very survival. Here 'activities supporting the survival of social systems' basically refers to activities that attempt to secure contexts enabling the reproduction of interpersonal communication. This is something that is conceivably always being attempted in some form, even when people are not clearly aware of the existence of the social systems that are supposed to be at issue.

At the same time, this chapter discusses social systems that enable the reproduction of interpersonal communication, such as those described here, as well as activities that consciously attempt to address the problem of their own reproduction based on a clear awareness of their existence. Specifically, I consider cases of social systems whose continued survival is being threatened by their own choices—choices made in the context of dealing with collective problems—and in which internal states have been produced that attempt to secure the survival of those systems by controlling the movements occurring therein. As a result, with changes in the systems' state of readiness, it appears that attempts to deal with their problems are themselves reconsidered under new conditions, so that they can be reproduced more reliably.

Although the double selection described above is inevitably manifested in the cases that follow, this seems to be due to situations in which the communication among fellows (i.e., acquaintances) (which should be the most fundamental event in terms of linking humans imparted with sociality) and fellowship (which represents the basis of such communication), along with the larger social framework that supports that relationship, are all considered to be uncertain and unreliable.

In contemporary Japan, it has been widely reported in recent years that a declining birthrate, an aging society, and the increasing concentration of the population in Tokyo are all gradually contributing to the realization of an extreme situation in terms of the disappearance of local communities. Beyond simply a decreasing population, the concern here is that this situation will lead to the loss of the social bonds that link people in local communities. Particularly in the case of the small island communities that I have studied, it seems that the overwhelming presence of the sea, which circumscribes the extent of local communities, means that they must be keenly aware of unsettling factors that threaten the survival of their highly independent and unified way of life.

I first take up instances of communication to deal with collective problems in Japan's island society to examine their characteristic features. Specifically, I draw on the case of Shiraishi Island in Okayama Prefecture, where I conducted a previous study (Kitamura 2014). On tiny outlying islands like Shiraishi, agriculture is of limited importance due to the fact that there is little land available for cultivation, and in the absence of any special conditions, a mix of subsistence farming and fishing is the norm. Accordingly, it is unlikely that large economic disparities or differences in social status will become entrenched among the islanders, and it seems that it is common for communication to take place on equal terms and without role differentiation.

On the other hand, in terms of relations with the mainland, alongside the recent shift to the 'borderless' circulation of goods and populations, reliance on the outside world is expanding widely, and traditional local communities are reportedly weakening apace. In this chapter, I discuss the ways in which people endeavor to continue living under such circumstances, even as they are aware of the extreme situation represented by the disappearance of their local communities as a possible future.

In the latter half of this chapter I develop this discussion with an attempt to approach the issue of social reproduction (survival) in human history. The key to this discussion is to concretely elucidate the process by which the dimension of 'society' appears as a context in communication interactions in societies where the power relations and institutionalization that structure society as something tangible are still undeveloped. In societies like this—as in those of hunter-gatherers or pastoralists whose lives are strongly dependent on nature—people pursue their livelihoods by making use of resources distributed over a wide area; in order to survive, they sometimes must live in small groups that engage in repeated migrations. Accordingly, the inevitable task here would be to ensure that society is able to regroup where necessary to deal with the problems it faces in a collaborative fashion.

In terms closer to the 'extremes' denoted in the title of this volume, it seems to me that extreme situations like the disappearance and demise of the social systems supporting the reproduction of communication among fellows, in the context of societies taken up for discussion here—namely, primitive societies and the peripheral island societies of contemporary Japan—must be envisioned as realistic issues. At such times people do not merely attempt to ensure the more reliable reproduction of communication in those settings; while concretely envisioning the existence of the underlying social systems that support that communication, they also endeavor to reproduce these as more reliable in themselves (see the frontispiece to this chapter).

Collaborative communication on Shiraishi Island

Overview of Shiraishi

The island of Shiraishi is one of the Kasaoka Islands, a chain of islands in the sea south of Kasaoka City in southwestern Okayama Prefecture that fall within the city's municipal jurisdiction. The island is connected to the Port of Kasaoka by regular shipping routes, and the fastest high-speed boat can reach the island in about twenty minutes. The island has an area of 2.86 square kilometers and a population of 672 (as of the 2005 census). In recent years, like other outlying islands, Shiraishi has been experiencing consistent depopulation compounded by an aging population and declining birthrate, with an aging rate which is specified as more than 55% of the population being aged over sixty-five years. The number of children is devastatingly small. The only elementary school on the island has double-cohort classes. Similarly, at the island's only junior high school, the possibility of the school's disappearance from the island as the result of amalgamation has been discussed as a real problem.

A variety of crops are grown in fields scattered here and there throughout the island, but these are mostly for private consumption. Although the fishery is not as active as in its heyday, about twenty households still engage in fishing, primarily bottom trawling. Other than these, the stone quarrying industry has been in decline since the 1990s, and only two households in this industry remained active at the time of the fieldwork survey on which this chapter is based. *Ryokan* (Japanese-style inns) and *minshuku* (guest houses), which are a mainstay of the tourist industry, are discussed separately later. The research that forms the basis of this chapter was carried out with students over three years from 2006 to 2009 within the framework of a research methods unit in a course offered through the Department of Behavioral Science in the Faculty of Letters at Okayama University (Setonai-kai ni okeru seikatsu to bunka kyōiku kenkyū gurūpu 2009). Although the following description of Shiraishi Island characterizes the time of our survey in the present tense, the reader should understand that our fieldwork and the writing of this chapter are separated by the span of almost a decade.

Islander responses to the revitalization of the island community

At the time of our fieldwork in the late 2000s, the Kasaoka Islands, including Shiraishi, were actively engaged in community revitalization projects centered on an Island Community Revitalization Working Group organized by the Kasaoka municipal office, as well as on local auxiliaries organized by island residents that were incorporated as a non-profit organization. The residents' organization ostensibly counts all residents of the Kasaoka Islands as members and is positioned as a group in which the residents themselves take the initiative in the revitalization of the island community. In reality, however, I got the strong impression that the Island Community Revitalization Working Group, which plays a role in the municipal administration, guided the residents in the planning and execution of various projects.

Shiraishi has long been known for having one of the best beaches in the prefecture. In the 1970s, when the number of tourists was at its peak, the island counted seven *ryokan* and eight *minshuku*, but at the time of our survey only a combined total of seven such facilities were still operating. In the case of local revitalization projects like the one described here, one of the main goals is to increase the number of people interacting with the urban district. This produces a tendency to plan more sightseeing events, and the proprietors of these accommodation facilities are expected to actively participate in these types of projects and join forces in making the events more exciting. In reality, however, although these people will cooperate when asked, their joint participation in such events seems to be far from active. Rather, the feeling of most is that they are too short-staffed to be able to help, or that joint initiatives are too difficult due to differences in their respective management policies.

In the face of such responses by island residents, some reservations are necessary in supposing that they are reluctant to participate in community-oriented projects, considering that in the context of a community revitalization project it would be only natural for them to work together to realize their goals. Despite calls for revitalization owing to the weakening of the local community by population outflows due to increased dependence on the outside world, it would be simplistic to expect that simply taking the initiative to launch a convenient project would bring about the return of vibrant cooperation on the part of residents. In this case, it seems more likely that residents have an aversion to the idea of forcibly mobilizing people for a project that could aptly be described as 'island tourism promotion.'

Instead, while building their own ways of life in their own fashion, residents are collaborating with those close to them, where necessary, in the attempt to launch activities involving those around them. One initiative that might merit being described as a tourism project involves welcoming students on school trips from the mainland to the island. This program, which began in response to an inquiry by a junior high school in a landlocked community, is fast becoming a major event that involves not only those who run accommodation facilities, but also the entire population of the island. As the scale of the program has expanded, it has gradually involved more and more people, with many accommodation facilities joining in to house students in separate groups and building a system by which the entire island is able to host students from a single school. As well as offering a full menu of hands-on activities such as marine sports and fishing, the program also includes exchange events involving local students and other islanders.

In areas where community revitalization projects have made the most progress, it is only natural that the projects should be advanced through collective efforts and consensus-building premised on an organizational approach. On Shiraishi, by contrast, even though the ultimate result differs little, people do not try to adopt the approach of achieving their goals by positioning themselves inside an 'organization.' To frame this difference more clearly, I mean that they do not try to create cooperative systems by imposing a view of collective efforts toward regional revitalization as being an unconditional good. Naturally, faced with problems that rebuff the efforts of individuals working separately, they will occasionally attempt to address these collectively by working together. But how they do so is left to the mutual coordination that takes place spontaneously in that time and in that place. In other words, rather than attempting to control the connection of actions toward an agreed-upon end on the basis of shared values, they make room for individual autonomy and exploratory back-and-forth, and then move the collaboration forward by sharing a basic general direction and flexibly incorporating one another's contributions.

So what we need to confirm here is that they do not think that choosing a more effective way of dealing with this kind of regional revitalization project will solve the problem all at once. They believe that when dealing with the future of the region, misguided efforts risk the 'dismantling of the region as a whole,' and they are skeptical of the idea that 'it is enough simply to formulate an appropriate plan and take steps under its supervision' (see Sasaki 2000). That is, they are trying to engage with the difficulties they face collectively through collaborative communication with their fellows, not with the expectation of immediate results, but by trying again and again without getting tired.

Transmission of the Shiraishi festival dance

Originally known and performed by almost everyone on the island, the *Shiraishi odori* (Shiraishi festival dance) is a local version of *Bon odori* (a style of dancing performed during the summer Obon Festival honoring the spirits of the ancestors). It clearly differs from typical *Bon odori* in that there are several types of dancing, and multiple dances are danced simultaneously to the same tune. In the past, it was not uncommon for the circles of dancers to be two or three deep, and the dances are said to have been performed throughout the night, interspersed with breaks. Owing to the high value attached to the dance itself, the *Shiraishi odori* has been exhibited off the island as a local performing art, and in 1976 it was designated as an Important Intangible Folk Cultural Property of Japan. In recent years it has taken on importance as a highlight for attracting tourists to Shiraishi. Aside from *Bon odori* events for the islanders, there are opportunities for tourists to dance on weekends as the Obon Festival approaches, and planned tours using chartered boats to bring visitors to the island to experience the dance.

For the people of Shiraishi, beyond being a familiar annual event, the *Shiraishi odori* is a cultural asset that is highly valued by the outside world, something special that they recognize as being the 'pride of Shiraishi Island.' At the same time, however, it is also clear to anyone that the dance circles have become smaller and smaller with the island's ongoing depopulation and the decline in the number of participants as entertainment options have diversified. This situation seemed to have been enough to elicit a sense of crisis on the part of the islanders, who feared that if things continued as they were going, the transmission of this traditional observance—whose performance on the dancing grounds of the Obon Festival had always been taken for granted—might one day cease.

In the face of such circumstances, one measure adopted was to teach the dance to children at school. For the past twenty years or so, members of the Shiraishi Odorikai (the Shiraishi Festival Dance Association), a group that is responsible for the preservation and transmission of the *Shiraishi odori*, have been visiting elementary and junior high schools to teach the dance to students. This is done not simply as a means for children to gain personal experience with the traditional culture of their home community, but with the aim of bringing students to the level where they will be able to perform the dance in front of others, and various opportunities are provided to show off the results. Meanwhile, when members of the Odorikai perform the dance for people who are not from the island, they also offer an accompanying lecture with opportunities to engage the audience's participation so that they can experience the dance together. Rather than simply a direct attempt to increase the number of dancers who will carry on the tradition, this is also a means of raising the profile of the *Shiraishi odori* as a traditional performing art by fostering supporters who will feel an affinity with it.

Islanders, however, would deny that passing on this traditional performing art is something they feel they must do because it is valued by outsiders. Far from it, they deliberately mention the dance's national designation as an important intangible folk cultural property as though it were an afterthought and repeatedly reject the idea that this is why they are passing it on. At the same time, while the Odorikai is solely responsible for the various activities involved in passing on the tradition, other islanders are, in many cases, unexpectedly critical in their evaluations of these efforts.

In dances performed for the benefit of tourists, the dancers dress in full costumes, with orderly dance circles and dancing styles. In contrast, islanders will point out that, in the authentic tradition, local people should be able to take part in the dance in their regular attire and to join in the circles freely and dance however they please. Moreover, some also express their dissatisfaction at how the dance is often interrupted by commentary on occasions when the *Shiraishi odori* is performed for tourists, noting how the atmosphere of the dance has been changed by the increased participation of tourists, to the point that they no longer know who the dance is really for. With regard to how the dance is taught in school-based education, as well, from the perspective of what the dance *should* be like for islanders, negative views have been expressed about how it has become something that students have to learn whether they want to or not.

Such responses on the part of islanders are quite surprising to outsiders like us. However, such feelings of surprise were already apparent in answers to a question we put to residents as an icebreaker when we first started our research on the island; namely, 'What are your thoughts on the *Shiraishi odori*?' Although they referred to it as their pride, instead of trying to talk it up by boasting about it to outsiders, it was usually the case that they made excuses for not being more enthusiastic participants in the dance themselves. In other words, rather than trying to tell outsiders about the value of this island tradition, their responses were tinged with regret about what it was they were actually doing as people ostensibly responsible for carrying on this annual observance.

Specifically, residents would say things like, 'I'm not a very good dancer, so I end up feeling awkward and just don't dance anymore,' or 'I came to the island as a bride, so it would be too much trouble to learn it from scratch—I've gotten by without dancing so far,' thus offering a string of responses that seem to explain the background to the dance circle's waning popularity. In other words, they were not trying to address the issue of the transmission of the tradition on the premise of the idea that all residents should actively cooperate to ensure the survival of a traditional observance as the pride of the community. Rather, it seemed that each individual was engaged in repeated trial and error as they reflected from a personal standpoint (as someone contributing to the life of the community) as to how to be involved in this event, which represents an important annual observance embedded in their daily lives.

This surprising response by people can be resolved by considering that their efforts are aimed not merely at the transmission of the event as such, but as a means of maintaining the viability of the life of the community. This sense of crisis concerning the transmission of this observance has resulted in various attempts to address the problems currently facing the community. Conceivably, however, these attempts should themselves be checked to see whether they might not work to dismantle the premise of the local community solidarity that enables people's collaboration at such times. This seems to show the basic attitude of the people of Shiraishi, who are attempting to ensure the reproduction of collaborative communication among people without resorting to social coercion and without forcing people's loyalty to their local community.

Dealing with issues concerning the survival of island society

Value in terms of the survival (reproduction) of local communities

Past studies suggest that although the residents of Shiraishi Island occasionally work together to deal with problems that rebuff the efforts of individuals working separately, even in these cases they try to respect each other's autonomy while coordinating their individual choices to make collaboration possible. In other words, they abstain from the method of trying to realize goals through collective efforts and consensus building premised on an organizational approach.

Accordingly, even in the context of activities aimed at the revitalization of local communities or the transmission of local traditional observances, they are not engaged in the creation of a cooperative system in the view that collective effort on behalf of such activities is an unconditional good. To put it another way, this is not a situation in which people make a concerted effort to realize the 'value of the survival of local communities,' in the manner of the 'value of the maintenance of the organization' for organizations. To clarify the difference with respect to the premise in the case of organizational systems, first of all, the relationship between a notional 'local community' and the residents who are its members is different in the sense that residents do not always have a clear sense of belonging to a 'local community.' Nor can they be assumed to share loyalty to their ascribed identity or the values that pertain to it. Therefore, even given what they are putting into practice, they would not consider this as being oriented to the realization of a particular value shared with others.

The first reason that we can think of as underlying such a difference is that in an organization, membership is clearly defined. This is why the coherence of the whole appears so solid, allowing individual members to have a sense of belonging to the organization. By contrast, the boundaries of a local community are not only indistinct, but can only ever be vague and hazy, rather than a coherent whole that can be pictured in the mind's eye at any given time. Accordingly, when the people of Shiraishi engage in activities aimed at community revitalization or the transmission of traditional observances, it may be that rather than a proactive motivation to secure the positive future of their community's survival, they are prompted by the reactive motivation of somehow avoiding a negative future that would rouse people's anxieties, such as the disappearance of the local community.

However, when local community becomes an issue in relation to the outside world, as in the case I examine below, the goal of the community's survival sometimes appears to take on a more proactive value. The records of historical events on Shiraishi include cases that can be regarded as attempts to define a vision of the ideal form of the local community. I would like to discuss these cases to deepen our understanding of local communities at the conceptual level.

The movement to prevent the private acquisition of forest lands on Shiraishi in the Meiji period

The southwestern reaches of Shiraishi Island are home to communally held forest lands that have been designated as a *zaisan-ku* (property ward), and which are also a source of high-quality granite. These woods were used communally by the islanders as a source of firewood and charcoal until the Meiji era (1868–1912) but were designated as a national forest in conjunction with land tax reforms following the Meiji Restoration. When the government decided to sell the land to the private sector in 1901, however, rumors began to circulate among the islanders. To provide against the possibility that the forests would be acquired by off-island capital for the purpose of quarrying, the chief priest of Kairyū-ji Temple, the island's only temple, which was supported by almost all the island's households, launched a campaign with volunteers from the island to prevent the forests from being sold to outside parties. The series of events that began with this movement were aimed not only at preventing the island's forests from falling into the hands of outsiders, but at forestalling private ownership outright, even by residents of Shiraishi, so that ultimately they could be jointly owned by the islanders as a common trust (Shiraishijima zaisanku kanri iinkai 1957).

This movement faced many difficulties, including fierce opposition from the majority of islanders. Nevertheless, after a long and convoluted process, it managed to achieve its original goal. In accordance with the government's policy of divesting its lands to locals on Shiraishi, ownership of the land was transferred, and the cost of the transfer was covered by funds loaned by Shiraishi volunteers. The debt arising from this borrowing of funds was paid from the sale of standing trees in the forest and licensing fees for the use of the quarry site by stone merchants. Afterward, once the debt was successfully cleared, the licensing fees were applied to various projects benefitting the islanders' welfare.

This movement sought to safeguard the lives of people living on the island by blocking the acquisition of the forests by off-island capital, and—for the people involved in the movement, at least—the goal of the local community's survival was regarded as having a positive value. Moreover, when we look back with the benefit of hindsight, it was important for the islanders to see the path to the survival of the local community, in the sense that in the midst of fierce competition by many stakeholders in pursuit of their own interests, finding a solution in the vein of having the land owned jointly by the islanders as a whole meant that once the lands were sold off, life on the island could go on much the same as before.

However, what is noteworthy about this course of events is that this goal was not initially shared by the majority of islanders. If more than a few people were planning to take advantage of the move to have the forest lands acquired by off-island capital and to cannily find ways to earn individual profit, it is hardly any wonder that they took a stand to try to block the movement. Such moves by the islanders themselves plunged the local community into a state of confusion. If we follow the process by which the movement was able to cut through such confusion and reach a settlement in accordance with its original goal, it seems that it was in some ways blind luck that the islanders ended up siding with the movement, despite its rather high-handed approach. Even so, the fact that this settlement was accepted by people at all was due to the fact that the islanders themselves, faced with a situation that jeopardized the survival of their local community, thought long and hard on their own choices and coordinated to maintain the general framework of the society as it had existed up to that point. Thus the Meiji islanders' response to the crisis faced by the local community, which had been triggered by outside intervention, should not be anything special in comparison to the coping methods that contemporary people are using to deal with the challenges facing Shiraishi today.

The first reason for this is that both cases have in common the point that, rather than people attempting to deal with problems by sharing the larger goal of 'the local community's survival' from the outset, there are, first of all, specific problems to deal with for each individual, and it is from responses to these that engagement begins. Second, both cases also seem to be alike in that even while endeavoring to respond to the needs of the moment in a certain kind of way, they also seem to have in common the fact that they are adjusting their behavior by reflecting on whether they are appropriate in light of the larger goal of the survival (reproduction) of their home community.

Returning to the question of understanding local community at a conceptual level, although a starting point in both cases is that this can be envisioned as something distinct from the world beyond, it is not necessarily the case that they are 'things' imbued with a positive value that people seek to perpetuate, in the sense that they are not definitive 'things' that can be concretely envisioned at any given time. However, when even this 'loose collective' of community, whose character can only be ambiguous, encounters a situation that plunges its very survival into danger, people become aware of it as something whose survival is seen as positively valued relative to the extreme eventuality of its future disappearance. Correspondingly, there may sometimes be cases where people start to be able to think of the local community as something more stable, and which exists within a certain range of time.

Another matter concerning the survival of local communities as a goal is that, in both cases, people are not simply striving toward it, but are cognizant of it as a value that points to a positive future. While respecting the respective autonomy of each individual in addressing more specific challenges, people try to coordinate their collaboration in a mutually flexible fashion. At the same time, taking their cue from these objectives, they reflect on the possibility that what emerges as a result might be misguided in terms of the bigger picture.

In this fashion, it seems that people will be better able to concretely engage with the issue of the survival of the local community, in the sense that they will be able to repeatedly reproduce collaborative relationships among people living in the same community without resorting to the use of social coercion or enforced loyalty to an organization.

The issue of social reproduction (survival) in human history

What does 'society' mean for interactions in primitive societies?

It is difficult for people living in the contemporary world to give a concise answer to the question, 'What is society?' Even so, as a citizen of the state of Japan, for example, or as the resident of a given municipality, we are subject to various obligations, and thus feel the tangible reality of 'society' through universally imposed social constraints. The source of this sense is likely our subordination to social authority based on our perception of social coercion. Moreover, though we may believe that we share with other members of the same society a certain way of seeing things that provides us with criteria for making judgments about meanings and values that serve as touchstones for making choices, including choosing the actions that individual people take to achieve certain results in various situations, the very thing that prepares us for such a state is the mechanism that we call 'institutionalization.'

In this section, I turn our discussion to what I refer to as 'primitive societies' in the sense that the societies in question have not yet developed the power apparatus or institutionalization that structure society into something tangible. These societies have no special means of ensuring the exercise of effective social coercion or the sharing of values for dealing with collective problems. In these circumstances, I would like to think about the question of what is taking place when the dimension of society appears as a context in interaction communication carried out to deal with the problems we face (Sugawara 2010). This should also be a crucial point in the discussion of a presumptive starting point for a consideration of the evolution of society in the context of human history. Here I develop my thoughts on this issue in relation to the idea of *double selection over two levels* of social activity, which I have discussed previously elsewhere (Kitamura 2015): that is, first attempting to achieve the desired outcome in the interaction at hand and, second, attempting to ensure the continued survival of the social systems that enable stable activities. By doing so, I expect that we might find a connection linking the relatively abstract problem of arriving at an evolutionary understanding of human history with challenges we face in contemporary Japan, thereby deepening our understanding of each in turn.

When considering the challenges facing communication to deal with collective problems in the context of primitive societies, the first thing that must absolutely be addressed is deciding how to deal with problems in which the interests of those party to the interaction are in direct conflict, and how to ensure their subsequent coexistence. Given the frailty of social control based on power relations and institutions, it is not possible in this case to rely on social authority to offer reasonable solutions to resolve conflicts, nor to give orders forcing either side to compromise. Likewise, the idea of building consensus based on shared values is completely unfeasible. Basically, people will have no choice but to think that the only way forward is to deal with the problem through collaboration between the parties concerned. Moreover, since there will be no difference in social status among the parties involved in this communication, each will be trying to behave autonomously. Thus they will only be able to enter into a collaborative relationship by means of a mutual respect for each other's autonomy.

When attempting to deal with problems by means of such collaborative communication, the key to the effectiveness of such an approach is the double selection over two levels that I have been focusing on here. In other words, this is more than just a matter of making choices and adjustments at the *level of interaction* that are aimed at the realization of some immediate goal. By simultaneously establishing more easily agreed-upon higher-level objectives at the *level of society* as the context of communication, with a view to making the reproduction of communicative activities easier and more desirable, it becomes possible to move ahead with coordination toward consensus at the level of interaction from this broader perspective. By incorporating such a process, collective problem solving can conceivably then become more substantial.

Below, I position this double selection over two levels as something found at the overlap between selection made at the *level of interaction* and selection made at the *level of society*. Now, to avoid ambiguity about what each of these two levels refers to, I want to advance my discussion with reference to Niklas Luhmann's theory of social systems. One of the twelve chapters collected in *Social Systems*, one of Luhmann's (1995) major treatises, is devoted to a discussion of the difference between society and interaction. I would like to use this distinction to reorganize my discussion of the problem of double selection over two levels.

Selections at the level of society, as addressed in this chapter, are basically choices pertaining to contexts that enable the *reproduction of communication* at the level of interaction; they are not choices pertaining to contexts that enable the *renewal of the process of making concrete choices* at the level of interactions, such as, for example, choices that instruct us to follow certain rules or norms inhering to society. However, when the reproducibility of society itself comes into question, as in the case of local community (discussed above), there will inevitably also be cases in which the renewability of specific processes of interaction in the context of the reproduced society also comes into question. As to how following rules or norms differs from the example here, the kind of society that is envisioned as a context is left up to communication interactions in the moment, and this also represents a part of collaborative problem solving between the parties involved, which can clearly be distinguished by following rules and norms put in place prior to the communication in the moment.

The difference between society and interaction in social systems theory

For Luhmann, the idea of social systems is the basis of his theory of society. He distinguishes three types of systems; namely, interactions, organizations, and societies. First, he regards social systems as clusters of communications (communication systems) that are created based on the appeal to past communications and the anticipation of future communications within networks created by connections between instances of communication. In order to sharpen this theory of social systems into a universal social theory, he believe it is essential to incorporate 'society' into this theory as an overarching social system that includes all social systems, as well as its inverse in the form of a system of elemental interactions, and to discuss each of these, and their interrelationships, in greater detail.

Out of these, the social system of interaction may be thought of as something that appears at any time, under conditions where behavior exhibited by one party (in a situation involving many) that exerts a direct effect on another becomes an 'action' for the other party to connect with their own behavior, including the response of ignoring it. On the other hand, this system is unreliable in that it ceases to exist if one of the participants refuses to initiate or continue it. That is, it seems that increasing the feasibility of the reproduction of interaction systems, which are constantly being halted and restarted, inevitably entails the foundation of a society that transcends individual interactions.

Luhmann (1995: 424) discusses the difference between societies and interactions at the stage of primitive society in the context of a consideration of the evolution of societies in human history:

> The difference between the societal system and interaction systems... presupposes itself in a rudimentary form and can then augment itself as difference. The horizon of meaningful action and experience always extends beyond those who are present to one another. No society arises as a single interaction system. But primitive societies are formed in a way very close to interaction.

Luhmann's (1995: 406–07) conception of selection at the level of a society that transcends individual interactions is concisely illustrated in the idea of understanding interactions as episodes:

> Interactions are episodes of societal process. They are possible only on the basis of the certainty that societal communication has been going on before the episode begins, so that one can presuppose sediments of earlier communication; and they are possible only because one knows that societal communication will still be possible after the episode concludes.

However, he also adds the following:

> Society, however, is a result of interactions. It is not an authority set up independently of what it selects...One can emphasize that society selects interactions, [and that] interactions select society... (Luhmann 1995: 433)

Thus, especially in the primitive society that I propose to discuss here, it seems we must consider that it is interactions themselves, rather than anything else in particular, that are actively involved in securing some kind of order at a broader scope that transcends individual interactions.

Below, to illustrate what I mean exactly by the overlap between selection at the *level of interaction* and selection at the *level of society*, I draw on the case of the Turkana, the nomadic pastoralist people with whom I have the most extended experience of fieldwork (Kitamura 1996b, 2008) and whose way of life, which lacks a governance system that could ensure the exercise of social coercion and the sharing of values, seems to correspond directly to the way that societies were maintained for more than 95% of the time since the emergence of modern *Homo sapiens* on earth.

A decisive step in overcoming what Thomas Hobbes (2017 [1651]) called the 'war of all against all' (*bellum omnium contra omnes*) would be, in the face of problems where the parties involved have conflicting interests, to be able to resolve conflicts peacefully through collaborative communication achieved solely by dint of sharing of information and leadership in the moment. I would like to deepen our understanding of a critical juncture in the evolution of human society by using the phenomenon of 'begging communication' among the Turkana to elucidate a way of establishing collaborative communication that offers the possibility of collective problem solving without inviting subordination to social authority.

Begging communication among the Turkana

I have been conducting fieldwork among the Turkana, a group of highly mobile pastoralists inhabiting the arid expanses of northern Kenya, since 1986. Even so, I have struggled to arrive at a clear understanding of how the Turkana themselves experience what I call 'begging communication,' which is a phenomenon that leaves a strong impression on outside observers like myself. A characteristic element of interactions among the Turkana is that they take advantage of various opportunities to demand that their fellows give them things. However, the importuned party in such a transaction cannot possibly give in every time such a request is made. Accordingly, in the act of refusal, these interactions unfold in such a way that the interests of both parties are transparently opposed. In the following, I reconsider this issue in terms of how choices in communication interactions constitute an overlap of selection at the level of interaction and selection at the level of society (which relates to the context that secures the reproducibility of subsequent communication), and endeavor to show that this represents an important step in substantiating communication to deal with collective problems.

Two points can be highlighted as reasons why these interactions are so difficult for us to understand. The first is a problem for the party being importuned. Namely, even if he refuses the other party's request, the begging party does not give up attempting to acquire the thing in question, so that the exchange of request and refusal threatens to persist indefinitely. This makes it impossible to understand how things might develop in future. The second is a problem for the begging party, in that in the context of this seemingly endless exchange, it seems impossible to understand what kind of relationship the begging party is trying to establish with the other person by being so unreasonably insistent when it seems clear that the other person is, for the time being, trying to refuse.

After struggling to find an answer to this challenge, I came up with the following solution. In the beginning, at least, both parties may be thought of as simply acting in such a way as to secure their own interests. As the exchange progresses and seems to drag out with no prospect of conclusion, however, it is confirmed that they are dealing with a special problem that cannot be solved by ordinary means, a problem that, in their experience, can only be solved by the people involved through collaboration with others. In addition, it seemed that the possibility of a settlement acceptable to both parties would be opened up if the parties were to approach the problem with a common motivation to 'resolve the stalemate' (Kitamura 2008).

The shortcoming in this understanding is that in order to promote this approach, parties who initially sought to secure their own interests are made to turn their attention in midstream to a completely separate problem. However, it is unnatural to assume that such a development could not have been anticipated at the outset. This difficulty can be resolved by adopting the idea of *dual selection over two levels* discussed in this chapter. In other words, these interactions are not only attempts to solve the problem at hand, but are also attempts to realize in that instance a broader order that transcends individual interactions and which makes it easier and more desirable for such interaction-type attempts to be reproduced on a repeated basis—and, moreover, should be considered as such from the outset.

In the context of this communication, each party respects the autonomous choices of the other, and will not abandon the communication even if it is not conforming to their expectations; the idea is simply to find a common ground to solve the problem by connecting instances of communication. Nevertheless, such efforts do not always lead to an easy resolution, as, for example, when dealing with a problem in which the interests of both parties are in direct conflict. At such times, the parties involved will pause and stop striving toward the immediate goal of solving the problem, attempting instead to put more emphasis on the higher-level goal of creating an order that supersedes individual interactions. And by being aware of the existence of this underlying 'social system' and working consciously to reproduce that order, while removing the obstacles to consensus, they should be able to reconsider their attempts to settle the immediate problem from the broader perspective.

This kind of approach is precisely the kind of communication technique that is essential to overcoming the war of all against all. In order to enable collective problem solving without inviting subordination to social authority, it will be necessary to implement what I have termed 'collaborative communication,' and the likelihood of this is heightened by the approach that I have called *dual selection over two levels*. In this communication, higher-order objectives that transcend the level of problem solving provide a mutual touchstone in the attempt to realize a state in which they will be able to deal all at once with issues in the moment by means of a flexible coordination characterized by mutual respect for each other's autonomy.

Where does society come from?

Finally, I would like to address the question of how the dimension of society (which we have considered as being more than the sum of communication interactions) should be understood to exist in societies where the power relations and institutionalization that structure society as something tangible remain underdeveloped, such as those of San hunter-gatherers in Botswana (Kitamura 1996a) and the Turkana pastoralists in the arid expanses of northern Kenya, who have been the focus of my own research. In particular, I approach this question by inquiring into the route by which society, as something that constrains and guides communication interaction systems from the background, is produced in the settings where communication takes place.

In terms of our daily lived experience, society would seem to be something that acts as a constraint on individual action selections in interactions. This does not mean, however, that we should think of society as having been present as such in the sites of our interactions from the very beginning, even in terms of evolutionary history. Were that the case, unless we ceased to be social beings entirely, we would have been unable to even imagine our liberation from subordination to social authority, or from our captivity to institutionalized ways of seeing. In terms of the order by which this would have taken place, the initial establishment of the context of society would have been followed by the infiltration into people's daily lives of a subordination to social authority and a surrender to institutionalized ways of seeing.

First, I would like to think about how the most fundamental events in our day-to-day social lives are interaction communications with the acquaintances with whom we live in geographic proximity. Since such interaction communications are destined to be constantly halted and started anew, for their repeated reproduction as such to be feasible will require a larger social framework that transcends individual interactions—namely, the foundation of society.

On the other hand, the respective horizons of meaningful experiences and actions of the parties to the actual communication extend far beyond the scope of what is actualized in a particular instance of communication. Therefore, in their appeal to past communication and anticipation of future communication in the context of connecting communication, people will refer to a larger network that transcends the interactions in the moment. In other words, they treat communication in the moment as though it could be reproduced beyond disconnections in the sites of interaction, and beyond the scope of the people present in those places. Thus the ways in which each of the parties to the interaction tries to deal with communication as something that can be reproduced in this sense are what manifests the foundation of society in the background of communication.

However, this does not necessarily mean that society is some divine phenomenon that can be put in place without being influenced by how communication takes place in these interactions; in other words, it is not something that exists independently or which we must accept unilaterally. Furthermore, we need not think that society is always conceptualized as being the same for each individual who inhabits it, and nor should we think that it must have any substantive existence. At its most basic, society, which appears as the background of communication, may be thought of as a kind of framework that constitutes a foundation for fellowship, as well as something which ensures that communication among fellows is a reproducible social event linking people imparted with sociality. And something that deserves special attention in this regard as a form of communication among fellows is communication manifested as 'collaborative problem-solving.' The reason I believe we should pay attention to collaborative communication here corresponds to what we have been considering as the most fundamental premise for our discussion in this chapter so far. In other words, I believe that in a primitive society, people must be able to secure a loose form of social organization characterized by fission and fusion in order that, when necessary, they will be able to regroup to collaborate and deal with the problems they face (i.e., reproduce a system that encompasses the extreme of extinction).

In this kind of communication for dealing with collective problems, it is not uncommon for people to be unable to draw conclusions that can win the agreement of the majority, or, once conclusions have been drawn, for doubts to well forth anew. In such cases, however, the question that emerges is how to ensure the survival of the underlying foundation of society that supports these communications in the background. Here, while concretely envisioning social systems that correspond to the frameworks distinguishing 'fellowship' from what it is not, people endeavor to ensure the survival of social systems themselves by making choices that reiterate the form of the ideal order therein. At the same time, this manner of trying to reproduce the social system itself seems to represent a touchstone that makes it more likely to achieve the reunion of fellows where there is a need to deal with problems through collaboration.

However, as societies evolve and grow more complex, more clearly defined social frameworks are brought into being, like 'organizations' or 'states,' which have a powerful governing structure. Once this happens, instead of some vague 'society' in the background, society ends up infiltrating communication with fellows as an agent that imposes specific standards of judgment and enforces choices. Nevertheless, as is clear from the various concrete examples discussed in this chapter—namely, the descriptions of a peripheral island society in contemporary Japan and an African pastoralist society whose lifestyle is strongly dependent on nature—no matter the time or place, collaborative communication based on social frameworks that form the basis of fellowship has always been the most fundamental social event linking people imparted with sociality to each other, and continues to be a means of supporting people's shared existence.

Even though the context of society gives form to the framework of communication in communal living, it cannot be considered to correspond to some substantive 'thing' inherited from the societies of pre-human animals. Nevertheless, it seems to be something that has been with us consistently from the very beginnings of the evolution of human society, or perhaps even much earlier, as something supporting the reproduction of communication in interaction systems. As has been made clear in the preceding discussion, this is the product of our own desire that the spatial clustering of our way of life should be solid, on the presumption that our way of living together with our fellows should be something that can be sustained across generations. Assuming the logical extreme of the transition of this spatial and temporal clustering to be the 'extinction' of such clustering, to avoid that extinction we have positioned the continued reproduction of such clustering as the basis of fellowship as a way of living together. It is almost as if we, as individuals, have positioned the continued reproduction of our own alter egos as the basis of our way of life.

References

Bateson, G. (1972) *Steps to an Ecology of Mind*. Ballantine Books.

Hobbes, T. (2017 [1651]) *Leviathan*. Penguin Classics.

Kitamura, K. (1996a) 'Byōdō shugi shakai' to iu nosutarujia: Busshuman wa byōdō shugisha dewanai' (Egalitarian society and the notion of nostalgia). *Afurika kenkyū* (Journal of African studies), 48: 19–34.

Kitamura, K. (1996b) 'Shintaiteki komyunikeishon ni okeru "kyōdō no genzai" no keiken: Turukana no "kōshō" teki komyunikeishon' (The experience of the 'collective present' in physical communication: 'Negotiative' communication among the Turkana). In K. Sugawara and M. Nomura (eds), *Komyunikeishon to shiteno karada* (The body as a means of communication). Taishūkan Shoten, 288–314.

Kitamura, K. (2008) 'Shakaiteki narumono' to wa nani ka? Tasha to no kankeizuke ni okeru 'ketteifukanōsei' to 'sōzōteki taisho' (What is 'social'? Indeterminacy and creative coping mechanisms in the process of making relations with others). *Reichōrui kennkyū* (Primate research), 24(2): 109–120.

Kitamura, K. (2014) 'Shima ni kurasu hitobito ga taisestu ni shiteiru koto: Okayama-ken Shiraishijima no jirei kara' (Communication in interaction systems among islanders: The case of Shiraishi island, Okayama Prefecture). *Bunka kyōseigaku kenkyū* (Studies in cultural symbiotics), 13: 43–60.

Kitamura, K. (2015) 'Sougokoui no komyunikeishon: Hito to doubutsu wo tsunagi-tsutsu hedaterumono' (Communication in the interaction systems that link and segregate between human and animals). In Daiji Kimura (ed.), *Doubutsu to deau II* (Encountering animals II: Emergence of mind and society).

Luhmann N. (1995) *Social Systems*, J. Bednarz Jr. and D. Baecker (trans.). Stanford University Press.

Sasaki, M. (2000) 'Kaisetsu: Edward S. Reed no shigoto' (Clarification: The work of Edward S. Reed). In *Afōdansu no shinrigaku* (The psychology of affordance). Japanese translation by N. Hosoda of E. S. Reed (1996) *Encountering the World: Toward an Ecological Psychology*, Shinyosha, 399–424.

Setonai-kai ni okeru seikatsu to bunka kyōiku kenkyū gurūpu (Education and Research Group on 'Life and Culture in the Seto Inland Sea') (2009) *Report of Social Research on Shiraishi Island*. Faculty of Literature, Okayama University.

Shiraishijima zaisanku kanri iinkai (1957) *Shiraishi-jima sanrinshi* (History of the mountain forest in Shiraishijima). [Privately printed book]

Sugawara, K. (2010) *Kotoba to shintai: 'gengo no temae' no jinruigaku* (Language and body: Anthropology of communication 'at the advent of language'). Kodansha.

CHAPTER 15

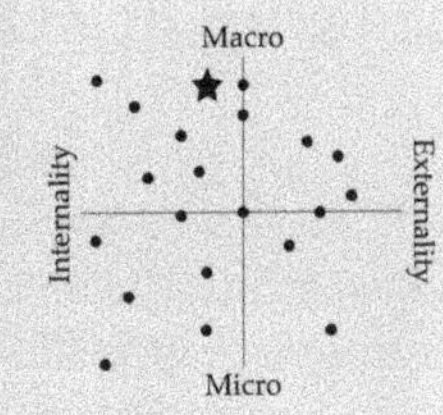

Extinction as an Extreme:

Its Entitative and Categorical Phases

Motomitsu UCHIBORI

Keywords: entitative extinction, categorical extinction, culture as ecological niche, Tatau, Neanderthal

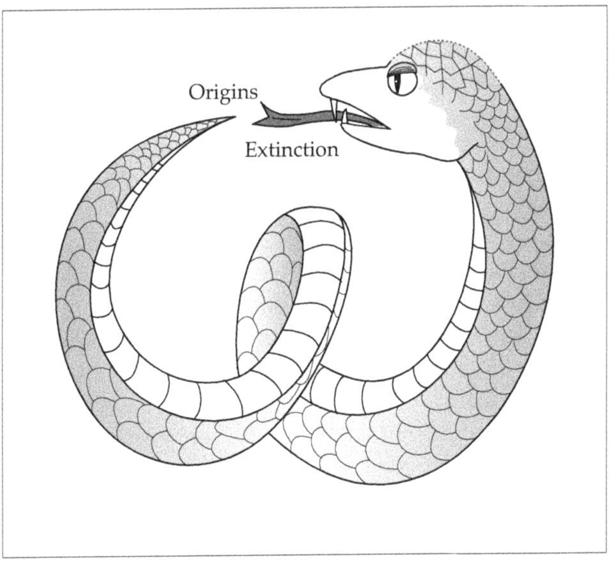

'A snake that cannot become an ouroboros'

This snake is attempting to return from its 'extinction' (the head and mouth) to its 'origins' (the tail) but cannot seem to manage it. The snake's body seems to be stuck in the shape of the Greek letter omega (ω). The snake seems slightly nonplussed at its inability to return.

Speaking of extinction

In this chapter, I would like to talk about the content and form of narratives about people who are recognized as sharing some kind of collective identity—whether kinship, ethnicity, nationality, or religion—and who are situated at the extreme limit of that collectivity. I also want to ask what happens when they have passed beyond that limit; that is, after that collectivity to which they belong has disappeared? At stake here, in addition to the mode of existence (a prerequisite for extinction) in terms of the nature of the (erstwhile) collectivity, are narratives about the process of its disappearance. In fact, the mode of existence and the process of extinction are two closely intertwined situations or phases within the same 'ontogenetic object,' which makes it difficult to separate them cleanly.[1] This is because these two phases are not simply demarcated as contexts for fixed temporal transitions, particularly from the privileged point of view of the present moment when people recall and discuss them. Rather, by including this very inseparability in the discussion, I want to think about extinction as an extreme, while paying attention to the ways in which memories of what once existed, 'that which was,' are created, as well as to the creeping oblivion that often accompanies them.

Let us differentiate extinction into two modes; namely, entitative extinction and categorical extinction. Whereas the former is usually mentioned with reference to living things, in connection with the extinction of individual entities that belong to a collectivity, the latter includes things that are not only living, but also non-living or abstract and, moreover, concerns the respective categories to which these phenomena belong. While this conceptual distinction may sound overly general, it is worthwhile noting that its appropriate use does, through differences in connotation, allow us to see certain things in a new light.

It is safe to say that extinction has become a prevalent theme in the popular discourse of the early twenty-first century.[2] Underlying this is a societal awareness of imminent crises facing our way of life as a species, as represented in the terms 'risk' and 'hazard.' The nature of this concern is prospective, heading from the present into an indeterminate future. Narratives concerning what, in formal terms, would be the temporal inversion of this—that is, about extinction as a past event—are apparently a reflective aspect of this concern. At the very least, some new significance is, seemingly, being discovered in such reflection. Regardless of whether the past really offers a salutary lesson for the present or future, the attempt to bring several possible forms of extinction (including those that exist only in the imagination) up for discussion as something that has already come to pass can only aid us in acquiring empirical criteria for measuring the urgency of crises in the present and future.

The most prominent of these past extinctions are the events gradually being elucidated as the 'mass extinctions' of organisms that evolved and developed after their appearance on Earth 4.3 billion years ago. These are extinctions that occurred in conjunction with great cataclysms in the global living environment that far exceed the scale of natural selection in the evolutionary context. The latest theory holds that five such events have taken place to date, with the most recent being the extinction of the dinosaurs 65 million years ago. The next mass extinction, whatever its cause, will be the sixth. This sixth mass extinction, unlike those that took place prior to the emergence of the primates, may be caused or triggered by environmental upheavals brought about by the activities of human life on earth (Kolbert 2014). One could go so far as to say that such a mass extinction, for the first time in the history of the planet, would be an intrinsic consequence of the presence of a particular organism. Certainly, there is something in this that arouses deep fear and misery in us as the agents of such an event, a fact that brings added tension to the current debate.

However, there must have been countless times in the past when the elements inherent in the activities of a particular organism paved the way for its own extinction in small, medium, or minute extinctions rather than on a mass scale. If we broaden our imagination a little, we might even say that such situations were probably closer to the normal state of affairs. For instance, we might consider individual cases of the extinction of regionally limited human populations, or in thinking of the Neanderthals (*Homo neanderthalensis*), the closest relative of modern humans, we might consider the extinction of the entire population. Discussing these extinctions entails exploring their intrinsic causes with varying degrees of uncertainty.

The case of the Bukitan and the Tatau

The western part of the island of Borneo, in what is now the Malaysian state of Sarawak, is home to a people called the Bukitan. Although the local people pronounce their ethnonym variously as Begetan or Beketan, the name is written as 'Bukitan' in Sarawak's local laws and ordinances. According to administrative documents and other written records dating from the mid-nineteenth century that have used this word, as well as in the living folk tradition of their own and of neighboring peoples, they were originally a people who lived by hunting and gathering. In 1981, early in my career, I conducted several interviews and field observations while staying for the better part of a month in a government-built hut constructed for the use of various survey personnel near a communal Bukitan settlement (a longhouse) along the Rajang, Sarawak's longest river. From those passing by the hut (mostly elderly women) and younger people frequenting the nearby Chinese-run shop, I heard somewhat fragmentary information about the current situation for the Bukitan and, from an old man who claimed to be familiar with the old stories, I heard that their tradition as it was passed down was chiefly concerned with the relationship between the

Bukitan and the Iban, a swidden-farming people distributed through the areas around them. While our conversations were conducted exclusively in the Iban language, all the Bukitan people spoke Iban without an accent.

The reason I am raising this topic now is that it already seemed to me in 1981 that the Bukitan as an ethnic group would disappear in the near future. In fact, I have elsewhere written a general interest article on the basis of this prospect (Uchibori 1994). To summarize that article, first, the Bukitan were considered to have accompanied the expansionist movement of the swidden-farming Iban, who, in the eighteenth century and earlier, were rapidly expanding their settlements northward from southern Sarawak. Second, the relationship between the Iban and the Bukitan was generally one of amicable coexistence, with violence (in the form of headhunting) occasionally being visited by the former on the latter. Third, this coexistence led to what might be called the 'Ibanization' of the Bukitan. Finally, as a consequence, the set of people referred to by the name (i.e., category) of Bukitan was on the verge of becoming what in mathematical terms would be called an empty set. While I regarded this last as

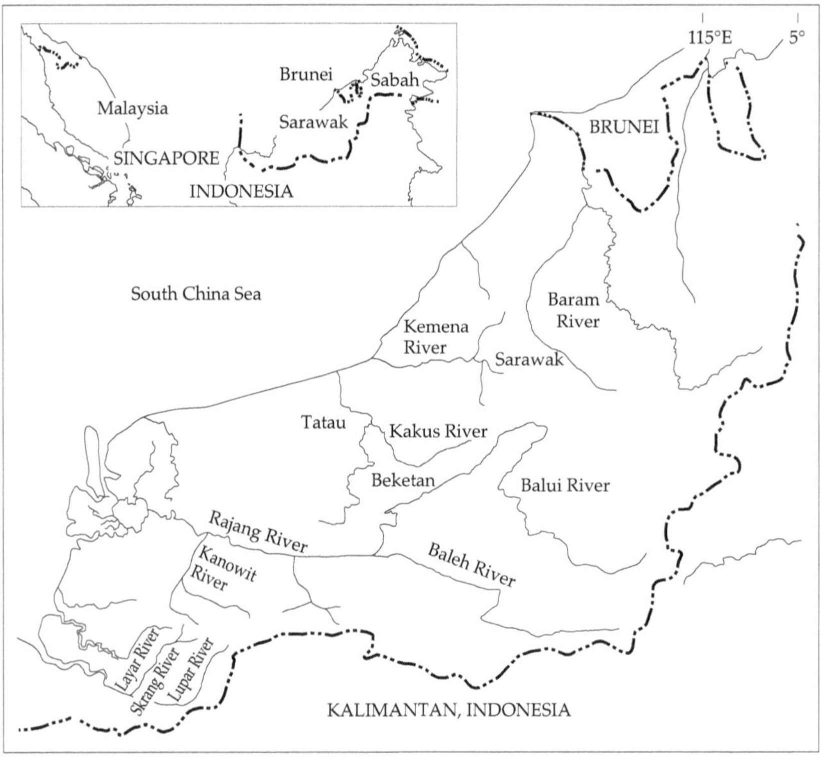

Figure 15.1: Northern and central Sarawak, home territory of the Bukitan (Beketan) and Tatau

an instance of categorical extinction, naturally such an extinction did not signify the deaths of individual people, nor did it entail that their futures would be dramatically disrupted. Rather, it simply meant that individuals would cease to fall under the general term 'Bukitan.' The Bukitan I met were cognizant of this fact, often remarking to me that 'one of these days, we'll all be Iban.' Nevertheless, what was somewhat unique about the case of Sarawak, which had been a semi-independent monarchy up to the Second World War, was that, even were such a situation to occur, the term 'Bukitan' would persist as a valid legal ethnic identity in the context of the laws of the nation-state of Malaysia, which is why I felt it was appropriate to use the term 'empty set.'

I confess now that my view of the Bukitan's extinction was actually mistaken. Even in the time of my fieldwork, the existence of four more Bukitan longhouses had been known along the tributaries of another river system, the Kakus, which runs almost parallel to the Rajang to the south, across a low divide. In the 2010s, these longhouses subdivided to become eight in number, so that (nominally, at least) the Bukitan today no longer seem to be in danger of extinction. But what is the reality of the situation?

Since 2005, I have had opportunities to carry out intermittent fieldwork in one of these longhouses. The Bukitan here, like those I met during my fieldwork on the Rajang, speak Iban as though they were native speakers of that language. Among themselves, however, they continue to speak Bukitan, and the non-Bukitan villagers living in the longhouse (peoples called Iban, Punan, and Melanau) also have some facility with Bukitan in their own way.

The Bukitan longhouse in question is a mixed community in which almost half the residents belong to ethnic groups other than the Bukitan. And even though the headman of the longhouse is Bukitan, this does not mean that Bukitan residents enjoy any particular privilege in the context of daily living. Basically, the current situation facing the Bukitan suggests that of a gradual dissolution of social cohesion, even as the population numbers justifying the ethnonym are maintained.

Because life for the Bukitan was once chiefly devoted to hunting and gathering, day-to-day mobility was an inherent part of their lives. Oral tradition and records by outside observers confirm that, as a result, Bukitan movements and migrations were carried out over a large spatial and temporal scale. In Sarawak, however, a significant proportion of swidden farmers are also a people with a high degree of mobility. In this part of western Borneo, relationships among ethnic groups and subgroups have been largely defined by the various migrations that have taken place, in parallel and intersecting with each other in both time and space. When the Bukitan first migrated to the tributary in question in the middle reaches of the Kakus River basin, likely in the late nineteenth or early twentieth century, the area would have been inhabited by a farming people called the Punan, but it is said that before them the area had been home to the Tatau, a people whose livelihood was based on the cultivation of the sago palm. Although limited to a portion of the middle reaches of the Kakus, what seems to have been happening here for more than two centuries is a successive—even violent, depending on one's point of view—series of ethnic replacements or transitions, in the progression from the Tatau to the Punan, then to the Bukitan, and most recently with the influx of the Iban. In the course of these replacements, the Tatau, the original inhabitants of the region, were actually the ones who faced an even more critical situation than the Bukitan, in terms of a discussion of 'ethnic extinction.' Very suggestively, they were arguably also more cognizant of the extinction crisis than the Bukitan in that they had their own unique way of narrativizing it, as shown below.

According to a Bukitan informant living in the middle reaches of the Kakus, the Tatau left the tributary in question as the result of an incident in which, facing starvation, they had slain a half-legendary serpent known in Iban as the *nabau*. Eating the *nabau* had always been forbidden, but they had shot it out of the trees with blowguns, and had then boiled it in a pot and shared it among all the villagers. Because of this, many people, adults and children alike, began to die of disease, so that the population declined. To avoid extinction, the villagers had been forced to leave the land that had been defiled by the violation of the taboo. In fact, only two Tatau longhouses are known to be extant throughout the entire Tatau River basin (the lower reaches of the Kakus River are known as the Tatau). Moreover, in one of these longhouses, fully half of the eight constituent households are Iban families who migrated from southern Sarawak about thirty years ago (in the 1980s).

At one point, the headman of this longhouse, who is still young, bemoaned the deplorable fact that, quite apart from the fact of the diminished Tatau population, the Tatau language was not even being used in their own longhouse, and that it was commonplace for Tatau residents, especially children, to speak to each other in Iban, whereas even Iban spoke Bukitan in Bukitan longhouses. The tale he told me about the cause for the overall scarcity of the Tatau people was similar to the story of the great *nabau* serpent. It is said that when the ancestors of the Tatau had once lived at the mouth of the Tatau River, someone caught an enormous caterpillar—as big as he was able to hold—which was sliced up to roast and eat. As a result, it is said, many of the children of the Tatau died in infancy, which is why their descendants today have almost entirely died out.

The tales of the near-extinction of the Tatau told by the Tatau and the Bukitan are so-called 'just-so stories'—etiological fables that explain the present as the effect of the past. However, the relationship between the ultimate cause (i.e., violating a taboo) and the proximate cause (i.e., epidemic disease) remains mysterious and completely incomprehensible outside of the symbolic and semiotic systems of the Tatau. The connection between a food prohibition and disease is likely specious. However, with regard to creatures of legend such as giant serpents, or even giant caterpillars, it would be necessary to explore the history of Tatau culinary semiotics, which is beyond the scope of what can be dealt with here. On the other hand, the direct cause is relatively straightforward. In this case there is a straight trajectory toward extinction, which shows a very different view of fate and causality than the Bukitan's own narrative of a one-sided and violent disruption based on the memory and tradition of their factual past with the Iban. To exaggerate the point slightly, one could characterize the Bukitan view as history, and the Tatau view as myth. Despite the fact that the situation currently facing the Tatau is the more precarious, the fact that their etiological myth sounds strange, even comical, is likely due to its being a literal ahistorical anachronism, which is to say, a confusion of temporality.

Whether for the Bukitan or for the Tatau, or for some other ethnic minority not mentioned here, the present state of affairs is the result of the intermingled movements of people in northwestern Borneo. These activities are kept in the memories, or made into memories, sometimes by historical narratives and sometimes by mythological narratives. Either way, both kinds of narratives subtly or implicitly presuppose the existence and activities of other human groups (ethnic groups) that differ from themselves, suggesting that they do not by any means see themselves as isolated entities. The region that serves as the stage for these activities is an ecological space that includes humans as principal, if not the most privileged, actors. Typically, ecological spaces have two phases: one of which is relatively stable and another that changes dynamically over time. However, following the Iban migration and expansion out of southern Sarawak, this region seems to have entered a particularly volatile stage of the latter phase. (For more information on the ethnic situation in this region, see Kato et al. 2020.)

In cases of swidden farming, there is a wide range in the degree of human migration, in terms of both frequency and distance, associated with the shifting of lands under cultivation. Among the swidden peoples of upland Borneo, the high degree of mobility (frequent migration) practiced by the Iban is particularly notable and has attracted considerable attention and debate from numerous researchers (e.g., Freeman 1981, most notably). I have elsewhere discussed this in the form of a rough-hewn hypothesis concerning the ethnogenesis of the Iban (Uchibori 1984, 1994). This hypothesis, which I would dare call the 'Iban Movement' Hypothesis, has two principal factors (while the word 'movement' includes the sense of geographical mobility, my usage extends beyond this to emphasize the additional sense of a social campaign). The first is the possibility that the ancestors of those who later came to form the strongly egalitarian Iban society were people who had broken away (or run away) from another agrarian society with a hereditary status hierarchy, while the second concerns the strong proclivity for hunting activities that the Iban still exhibit today. However, needless to say, for these factors to be combined to achieve high mobility requires that a space for swidden cultivation will need to be available, or created, at the destination locale. In fact, the area from the southern to the central part of the present-day state of Sarawak, which is blanketed with tropical forests and has a very low population density, still satisfied this condition as late as the early twentieth century. The Iban custom of headhunting, at least in view of its outcomes, may be said to have further prepared conditions at the destination locale in this vein. Rather than actual acts of killing, headhunting had the effect of spreading fear (terror) in the area.

For a group of people with a strong desire for cultivatable land to move into newly vacant lands in the gaps between those who do not have as much of a desire to do so can be rephrased as their establishment of an ecological niche in that locale. What made this expansion successful was culture—first, as a subsistence technology in terms of the cultivation of upland rice using the slash-and-burn method of swidden farming and, second, as a social customs value system in terms of the practice and appreciation of headhunting itself. In the sense of forming and excelling in a new ecological niche with the power of culture as a resource, such a niche is not purely ecological but, rather, a cultural-ecological niche made up of composite factors. In short, the process of competition in such a niche constitutes the migration of the Iban, and it is through the accumulated experience of migration that they have expanded their home territory. And, almost naturally, this accumulation has progressed in step with the growth of their population.

The concept of a cultural-ecological niche, if it is applied not to the advancing group but to those advanced upon (i.e., the indigenous human population), could shed a more appropriate light on the subject we are dealing with here. After the arrival in the Tatau lands of the Punan agriculturalists and then the Bukitan hunters (as though they were being pushed out of their own territories by the Iban), with their larger population, the Iban themselves began to move in rapidly and in strength at the beginning of the twentieth century. At present, the Iban constitute the overwhelming majority of the population across the overall basin of the Kakus River. The consummation of the shift toward becoming an empty set—an ethnicity in name only—is much more imminent for the Tatau than in the case of the Bukitan. In other words, this means that their cultural situation, which centers on language, and their ecological situation, which focuses on subsistence, have almost completely lost their unique niche (i.e., status). My use of the word 'unique' here denotes a clear and mutually acknowledged difference with respect to the niches of other groups around them, and it is this difference that has now largely been extinguished. When the imminent extinction of what might be described as the comprehensive difference between self and others becomes immediately apparent, and when people exist who are capable of acknowledging this, it becomes possible to speak of the extinction of one's own ethnicity as a collective identity. While this might be described as the recognition of a collectivity poised at the extreme limit of existence, it remains within the realm of rationality, unlike the already-dead speaking of their own death as a memory of that extreme.

In the usual sense, 'entitative extinction' is not something that can be remembered as such internally. After a 'categorical extinction,' there may well be individuals who remember and record it, but the fact of their extinction as a group will be known as a matter of course by those who go on to read the records after the fact. In the history of our species, there have been countless extinctions, both large and small. In the next section, I would like to step beyond the usual timeframe of history to cast an outside eye on narratives about large extinctions that are far from written or even oral records.

Discourses of the Neanderthal extinction

The Neanderthals (*Homo neanderthalensis*) have been discussed in the position of being the 'closest' of the past relatives of modern humans (*Homo sapiens*). This position itself has remained unchanged for more than one-and-a-half centuries, ever since the confirmation in 1857 of fossil bones as belonging to a past human species. Apart from discussions over finer points, the principal debate fought within the context of this positioning has chiefly concerned whether the Neanderthals were the ancestors of modern humans or whether they were an offshoot from our own ancestors—one that had no surviving descendants. Until the last decade of the twentieth century, this controversy was not one that would be settled easily, and in that it was variously linked to both monogenic (single region) hypotheses and polycentric (multiregional) hypotheses concerning the evolution of modern humans, it appeared to have the character of an ideological struggle quite removed from empirical evidence. Any serious discussion of the 'extinction' of the Neanderthals was associated with the positing of monogenic hypotheses of the evolution of modern humans.

Discussions about how the Neanderthals went extinct are generally premised on the fact that they were contemporaries with (the ancestors of) modern humans, coexisting with them—sympatrically on a scale of thousands or tens of thousands of years—in various regions from the Middle East into Europe. In essence, to talk about the extinction of the Neanderthals is to talk about the nature of this coexistence, and ultimately the factors that led to its eventual demise. Viewed in this way, the sketch of the existence of the object in question bears several surprising similarities to the situation on the Kakus River in the twentieth century. Naturally, there are overwhelming qualitative differences between the two in terms of their temporal and spatial scales, and these differences are precisely what characterizes these respective narratives of extinction.

One of the most popular, though old-fashioned, images of the Neanderthals is that they were a people adapted to a cold climate, something that would have had a bearing on their mode of survival and the survival strategies they employed. They are imagined as having had the robust physical strength that would have been necessary to overcome the harsh ecological environment of the Ice Age by hunting mammoths and other large animals, something that came at the expense of having a rather dull appearance. In some non-academic but nevertheless unique and well-thought-out fiction, Neanderthals are portrayed as lacking articulated language skills in comparison to modern humans, communicating with each other by means of empathic ability, with some being able to achieve a kind of shamanic spiritual communion.[3] The attributes implicit in these popular images, many of which have now been proven to be fanciful, would not logically have led to the Neanderthals' extinction in and of themselves. But in another once-popular but evidently false scenario, the attribute of cold-climate adaptation was thought to imply a maladaptation to the warmer environment that arrived in conjunction with the end of the Ice Age, while empathic communication was seen as inferior to articulated language, which was regarded as being more associated with reason. In comparisons with modern humans, this supposed inferiority of Neanderthals in terms of their abilities is tied up with environmental changes and is seen as the ultimate cause of their entitative extinction.

Within this supposed relationship of superiority and inferiority, the tale is told of how modern humans overwhelmed the Neanderthals. Though they existed contemporaneously, did their relationship include direct negotiation, or were the two populations socially isolated? If the former, was this violent in nature, or might it have been essentially peaceful? Such an intergroup relationship would constitute a direct cause in the context of narratives about the extinction of the Neanderthals. In narratives that presuppose the most violent relationships, the cause is the dramatic mass death of the Neanderthals, caused by what we would call wars or genocide wrought by modern humans. In narratives at the opposite extreme, extinction is suggested or referred to as the result of a gradual decline, or self-destruction, without the direct or intentional involvement of modern humans. In this case, the ultimate cause and proximate cause are discussed as being identical or having taken place in virtual concurrence.

It is important to note here that, seen in very formal terms, when a violent relationship is seen to be the proximate cause, there will ultimately be a need for some kind of physical evidence (i.e., remains), the existence of which will be able to support the narrative. Conversely, in the case of a narrative of self-destruction, the decisive factor of physical evidence is inherently impossible in any positive form. In order to think about this, it is instructive to envision the defined aspect of 'marginal settlements' in Japan today, where the possibility of their continued viability has been weakened to an extreme degree. In the distant future, the cause of the extinction of those settlements will not be able to be gleaned from physical evidence. In this sense, these two narratives are not equivalent in evidentiary terms. More simply, whereas the former eventuality is accompanied by an incident, the latter is not of that nature. This is not to say that the latter necessarily prevails in the absence of any clear physical evidence left behind. Either way, the choice of which narrative to adopt depends largely on indirect corroboration, context, and analogy with intergroup relationships in extant human societies and will also be influenced by current trends surrounding and informing both those who propound these narratives and their audiences. The relationship between the Neanderthals and modern humans is, on the one hand, a matter that bears on aspects of coexistence and, on the other hand, a matter that bears on the evolution of humanity as a whole. And with regard to current trends, as I discuss in the next section, the direction of narratives concerning the evolution of humanity as a whole is also subject to their influence. The configuration of social trends in the West, particularly in the United States, with the evolution of humanity as a whole, and the positioning of the Neanderthals and narratives about their extinction, seem to be progressing with an interesting synchrony.

In the final decades of the twentieth century, there was a widespread view that the differences between Neanderthals and modern humans were greater than had previously been understood. While this was due to a number of reasons, the decisive factor was the success of deciphering human mitochondrial DNA. As far as is evident from the commonality of the mitochondrial DNA of modern humans, it can be calculated that all humans now living on Earth can trace their maternal lineage back to a single female ancestor who lived around 200,000 years ago, an era that is much more recent than the 500,000 or 800,000 years ago that constitute the period when Neanderthals and modern humans are thought to have diverged. This led to the conclusion that the Neanderthals were a distinct species from *Homo sapiens*, but the important thing, in terms of an ideological trend, is that around the same time, the Neanderthals began to be placed at a remove from modern humans, not only in terms of genetic distance but also in terms of capabilities connoted by such words as 'spirit' and 'culture.' In other words, the capabilities of modern humans in these regards were now being emphasized more than ever before as something that set them apart from their predecessors.

The attendant academic narratives (i.e., discourses), needless to say, are based on the proofs mustered in their respective fields. As I am not qualified to go into the evaluation of these proofs themselves, my focus here will be limited to a discussion as to which topics came to be framed as discourses using what concepts.

The most comprehensive concept mobilized in this discourse is that of the capacity of *Homo sapiens* for 'symbolization.' This comprehensive concept encompasses all activities and outcomes not considered to be directly practical for survival, including everything from artifacts, such as art objects and ornaments, which can survive as relics to posterity, to ritual acts and language, which leave no physical evidence in their own right.

In fact, the interpretation of what these seemingly impractical objects might have been and what they were used for is essentially accomplished by analogy with contemporary human societies. It is estimated that the emergence of culturally and cognitively modern humans took place about 100,000 years after the emergence of anatomically modern humans. The prevailing theory to date is that both of these emergences took place in Africa. And while there are considerable gaps in the chronology of these two phases of 'modernity,' what is certain is that humans were already culturally and cognitively modern by the time they reached the final stages of the expansion from Africa into Eurasia (i.e., into Europe), when they began to coexist with Neanderthals. As argued by cognitive archaeologists, the highly developed capacity for symbolization that the Neanderthals are presumed not to have shared with modern humans, in as much as it expresses the key distinction separating the two, is to be more appropriately termed as a human 'revolution' than a human 'evolution' (Mellars et al. 2007).

The extinction of the Neanderthals is thus framed as their revolutionary 'replacement' by modern humans, and in these discussions they came to be imparted with a heterogeneity in relation to the latter. Their erstwhile image of cold-climate adaptation as a mode of survival had been out of the question. And yet simultaneous with the subtle change of their scientific name from *Homo sapiens neanderthalensis* (a subspecies of *Homo sapiens*) to *Homo neanderthalensis* (a species in their own right) in the late twentieth century, an active body of discourse arose that included the rejection of the reputation accorded to the Neanderthals of the Shanidar Cave in the 1970s as the 'First Flower People'—an assessment that hinted at the efflorescence of a capacity for symbolization—as well as a focus on the difference of modern humans in terms of their flexible mode of learning, which was capable of innovation. In light of research progress as it stands today, however, these discourses may have emphasized heterogeneity more than was necessary. The question remains of whether the searchlight may have been focused too narrowly on the extreme event of entitative extinction, at the expense of other possibilities.

In the 2010s, the course of this debate took a very noticeable and significant turn. Most decisively, Neanderthal genes were found to have been transmitted, even if only by a few percent, to modern humans living outside Africa. In the sense that we now know that they have surviving descendants, and that these number in the billions, moreover, we can say that the Neanderthal did not undergo extinction in the actual entitative sense. Even if we do not know how the ancestors of the modern humans who came out of Africa interbred with

Neanderthals (see Terashima's discussion in Chapter 21 of this volume), some of these genes are definitely those of Neanderthal males. We see no recognizable link between sexually asymmetrical relations and hierarchies of social—racial or ethnic—dominance in the background of this crossbreeding, as we find often in modern society. Moreover, given that such a significant percentage of genes remain after a thousand generations, it also seems likely that the initial rate of crossbreeding may have been quite high for generations. It is even possible that instances of such crossbreeding did not constitute an exceptional case.[4] And yet, given that we can still talk in terms of their extinction, what does this even mean?

Here, once again, the problem is actually the name of the set (collectivity). In the case of the Bukitan and Tatau, regardless of whether or not they originally went by those names themselves, after a certain period those were the names by which their respective sets came to be recognized by themselves and others. But this would not have been the case with the Neanderthals, of course. In this case, rather, we should say that the people regarded as belonging to the named set were created as a bounded, classified group in the material record explored by contemporary observers, and in the memories they created—and went extinct as such a group. This is precisely why Neanderthals can be said to have both gone extinct and not gone extinct, even in the context of a single narrative.[5] The word 'extinction' here begins to look more like a brush depicting the image of the person who uses it, rather than something that indicates an objective reality. References to the extinction of the Neanderthals even seem to derive from the mental landscape of modern humans—especially people living in Europe today—who are concerned with their own existence.

Neanderthals are a popular part of the European imagination. Although the time scale differs, this is something akin to the popularity enjoyed by the prehistorical Jōmon people in Japan, with their distinct pottery style and pre-agrarian sedentary settlements (Katayama 2015). In both cases, no matter how their relationship with contemporary people is described, there exists a sense of intimacy in terms of having shared the same living space or land, separated by time, and their name, which today is a name whose substance has disappeared, but, at the risk of exaggeration, perhaps, can even be imagined as the source of their identity. Recent DNA studies have indicated that European traits such as light skin, blond hair, and blue eyes may possibly derive from the Neanderthal population, as might the origins of certain diseases. The differences between Neanderthals and modern humans acknowledged today have become much narrower than in the past, and it seems likely that they will once again go back to being treated as two subspecies of *Homo sapiens*. However, whether as a species or as a subspecies, when we speak of the disappearance of a unit's boundaries in terms of its extinction, it is precisely because of this emphasis on extinction that we paradoxically become aware of the timeless connection between the two.

This awareness seemed almost to reach its zenith when Clive Finlayson (2011), the director of the Gibraltar Museum, in a region that may have been the last bastion of Neanderthals in Europe, wrote that the Neanderthals' extinction was not a matter of their abilities, but simply an accident of luck and circumstance, a matter of being 'in the wrong place at the wrong time' as Europe began to cool once again. By this logic, the fates of the Neanderthals and modern humans could have traded place with one another by a stroke of luck and, in that sense, were on the same horizon of survival.

Ways of talking about evolution

The stories of the Bukitan and Tatau are narratives of *history* in relation to very recent communities that extend up to the present day, while the extinction of the Neanderthals is part of a narrative that is generally referred to as concerning humanity's *evolution*. These two narratives, of history and evolution, tend to be treated in the same way, since in both cases their plot composition concerns developments beyond the temporal scale of daily life. As a classic example, the paleontologist Stephen Jay Gould (1989), dissatisfied with the gene-centeredness and calculating hindsight rationalism of evolutionary biology, overtly declared that evolution was 'just history.' The rationale that prompted Gould to turn to history was the contingency of the instances of survival and extinction seen in the evolution of life on earth. Beyond this, he offered the critique that theories of evolution had not paid enough attention to this aspect of extinction. In this context, a narrative like Finlayson's (2011)—which attributes Neanderthal extinction not to differences in the ability to adapt to new environments but to the coincidence of their habitat at a certain point in time—would be a way of telling the story of extinction just as history. But is that really the way to talk about evolution?

The answer is both yes and no. Yes, because it is a narrative that provides a rationale for 'choices' that explain the course that humanity (including the Neanderthals) has ultimately followed—that is, the future as seen from that point in time. Even if these choices are contingent, this does not affect the quality of the future being explained. No, on the other hand, because this means that natural factors, in terms of the initial conditions (initial population or initial habitat) under which the ancestors of modern humans were driven along a narrow path to the final stages of contingency to occupy an ecological niche in western Eurasia, including Europe, end up receding into the background. This is because the narrative of evolution—or, to place a deliberate emphasis, the narrative of evolution as 'science'—seeks in the foreground of inquiry, as an inevitability, the logic that generated the contingency by which the group that remained after the boundary disappeared could only be one and not the other.

It goes without saying that the typological difference between history and evolution, and especially the narrative of evolutionary biology as represented by Richard Dawkins (see, for example, Sterelny 2004), is not simply a matter of temporal scale. All narratives concerned with change over time are activities that give form to the substance of those narratives as though they were memories. In the case of history, in the formation of this memory, it is likely that far more is left behind and forgotten. And while we could even say that ignoring this function of forgetting is precisely what makes a historical narrative, there are several ways of choosing what to include and what to dismiss, which is where conflicts arise in the context of historical narratives. In contrast, the formation of memory in the case of evolution (i.e., biology) is logically constructed from the details of the natural (i.e., physical) materials that are available at any given moment. To that extent, change over time is merely apparent, and should be said to represent the logical context of propositions that are included in formally causal explanations. More straightforwardly, we can contrast the existence of the agent in the context of history with its absence in the context of evolution. It follows that whereas history can be an object of 'understanding,' this condition is fundamentally lacking in evolutionary narratives. To put it plainly, evolution is devoid of *intention* in any sense.

In the narrative of human evolution, however, it is often the case that the subject who is not an actor is made into a pseudo-protagonist. This pseudo-subject is then given a position much like that of the hero in a hero tale (Landau 1993). Following Vladimir Propp (1958), who analyzed the structure of Russian hero tales (magical folktales) and who is also mentioned by Misia Landau (1993), the protagonists (i.e., heroes) of these folktales encounter various trials and tribulations, but manage to overcome them each time with the help of divine spirits or their own supernatural abilities and wiles, eventually succeeding in acquiring the treasures they seek or marrying the women they love. The heroic story of human evolution has exactly the same structure, and modern humans are successful in overcoming climatic changes, food insecurity, threats from predators, and competition with their kin until they finally flourish as *Homo sapiens*, their success being described as a series of fortunate events that unfold as if by magic.[6]

Narratives of evolution as they concern extant organisms tend to be success stories, something that is especially true in the case of humans (see Harari 2015). While this tendency is perhaps even more pronounced than in purely historical narratives, this is because the subject is a pseudo- or hypothetical agent and hence not subject to deep personal understanding, and the supporting role of the players of the story—the side of the losers, so to speak—is naturally not given any attention. Needless to say, however, in the sense that they have left no descendants, most of the organisms that have appeared on Earth have gone extinct, whether as individuals or as bounded sets. Accordingly, we cannot talk about the evolution of life as a whole without considering the significance of this (Sarashina 2018). In fact, this is something that Gould was keenly aware of, since, as a paleontologist, he was an expert in discussing organisms that no longer exist (Gould 1989).

The significance of extinction narratives

I hope to rehabilitate the topic of extinction, which is logically inherent in evolution as a fact, but which has only ever been dealt with in the background, aside from in the case of the dinosaurs. Could this revolutionize the overall discourse concerning evolution? In concluding this chapter, I would like to reaffirm the significance of discussing extinction within the frameworks of evolution and the present state of human existence.

As I mentioned at the outset, in the section 'Speaking of extinction,' extinction has become a prevalent theme in popular discourse. When surveying the future from humanity's present moment, a central proposition is that the possibility of extinction has begun to appear, or is fast approaching, and this tendency is most prominent in arguments being advanced in the field of popular science (e.g., Guterl 2013; Newitz 2015; Kolbert 2014). While each individual field has its own scenarios and factors leading to extinction, the discourses directly related to what I am considering in this chapter are those based on analogies between the extinction of past organisms and the extinction of humanity as a whole. Conversely, discourses on the extinction of specific human groups focus on the political violence implicit in the Holocaust and other instances of genocide. Whether in whole or in part, in both cases, these discourses seem to envision extinction as an annihilation in an aspect of extremity in terms of a catastrophe that occurs rapidly and within a relatively short period of time. There is certainly an irresistible basis for this sense of urgency and crisis.

However, the true aspect of extinctions that have actually taken place in the past and which may occur in the future are very different from these sudden catastrophes, both in terms of the scale of the time that elapses and the relationships between individuals and groups. Not only has this happened several times before, in terms of the mass extinctions of Earth's organisms, but it is probably no different for any human population, whether it is something that has happened to a regionally limited ethnic group or something that has taken place over a somewhat more extended period, as with the Neanderthals. Most processes of the extinction of a set take place in a temporality that cannot be seen and is not experienced as such by individuals. To that extent, extinction should not normally be a source of anguish in terms of the entitative and existential annihilation of individuals. This is because most (though of course not all) of these, rather than being entitative extinctions, are in fact the erasure of the boundaries of a given set—or, in other words, categorical extinctions.

The zoologist Hideo Obara distinguishes two categories of extinction. The first is the disappearance of an old species, and its emergence as a new one, while the second is the forced extinction of a given species by a higher-order species. In addition to these, he regards destructive extinction by humans as a third category, though he understands this as something that is isolated from the evolutionary process (Obara 2005). The extinction of the Neanderthals would fall into the second category, on the condition that we accept modern humans as a higher-order species. As we have seen in this chapter, however, the reality is not so simple. Nevertheless, to borrow Obara's terminology, I wonder whether this third category of extinction could be typologically similar to the process of extinction in the first and second extinction processes. If extinction is a problem for us, then perhaps the meaning of all questions comes down to this.

The term 'Anthropocene' was coined to describe the period from the mid-twentieth century, which represents the most recent decades of the Quaternary Holocene, which has lasted 10,000 years.[7] It is understood to refer to a geological situation that has been decisively influenced by human activities, and which is utterly unlike what has gone before, even from the standpoint of atmospheric movements and geology. More significantly, it is also extremely likely that the nature of the extinctions that many organisms will meet with in the Anthropocene will be more extensive and abrupt than those that have taken place in the past. According to a rough calculation, twenty-seven thousand biological species are dying annually of direct or indirect human impacts (Newitz 2015). The question now is whether or not this blade will be turned on ourselves.

Extinction as the result of some absolute extrinsic factor that humans will not be able to stand against is only a matter of time and, in that sense, inevitable. The question of when this will happen is more or less a matter of chance. However, among the six extinction scenarios outlined in Fred Guterl's (2012) *The Fate of the Species*—superviruses, the recurrence of mass biological extinction, sudden climate change, threatened ecosystems, synthetic biology (i.e., bioterrorism), and machines (i.e., computers run amok)—there is only one absolute external cause in which humanity has no possibility of being involved; namely, that of a catastrophic natural disaster such as a huge meteorite impact or a major volcanic eruption that leads to mass extinction. In other words, his argument is that if we can reduce our involvement in these other factors, we can prolong the life of humanity as a species, regardless of how much time we might have left.

However, the ultimate significance of thinking about extinction lies elsewhere. That is, it may lie in the idea that there is no point in such efforts to extend life. This is analogous to believing that it is pointless to measure the life expectancy of individuals and then to make technical efforts to extend their lives. Whether we are talking about the evolutionary possibility of a new human species that could branch off from modern humans—including the possibility of an artificial intelligence—or the idea of the cosmic extension of the niche within which humanity survives in the sense of our expanding our habitat beyond the Earth, these speculative scenarios still cannot be said to be disaster-proof. Survival as a new set in the context of evolution and the survival of extant individual organisms may be analogous, but it hardly bears mentioning that they are not the same in practice. Far from it, in fact, for in some cases they might be said to be in conflict. It is possible that a set may overcome the extinction of individuals or, conversely, that an individual may remain after the disappearance of the set.

I ended the article on the extinction of the Bukitan as a distinct people with the following words: 'I can envision a situation in which they become an empty set, so that only their name survives. At that point, I wonder if any (former) Bukitan would feel unhappy about their lot' (Uchibori 1994). I had then doubts about the credibility of the idea of future unhappiness that might be elicited by memories of the distant past, but the violence of the past in terms of the erasure of boundaries has certainly brought suffering and sorrow to the individuals who belong to that set. And yet, in the wake of the absolute oblivion that would follow the disappearance of even the empty set, even discussing this idea would probably be unreasonable, which is to say absurd. To that extent, this means that the use of the term 'extinction' in conjunction with 'oblivion' is synonymous with such absurdity.

Notes

1. I would like to use the term 'ontogenesis' to refer to the overall aspect of the emergence and disappearance of a unified entity, which I refer to as an 'ontogenetic object,' or an object to be seen in the context of its genesis, whether individual or collective. This term was coined by the zoologist Ernst Haeckel to refer to the process of an individual's development, but it can also be applied to a collective to encompass the entire process of existence. The term 'ethnogenesis' can be used when the genetic object is an ethnically distinct human group.
2. Extinctions in the context of regional human populations have been widely discussed, especially from the perspectives of ecology, nutrition, and demographics (e.g., Laughlin and Brady 1978; Eder 1988; Sodikoff 2011). For representative discussions informed by the sense of crisis in the twenty-first century, see, for example, the special issue of *Gendai shisō* that includes an article by the Japanese anthropologist Keiichi Omura (2015).
3. Many positive and tragic images of the Neanderthals are found in the popular epic novel *The Clan of the Cave Bear* (Auel 1980), whose main character is a modern human girl named Ayla, an orphan taken in by a clan of Neanderthals.
4. See Pääbo (2015) and Barton et al. (2011). In the latter, the Neanderthals move south to weather the cold climate. There, they encounter modern humans, with whom they mate and produce a mixed species. Based on the resulting genetic makeup, it is presumed that the genome of the overwhelmingly small Neanderthal population gradually diminished, eventually being absorbed into that of modern humans. Pat Shipman (2015) downplays the effect of mixing on the grounds that

the overall mating rate seems to have been low and that no Neanderthal genes associated with male fertility are found on the human X chromosome.

5. Both Pääbo (2015) and Finlayson (2011), for example, refer to 'went extinct' and 'not extinct' in different places in the same book.
6. There are a few exceptions with historical narratives. One classic example is historical materialism, which might better be described as a sort of subclass of evolutionary narrative. In fact, the case of Bukitan and Tatau in this chapter is an instance of relating a potentially historical story within the framework of an evolutionary narrative, while in the case of Neanderthals we might say that I mentioned talking about evolution as history. Although this twist was intentional, one could certainly also tell the story without the twist. The significance of introducing 'historical' elements like extinction and contingency into the evolutionary narrative is discussed by Gould (1989), as well as by Hiromitsu Yoshikawa (2014), who uses the word 'absurd' in relation to evolution, and Kim Sterelny's (2004) summary is also instructive. In any case, debates over the relationship between history and evolution depend on the way we interpret the concept of history. In Losos (2017) and Blount et al. (2018), experimental evolutionary biology is discussed as a way of weighing the validity of Gould's argument about the historicity of evolution based along an axis of contingency and predictability. I plan to discuss the significance of this debate in another paper. Moreover, new works—not only on the Neanderthals, but also on human evolution writ large—have appeared that use the concept of 'extinction' as a driving force for evolution (Sarashina 2018).
7. For a discussion of the validity of the introduction of the idea of the Anthropocene, see Walters et al. (2016).

References

Auel, J. M. (1980) *The Clan of the Cave Bear*. Crown.

Barton, C., J. Riel-Salvatore, J. M. Anderies and G. Popescu (2011) 'Modeling human ecodynamics and biocultural interactions in the Late Pleistocene of western Eurasia'. *Human Ecology*, 39(6): 705–725.

Blount Z. D., R. E. Lenski and J. B. Losos (2018) 'Contingency and determinism in evolution: Replaying life's tape'. *Science*, 362: 6415.

Eder, J. F. (1988) *On the Road to Tribal Extinction: Depopulation, Deculturation, and Maladaptation among the Batak of the Philippines*. University of California Press.

Finlayson, C. (2011) *Humans Who Went Extinct: Why Neanderthals Died Out and We Survived*. Oxford University Press.

Freeman, D. (1981) *Some Reflections on the Nature of Iban Society*. An occasional paper of the Department of Anthropology, Research School of Pacific Studies, The Australian National University.

Gould, S. J. (1989) *Wonderful Life: The Burgess Shale and the Nature of History*. Norton.

Guterl, F. (2012) *The Fate of the Species: Why the Human Race May Cause Its Own Extinction and How We Can Stop It*. Bloomsbury.

Harari, Y. N. (2015) *Sapiens: A Brief History of Humankind*. London.

Katayama, K. (2015) *Hone ga kataru Nihonjin no rekishi* (What the bones tell us about Japanese history). Chikumashobo.

Kato, Y., J. Langub, A. R. Abdullah, H. Samejima, R. Soda, M. Uchibori, K. Okuno and N. Ishikawa (2020) 'Multiethnic society of central Sarawak: An ethnographic analysis'. In N. Ishikawa and R. Soda (eds), *Anthropogenic Tropical Forests: Human–Nature Interfaces on the Plantation Frontier*. Springer, 85–109.

Kolbert, E. (2014) *The Sixth Extinction: An Unnatural History*. Bloomsbury.

Landau, M. (1993) *Narratives of Human Evolution*. Yale University Press.

Laughlin C. D. Jr. and I. A. Brady (eds) (1978) *Extinction and Survival in Human Populations*. Columbia University Press.

Losos, J. B. (2017) *Improbable Destinies: Fate, Chance, and the Future of Evolution*. Riverhead Books.

Mellars, P., K. Boyle, O. Bar-Yosef and C. Stringer (eds) (2007) *Rethinking the Human Revolution: New Behavioural and Biological Perspectives on the Origin and Dispersal of Modern Humans*. McDonald Institute for Archaeological Research, University of Cambridge.

Newitz, A. (2015) *Scatter, Adapt, and Remember: How Humans Will Survive a Mass Extinction*. Anchor Books.

Obara, H. (2005) *Jinrui wa zetsumetsu o sentaku suru no ka* (Will humanity choose extinction?). Akashi-shoten.

Omura, K. (2015) 'Jinrui de nakunaru tame no jinruigaku: Judōteki na nōdōsei ga hiraku ozomashiku mo utsukushiki mirai' (Anthropology for becoming post-human: Clearing the way to a dreadful yet beautiful future through passive activity). *Gendai shisō* (Contemporary thought). Seidosha, 2015–9: 227–245.

Pääbo, S. (2015) *Neanderthal Man: In Search of Lost Genomes*. Basic Books.

Propp, V. A. (1958) *Morphology of the Folktale*. Mouton.

Sarashina, I. (2018) *Zetsumetsu no jinruishi: Naze 'watakushitachi' ga ikinobita no ka* (The human history of extinction: Why 'we' survived). NHK Publishing.

Shipman, P. (2015) *The Invaders: How Humans and Their Dogs Drove Neanderthals to Extinction*. Harvard University Press.

Sodikoff, G. M. (ed.) (2011) *The Anthropology of Extinction: Essays on Culture and Species Death*. Indiana University Press.

Sterelny, K. (2004) *Dawkins vs. Gould: Survival of the Fittest*. Icon books.

Uchibori, M. (1984) 'Aru yakihata kōsaku-min no rekishi: Sarawaku iban no baai' (History of a swidden people: The Iban of Sarawak). In T. Ōbayashi (ed.), *Tōnan'ajia no minzoku to rekishi* (The peoples and history of Southeast Asia). Yamakawa Shuppansha, 416–436.

Uchibori, M. (1994) 'Minzoku no shōmetsu ni tsuite: Sarawaku bukitan no jōkyō o megutte' (On ethnic extinction: The situation facing the Bukitan of Sarawak). In E. Kuroda (ed.), *Minzoku no deau katachi* (Forms of ethnic encounter). Asahi Shimbun Company, 133–152.

Walters, C. N., J. Zalasiewicz, C. Summerhayes, A. D. Barnosky, C. Poirier, A. Gałuszka, A. Cearreta, M. Edgeworth, E. C. Ellis, M. Ellis, C. Jeandel, R. Leinfelder, J. R. McNeill, D. Richter, W. Steffen, J. Syvitski, D. Vidas, M. Wagreich, M. Williams, A. Zhisheng et al. (2016) 'The Anthropocene is functionally and stratigraphically distinct from the Holocene'. *Science*, 351: 6249.

Yoshikawa, H. (2014) *Rifujin'na shinka: Idenshi to un no aida* (Absurd evolution: Between genes and luck). Asahi Press.

CHAPTER 16

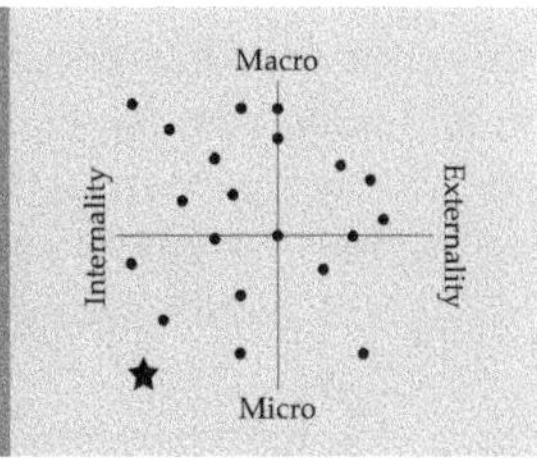

Living at the Extreme:

A Case of Sex Workers in Mumbai, India

Masakazu TANAKA

Keywords: prostitution, sex work, devadasi, Hinduism

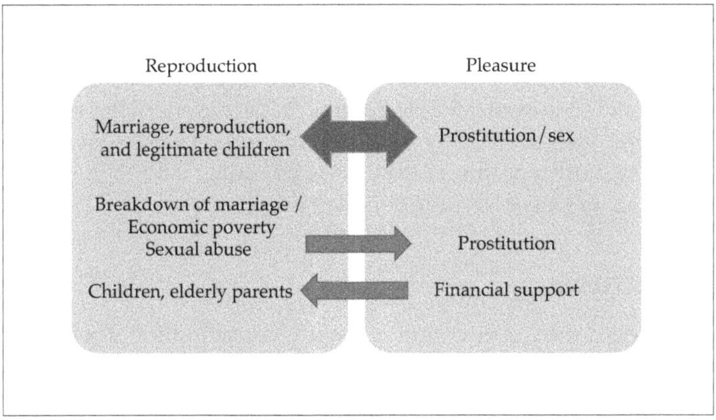

As sexual motivation can be broadly differentiated into pleasure and reproduction, prostitution and marriage have been institutionalized accordingly. Many women in Mumbai became involved in prostitution due to impoverishment resulting from the breakdown of their marriages. These women need to support not only themselves, but also children and elderly parents back in their home villages. In these circumstances, these women are stricken by a sense of hopelessness and helplessness. In this chapter, I consider these emotions as constituting an extreme in the existential dimension in their own right.

My aim in this chapter is to discuss the realities of prostitution / sex work based on an Indian case study. For many Indian women, marrying the man of their parents' choice, bearing (legitimate) children, raising a family, and being cared for by their husbands and children represents the ideal life. However, even marriage does not necessarily guarantee that a woman will be happy in life. It is not uncommon for women to be widowed, or for their marriages to break down due to separation or abandonment by their husbands. Remarriage is also difficult for women in the Indian middleclass, which places importance on a bride's virginity. When the survival strategy represented by marriage is no longer available, women will be faced with extremely straitened economic circumstances. What existing paths to self-sufficiency are open to such women? What ways of life will they end up turning to? In addition to their economic circumstances, something else that I would like to consider as an 'extreme' in this chapter is the type of situation that elicits a sense of fatalistic resignation in the person concerned. This is because 'habituation' may also be considered to be a most dire condition in the sense that it coaxes the living to abandon hope about how to live their lives, transforming them into a kind of living dead.[1]

Given this understanding of the issue at hand, I would like to discuss sex work as one form of extreme situation, with a view to clarifying some of its inherent aspects. In what follows, I begin by describing Kamathipura, the site of my fieldwork in Mumbai, and the general characteristics of the women who work there, and then proceed by introducing specific cases on an individual basis. In closing, I discuss some of the themes informing the struggle to survive for women who no longer have the option of marriage.

Kamathipura

One of the largest red-light districts in India is Kamathipura, a neighborhood in central Mumbai that comprises more than a dozen lanes about 120 meters in length. The name derives from the textile workers from Andhra Pradesh belonging to the Kamathi caste who originally settled the area in the eighteenth century. The neighborhood's roots as a red-light district date to the subsequent establishment of brothels for British troops stationed in India during the colonial period.[2]

In 2003, the area was home to approximately 30,000 female sex workers, but as a result of the spread of HIV / AIDS, over the next decade to 2012 the number plummeted to only 2000 female sex workers (as communicated by an older informant). Up to 40% of these women are reportedly homeless. Many have migrated not only from elsewhere in India, but also from Nepal, Bangladesh, and other South Asian countries.

Although there are fifteen lanes in Kamathipura, my survey in the area has not been exhaustive. The following case study concerns only one of these lanes (referred to here as 'the Lane') and does not presume to offer any conclusive insights into payment methods or the characteristics of female sex workers in the other lanes. Although the existence of a lane populated by *hijra* (a 'third gender' described below) came up in conversation on several occasions, it seems that the area had already disappeared, as I was unable to confirm any further details.

The entrance facades of about thirty buildings line both sides of a lane approximately three meters wide (Photograph 16.1). The ground floors of about half of these buildings are occupied by shops selling general merchandise and goods. Women do not stand in the Lane; rather, they stand at either entrance of the Lane where it joins the main thoroughfare. After haggling with their clients, the women will bring them into the Lane to their own rooms, or else to rooms that have been rented for the purpose. To kill time, women stand near the entrances to buildings in the Lane, sometimes bringing their cots outdoors or chatting and sharing light meals with each other.

The entrances to the buildings are narrow, about a meter in width. Inside, there will be a small room or workshop that serves as a foyer. Deeper into the building are other rooms, some of which are used by the women as living quarters (and sometimes as places of work), while others will be work rooms furnished only with a bed (Photographs 16.2a, b, c). Even though their narrow entrances are separate, the buildings have a complex structure, interconnected on the inside with passages to other buildings on the second floor. The buildings are not particularly specialized for prostitution, however; although this floor might be occupied exclusively by female sex workers, the third floor might be used as a workshop or as the residence of a family who has no connection to the sex industry. So, while the area is a red-light district, it would be more appropriate to think of it in terms of a mixed residential area.

Fieldwork

After some preliminary research, I visited the Lane to carry out interviews on two occasions in 2015 and 2016. These interviews were conducted in Hindi through an interpreter, a (female) graduate student from Tata Institute of Fundamental Research.

Photograph 16.1: The Lane

All interviews were recorded, after which I requested a Hindi transcription that I then had translated into English. The Hindi transcriptions were also checked by a Japanese graduate student and an Indian national from Mumbai.

I conducted interviews with forty people ranging in age from their twenties to their fifties.[3] More than half of the interviewees hailed from the neighboring states of Maharashtra and Karnataka, and West Bengal, where Kolkata is the state capital. Others came from states such as Tamil Nadu, Madhya Pradesh, Andhra Pradesh, Bihar, and Assam (Figure 16.1). They came from a diverse array of castes, ranging from Brahmins to untouchables (Dalits), while in terms of religion I counted faithful from various religions, namely Hinduism, Islam, Buddhism (Dalit), and Christianity (to which some converted after turning to prostitution). The following cases constitute a summary of some of these interviews.

The lives of women

Women who lack the financial wherewithal to rent a room in a building in the Lane live on the streets, while those with small children rent houses outside of Kamathipura and commute, out of concern for the impact on their children. Some women work from around noon until evening, while others arrive round three o'clock in the afternoon and stay until midnight, even if they have children. Daily rent for women living in the Lane amounted to 220 Indian rupees (in 2015, 100 INR was equivalent to about 1.50 USD). The prostitution price was generally set at 250 rupees ($3.75) but could range between 100 and 400 rupees

Photograph 16.2a: A woman's room (which doubles as a work space; the foreground is a common area)

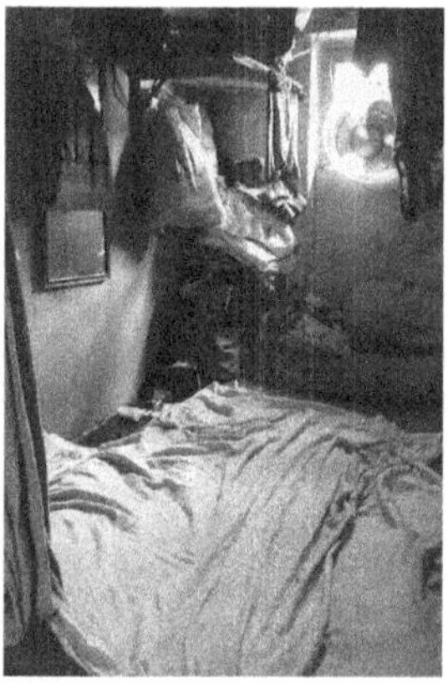

Photograph 16.2b: A woman's room (interior of Photograph 16.2a)

Photograph 16.2c: A woman's room (which doubles as a work space)

depending on the person. Some women said that they were willing to offer their services for as little as 40 rupees. On a busy day, a woman might earn in excess of 2000 rupees. Something that was repeatedly pointed out to me was that earnings were not consistent. Nor was there any set time for a work session, which only ended with the man's ejaculation. A work session is usually quite short, lasting between five and fifteen minutes per customer. The women bathe only once a day, rather than every time they have sex. They say that between 500 and 700 rupees a day is necessary to cover living expenses. Although women

Figure 16.1: India (adapted from International Energy Agency 2002: 15)

stand at the entrance to the Lane, soliciting customers almost every day from late morning until evening or into the night, they will take anywhere from five to ten days off during their menstruating period. Some women affirmed that they needed seven days. If a customer were to discover that a woman he had engaged was menstruating, he might get angry and refuse to pay her.

In the course of my research in the Lane, I found no evidence of the involvement of intermediaries or procurers like pimps. The absence of procurers meant that women were entitled to keep all the money they were paid by customers.[4] The buildings on both sides of the Lane, in addition to the women's living quarters, also contained several work rooms available primarily for the use of the women who commuted from places outside of the Lane. For the use of these rooms, women would pay the landlord a fee of about 20 rupees ($0.30) per customer.[5] While some landlords also live in the same building, they do not directly interfere with the women's line of work (*dhandhā*; i.e., prostitution). By contrast, women's share of the takings at brothels that face the main thoroughfares, where men sit at the entrance, range in the vicinity of 50% to 70%. The absence of intermediaries in the Lane is important, and several women said to me that they feel a greater sense of freedom there.

—— *Will any Mafia-like men come to skim off a large part of your gain here?*[6]

> No such men come. They know that we are very poor. Why do they come all the way to the Lane, where they can find no big gain? Such exploitation usually happens to the high-profit prostitution in other parts of Mumbai. They know that we make a profit of ten rupees only.

Although they enjoy their gains without any middleman's exploitation, women in the Lane have to deal with troubles with customers for themselves.

—— *Have you got any man who can deal with the troubles and control the situation?*

> No, we haven't. We deal with them for ourselves.

The women endeavor to avoid customers intoxicated by alcohol or drugs as much as possible, as such men are more likely to refuse to pay or to resort to abuse. According to one woman, drunks are handled by female sex workers who perform their services on the street, and who are likewise intoxicated. No one in the Lane will rent a room to a drunk woman who wants to bring in a drunk customer. While the women require payment up front as a rule, if they somehow miss the opportunity to take payment and the customer refuses to pay after sex, they have no recourse, and can only accept the situation. Not wanting to bear the brunt of a customer's violence, they take no action. Other than drunks, they also tend to avoid men they described as 'Arabs' and 'Blacks' (i.e., Africans). Although I did not inquire into the reasons for this, there is likely a language barrier. Sikhs are also avoided because they sometimes carry *kirpans* (daggers) for religious reasons. They also have a reputation for being belligerent and quarrelsome.

The women agree that their greatest fear is of arbitrarily being hauled off by the police, as prostitution in India is illegal. Usually, the fine can be settled for 1250 rupees, but this still represents a significant loss. I will return to the matter of police harassment below.

Two brothels I visited (one near the Lane and the other some distance away in the downtown core, at which I interviewed a total of three women) were both cleaner, as well as nearly ten times more expensive, than the rooms on the Lane. The women came from elsewhere and performed sexual services in the brothels' rooms, or were dispatched to hotels where their clients stayed. Some of the women in the Lane said that they would not work in such establishments due to the dangers associated with going to hotels.

> If you accept a guest's invitation to accompany him to a bungalow or lodge, you will be raped. So we try not to go.

While it is said that there is no conflict among the women, this is not to say that they have a strong solidarity. They occasionally go shopping together, or accompany each other on outings to temples and shrines, for example. It was also pointed out to me that although a building's fellow residents will help each other when there is a need, such as taking them to a hospital when they get sick, they generally prepare meals only for themselves, since their meal times are so varied. When someone who lives in the same lane as them dies, they will fast in mourning (though this may be a general practice, not limited to the Lane). Likewise, while some women call older women 'mother' or 'elder sister,' as though they were in the same family, it is unclear how strong this awareness is.

> Yes, yes, we help each other. We are living together like a family. This is not our home village. Who would help us?

The women will give between 50 and 100 rupees to women who are in need of cash. Some women also take part in a kind of rotating savings and credit association, saving money communally to cover individual emergencies. Each woman pays 10 to 20 rupees into a pot each day, and they can withdraw money to cover expenses when they go back to their home villages or when they get sick. However, this alone is often not enough. If they borrow money from moneylenders in the area, they will pay 2000 rupees of interest for every 10,000 rupees they borrow (some even say that a loan of 5000 or 10,000 rupees will incur an interest charge of 5000 rupees). Also, failure to repay the loan by the date it comes due means having to repay double the amount.

While women from the same region do sometimes tend to live in the same building, such cases should be considered an exception. The same applies to religion and caste (i.e., occupational and ritual status groups). It would be easier to think of prostitution as an industry that operates outside considerations of religion or caste. Customers appear to show no preference for women based on their place of origin. It is also interesting to note that customers do not ask about the caste of the women whose services they engage. Lower caste men do not necessarily have a preference for higher caste sex workers, nor does higher caste necessarily correlate with higher price.

The lanes of Kamathipura are regularly visited by members of non-governmental organizations (NGOs) who distribute medicines and condoms, and the women help themselves to condom supplies kept at the front door to ensure that their customers wear them.

Occasionally, a customer will refuse to wear a condom after entering the room, which leads to arguments. Then, a woman drives him out. She never accepts him without a condom: 'Let him leave, another man will come.'

Condoms have been distributed for more than two decades. I would like to revisit the question of NGO activities another time, but I should also add that former female sex workers are employed to conduct instructional activities. Free sexual health clinics are offered every three months at a nearby hospital. In our interviews, one woman with HIV was the only one to state that she had contracted a sexually transmitted disease. Also, in the event that a woman becomes pregnant as a result of a liaison with a client, she will not have an abortion. There are several facilities that specialize in taking care of such children.

So what sort of circumstances might lead to a woman ending up in Mumbai's red-light district? A list of reasons would include the fact that many women marry and start families in their own hometowns, but then leave to come to Mumbai after a bereavement, after being left by their husbands, or when they are unable to cope with domestic violence. While most of my informants' marriages had broken down or ended for these reasons, several were still married. One woman was keeping her work secret from her spouse, and two women were working in the full knowledge of their husbands—who had been regular customers of theirs. All the others were single mothers or were living in Mumbai on their own, having left their children in the care of their own parents or in-laws. Those with elderly parents or children back home send remittances for support. With some exceptions, as in the case of devadasis (which I discuss below), the women keep the nature of their employment a secret from their families and relatives. They did not necessarily come straight to Kamathipura (whether they arrived on their own or were brought there) to start working in the sex industry. They might have taken other employment, perhaps as domestic help in other parts of Mumbai. In some cases, they were sold to a brothel after being told they would be referred to employers as domestic help. However, even those who were indentured into prostitution were able to leave somehow—either by paying cash or finding assistance from a man (generally a customer) or running away—before coming to live in the Lane.

Women's narratives

In this section, I would like to introduce some cases.

AX (age twenty-six, Muslim, Sheikh,[7] originally from Bengal, seven years of schooling)

AX is a twenty-six-year-old woman from Bengal. She is Muslim and belongs to the Sheikh caste. She went to school for seven years. She came to Mumbai in 2002 when she was fourteen or fifteen years old. A man she knew from her village approached with an offer for a job in Mumbai as a domestic helper, and she accompanied him to the city. However, she was criticized as being too petite, and was sold to a brothel ('bungalow'). She had been a virgin until then. Her father died when she was only one year old and she is an only child, so she supports her mother with cash remittances. Her mother used to earn a living as a domestic helper, but no longer works. AX has apprised her mother of her work in Mumbai.

At first prostitution was a harrowing experience, but now she works with enthusiasm. At the brothel where she was first indentured, she gave all her money to the manager. Her daily take ranged from 2000 to 3000 rupees, but this was all taken away. Nor was she free to leave. She worked there for about a year.

She made the acquaintance of a Muslim man working in a cafeteria and succeeded in leaving the brothel with his aid. She lived with him for about two-and-a-half years, and he never said anything about her decision to continue working as a prostitute. However, this arrangement ended when he returned to his hometown to marry at the insistence of his parents. Now she lives in the Lane on her own. She pays a daily rent of 220 rupees (on average). She charges her clients the same amount for sex (220 rupees per ejaculation). Occasionally she receives presents or *bakshis* (small gratuities). She returns to her hometown each year for the Islamic festival of Eid al-Adha (Bakr Id). However, even if she were to quit her job and return to her hometown, she would not be able to marry or go to school. In her own words, she feels that 'My life is worthless…so long as I bear hatred in my heart, it will never be healed.'

What was striking about my conversation with AX is that although she was initially forced into prostitution, she continues to engage in the work willingly now that her freedom has been secured. As we shall see, this is an exceptional case. The timing of my second fieldwork period coincided with a period of heightened police presence, partly due to the influence of property developers. While many other sex workers were suffering, since they were unable to solicit and lacked customers, she was attracting five customers a day. Although many of the other women were unaware of the reason for the police crackdown, AX explained to me succinctly how the developers had lobbied the police not to allow prostitution in the area, which they were trying to clear for redevelopment. Although I did not attempt to verify her explanation, I was impressed that she was able to give such a coherent explanation.

I should add that it is not necessarily the case that the women working in the brothels are completely without freedom. After the first few years, they are entitled to take half, if not more, of their earnings and are free to live off the premises.

BY (age fifty-two, Muslim, Sheikh, originally from Uttar Pradesh, no formal education)

BY married an agricultural laborer at the age of seventeen or eighteen. Her father was also a landless agricultural laborer. Having come from a poor background, she was once again in financial distress after the death of her husband when she was in her mid-thirties. By then her parents and siblings were also dead. She contracted an eye disease and was no longer able to work. A woman from the village introduced BY to her son, who was working in Mumbai (though it is unclear whether or not she was able to meet this man). After arriving in Mumbai, she lived for a time as a beggar. One day, while offering *namaz* (i.e., praying) at a mosque, she was approached by a bearded mullah (i.e., a Muslim priest) with whom she became intimate. He gave her a promise of marriage, but then brought her to Kamathipura. Because of this, she and the mullah quarreled and fell out. He died soon after, but not before cursing her name, promising her that 'Your prayers shall not reach the ears of God!' After living on the streets in a primarily Muslim neighborhood, she came to the Lane two years earlier and decided to rent a room.

She engages only one or two clients a day. On the day of our interview, she said that she had been standing on the street since the morning, but that no one had come. Since she cannot pay for her room unless she finds a customer, she was going to have to ask to postpone payment until the next day. When she needs money, she finds someone to lend it to her. She does not cook for herself, choosing to eat her meals out.

CZ (age forty, Hindu, originally from Bihar, caste unknown, no formal education)

CZ's father was an agricultural laborer and her mother died when she was young. Her younger brother's family now lives in a village in Bihar. She married when she was sixteen or seventeen and moved to Mumbai with her husband around ten years before the time of our interview. He became a drunkard and ran out on her two or three years ago. She does not know what he is doing now, or even if he is still alive. While they were married, her husband worked at a teahouse, while she worked as a domestic helper. Now she lives with her children in a slum area. Her eldest son is twelve years old, and she has another son and two daughters, all in elementary school. Although they return home at four o'clock, she works in the Lane until midnight. While she is at work, her landlord takes care of the children. The steeply rising costs of childcare and education are her most pressing concern. About a year ago, she had learned about this place from a woman she knows (and who it seemed had also worked in Kamathipura), who told her that there was a job for her in a company, but when she arrived she knew what she would be doing. Since she had no choice, she turned to prostitution. She is uneducated and cannot even write her own name. She has a debt of 10,000 rupees, which she used to rent her room, and is now in the process of paying off.

DU (age forty-five, Muslim, Qureshi,[8] originally from Mumbai)

DU's father was a taxi driver. She married a beads seller when she was eighteen or nineteen, at which time she moved to a suburb of Mumbai. Six months later, however, her mother-in-law cast her out, sending her back to her own parents' household. But her elder brother told her that she no longer had a place there and sent her back to her matrimonial home. At that point, her mother-in-law dragged her to the train station and abandoned her there. She was living as a beggar near the station when she was raped.

> I was too young to know that I got pregnant. Somebody dragged (and raped) me, then everything had gone from my head. Still, I can see nobody around but the God above me and the Earth below me.

A man stole her bag, which contained all her valuables, and put her on a bus to Pune. There she met another man, who sold her for 40,000 rupees. However, she managed to escape, and married a Muslim man from Gujarat, where she went to live. She had six children with this man. Four of her children died, and she does not know the whereabouts of two others—a daughter, who was eighteen years old and in the tenth year of her schooling, and a son who was ten years old. Fearing that someone had kidnapped them, she went to the police. She was with her husband for ten years, and he is still in the village. He used to work as a newspaper reporter in his hometown, but since he was no longer able to work after he developed a tumor in his back, she came to Mumbai on her own ten years ago.

DU had been sleeping at a mosque. But since she was originally from Mumbai, she was vaguely familiar with Kamathipura. It had also come up in conversations with her parents, so she made her way there. She was in good health at the time and was able to earn a lot of money by taking many customers. She fell in love and spent between two and two-and-a-half years with this person, but on the day of our interview, she had quarreled with him. The man in question had originally been a client of hers. She had answered his phone and heard a woman's voice. The woman claimed to be his wife. In response, DU also claimed to be his wife. This was when she learned that the man had a grown-up daughter, so she had kicked him out. This man had given her between 500 and 1000 rupees on a monthly or bi-weekly basis. Sometimes he had paid her rent (though she also stated that she had been giving him money). She no longer sends money to her husband in Gujarat. She still takes two or three customers a day, but that is all she can take, as she has contracted tuberculosis and her body is weak.

EK (age fifty, Sikh, originally from Delhi, ten years of schooling in Madras)

EK's father was a Sikh official in Madras (i.e., Chennai, the capital of the state of Tamil Nadu). When she was thirteen years old, she married and moved to Punjab. Her husband worked for a company that made sweaters and vests. However, they were unable to have children, and after ten years they divorced. She married again, but her husband died only a few days later. She tried returning to her parents' household, but there was no longer a place for her there. So she began working as a dancer at a cabaret in Mumbai. When the cabaret shut down, she came straight to Kamathipura. By then, she was twenty-five years old. Her room rent was 150 rupees a day. She had a debt of 80,000 rupees. She had borrowed 60,000 rupees from one person, and the remaining 20,000 rupees from several others. Her son (by one of her clients) now works in a shoe store. When he was five or six years old (i.e., about twenty years ago), he suffered a fall down some stairs. At the time, she had taken him to a hospital in Thane (another city in Maharashtra). It was because of this accident that she took on so much debt, but the repayment schedule was not onerous. Her son studied until the seventh year of school. He was unaware of what his mother did for a living until he became an adult, but now he knows. He used to tell people that she worked at a hospital.

FS (age twenty-eight, Hindu, Choudhury caste, originally from Karnataka, no formal education)

FS was the only daughter in her family. Ten years ago, she married a man in Mumbai, but he was prone to violence. He worked as an odd-job man at a company, earning between 200 and 300 rupees a day, but he died in an accident four or five years ago. Afterwards, a woman she met referred her to her current line of work. She was told that this job would be enough to support her. She has left her thirteen-year-old son and nine-year-old daughter in the care of her parents. She returns every few months to visit them. Both her parents are engaged in agriculture. If she were to go back to live with her parents, she would only be able to earn a daily wage of 100 rupees, even if she worked all day, whereas in the Lane she can earn two or three times as much. She sends her parents remittances of a few thousand rupees now and then.

GR (age forty-five, Buddhist, originally from Maharashtra, no formal education)

GR's parents died when she was a child. Thereafter she lived with her uncle, but when he lost patience with her and asked how long he would have to keep feeding and clothing her, she ran away. When she was around fifteen years old, she came to Mumbai on her own, with only a sari to her name. She was sleeping on the street when a woman came up to her and asked what she was doing in such a place. The woman introduced her to a house where she worked for three years as a domestic helper, after which she moved to the Lane.

LE (age twenty-five or twenty-six, Muslim, caste unknown, originally from Kolkata, no formal education)

LE married at the age of thirteen and has one son. She was thrown out by her husband after five years of marriage. Although she worked for a time as an agricultural laborer, she came to Mumbai when an old man, a relation of her in-laws, said that he would get her a good job. He sold her to a nearby brothel. She is still paying off her indenture. She borrowed 10,000 rupees and has somewhere between 3000 and 4000 rupees left to repay. Her seven-year-old son is now in an orphanage. She visits him every two weeks. When the shop shutters close each night, she sleeps in front of them on the street with some other women. Sometimes she is abused by her customers when she is working, and she is chased away by the police when she sleeps on the street. She is afraid of drunks, as well as the police.

—— *You were cheated and sold. Do you want to go to the police?*

> I have been working in the trade for many years. I am standing on the street. Thus, there is no point going to the police. Do you think that they will feed me?

Finally, I would like to mention *hijras*. *Hijras* are people in India and Pakistan who, although biologically male, are considered female and dress as women. In some cases, they are castrated as eunuchs. Although they have achieved political recognition, some *hijras* make a living by prostituting themselves to male clients. They live in communal groups headed by a leader called a guru and celebrate weddings and the birth of children as a group, for which they receive cash payments (Kunihiro 2009).

Hijra (age nineteen, Muslim, originally from Mumbai, nine years of schooling)

At the age of eight, she was taken away by the *hijras*. She says this was because she was judged to be a born *hijra*. Her parents were terribly saddened by this. She visited her parents every Sunday. The guru was both mother and husband to her. She has been taking men as partners since she was twelve years old. She lived in a lane with other *hijras* and entertained customers there. She has left the guru and now lives with her husband. Her husband, who had been one of these customers, pursued her affections for four years before she finally consented to marry him. She met his parents, and they married in a court marriage.[9] This was a year before our interview. Her husband drives a car for the municipality. Nowadays she herself does nothing in particular, and on the day we met said she was simply paying a social call to Kamathipura.

Devadasis

In 2015, when I first entered one of the buildings to conduct an interview, I vividly remember encountering a woman who told me that she was a devadasi. This was a word I never thought to hear in a red-light district in Mumbai. Although I was intellectually aware that there must be many devadasis working as prostitutes in Mumbai, I never expected to meet one in person.

The term devadasi (Sanskrit *devadāsī*, meaning servant of a god or goddess) refers to a woman who has been dedicated to a deity in a temple. Because such women are considered to have been wed to divine beings who transcend death, they will never be widowed, and will thus remain auspicious. In states like Odisha in eastern India and Tamil Nadu in the south, the word denotes women who work in the great temple complexes, singing and dancing during ritual performances (Marglin 1985).

In Maharashtra, Karnataka, Andhra Pradesh, and other areas southeast of Mumbai, however, the term has taken on a different connotation. There is a custom among the Dalits of these regions for each household to present one daughter to the temple before menarche to be wed to the goddess Yellama (Renuka), and these girls are also known as devadasi (Photographs 16.3 and 16.4).[10] They sit before the goddess, their necks adorned with necklaces called *modi* (Photograph 16.5). The marriage is signified by their foreheads being painted with red powder and their necks adorned with a garland of flowers. However, Yellama is also revered by higher castes, and her priests are accorded high positions in important temples. When the girls begin to menstruate, they become the sexual partners of influential men in the village, from whom they receive financial support. Some devadasis maintain relationships with one man and play a religious role as a kind of shaman, and they may be given land to stay in the village (Photographs 16.6 and 16.7). However, their relationships as lovers do not last indefinitely, since the attentions of these influential men will inevitably turn toward younger women. Many end up migrating to large cities like Mumbai, where they find work in the red-light districts.

The term 'devadasi' relates to the marriage of the woman in question, but this marriage is itself both auspicious (in that the deity will not die) and dehumanizing (in that she can never marry again). Underlying this situation is a strong belief that the ideal marriage is one in which the wife should die before the husband, which is considered a sign of chastity. Ironically, however, the devadasi's sexual relationships are anything but chaste.

Photograph 16.3: Yellama

Photograph 16.4: Temple to the goddess Yellama (Renuka)

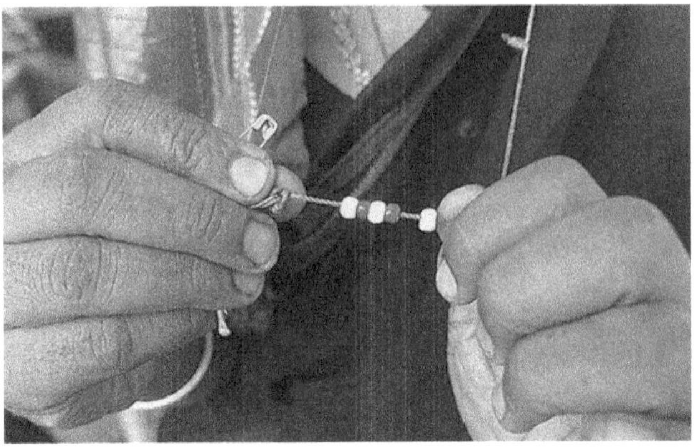

Photograph 16.5: *modi*

Photograph 16.6: Devadasi performing devotions at a temple

Photograph 16.7: A devadasi ritual

From the perspective of an outsider, the situation of the devadasis appears as follows. Dalit families deep in rural India offer innocent girls to the gods or goddesses (in marriage). Then, without ever marrying, they become the victims of sexual exploitation by the upper-caste men of their villages and, furthermore, end up selling their bodies in the city. Their remittances provide support for their parents and siblings. Consequently, the devadasis indirectly support the political and economic systems of their home villages. This sexual exploitation is never rejected or criticized because it is imbued with religious meaning.

I should add that the devadasi system also includes *jogappas*, men who marry the goddess Yellama and then wear saris and live as women. This world gives rise to a complex gender fluidity, but I will leave discussion of that for another paper. Below I introduce the case of the devadasis I encountered in Mumbai.

Devadasi narratives

MN (age fifty, Hindu, originally from Karnataka, Maddur caste)

Around the time that Indira Gandhi was assassinated (in 1984), India's economy took a downturn and MN's entire family left the village for Mumbai. MN was working as a domestic helper. Her parents eventually returned to their village, where her father subsequently died. Since she was an only daughter, she had been offered to Yellama while she lived in the village. After working as a domestic helper, she followed another woman she knew and began to work as a prostitute. Devadasis are not forced into prostitution but, rather, do it of their own free will. MN asserts that she has never been sold. She does not remember how many years she has been working as a prostitute, but given that her son, whom she had with one of her clients, is now twenty-five years old, it must have been longer than that. She was a virgin when she began working as a prostitute (see AX's narrative. I occasionally heard of similar cases). Before she came to her current quarters, she lived in a different house in the Lane but moved after the death of her landlord. She has children (a son and a daughter), as she did not use condoms in the past. Her son lives at her mother's house and is a student. Her daughter is married with two children. She sends 2000 or 3000 rupees to her mother and her son. Her room is not in the Lane, so she pays 20 rupees (as a room fee) each time she entertains a customer. However, she does not sleep outside, but in the corridor of a house in the Lane.

NR (age twenty-eight, Buddhist, originally from Karnataka)

NR's father was a farm laborer. She came to Mumbai three years ago, at the age of twenty-four or twenty-five. Before that, she had busied herself with chores at her parents' house. She explained to me that a girl is offered to the goddess only if there are no sons in the family (see the case of MN above).

> In households with no sons, the daughter is given to the deity. Thereby, she becomes a son, with the idea that she can then earn a living and give what she earns to her parents.

Once a woman is married to Yellama, she may never marry a man. However, NR says that secular civil unions are still possible. Even so, she says that rather than marry, devadasis generally only cohabit with men for financial support.

It is also worth noting that in contemporary India, the term 'Buddhist' denotes Dalits who converted to Buddhism after India's independence, or their descendants. Accordingly, it seems quite likely that NR was originally a Dalit—and was treated as such in her home village—and was presented to the goddess as a matter of custom, as in other Dalit households.

The hopelessness of female sex workers, client relationships, and family relationships

Above I introduced the women I met in the Lane, a site located in the Mumbai red-light district of Kamathipura. Below I would like to take a closer look at the reality of life for these women by teasing out some of the common threads that emerge from their narratives. In a previous study I conducted with female sex workers in Japan, I found that many had positive views about prostitution, in part because they regarded prostitution as sex work (i.e., a job) and because many of my informants were women with a high work ethic (Tanaka 2013). I admit, however, they are not typical among Japanese prostitutes. In India, conversely, I found that prostitution was seen in a very negative light—although not without exceptions. There was also a prevailing sense of resignation and hopelessness. In the following, I would like to highlight several narratives that touch on this sense of hopelessness, as well as on relationships with clients and family members, to bring these matters up for further consideration.

A sense of resignation and hopelessness

The women I interviewed often used the word *majboori*, which connotes ideas like inevitability, helplessness, and the absence of choice. Their usage derives from their sense that they have no choice but to live the way they do (see AX's narrative above).

GR (age forty-five):
Nothing good here. Because of helplessness (*majboori*), I am living here.

GE (age forty-five):
I wish to leave here. If I can make a lot of money, and find a good job, then I could leave this Lane. As there is no such job, I will have to keep staying here.

FS (age twenty-eight):
I have no choice but to stay here. I have to take care of my children.

NR, on the other hand, laments over the fact that she fails to marry, a major purpose of life.

NR (age twenty-eight):
My life is finished. Who would marry a woman like me? I have lost everything in this Lane. Even if I've got a man, he would not marry me working in this trade.

In more specific terms, this sense of hopelessness manifests itself as fear of customers.

FS (age twenty-eight):
Sometimes I feel terrified by their eyes and beard, too. Our life is always at risk. Sometimes we meet good customers and sometimes bad customers. Some beat us, others treat us with care and respect. Any customer is fine with me. Without him I cannot take any meal. Even if I feel bodily pain, I have to work. Sometimes the police come to arrest us. We are living with fear.

HH (age twenty-six):
I don't like this place. It is our destiny. We have to accept a customer, whether a thief or alcoholic. Sometimes a customer beats us and runs away. Sometimes the police come to arrest and beat us.

IF (age thirty-five):
To fill my stomach, I am happy to accept any customer.

LE (age twenty-five or twenty-six) explains this impotence in connection with the abandonment of religious practice.

—— *Do you observe the practice of fasting during the period of Ramadan?*

> Yes, I did many years ago, but I stopped it after working in the Lane. Where can I sleep and pray? I have no water for bathing during Ramadan. Where can I go and pray here?

Getting older is yet another important concern that intensifies the sense of hopelessness.

> JB (age forty-five):
> Sometimes I cannot make money because customers prefer young women to old women. Who would like old women like me?

However, they are not necessarily completely without hope. EK, who is fifty years old, puts it thus:

—— *Do some young men come to women over forty years old?*

> Yes, certainly they do. A young woman is fussy, and does not allow a man to touch her body. This is not the case with old women. Old customers call us old bats, but not young men.

Client relationships

What kinds of relationships do these women have with their clients? In the Japanese context, the relatively long duration of the engagement (sometimes two hours) means that conversation is important. Moreover, in order to secure regular customers, women will try to entice their clients with affectionate words and conduct other than sex. In Kamathipura, on the other hand, the duration of the interaction is extremely short, with the consequence that there is little time for anything other than sex. This serves to reinforce the feeling that one is being treated like an object (Tanaka 2013).

> JB (age forty-five):
> Nobody expresses any emotion to us. They ask nothing about us. When sex is over, they just leave.

—— *Don't you talk to a customer?*

> No, I don't. I do only my job.

If a customer ever does talk to her, it is generally an admonition that she should give up her line of work. KM, a twenty-eight-year-old devadasi, had the following to say:

—— *Do you talk to a customer?*

> Yes, of course. We talk and joke to each other.

—— *Can you share feelings with him?*

> No, I don't. I never share feelings.

—— *Do they try to share them with you?*

> No, they don't. Neither do we ask about their feelings. They just come to enjoy sex.

Nevertheless, they need to do something to entice the customers on the street. FS (twenty-eight years old) states:

> Yes, we need to attract men. I do my best to do everything like smiling, winking, etc. to make a living. This is a strange place for us. Having left our homeland, we came to make money here. Customers come to meet good girls.

I managed to interview one man who was a regular customer. Close to forty years old, he has been visiting the same woman (from Kolkata) once or twice a week for the past year. He is a civil servant with a wife and child and has been frequenting Kamathipura for almost twenty years now. When he was younger, he says, he rarely came more than once a week, if at all. According to the women, this man has not caused any trouble so far, but regular customers are notorious for being bad at paying and not wearing condoms. For this reason, the women do not seem concerned about developing regular clientele. While they sometimes give out their phone numbers, they say that when they do so they end up being harassed in the middle of the night.

> GE (age forty-five):
> Regular customers tend to try to have sex without a condom. Even if we ask them to use it, they refuse. Thus, we never accept them.

> FS (age twenty-eight):
> I do not make regular customers. Everything is fine without them. They come to give something to us, and grab our heart and sweep us away. Then, they beat us, take money from us and never pay us. I don't want such a tense relationship. Making money on our own, enjoying good food and relaxing. This is the best way of life.

On the other hand, from the narratives introduced above, we know that some women have married, or at least cohabit, with their regular clients.

Family

Overwhelmingly, the answer to the question of why the women continue to work in the sex industry concerns economic deprivation. However, this is not purely a personal issue, as financial support for elderly parents and children also represented an important motivation.

> AX (age twenty-six):
> My humble wish is that my mother will die a peaceful death. I wish this dream will come true. I thank her for being nobody but my mother, giving birth to me, and taking care of me.

I would like to focus on the relationships that AX and others have with their home villages, particularly the remittances that they send to their parents or children (who are sometimes left with the women's parents). In order not to fall behind on their remittances, the women end up taking out loans. The fact that these women carry significant levels of debt, sometimes running into the tens of thousands of rupees, and must continue to work to be able to repay the debt is also related to the support they provide to their parents and children. Most sex workers endeavor to maintain their family ties even while working in the sex industry. They were cheated by their marriages. For those with children, however, they do not abandon them but find fulfillment in supporting them. Prostitution is thus involved not only in these women's own livelihoods, but also in the lives of their children and elderly parents.

To sum up, extremes presumably include situations in which people's physical and economic survival is put at risk—such as war, disasters, or poverty, for example—but we can also imagine that they include more existential situations, such as those that sap people's will to live. The life of a sex worker could be described as an extreme situation in which people are denied the hope of living, even if they manage to secure a certain level of economic security and freedom. This is because prostitution entails social and ethical death for these women. What I am referring to here as social and ethical death is also something cultural. With rare exceptions, prostitution is surely the profession that women most want to avoid, one which they do not want people to know about. Accordingly, it is not surprising that women who engage in prostitution feel a certain amount of resignation.

Separating sex for reproduction from sex for pleasure

Prostitution is an institution in which clients enjoy sexual pleasure in exchange for some form of paid remuneration. Most clients are men, and it is mostly women who fulfill their demands. It is owing to this asymmetry that prostitution has been critiqued as a classic form of sexism. In a patriarchal system, it is clear that prostitution is inextricably linked to marriage. From the male point of view, marriage is an institution that guarantees the legitimacy of a man's own offspring and enables the continued survival of society. In traditional societies, marriage partners are decided according to the wishes of parents and societal rules. In contrast, sexual pleasure is sought from female sex workers via paid remuneration. Prostitution could thus be said to be a world that is premised on the idea that sex can be separated according to whether it is pursued for reproduction or for pleasure. Moreover, prostitution is regarded as an occupation that is arrived at by women in straitened economic circumstances whose own marriages have broken down or who have lost their virginity by sexual violence or through love affairs, and are unable to marry.

So where does this separation between reproduction and sexual pleasure come from? To put it another way, when—in the long sweep of human history—did sex become an autonomous act pursued for the sake of pleasure alone? What I would like to focus on here is the disappearance of a breeding season and estrus in the context of human history.[11] This development means that we can have sex at any time, and that if we want to have sex, then we must draw attention to our sexual appeal. Since the disappearance of a breeding season and estrus, humanity has acquired a culture of sexual appeal. This disappearance meant that for humans, sex not only became an activity that transcended biological time, uncoupled from both the seasons and developmental age, but was also liberated from reproduction. Humans can now enjoy sex purely for its own sake. As a result, it has become possible to have inventive forms of sex that are not limited to genital copulation. Humans are now able to engage in sex even before sexual maturity or after menopause.

A consequence of this separation of the sexual act from reproduction has come to be known as the commodification of sex. However, the enjoyment of sex has also brought about it sacralization. This is because whereas procreative sex is animalistic, sex for the sake of pleasure itself is a special activity unique to humans (Graeber 2011: 181). A classic example of this is found in the holy prostitutes who resided in the temples of Mesopotamia. These women excelled not only in the sexual arts, but also those of civilization, such as song, dance, and music. And it is in the devadasi, who also perform dances in temples, that we find their distant descendants.[12]

Of course, these facts alone cannot explain the commodification of sex or the extreme of prostitution that it has brought into being. There remains an enormous gap between the phase of human evolution and the social conditions that gave rise to prostitution. Nevertheless, something I have deliberately sought to confirm at the end of this chapter is the perspective of the fundamental principle of sexual equality between men and women, which is currently obscured from view. The basis for criticizing prostitution's inherent sexism is the inequality of men and women in the context of the sexual act. Any attempt to criticize prostitution without starting from this fundamental principle would risk preserving traditional views of gender (e.g., arguments that sex should not be considered separately from reproduction, or that marriage is a source of happiness for women in its own right).

In this chapter, I have used a case study of sex work in Mumbai to discuss the idea of the extreme. The hopelessness and isolation that are so evident here should not be reduced to mere economic factors. They are underpinned by a system and discourse that recognizes sexual pleasure solely on the part of men, which denies women's desire and which refuses to acknowledge their sexual agency. With this in mind, it will be necessary to bring a more comprehensive and fundamental approach to the study of prostitution.

Postscript: This chapter represents part of the findings from the Japan Society for the Promotion of Science Grant-in-Aid for Scientific Research (B) on 'An Anthropological Approach to "Honor-based-Violence" in the Area Spanning the Mediterranean, Middle East, and South Asia' (PI: Masakazu Tanaka, 2013–2016, 25284176) and the Japan Society for the Promotion of Science Grant-in-Aid for Scientific Research (A) on the 'Cultural Anthropological Study of "Gender-based Violence Complex"' (PI: Masakazu Tanaka, 2016–2020).

Notes

1. A classic example would be the concentration camps set up by the German Nazi government during the Second World War to house Jews and other declared enemies of the state. Inmates resigned to their impending deaths in the gas chambers were called *Muselmanner*—in other words, the 'living dead.'
2. Shah (2014) offers a detailed account of Kamathipura, although her analysis privileges the perspective of women's internal migration rather than prostitution itself.
3. This number does not include the three female sex workers who worked with so-called pimps.
4. This situation differs significantly from that found in a report of a study conducted in the same area in 2006 (Karandikar et al. 2014), in which violence by regular customers, pimps, and husbands was a major concern for female sex workers.
5. The price the women charge for services and the price of rooms fluctuate considerably throughout the year. The prices discussed in this chapter were current in the spring of 2016 as a rule.
6. My questions are indicated with two dashes (——), while quoted statements with no mark relate to the words of my interviewees. I should also note that to avoid confusion, I have refrained from necessarily describing all of the attributes of my interviewees.
7. Groups analogous to castes are also found among adherents of Islam. These exist in a hierarchical relationship determined by several factors, such as their relationship with the Prophet Muhammad and whether or not they are of Middle Eastern or Indian origin. The Sheikhs (or Shaikhs) are believed to descend from low-caste Hindu converts to Islam.
8. Qureshi are regarded as a Muslim population that migrated from the Middle East to South Asia but occupy a lower status than the Sheikhs due to their involvement in slaughter and other ritually polluting activities.
9. A form of marriage performed in a government office in the presence of three witnesses under the supervision of a marriage officer, and which has no accompanying ceremony.
10. For a detailed discussion, see Arun Jaganathan (2013) and Ramberg (2014). As many as 15,000 girls are reported to have been offered this way every year (Kermorgant 2014: 258).
11. See Tanaka (2019) for a discussion of reasons why a human breeding season and estrus might have disappeared.
12. Strictly speaking, however, the term 'devadasi' should be thought of in reference to the women who sing and dance for ritual performances in the great temple complexes, who do not feature in the discussion in this chapter.

References

Arun Jaganathan, V. R. (2013) 'Yellamma cult and divine prostitution: Its historical and cultural background'. *International Journal of Scientific and Research Publications*, 3(4): 1–5, http://www.ijsrp.org/research-paper-0413.php?rp=P161051 (accessed October 10, 2017).

Graeber, David (2011) *Debt: The First 5,000 Years*. Melville House.

International Energy Agency (2002) *Electricity in India: Providing Power for the Millions*. Organisation for Economic Co-operation and Development, http://www.rist.or.jp/atomica/data/fig_pict.php?Pict_No=14-02-11-04-10 (accessed February 1, 2019).

Karandikar, S., C. Frost and L. Gezinski (2014) 'Patriarchy and gender-based violence: Experiences of female sex workers, India'. *International Journal of Social Work*, 1(1): 1–14.

Kermorgant, C. R. (2014) *Servants of the Goddess: The Modern-Day Devadasis*. Vintage Books.

Kunihiro, A. (2009) *Hindū megami no kiesha hijura: Shūkyō jendā kyōkai-iki no jinruigaku* (*Hijra*—devotees of a Hindu goddess: Anthropology at the boundary of religion and gender). Fukyosha.

Marglin, F. (1985) *Wives of the God-King: The Rituals of the Devadasis of Puri*. Oxford University Press.

Ramberg, L. (2014) *Given to the Goddess: South Indian Devadasis and the Sexuality of Religion*. Duke University Press.

Shah, S. P. (2014) *Street Corner Secrets: Sex, Work, and Migration in the City of Mumbai*. Duke University Press.

Tanaka, M. (2013) '"Yatto honto no kao o misete kureta ne!": Nihonjin sekkusu wākā ni miru nikutai kanjō kan'nō o meguru rōdō ni tsuite' ('You've finally shown your true face!': On physical, emotional, and sensual labor from the perspective of Japanese sex workers). *Kontakuto zōn* (Contact zone), 6: 30–59. February 1, 2019).

Tanaka, M. (2019) *Sekushuariti jendā taisei to sono shūkyō-teki kakuran: Devadasi to Shikyū iinchō Haru o megutte* (The sexuality-gender regime and religious subversion: On Devadasis in India and Shikyū-iinchō Haru in Japan). *Shūkyō kenkyū* (Journal of Religious Studies), 395: 107–134, https://www.jstage.jst.go.jp/article/rsjars/93/2/93_107/_article/-char/ja/(accessed February 1, 2019).

CHAPTER 17

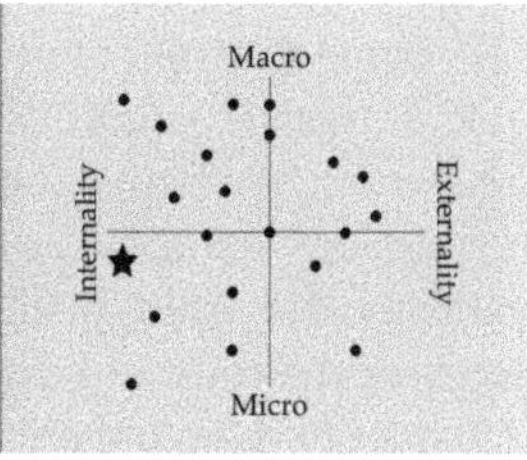

The Inevitability of Extremes:

Thinking about Initiation in Papua New Guinea

Naoki KASUGA

Keywords: matter and mind, initiation, procreative power, magic

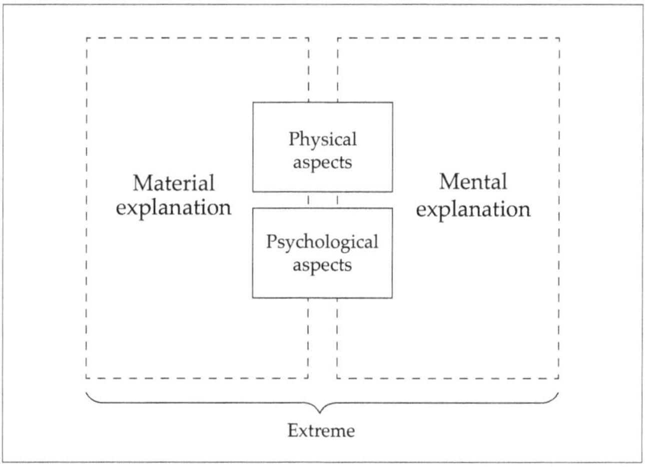

There are both physical and psychological aspects to extremes. These are extreme states in terms of our bodily limits and extreme states of consciousness or psychology at the limits of our endurance. More often than not, these two aspects coincide to trouble the acting subject. Also, in contrast to studies that place their focus on material logic and causes, there is another style in which this analysis of extremes unfolds as a reflection of the mind or thought. Anthropology tends to favor approaches that have the latter character. This chapter strongly pushes this anthropological tradition in depicting both physical and psychological extremes as mechanisms of the mind.

The theme of extremes permits us to focus on their material aspects, as well as to discuss them from a mind-oriented perspective. Notwithstanding the problems inherent in distinguishing between mind and matter, the question of whether or not humans are in a state of physiological or psychological extremity is a highly valid one. Nor is this distinction completely irrelevant to primate extremes, as we see from the contributions in this volume by Nakamura (Chapter 6) and Nishie (Chapter 8), who intentionally discuss psychological aspects. It should also be emphasized that this dichotomy not only characterizes extremes, but also serves as a constitutive principle that forms the essence and cause of extremes. For example, mental extremes may be explained by reducing them to their physical dimension and, conversely, physical extremes might be understood as products of the mind.

Many traditional anthropological studies have also advanced arguments in line with this mind–matter dichotomy, discussing how specific people in various parts of the world create their own unique material and mental environments. However, with the exception of specialized subfields, like ecological anthropology, the material environment is often regarded as either a precondition for or a reflection of the mental world and tends to be relegated to a secondary role in relation to the mind. This chapter is no exception, as I rather intentionally bring this tendency to the fore. In sum, this chapter examines the situation whereby the material and psychological extremes of humans are created in response to their mental mechanisms by way of a discussion of initiation in Papua New Guinea.

Discussion of the physical and psychological extremes experienced by those undergoing initiation is commonplace in ethnographies of societies from all over the world. Although initiation may require the staging of death and rebirth, this can be performed symbolically, and hence there is no necessity that it should reach an extreme. With reference to Victor Turner's (1974) theory of ritual, we can say that extreme states produce the effect of acquiring lifelong friends through an 'existential' experience and gaining a proper sense of the constructed nature of society (to use a contemporary expression). In the case of Papua New Guinea, however, extreme states are not necessarily relevant to existential experience or to gaining a sense of the constructed nature of communal life. Initiation has been carried out on the premise that, more than anything else, it is essential for the sexual development of young men, and that it is necessary for initiates to be placed in an extreme state, both for their own sake and for that of the group.

Initiation in Papua New Guinea, similar to local rites of food prestation, is based on an indigenous way of thinking about the body. Contrasting it with African initiation, Ian Hogbin (1970: 103) wrote that it is typically carried out based on the understanding that whereas girls' breasts swell naturally and their menstruation begins as they walk with their mothers, boys' growth is delayed and it is not certain that they will become men without such intervention. However, the question of how the necessity of their sexual growth should demand such an extreme state is one that merits investigation in detail.

This chapter introduces and discusses initiation based on data collected by the American anthropologist Fitz Poole (1981, 1982), who conducted fieldwork in the early 1970s among the Bimin-Kuskusmin in West Sepik Province, now Sandaun Province. Details of initiation rituals have been reported from the northern coast, the Sepik River basin, the interior, and the eastern highlands, but Poole's description may be judged as the most suitable for discussing extreme conditions. Below, in the section 'The initiation process,' I present an overview of the communal life and initiation processes of the Bimin-Kuskusmin and the harrowing experiences undergone by the boys who become initiates. Then, in 'Initiation from each standpoint,' after confirming that their situation is appropriately referred to as a boundary state, I want to examine each participant's construction of reality in terms of how this boundary state was rendered as inevitable. Finally, in 'Female capacities versus male control' and 'Special female priests,' I want to try to highlight the necessity that sustains initiation by re-examining the relationship between gender and communal life.

The initiation process

The communal life of the Bimin-Kuskusmin has many aspects in common with that of groups in many other parts of the New Guinea Highlands. Although they emphasize patrilineal descent, they form exogamous clans that occasionally also accept matrilineal kin and even people from outside their kinship networks, which in turn form the nuclei of their ridgetop villages. Adult men live in the higher-situated 'men's houses,' while women and children live in the lower-situated 'women's houses.' They draw a distinction between men's food and women's food, which they are careful to strictly maintain. In this region, men grow taro in the fields and hunt animals in the forests, while women gather plants, but the sweet potato is eaten by both men and women as a staple food.

Bimin-Kuskusmin children are considered to be the product of the sperm of their father and the bodily fluids and menstrual blood of their mother, and to inherit blood from both parents. Whereas uterine (matrilineal) blood is considered to disappear after about four generations, the patrilineal descent group is based on the perceived strength of agnatic (patrilineal) blood. However, because agnatic blood is weakened by exposure to menstrual blood and other female substances, men are obliged to consume men's food. The father's procreative power is inherited by both sons and daughters in the form of sperm for sons, and of vaginal secretions and breast milk for daughters.

A problem arises with the inheritance of menstrual blood from the mother, which is passed to both sons and daughters. For girls, this represents a part of their own fertility, but for both sexes it is something that is dangerous and needs to be handled with care. Taking precautions against menstrual blood through its associations, the idea of ritual pollution is found in most parts of New Guinea. Characteristics specific to the Bimin-Kuskusmin are that vaginal secretions and menstrual blood are treated as separate, and that control of menstrual blood is placed under the control of the patriline. When a girl experiences menarche, a rite that corresponds to initiation is performed by a female priest. The rite, however, is completed with a short ritual in which the menstrual blood is stopped by the elder brothers of the lineage. In effect, the girl's menstrual blood, inherited from her mother, can be stopped by being controlled by the patrilineal clan, thereby preventing the spread of pollution.

Menstrual blood also remains in boys' bodies. Unlike in the case of girls, the procreative power of boys remains latent, so it is little wonder that the Bimin-Kuskusmin (like so many other communities) came up with the idea of associating the removal of menstrual blood with the generation of virility on the part of men. The removal of menstrual blood also signifies the eradication of the mother as a woman, and the generation of virility is the gushing forth of the patrilineal semen. On the basis of these paired themes, boys traverse ten intermittent stages of initiation over a period of at least ten years to become full-fledged men qualified to marry and perform divination or sacrifice.

Initiations are led by elders of the two major clans that represent the Bimin-Kuskusmin and involve movements on the part of many actors. One of these two major clans is a representative of the clan group with mythical origins, while the other is a chief of a clan group regarded as being historical immigrants, and all recipients and everyone involved in the right, whether as initiates or as initiators, belong to one clan group or the other (which constitute the ritual-moieties, in anthropological terms). Although this is a male initiation ritual, a significant role is also played by female ritual leaders.

As a stage for the initiation, several houses are erected in accordance with each clan on the same forested slope. Taro gardens are tended at the upper reaches of the slope, and bamboo tubes containing the semen of the clan elders are buried in the soil. At the time of Poole's fieldwork, several houses were prepared that would be able to accommodate 186 initiates (i.e., boys), alongside shelters for around 100 initiators (a majority of whom were men).[1] Once the elders confirmed that all was in readiness for the ritual, they brought the ancestral skulls from the clan cult houses to be enshrined in the forest houses as sacra, along with cassowary and spiny anteater bones, which are respectively associated with the mythical ancestors Afek and Yomnok.

Poole's description concentrates particularly on the first stage of initiation. As in other regions, the first stage is the longest and most intensive part of the ritual. It involves the greatest number of tasks and pushes the boys to an extreme state to realize the elimination of menstrual blood. I give an overview of this process below, summarizing the salient details of Poole's (1982: 120–34) detailed account.

The *kauun* pandanus rite

The initiation begins in the predawn darkness with the chanting of the male initiators. The chant summons the boys who have all gathered in a single settlement to the side of a huge tree where the great ancestress Afek is said to have planted her walking staff. The chant repeats how the boys, polluted with their mothers' secretions, are to be abominated. The boys, bewildered, take the first steps toward fear while laughing nervously. They are handed cucumbers and sugarcane by the female initiators, with which they are instructed to quench their thirst, as they will no longer be able to drink water. The women strip the boys naked and wash their bodies while mocking their tiny genitals, and then cover their heads with yellow funerary mud. They reiterate a rumor that the boys have already heard, namely that the men 'will kill them.'

The chant also calls the boys' mothers to listen to the speeches delivered by the elders on the other side of the stream, which detail how they have polluted and weakened their sons and why their sons must now be killed. The sons are made to bathe in sow's blood, which is hinted to be human blood, and are now beginning to wail and scream. When the mothers see their sons, they rush to the bridge but are restrained by male initiators. Even though this is a ritual scenario, it ends up turning into an unexpected bloodbath.

The various steps by which the boys and their mothers are separated at the beginning of the initiation can be found in other ethnographies as well. Perhaps the most detailed of such accounts is found in Gilbert Herdt's (1986) ethnography of the Sambia of the Eastern Highlands, which describes how hatred erupts between the mothers and the young men as initiators, requiring the intervention of another man. The details of the initiation are a secret known only to those who have passed through it, and this knowledge, which is constitutive of male adulthood, grows as the initiates rise through the stages of the ritual. However, since the intensity of the rite is perceived by women and children, boys about to undergo initiation will be nervous, and their mothers must suppress their resentment and sadness to attend the rite.

While the boys are being taken into the forest for their initiation, the waiting initiators sacrifice *kauun* pandanus (classified as a female plant) and marsupials to the ancestors. They lure the boys from behind a blind of boughs, then leap on them and beat them soundly with their switches. The boys cry out and run away, covered in blood. Then men drive them into the low root house that has been prepared for them and address them. The speeches describe how the boys have been soiled by women, how defiled and disgusting they are, and why they must be subjected to such abuse. Hearing the vivid imagery described by these loud voices, the boys panic at the idea that they will be attacked again.

This, the *kauun* pandanus rite, is the first part of the ritual. For the next twenty days, they will spend their nights in a state of confinement. Throughout the initiation period, female initiators will supply them with cucumbers and raw sweet potatoes.

The sweet potato rite

From the following morning, a new rite begins that will continue over four days. Each morning and evening, the men will address the boys with loud chants detailing their bodily defectiveness and abhorrent female nature, stamping their feet violently all the while. The idea is to humiliate the boys, who must have 'shame' for their 'softness.' By now, the boys are terrified, not knowing what to expect next.

After the men build a screen of boughs near the entrance to the house, the boys are told to stick out their heads. Male initiators, their foreheads painted with red pigment, await on the right side of the boughs, while female initiators, their foreheads painted with black pigment, are waiting on the left. The forehead is the site where knowledge is imparted. Since male knowledge cannot be imparted as long as the female element remains there, the boys are made aware that they must change from the state of a woman with a black forehead to that of a man's red forehead.

With their heads sticking out, the boys are forced to eat five raw female foods in succession; namely, frogs' eggs, frog tadpoles, black mushrooms, adult frogs, and crayfish. In each instance, a boy ingests one of these foods and then eats a piece of raw sweet potato and ginger, then is rubbed harshly with nettles and forced to vomit. In the event the novices cannot vomit, they are made to do so by being forced to drink a mixture of pig blood and urine. Women's food appears in different forms over the next three days, and the repeated speeches of the men instill in the boys the contrast between the superior red blood of men and the black and ugly blood of women (reminiscent of menstrual blood). By now their entire bodies are swollen with wounds, and their vomitus, incontinence, and diarrhea combine in a foul-smelling miasma. The boys are no longer able to ingest the food delivered in the evenings, and their moaning and sobbing can be heard throughout the night.

The domestic sow rite

After the women's food, a rite using domestic pigs follows. The blood of sows shot with arrows is collected in gourds and smeared on the boys' chests and navels. This is done so that female knowledge will not pass through the navel to re-contaminate them, and so that the residues of breast milk can be expunged. The initiators tie the sows' genitalia to the entrance of the house and allow the boys to eat a bite of parts of the sows associated with the animals' female anatomy. The boys are ordered not to eat these parts for the rest of their lives and told that disobedience will be met by a beating from their elders and vengeance by their ancestors. The boys are again made to eat the female parts of the pigs while listening to other initiators chanting to the ancestors of the pigs to take vengeance. Feeling that they are doomed no matter what they do, the boys begin to cry as they eat the female parts. This ritual is repeated the following day.

The great ancestor cassowary/spiny anteater rite

The most important rite of the first stage begins at night, purportedly at the same time as a parallel rite conducted in the ancestral underworld. After the adult men demolish the roof of the low root house to frighten the boys inside, they set the roof alight and begin a chant to summon the great ancestors Afek (as a cassowary) and Yomnok (as a spiny anteater) from the ancestral underworld. Emerging from the now-roofless low root house, the boys are escorted to another building where a fire has been kindled in the hearth. As they enter, the novices are made to pass between the legs of a male initiator of their own moiety. Throughout the night, they listen to speeches and chants, but most of these are in a secret ritual language that they cannot understand clearly.

Men who play the role of the boys' ritual mothers' brothers appear and make incisions on the boys' navels and temples, telling them that they are no longer 'people of women's houses.' Nevertheless, once blood from the incisions is collected and applied to the penises of boys of the opposite moiety, the boys are given cause to fear that the blood is destroying their penises. Subsequently, female initiators bring out special net bags decorated with the design of a red circle inside a black circle. They are told how this indicates that their 'male life' is still weak, surrounded by the residues of menstrual blood.

On the fourth night, the adult male initiators perform a sacrifice to the ancestress Afek. The ancestral siblings Afek and Yomnok are both hermaphrodites, but it is said that Afek, who is more feminine, married Yomnok, who is more masculine, to give birth to the ancestors of the clans, who were the progenitors of humans. When the sacrifice to Afek, who is invoked in the form of a giant cassowary, is complete, the boys are grasped by the earlobes by their ritual fathers while their mothers' brothers quickly pierce the nasal septum from the right side with a cassowary-bone dagger. The boys become hysterical, and it is not uncommon that some faint. They are actually told that they are now dying.

On the fifth night, the boys' bodies are smeared with hot marsupial fat and dew water on the inside of their forearms, eliciting shrieks of pain as large blisters form. When the initiators resume their chanting to invoke Afek, two paramount female ritual leaders appear to impersonate Afek. Dressed in cassowary feathers, with exaggerated breasts, and red pandanus fruit worn in the position of an erect phallus, the two women perform a dance. The men lance the boys' blisters, collect the pus, and apply some of it to the dancers' phalluses. The pus is a manifestation of semen, which Afek is said to eat.

Pig fat and red pigment are then smeared onto the boys' chests, whereupon they are called over to the fire. They now listen to a lengthy list of sacred myths from the paramount male ritual leaders, representatives of the ritual moieties. They are told how Yomnok gradually gained ascendancy over Afek, how men gained control over the initiation rituals (which were originally created by the women), how this is not a complete victory, and how men remain largely ignorant of Afek's ritual power. When the myths are finished, the men wipe away the boys' sweat and tell them that they are about to 'become new men.' Yomnok's appearance in the ritual, however, has yet to take place. The part of the hermaphroditic Yomnok, like Afek, is danced by the female ritual leaders, but this time the boys are bound to drink their own pus.

On the sixth night, the boys are required to make further sacrifices to Yomnok, and their ritual fathers and mothers' brothers pierce their nasal septa, this time from the opposite side. Then, on the seventh night, after further incisions are made to their foreheads and left temples, the female initiators bring out net bags featuring the design of a red circle inside a white circle. They are at last given to understand that their red blood has become dominant but must be protected with white pigment. For the rest of their lives, they will use this white pigment to protect themselves when they become vulnerable.

On the final night, for the first time, the men's chants now depict the boys as growing in strength and manliness. Further chants bid Afek and Yomnok to return to the ancestral underworld. The boys drag their wounded and swollen bodies to a new building, where they lie down to sleep.

The wild boar rite

A rite using wild boars is performed over the next two days, following the same procedure as that for the female pig. Wild boar genitalia and skulls are hung at the entrance of the building, and the boys are given portions of hot boar meat. Their meals of cucumbers and sugarcane are now supplemented with cooked taro. The demeanor of the initiators eases up somewhat, but many boys remain suspicious, displaying panicked reactions to the shouting and stamping that accompany the chants.

The taro rite

Next, another four-day rite takes place involving the replacement of women's food with men's food. The boys are ordered to step through a screen of boughs around the entrance of the low root house built by the initiators. Like the previous time, male initiators whose foreheads are painted with red pigment stand on the right, while female initiators with black pigment painted on their foreheads stand on the left. This time, however, all of the five foods the boys must eat are already cooked. Some of the boys cry out and try to flee, recalling the previous stages of the ritual, but many exhibit the stoic masks of men in pain.

The *bokhuur* pandanus rite

The ritual that concludes this first initiation, like the *kauun* pandanus rite, takes an entire day. The initiators begin a loud chant that echoes through the forest and nearby villages, summoning the 'becoming new men' and their fathers. The boys don new phallocrypts (penis sheaths) and net bags and their skin glistens with the pig fat and white pigment that adorns their chests. For the first time, they sacrifice marsupials alongside *bokhuur* pandanus, which is classified as a male plant. This time, it is the men, anointed with pandanus, who are confined inside a high root house. The boys then demolish the roof and give short speeches in turn, telling the initiators about why they had to endure what they endured, and how strong their bodies have become.

A long procession, led by the boys (now the 'becoming new men'), forms with the female initiators in the rear, marching silently toward the great tree associated with Afek. There the boys find their fathers assembled on the opposite bank of the river. The boys stand (or are held up) to listen to lengthy speeches by the elders. Proudly and rather self-consciously, even in their debilitated and ill state, they listen to their elders explain how their own patience and their fathers' sperm have combined to create their strong bodies. When the speeches are over, boar's blood is poured over the boys, and they finally relax and sit on the ground, where they are embraced by their fathers, who cross the bridge to join them.

Initiation from each standpoint

Although substantially abridged, the above has roughly summarized the procedure followed by the initiation and the intense experience undergone by the initiates. To start my discussion, I want to confirm that this situation is not merely a physical extreme, but also a mental extreme—that is, an extreme that encompasses both mind and matter. According to Poole (1982: 143):

> Throughout the performance, I noted many times how much attention was devoted to the body, corporeal changes, and bodily effluvia when the boys were in the root houses. All lost weight rapidly, and many suffered from nausea, diarrhea, uncontrollable urination, abdominal pain, bladder infection, constant colds, and suppurating ritual wounds. Ritual stress aggravated most of these physical problems...At first, forced to live with their own wastes in the root houses, they often joked about their situation. Later, some became nauseous from the stench and blamed or ridiculed others for fouling the root house. By about the third phase, most were too ill, tired, hungry, and terrified to mention what seemed to have become a relatively minor discomfort.

It was in this state they were moved into the high root houses to undergo the great ancestor cassowary/spiny anteater rite that forms the core of the initiation sequence. Then, as they moved into the latter phases of the ritual, they undoubtedly approached the extremes of harshness and intensity. The question is, as I noted at the beginning, how the boys' extreme state was constructed as something inevitable.

There can be no doubt that the Bimin-Kuskusmin initiation was painstakingly and thoroughly carried out with the understanding that the boys' sexual immaturity was linked with their mothers' menstrual blood, and that the menstrual blood in their bodies had to be expunged in order to activate the agnatic blood they had receive from their fathers. However, as far as this explanation makes a direct correspondence between belief and behavior, it must therefore accept being subjected to critiques that have been advanced in the anthropological study of religion and magic. The relationship between belief and behavior is endlessly complex; the certainty of belief does not necessarily promote behavior, while, conversely, in some cases it is actually uncertainty that may have the effect of eliciting behavior.

We cannot deny that Poole's analysis, as a product of the anthropology of the 1980s, explains behavior in terms of belief and describes the rites in terms of myth, but (inspired by Victor Turner and Clifford Geertz) he provides us with a valuable description by thematizing the subjective experience of the rite. That is, his ethnography of the Bimin-Kuskusmin initiation rites allows us to develop our thoughts on the certainty and uncertainty respectively taken on by the boys who undergo the ritual as initiates, by the adult men who serve as initiators, and by the elders who serve as its leaders.

The initiates, as shown in the above quote, are placed in a state in which their own physical bodies become objects of their consciousness. While the strain of the unpredictable commands given by the initiators one after the other drives their bodies even harder, as far as we can determine from Poole's report, they never seem to doubt the intent of the orders themselves. As their physical and mental debilitation and distemper grow worse, they endeavor to take pride in expunging the female substance from their bodies to become men. Though they may feel resentment at their unexpected and harsh treatment, they do not direct it at their elders, whose power they fear and whose authority is overwhelming. Rather, the brunt of their anger is directed at lower status men with whom they have less of an age difference and other adults in general, and particularly toward their mothers and other women. They resent them for not warning them, for not protecting them.

In this respect, the ideas of male pride and female pollution that the various initiation rites inculcate in them are implanted as assured notions. The 'something' that they felt had been cut off and ruptured inside their bodies through their intense experience *must* have been the female substance, represented by menstrual blood. Even so, in interviews conducted with 'becoming new men' who had passed through this first stage of initiation, Poole found that despite the praise lavished on them, some boys had the sense that their female substance remained with them in some unrecognizable form. That is, the first stage of initiation elicited doubts and anxieties about their own bodies and, in doing so, prepared them to undergo rites in the next stage in which their masculinity would once again be denied.

How confident are the initiators of the rites' effectiveness that they would subject the initiates to this harrowing ordeal? It seems as though they share the inclination that they must inflict great suffering on the initiates because it is something that they themselves experienced and, in doing so, have acquired (or are gradually acquiring) the bodies and souls of men. However, while most seem to express sadness—as, for example, during the ritual in which blisters are raised on the initiates' skin—there are nonetheless some who are so enthusiastic that the elders are careful not to allow them to lead the rites (Poole 1982: 128). Such individual differences are also noted with regard to initiation in the Arapesh region of the East Sepik, but this is not necessarily in proportion to the strength of the belief. The trust that the initiators place in the rites seems to be supported by the perception that, as far as they know, the rites have been established in this way and that a man's maturity does, in fact, increase as he progresses to higher stages and grows older. Hence, younger initiators know that they, like the initiates, still retain female substance in their bodies, and that they, too, will experience the trials faced by initiates in the rites at the next level. In other words, in the bodies of the initiates before them, they should see themselves not only in the past, but also in their imminent future.

The elders lead the rites with detailed knowledge concerning each stage, and they personally handle the sacra and pass on their myths. Poole describes the difficulties they face: the problems of how to interpret the details of rite, how to clarify uncertainties, and how to deal with new situations without being able to know the intentions of their ancestors, the founders, are intractable.

Nevertheless, they supervise the initiators and subject the initiates to the test, as though the format of the initiation has been settled since the time of the ancestors: 'Bimin-Kuskusmin elders express ambivalence and even cynicism about the ultimate truth of what they render ritually of such elusive (and possibly illusive) understandings of ancestral knowledge' (Poole 1982: 111). For that reason alone, they are careful not to push the limits of the ritual, especially the boys' anger and despair, and the mental and physical illness they face. Poole witnessed many boys faint or fall into hysterics, and the elders stressed that in the past the rites had resulted in deaths, cases of insanity, and maiming of the boys. Yet no matter how harsh the experience, they maintain that initiation is absolutely necessary to achieve the desired efficacy, as long as it is accompanied by calm discipline. They are firm in their belief that the dramatic transformation of the body and the emotional control required of a man can all be obtained through this series of rituals.

Beyond the differences in the respective positions of the initiates, initiators, and elders or any differences in their individual personalities, and despite the uncertainty of their positions, there exists nevertheless a shared logic that extreme states are created because they are essential to adulthood, and it is by this logic that this practice can be seen to be justified.

Female capacities versus male control

The logic on which the initiation relies, of course, is based on the gender dualism of male and female.[2] In my own view of Papua New Guinea, the relationship between men and women is generally defined by five capacities possessed by women; namely, sexual growth, menstruation, fetal development, childbirth, and lactation—when a woman grows, her breasts swell and she experiences menstrual bleeding, and then, when the bleeding stops, the fetus grows and she eventually gives birth to a baby, who is nursed and then weaned before the mother's menstruation resumes. This cycle is repeated until menopause takes place. In contrast, men are linked with the female component through their participation in fetal development through intercourse—involvement that also includes sexual abstinence and allowing men of other groups to have intercourse with women. Their idea is that the fetus is nurtured when the woman's blood, which is related to menstrual blood, mixes with a man's semen. Thus there is no concept of pregnancy, and intercourse continues until the woman's belly has swollen (that said, where there is the idea of an ancestral spirit entering into the woman's body, this could be regarded as a kind of pregnancy (Malinowski 1929; Weiner 1976).

The reproduction of society might seem to rely heavily on these female capacities, but since men are necessary for any fetal development, for children to be born and grow up, it is perhaps not that strange that men have been able to bring female capacities under male control. In general, Papua New Guinean societies place emphasis on the inheritance of male blood, while female menstrual blood is regarded as tainted and something to be avoided. There is a tendency to see not only menstrual blood but all vaginal secretions as dangerous, and the bodies ideally subject to control extend not only to women, but also to men who engage in intercourse with women. In many areas, it is reported that while a wife sequesters herself in the menstruation hut, her husband will secretly go into the forest to cleanse himself by bleeding his nose and penis.

The point to remember here about menstruation, sexual intercourse, childbirth, and child rearing is that these are all fundamentally premised on men's lack of knowledge. A woman's inherent capacities may be beyond her understanding, but she knows things about her body that a man could never know. A woman's body should be visible to both men and women, but it also creates male ignorance. In addition, the visible female body wields a power commensurate with the principle of the invisible spirits. Since men's magic and witchcraft are neutralized by women's secretions, it follows that mediums, hunters, warriors, and ritual performers must all break off their contact with women.

Whereas the environment of magical power is given shape by the invisibility of the spirits, as well as by spells, sacred objects, and other secret things and knowledge, female capacities are imbued like laws in women's bodies themselves. Although magical power has a secret nature, female capacities are a physical and obvious necessity. In contrast to the imponderable character of magic, wherein 'anything is possible,' female capacities are unknowable in the sense that they 'necessarily go on and on.' The relationship between male and female in Papua New Guinea is, to some degree, commensurate with a scheme in which men make use of magical power to resist women's capacities. A typical example of this is how men will draw on the power of ancestral spirits to purify the polluting influence of the menstrual blood of women marrying in from other groups and, in particular, how those in positions of authority will exercise powers that depend on ancestral spirits over both men and women. However, cases of magical powers being levied directly against women's inherent capacities are rare. A man might entice a woman with love magic, but after their marriage he might order her about or even beat her. He might shut her out of the house and devote himself to his cult or keep her in the home and make her bring food to the men's house. But, in principle, the woman's female capacities are beyond his touch.

Nevertheless, precisely speaking, it would be incorrect to posit a scheme wherein women have capacities similar to those that would be considered 'natural' in the context of Western modernity, and that men have 'supernatural' abilities in the sense that they are outside this realm of nature. In the first place, some magic is accessible to women, and there are mediums and witchdoctors who are recognized for their powerful magical abilities. Second, women's bodily capacities in many cases are considered to be imbued with 'supernatural' qualities. Third, although rare, cases of men being seen to exercise magical power to exert direct control over a woman's bodily capacities are not unknown.

Instances of magical powers wielded by women have been reported in many places, but it cannot be said that these powers increase their influence over men. Although their magic is limited to the female realm, women's power as sorcerers is limited even more when they have male clients. Second, 'supernatural' abilities attributed to female bodies may be mentioned in cases where, for example, vaginal secretions are believed to not only cause crops to wither but also work harm against the spirits—and, even more appropriately, in cases where they attract evil spirits. Whereas a sorcerer summons spirits by force of a strong will via spells and magical objects, a woman will exert her power over the spirits unintentionally (sometimes intentionally) using only her body. This developmental form is what anthropologists call witchcraft; that is, the unintentional use of power to inflict harm. In Papua New Guinea, we find a socially perverse image of the witch as an invisible figure who feeds on the viscera of others in his or her group. Although it is an extremely radical and dangerous evil, it does not have the same impact as sorcery; in addition to women, other agents accused of witchcraft include those who are somehow socially deviant, elderly people who have been neglected, and members of any other group that is considered to be weak or marginally valued.

Finally, I would like to consider the relative importance, if any, of cases in which a man's magical power exerts direct control over female capacities. As we know from the previous section, the Bimin-Kuskusmin offer one such case, so we shall return to them now to conclude this chapter's discussion of the extreme.

Special female priests

As already noted, among the Bimin-Kuskusmin, the menstrual blood that girls inherit from their mothers is countered by the power of their patrilineal male relatives. A girl inherits her procreative power from her father in the form of vaginal secretions and breast milk. The visible bodily products that are an essential aspect of her female capacities are thus attributed to a man's sperm and blood. While the menstrual blood that lingers in the bodies of boys poses a problem, this is expunged from the body in an initiation ritual led by the male elders, who drive the boys into an extreme state and cause them to experience a definite bodily transformation, thereby allowing the male procreative power they have inherited from their fathers to flourish.

Even so, this is but one aspect of the male–female relationship. Here female initiators participate in the male initiation on a scale that has not been observed in other areas. Chosen to become ritual performers since they were girls, these girls have been raised by female ritual leaders known as *waneng aiyem ser* and have mastered rituals of healing and divination. They marry at menarche, give birth, and raise their children, but they refrain as much as possible from eating women's foods, adhere to male taboos, and have nurtured the spiritual power of their male sperm by consuming men's foods such as taro and boar meat. When they reach menopause after bearing and raising many children (particularly sons), they now turn the procreative power they have inherited from their fathers to the reproduction of the agnatic bloodline. In particular, agnates of senior lineages of their clans play a vital role in the coming-of-age ceremony as *waneng aiyem ser* (Poole 1981: 151).

The *waneng aiyem ser* returns to her clan's territory and retires to a secluded life in a special house built in the forest that is subject to several taboos. She will be given a special taro garden and sacra to keep. In addition to the boys' initiation, she also takes on important roles in other important rites, including birth ceremonies and funeral rituals. In this chapter, I have already described how the role of Afek, the great progenitor of the Bimin-Kuskusmin and the founder of their initiation rituals, is performed by the *waneng aiyem ser*. In this part of the ritual (the great ancestor cassowary / spiny anteater rite), which may be regarded as the core of the initiation, Afek's hermaphroditism is accentuated by her large breasts and erect penis, while her femininity as a wife is affirmed by the behavior of imbibing the pus (i.e., sperm) of the boy initiates.

In the next myth related to the boys by the elders, they are told that although control of the initiation has been usurped by her husband Yomnok, Afek's ritual power remains hidden from the men. Indeed, neither the initiate nor the initiator knows why the pus from the painful blisters on the inside of the boys' forearms is seen as sperm for Afek. One would think that the elders would be privy to this information, but as I have already noted, this is not the case. The same is true for the design of a red circle inside a white circle which the female initiators show to the initiates; no man can know where or how Afek created or else found this design, which is so essential to manhood.

Among the skulls brought as sacra to the initiation site are those of the great and eminent *waneng aiyem ser* of the past, although it is disconcerting for the elders who are familiar with magic not to know the ancestors' intentions here. According to Fredrik Barth (1987: 4), who studied the Baktaman people in the nearby territory, his informants spoke with disgust and disbelief of a neighboring people who venerated the women's skulls. The Bimin-Kuskusmin's determined attempt to control the capacities of the female body with male bodily substances of blood and semen seems conversely to have led to men's dependence on women. If this point were to be explained as inevitable, then the background of the boys' extreme state can also be presented as more inevitable.

The attempt by patrilineal kin to assert control over menstrual blood and to attribute vaginal secretions and breast milk to themselves consists in making children heirs to both matrilineal and patrilineal blood, then giving men the power to counter the menstrual blood of women who bear the same blood from their father, while women combine menstrual blood from their mothers with the vaginal secretions and breast milk they inherit from their fathers to form their own procreative power. This is an ingenious logic that allows both male and female procreative power in the male line, and which reproduces patrilineal group members as both men and women while maintaining the dominance of the father's agnatic blood and semen over maternal menstrual blood. The price of this logic, however, is that the boys must face a situation in which the expression of sperm as procreative force is blocked by menstrual blood from their mother's side.

What is more, this logic cannot escape the contradiction that the matrilineal menstrual blood is simultaneously evil and essential for the production of children; in fact, it reformulates this contradiction within the opposition of the male line and the female line, of girls and boys. As a consequence, whereas a woman who bears and raises the children of her patrilineal kin without menstrual blood emerges, on the one hand, as an ideal to be sought out, on the other hand, a man who cannot fully exercise his patrilineal procreative power owing to menstrual blood emerges as a problem to be solved. The former is embodied in the *waneng aiyem ser*, while the latter is typified by the boys who undergo initiation. The *waneng aiyem ser* is hermaphroditic (i.e., a union of a patrilineal male and female), but plays the role of a female (i.e., a mother), and by passing under her genitals the male initiates are reborn as boys possessed of procreative power.

The *waneng aiyem ser* are often compared to witches known as *tamam*. These are women who are immune to the efficacy of rites performed by patrilineal male kin, and who attack the men by dripping menstrual blood indiscriminately. It is said that only *waneng aiyem ser* can resist *tamam*, and that male patrilineal kin secure their safety through her.

Conclusion

The extreme state experienced by the boys who undergo the Bimin-Kuskusmin initiation ritual is recognized as essential by the boys themselves as initiates, by the elders, and by the initiators, and it is something that is engendered by all of these groups in cooperation. In this, we can certainly acknowledge a correspondence between belief and behavior. Furthermore, when we examine the relationship between gender and communal life, which is the basic underlying premise of the initiation, it becomes clear that they have placed the capacities of the female body under the control of patrilineal blood and semen, which, in turn, has engendered a dependence on women. They use an ingenious logical construction to create the ideal woman—the *waneng aiyem ser*—through whom they generate male procreative power and protect themselves from the dangers of witchcraft.

The logic and practices of the Bimin-Kuskusmin represent their solution to the dilemma posed by the fact that the existence of women is essential for the production of children to reproduce the patriline, even while their menstrual blood, which is essential for their own procreative power, is seen to have an evil influence on men. The motifs of female menstrual blood and male agnatic blood, of the pollution of menstrual blood, and the dominance of patrilineal descent, which are also reported across Papua New Guinea, are here again situated as the basis of gender and communal life. The relative merits of this solution notwithstanding, both the performance of male initiation and the involvement of the female ritual leaders can be regarded as its inevitable consequences.

Given the foregoing, we can see how the harsh conditions experienced by the boys of the Bimin-Kuskusmin are likewise an inevitable consequence of the fundamental problems arising from the difference in reproductive roles between men and women, and differences in their affiliation to their own versus other collectives. The themes of reproduction and group affiliation, which are essential to human survival, will give rise to extreme situations in response to their particular ways of thinking, as we have seen here. As mentioned at the beginning, it would be nigh impossible not to emphasize the influence of the realm of the psyche over the realm of the material.

The necessity of emphasizing the role of the mind in discussions of extremes most certainly stems from the fact that our object of study is humans. This is not so much a function of the rich mental world that humans possess as due to the fact that it is academically difficult to discern details of the mind from nonhuman animals. If we allow ourselves enough scope of imagination, we might be able to depict extreme states among primates not only in terms of their own thoughts and emotions but, in some cases, even turn them into inevitabilities linked to their psyches. With the introduction of the mind, the evolutionary process would come to life in a much more immersive and immediate way. Alas, as this must remain an academic discussion, I shall end without stepping over into this extreme.

Notes

1. The term 'initiates' is used to refer to novices undergoing the initiation, while the 'initiators' are those who impart the necessary experience and knowledge to the initiates during the ritual and who have already undergone (a least part of) the initiation themselves.
2. As noted by Marilyn Strathern (1988), it is also true that initiations that create masculinity are made up of both male and female elements. However, whereas her discussion focuses on the aspect of how the two elements form analogies while continuously transforming into other analogies, the theme of this chapter is the construction of the adult male person as such.

References

Barth, F. (1987) *Cosmologies in the Making*. Cambridge University Press.

Herdt, G. (1986) *The Sambia: Ritual and Gender in New Guinea*. Holt, Rinehart and Winston.

Hogbin, I. (1970) *The Island of Menstruating Men*. Chandler Publishing Co.

Malinowski, B. (1929) *The Sexual Life of Savages in North-Western Melanesia*. Routledge.

Poole, F. (1981) 'Transforming "natural" women'. In S. Ortner and H. Whitehead (eds), *Sexual Meanings*. Cambridge University Press, 116–165.

Poole, F. (1982) 'The ritual forging of identity'. In G. Herdt (ed.), *Rituals of Manhood*. University of California Press, 99–154.

Strathern, M. (1988) *The Gender of the Gift*. University of California Press.

Turner, V. (1974) *Dramas, Fields, and Metaphors: Symbolic Action in Human Society*. Cornell University Press.

Weiner, A. (1976) *Women of Value, Men of Renown*. University of Texas Press.

PART • FOUR

Extremes Engraved

in the History of Human Evolution

CHAPTER 18

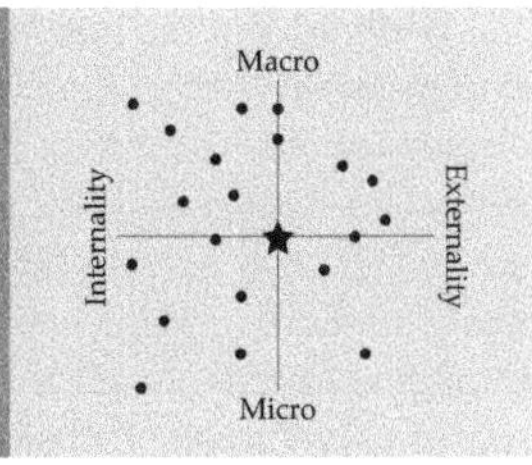

Childbirth and Birth as Extremes:

A Benediction by a Hard and Narrow Path

Suehisa KURODA

Keywords: obstetric dilemma, difficult births, cephalopelvic disproportion (CPD), midwifery care, oxytocin, stress tolerance, labor hypothesis, extremivora

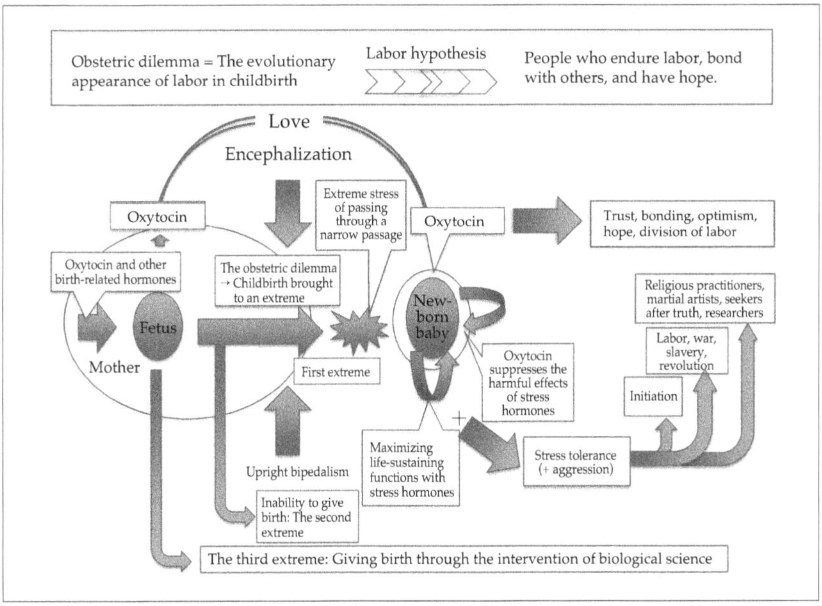

Childbirth for modern humans can be seen to have three extreme phases. First, there is the 'extremization of childbirth' in which the fetal brain grows to the maximum size possible for passage through the birth canal. Second, there is the appearance of mothers who go beyond this limit and who cannot give birth without a caesarean section, which brings us face to face with our reproductive limits as a species. Third, there is artificial childbirth thanks to modern medicine, which is far removed from normal mammalian childbirth. In the first of these extreme phases, the fetus develops a stress hormone system and an oxytocin system to withstand the extreme stresses caused by the pressure of the narrow birth canal and the change in its living environment. Stress hormones contribute by supporting the life of the newborn under severe conditions, such as strong pressure during passage through the birth canal, anoxia, and low body temperature, and by preparing it for the start of pulmonary respiration. When secreted in large amounts, however, they can cause serious damage to the infant's brain, internal organs, and circulatory system. Oxytocin suppresses pain perception, reduces stress response, protects the brain and nervous system from excessive stress hormones, and strengthens the immune system. Moreover, oxytocin is known to promote suckling response after birth, subsequent attachment to the mother, recognition of others, and the formation of bonds of trust and affinity. These hormone systems, which are triggered at birth, are activated as needed in later life to create strong stress tolerance and trust in others, which form the foundation of human society.

Do humans have difficult births?

It is thought that the cranial size of the human fetus in the final month of pregnancy is almost the same as the mother's birth canal, and that, as a result, births are likely to be difficult. Washburn (1960: 74) used the term 'obstetric dilemma' to characterize the contradiction between bipedalism, which limited the size of the birth canal, and tool use selecting for a larger brain, resulting in a larger fetal head—a dilemma that is resolved by producing the newborn in the fetal state.

In the Book of Genesis, the first book of the Hebrew Bible and the Christian Old Testament, it is written that when Adam and Eve were expelled from the Garden of Eden for deceiving God, the pain of toil was inflicted on man and the pain of childbirth was inflicted on woman (Genesis 3:16–19). Thus, for the ancient Hebrews, painful childbirth was a characteristic of being human. Nor is it unreasonable to think that humans have difficult births, or that only humans have difficult births, when compared with livestock and other animals that seem to have easy births.[1]

Difficult births were a commonplace threat. Judith Leavitt (1986), who studied childbirth and child-rearing in the modern United States of America, cites many diaries and letters in which women lament the deaths of sisters and female friends in childbirth and dread their own pregnancies. She writes that 'during most of American history, an important part of women's experience of childbirth was their anticipation of dying or being permanently injured during the event' (Leavitt 1986: 14). She even cites a man's letter from 1937 referring to 'the worst of all human Perils, that of giving birth to a child' (Leavitt 1986: 26).

Conversely, however, in pre-modern non-Western societies, many working women gave birth to their children without difficulty, and reports of unattended labor are not uncommon (e.g., Arriaza et al. 1988; Engelmann 1884; Goldsmith 1984). Even in the Old Testament, another passage describes how Hebrew women give birth to their children before the midwives arrive.[2] Likewise, in Japan, unattended labor was commonplace in some areas until around the Second World War (Matsuoka 1991, 2014; Nishikawa 1997).

As a child, I personally witnessed my delicate and petite elder sister give birth, and I am reluctant to say that people have difficult births. The day before, she had asked me to babysit my niece and run to my brother-in-law's office if anything happened, and the next morning she gave birth to her second child at home. Hearing a low moaning, I entered the room to find my sister sitting on a futon, with a diminutive midwife quietly rubbing her back. Without thinking, I knelt attentively to attend with my niece, but we did not have long to wait. I remember the relief I felt once the baby, after being washed in the tub beside me, was held aloft and began to cry softly, and how my sister, now cuddling the baby, thanked me for watching my niece.

So do we humans have difficult births or do we not? That there have been conflicting discourses on childbirth—the most important of all human activities—for thousands of years suggests that whether or not we have difficult births is neither a matter of definition nor a question that can be settled decisively. Yet this question is not easy to answer, in part because of the weight of faith placed in women's ability to give birth. Veteran midwives such as Fujie Sakamoto (2011) say that women have the ability to give birth and babies have the ability to be born. Nor does Sakamoto deny the pain of childbirth, instead finding significance in the fact that both mother and baby can overcome it. In addition, according to a midwife interviewed by Matsuoka (2014), 85% of births do not require special assistance. From a midwife's perspective, then, it must be difficult to ignore this reality and agree with the common sense view that 'humans have difficult births.' This is because tension arising from an expectant mother's fear may lead to the cessation (failure to progress) or prolongation of delivery—literally, to a difficult birth—and we should therefore refrain from potentially frightening words or actions. In childbirth, the pelvic ligament may loosen, and the bony birth canal may widen, but anxiety may bring childbirth to a standstill (Odent 2011). Conversely, a woman whose delivery has stopped may give birth safely with the encouragement of her family. Thus, in order to determine whether humans have difficult births, we must keep in mind that the mental state of the mother has a profound effect on childbirth (Sakamoto 2011).

The manner in which childbirth takes place and how it has been perceived changed dramatically in the twentieth century. Before 1950, the United States and Europe (and Japan) experienced a dramatic decline in childbirth-related deaths thanks to sweeping improvements in hygiene and disinfection and the introduction of sulfuric agents and antibiotics, as well as improvements in caesarean section anesthesia and techniques (Loudon 1986; Sewell 1993). Today, further improvements make childbirth safer and more comfortable, and medical techniques such as caesarean sections (Liston 2003)[3] and epidural anesthesia, which eliminate the unnecessary suffering and risks of vaginal delivery, are becoming more common and widespread worldwide. Conversely, studies of the endocrine system during the childbirth process and long-term follow-up surveys of mothers and children reveal the undesirable effects of unnecessary medical interventions (Ben-Ari 2018; Buckley 2015; Odent 2011). In this context, too, we find contradictory assessments with regard to easing of childbirth with medical technology, some in favor of such intervention and some against. In addition, while it is certain that, owing to the obstetric dilemma, childbirth is not without difficulty, on the other hand primatologists have pointed out that 'humans are not unique among primates' in terms of their mechanism of birth (Hirata et al. 2011).

So how can we explain this deluge of conflicting discourses around human birth? Do humans really have difficult births after all? In this chapter, I review human birth and childbirth in light of the concept of 'the extreme' and the results of endocrine research, focusing particularly on discourses concerning the difficulty of human birth, in order to reflect on what it means for humans to be born via the narrow birth canal. Thus this chapter constitutes a rereading or repositioning of the obstetric dilemma.

Extremivora: an animal that feeds on extremes

Perhaps the most typical image of the extreme is that of a situation in which the survival of the individual, the preservation of the spirit, or else some kind of social mechanism has reached or just started to cross over the brink of collapse (as discussed elsewhere in this volume; see the discussions by Kasuga in Chapter 17, by Kitamura in Chapter 14, and by Funabiki in Chapter 13). It is at the level of individual events that childbirth and birth show an extreme aspect, fraught with the risk of death. However, considered at the collective level, birth and childbirth cease to be extreme events, since they are essential for the survival of the group. Thus how extremes appear depends on the dimension in which they are viewed.

In this chapter, I want to consider the extreme separately from the catastrophes that attend them. When a human (or animal) individual or group survives an extreme situation in which life and death hang in the balance, it may begin to act or to become something different than it did or was before. While the organizing theme of this volume is oriented to seeking a new understanding of human society by paying attention to such 'metamorphoses in extremes' and to the unseen aspects of extremes that underlie social life, this consideration is made possible by the fact that while extremes are closely related to catastrophes, they do not mean the deaths of everyone and everything involved; that is, there is a discontinuity between extremes and crises or catastrophes. Discontinuities also appear as a result of timing differences or probability up to the point that a crisis manifests, even if the tipping point that invokes the crisis has already been crossed, or due to the postponement of the crisis by technology. Furthermore, they will not necessarily appear uniformly in all members. Individuality may be accentuated by differences in individual qualities, life histories, the robustness of the social system, or simply quirks of fate.

If the time lag between the onset of a crisis and its manifestation is large or probabilistic, its evaluation and any responses will be divided, as we see in the case of climate change due to global warming or internal radiation exposure from the Fukushima nuclear power plant accident (see the discussions in this volume by Nishii in Chapter 10 and Tokoro in Chapter 11). With regard to global warming, the earth's temperature has already risen by 1°C relative to pre-industrial levels, and omens in the form of destructive events (such as heat waves and aridification accompanied by massive forest fires, as well as torrential rains, hurricanes, and typhoons) are occurring with increasing frequency on a growing scale. In view of the resulting loss of ecosystems and the destruction of so many human lives, the current situation can be described as having reached a critical point—which is to say, an extreme state. However, the time lag relative to full-scale catastrophe is causing us to delay our response, which will likely result in an even greater catastrophe being imposed on future generations. In order to include cases such as these, I would like to define the term 'extreme' to mean 'some sort of critical point that threatens the survival, identity, or social existence of its participants and *their descendants*.'

However, there is something else pertinent to the purpose of this chapter that is made possible by the gaps and discontinuities between extremes and catastrophes—namely, the fact that humans are not only passively exposed to extremes, but also create and manipulate their own extremes, and seek identity by intentionally leaping into extreme situations fraught with danger (see Kasuga's discussion in Chapter 17 and Kawai's discussion in Chapter 9). Furthermore, rulers have started wars to gain the allegiance of soldiers in the extreme situations that follow, people have risen up in extreme situations to stage revolutions, and martial artists have pushed themselves into extreme states to improve their physical and mental functions, and it is in extreme situations that religious practitioners have attained enlightenment and established the foundations of culture. Humans are an animal that could be characterized as an 'extremivora': an animal that grows by feeding on extremes.

Three extreme aspects of childbirth and birth

Childbirth, in which the fetus is released to the outside world by the contraction of the uterus to become an independent organism, is a phenomenon that endangers the lives of both mother and offspring in animals in general. This is especially true with humans, for whom the birth canal has no room to spare. The fetus must squeeze its head and twist its body to pass through the birth canal. During this transit, it expels the last amniotic fluid from its lungs and switches over to the biological maintenance mechanisms of an independent terrestrial

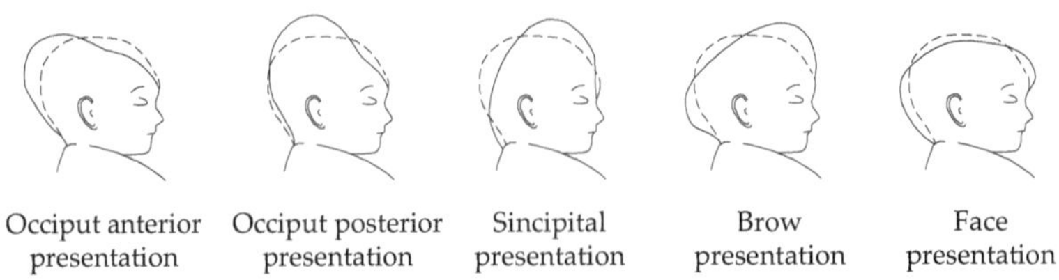

Figure 18.1: Cranial deformities for newborns immediately after birth: the shape of the cranium will vary depending on the position of entry into the birth canal (Kitagawa and Uchiyama 2013: 381)

organism, including respiration, blood circulation, nutritional intake, and body temperature maintenance. The harsh and extreme character of this process can be inferred from the fact that stress hormones like adrenalin, noradrenalin, and vasopressin seen immediately after birth are present at levels rarely seen at other times, even when exposed to maximum stress (Ben-Ari 2018), and the head of a compressed newborn is shockingly deformed (Figure 18.1). The newborn's transit through this narrow passage injures both mother and newborn in various ways. Until the development of antibiotics, this often led to deadly infections.

When childbirth is viewed from the level of the species, another extreme appears. The fetus is born at a stage of physiological prematurity that allows it to pass through the birth canal. On some occasions, however, a fetus will grow too large for the birth canal and be unable to be delivered naturally. In obstetrics, this is called cephalopelvic disproportion (CPD), and without medical care, both mother and child will die. The World Health Organization (WHO 2012) estimated that the worldwide rate of CPD among primiparous (first-time) mothers in 2010 was 3.3%–3.6%, accounting for 12% of all maternal deaths, and that this rate has increased by as much as 0.3%–0.6% in the half-century since 1960. In addition to heredity, other factors contributing to CPD include pelvic deformity due to disease, poor nutrition, obesity, and advance maternal age, but it is not clear how widespread CPD would be without these acquired adverse effects. When we simply consider evolutionary process, it would suggest that CPD due to genetic factors should be eliminated by childbearing failure, but a study by Mitteroecker et al. (2016) has found that this is not the case. On the basis that higher neonatal weight reduces the risk of postnatal disease and increases the chance of survival, the study's authors suggest that the tendency to have the largest possible number of offspring will continue to be selected for because of the reproductive benefits, even though the offspring may sometimes be too large, costing the lives of both mother and child. This explanation makes sense because mothers give birth to immature fetuses, and it is obvious that a slightly more mature fetus would be more advantageous for survival. However, considering that some mothers die as a result, cerebralization in human development trespasses into another of our limitations as a species; namely, the (second) extreme in which some women are no longer able to give birth. Today, rates of CPD are steadily increasing, not only because newborns are getting bigger due to improved nutrition (Japan being the only exception),[4] but also because the widespread use of caesarean section procedures saves the lives of very large babies, along with their mothers (Mitteroecker et al. 2016; WHO 2012). In situations where caesarean sections are not possible, mothers and children who cross this limit are doomed. In this sense, this is an extreme in which technological intervention has been able to shift the implications for catastrophe.

Caesarean section, which is performed not only for CPD but also for various other emergencies, is estimated by WHO to be used in no more than 10%–15% of childbirths if they do it for only necessary cases. In recent years, however, the rate of caesarean births has surpassed 50% in Latin America and has reached about 20% in Japan. And in the United States, it occupies 27% of even low-risk childbirths that are not inherently problematic (Boerma et al. 2018). This is because many obstetricians regard the caesarean section as a salutary method of childbirth that minimizes risk, is safe, and does not place a burden on mother and child (e.g., Liston 2003). While it is true that expectant mothers are also averse to the pain and genital deformation that accompany vaginal childbirth, many critics argue nonetheless that physicians are unjustifiably increasing the number of caesarean childbirths due to the convenience of managing the childbirth schedule, the high cost of care, and the ease of proving the absence of medical malpractice (Boerma et al. 2018; Matsuoka 1991, 2014; Potter et al. 2001).

In recent years, reproductive medicine and obstetrics have seen remarkable progress. As well as cesarean sections, techniques such as accelerating delivery times with oxytocic agents, in vitro fertilization, and epidural anesthesia have become standard medical practices, to the point where it is now possible to save extremely low birth weight infants weighing less than 1000 grams (BBC 2019).[5] Although not approved in Japan, techniques have also been developed to pre-select the sex of the fetus and to make surrogate motherhood possible. Furthermore, some are pushing forward to the stage of the creation of cloned babies[6] and genome editing of eggs. In a latter intervention, the American biologist Shoukhrat Mitalipov and his team successfully removed a mutation associated with hypertrophic cardiomyopathy, a common inherited heart disease, from fertilized human eggs (Blakemore 2017). According to the study, no unwanted changes were found at other sites in the edited DNA. Mitalipov's stalwart intention to 'push the boundaries' of science was echoed in an observation by the cardiologist Sakthivel Sadayappan that 'this [technology] is the future' (Blakemore 2017).[7]

Hence, in the world's most advanced fields of reproductive medicine, human childbirth and birth have already taken off from mammalian reproduction; human life in the flesh, so to speak, has been brought to a new limit—another (our third) extreme. This third extreme is the result of using the biological sciences to stitch together and repair catastrophes that take place at the individual level, but evaluating its relative merits is a difficult prospect. This is because so many people have gained life or have inherited life as a result of the 'gospel' of these advances, and the individual's desire to have their own offspring cannot be easily denied. Reproductive medicine is sure to continue to innovate in the future. Yet if fetuses with a short gestation time can now be saved, it seems likely that all abortions will eventually come to be seen as murder, and if gene editing leads to the birth of 'those bearing superior genetic material,' then what does this entail for their equality in relation to those for whom such interventions were not available, or for parent–child relationships? It is difficult to imagine that there could be ethics and laws for reproductive medicine that can be shared in a world that, first of all, has so many cultures, in which the disparity between rich and poor is so great, and where many people have no access to general medical care.

However, grappling with these difficult questions is beyond the scope of this chapter, and it would detract from our purpose of looking at childbirth and birth in terms of extremes and evolution. Below we shift focus to the discourse that humans have difficult births and a discussion of our first extreme.

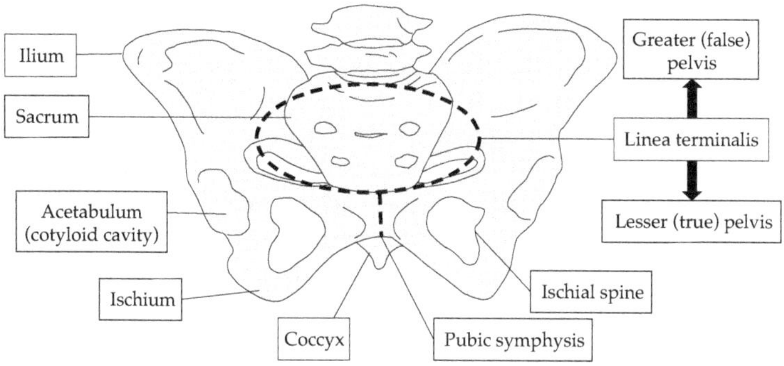

Figure 18.2: Structure of the pelvis. The cavity created by the true pelvis constitutes the bony birth canal. The entrance is the linea terminalis, and the exit is composed of the lower end of the pubic symphysis, the tip of the coccyx, and the ischium.

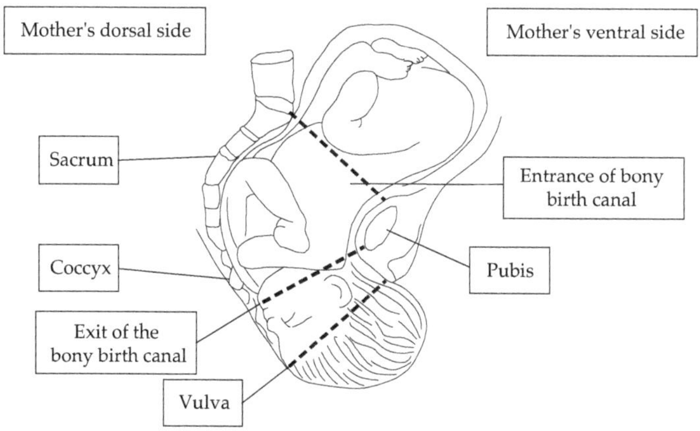

Figure 18.3: The 'crowning' fetus pokes its head out of the vulva. After this, the back of the head is turned toward the mother and the head bends backward (adapted from Kitagawa and Uchiyama 2013: 380).

The obstetric dilemma: childbirth brought to the extreme

The human pelvis is divided into an upper part (the large pelvis) that widens like a bowl to cup the internal organs and a lower part (the small pelvis) that is shaped like a cylinder. At the midpoint of the upper and lower parts, on both the left and right sides, is an acetabulum, the cup-shaped socket into which the femoral head fits. The cavity in the small pelvis is the bony birth canal, which consists of the sacrum and coccyx on the dorsal side, and the lower ilium, ischium, and pubis on either side. The pubic symphysis, which forms the ventral side, is vertically short, and the dorsal sacrum and coccyx curve gently forward behind, so that the bony birth canal opens anteriorly (ventrally) and downward (Figures 18.2 and 18.3). During childbirth, the lower part of the uterus and the cervix extend to cover the inside of the bony birth canal, and the exit of the bony birth canal leads to the vagina and vulva. These soft tissues that the fetus actually passes through are the soft birth canal. The structure of the pelvis is determined by its function of connecting the upper body and lower limbs, enabling humans to walk and run efficiently on two legs, and supporting the internal organs, so the width of the internal cavity (the bony birth canal) cannot increase beyond a certain point. Thus, the major axis from the entrance to the exit of the bony birth canal is only slightly larger than the anteroposterior diameter of the perinatal fetus head (about 11.5 centimeters).

Washburn (1960) speculated that the obstetric dilemma appears with the emergence of the *Homo* genus after brain volume expanded beyond the level of apes. The dilemma was solved by giving birth to an immature fetus with a small head, but this served to increase and prolong the mother's burden of caring for the child, necessitating a division of labor between the sexes and the emergence of the family.

As mentioned above, immaturity is precisely the reason the fetus stays in the mother's body until the head is almost too large to pass through the birth canal—why childbirth became extreme. A. Portmann (1956) explained that, based on the immaturity of the newborn's brain, humans experience early birth (which he calls 'physiological prematurity'), corresponding to being born about a year earlier than would otherwise be the case. He suggested that the blank slate of a newborn's brain is thus exposed to the culture of the group into which it is born, which in turn becomes a behavioral norm. In other words, physiologically premature birth could be characterized as the birth of a human as an (inevitably) cultural being.

However, the past half century has seen many objections raised against the obstetric dilemma. To start with, the underlying theories of human evolution have changed: in the 1960s, upright bipedal walkers were thought to have appeared a million years ago, and according to Darwin's theory of human evolution, tool use, upright bipedal locomotion, and the expansion of the brain were thought to have mutually facilitated each other. So, for Washburn, the perfection of upright bipedalism and brain expansion were thought to have evolved in parallel. Later, however, it was discovered that upright bipeds appeared more than six million years ago. Moreover, not only was their cranial capacity comparable to that of apes, but it remained so for another several million years. With the rejection of the Darwinian paradigm, since the late 1980s the origin of upright bipedalism has come to be debated in terms of changes in foraging ecology and lower energy costs of migration (e.g., Richmond et al. 2001).

Further, the birth canal is also narrow for fetuses in small to medium-sized primates, for which new data have been presented on neonatal head deformities, difficult births, and accidents during birth, confirming that the obstetric dilemma is not unique to humans (Leutenneger 1982). A recent review by Betti (2017) introduces a number of critical studies. To summarize, no differences are found in the energy efficiency of locomotion between males with narrow pelvic cavities and females with wide pelvic cavities, nor any direct association between pelvic cavity width and locomotory efficiency; the range of variation in pelvic morphology is considerable, and it cannot be said that the pelvis has been subjected to stronger selection pressure from upright bipedalism than other parts of the body; and the principal limiting factor of the birth canal is the visceral support function of the pelvis (women with a wide pelvic cavity—i.e., wide bony birth canal—are more prone to pelvic organ prolapse after childbirth). There are, moreover, additional factors unrelated to bipedalism, such as the fact that CPD increases with excess carbohydrate intake, that the bony birth canal changes over time, and that the probability of CPD is higher for childbirth at too young and too old maternal ages. Betti (2017) concludes that the female pelvis is basically adapted to childbirth, and that thermoregulatory body size adaptation to climate temperatures and genetic drift contribute significantly to the variation.

However, the gist of Betti's (2017) critique is directed at Washburn's overly simplistic description that the pelvis structure, which has multiple functions and changes its morphology over time, even for a given individual, was determined solely by its relationship to upright bipedalism. It is not incorrect to say that upright bipedalism is a major determinant of pelvic structure, nor that the enlargement of the brain has led to the birth of babies who are too large to fit through the birth canal. What Washburn (1960) emphasized in this context is that the birth of immature babies led to the emergence of social organizations such as the family and the division of labor. The accuracy and importance of these statements remain unchanged.

The argument, advanced by Wells et al. (2012), that CPD rates increased after the emergence of agriculture is important for examining the problem of difficult birth. Farming leads to the overcrowding and poor hygiene that accompany settlement and larger group sizes. Furthermore, a bias toward low-protein carbohydrates exacerbates physical deterioration, anemia, and tooth decay. As a result, women become smaller and the pelvis shrinks, whereas the head size of newborns will not decrease beyond a certain point. In addition, higher blood glucose levels on the part of the mother due to a greater carbohydrate intake will lead to greater obesity on the part of the fetus, resulting in an increase in difficult births and increased mortality for expectant mothers and newborns (Liston 2003; Wells et al. 2012). Although poor hygiene and nutrition raised infant mortality, the rapid growth of the population nevertheless took place due to the fact that carbohydrates cooked in earthenware vessels accelerated weaning and shortened the birth interval, while food stocks eliminated periods of shortage. We also cannot ignore the fact that most physicians who have lived with American Indian tribes for extended periods have reported the rarity of childbirth problems among women who live a traditional lifestyle, and associate increases in the duration of childbirth and the rate of difficult births with marriage to whites and the transition to modern lifestyles (Engelmann 1884). Moreover, an analysis by Loudon (1986) of detailed childbirth records in London from around 1900 found that the percentage of cases with no midwifery care was 97%–97.5%, and he used this figure as the value that could be expected for survival in unattended labor (see the section 'Midwifery care and medical treatment for crises in childbirth' below).

A hypothesis concerning the evolution of midwifery care and the history of unattended labor

The obstetric dilemma has shed light on the evolution of human patterns of childbirth. One development is a hypothesis by Trevathan (1988) that posits an evolutionary explanation for why midwifery care is commonplace in childbirth in modern humans (referred to here as the 'hypothesis of midwifery care evolution'). Compared to the bony birth canals of the other great apes, which are cylindrical in shape and have plenty of room for the passage of the fetus, those of modern human females are narrower and longer on the left–right axis at the entrance, and longer on the anterior–posterior axis closer to the narrow exit. Normally, the fetus goes into the birth canal head-first (called a cephalic or normal presentation). Because the head is longer front-to-back and the shoulders are longer side-to-side, the fetus enters the birth canal with the front and back of the head aligned with the transverse pelvic axis at the entrance, and then the shoulders are narrowed and tilted, and the body is rotated so that the width is now aligned with the transverse pelvic axis. The fetus turns the now-distended back of its head toward the mother's ventral side, turns its neck back along the exit curve to emerge head-first from the vulva (Figure 18.3), and then turns further to bring the left and right axes of the shoulders into alignment with the anterior–posterior pelvic axis. Another party catches the newborn while pulling it out. On the other hand, in many monkey species, in contrast with humans, a mother will pull her newborn offspring up into an embrace by reaching behind to catch the newborn as it emerges from her vulva with its head facing in the same direction as its mother.

Trevathan (1988) explained that by the time newborns came to be born in a position similar to that of modern humans, for mothers to pull out their newborns as monkeys do would bend their babies' necks backward, possibly damaging their cervical bones and nerves. She hypothesized that this put selection pressure on the provision of midwifery care by others, which was safer and more advantageous for reproduction. In fact, every childbirth care manual warns against damaging the cervical nerves when pulling the baby's head out of the vulva to facilitate delivery of the scapula. Grounded in this concrete knowledge of childbirth and midwifery, this hypothesis that the socialization of childbirth (i.e., midwifery) constitutes the way to deal with difficult births caused by the obstetric dilemma is quite compelling. Nevertheless, the fact remains that this hypothesis is inconsistent with ethnographic data that attest the practice of unattended labor in many human societies. There are also counterarguments to this hypothesis from primatology. Let us begin by considering these.

Douglas (2014) reports the case of a primiparous female wild bonobo who made up a bed and then gave birth in it in the course of less than an hour. During this time, two multiparous females stayed close at hand, frequently peering in, while another multiparous female came by after the birth had taken place. Furthermore, in thirty-four out of thirty-nine cases of births observed in various species of wild primates, other individuals were present, suggesting that humans are not unique as a species in having other females or males present at birth. Demuru et al. (2018), observing zoos in the Netherlands and France, found that multiparous bonobo females showed a high degree of interest in other birthing females and did not allow males or humans to approach them. This was particularly noted in the case of two births observed in La Vallée des Singes (Romagne, France), where the multiparous females removed flies that were pestering the birthing females. In each case, a different multiparous female reached under the vulva of the parturient female to catch the newborn at the time of delivery. In one case, a mother put her hand on the newborn's head when it was being pulled out of the parturient female's vulva, as though to help.[8] In the other case, another female provided support by gently holding the newborn's head from the side while the parturient female was climbing down from a rope with the baby's head still crowning from her vulva. Neither event caused any birth problems.

Solidarity among female bonobos is strong, and may be the basis for the act of (what approximates) midwifery care.[9] In both Douglas (2014) and Demuru et al. (2018), midwifery care is not so much an essential act—the 'obligate midwifery' that Trevathan (1988) argues has evolved as a result of the difficulty of childbirth—as the product of the involvement of social factors such as the high degree of solidarity and tolerance among females.

Objections have also been raised in the context of chimpanzee studies. Hirata et al. (2011) filmed detailed video recordings of the births of three chimpanzees in captivity. They report that 'In all three cases, the fetus emerged with an occiput anterior orientation, and the head and body rotated after the head had emerged'—just as in humans. In other words, the chimpanzees, whose bony birth canal was thought to offer a simple and easy birth, were found to have a delivery pattern that had been thought to be unique to humans, including rotation and presentation. With this, the premise of the evolutionary hypothesis of midwifery care collapses.[10]

More pertinent to our discussion here, however, is that 'in two of the three cases, the chimpanzee newborns landed on the ground without being guided from the birth canal by the mother' (Hirata et al. 2011: 686). In these two cases, the parturient chimpanzees simply squatted down and delivered the newborns onto the straw floor. In the other case, the mother caught the newborn when its entire body had been delivered. In the six cases of births by chimpanzees in captivity observed by Matsubayashi (2002), the mothers all gave birth in a half-sitting position to drop babies onto the floor. While cases of females giving birth on the ground have been observed in Mahale Mountains National Park in Tanzania (Nishie and Nakamura 2017), chimpanzees in the wild more usually give birth in well-cushioned beds in trees (Goodall and Athumani 1980; Fujisawa and Matsuzawa 2012; Kiwede 2000). From this it can be inferred that dropping a newborn baby is not a dangerous act (Matsubayashi 2002; Hirata et al. 2011).[11]

This brings us to an understanding of unassisted labor among humans, when the mother gives birth alone. If a woman squats with both hands placed on some object, or else with both hands placed on the ground and her legs apart, she will be able to widen her pelvis and focus all her strength, which facilitates delivery (Kitagawa and Uchiyama 2013; Wells et al. 2012). In the past, unassisted labor was experienced in a similar position, with the newborn being delivered onto a soft object. At such times, the woman would kneel with her heel against her anus to prevent perineal tears (Matsuoka 1991, 2014; Nishikawa 1997). This may also be a technique to reduce the speed at which the newborn baby emerges, allowing a safer delivery.

Interestingly, then, similarities to humans have been observed in (what can be likened to) midwifery care among bonobos and in unattended childbirth and delivery with chimpanzees. For now, I just want to confirm what we could say from the latter. Based on ethnographic data, it is clear that the practice of both assisted childbirth and unattended labor have a long and continuous history among humans, and these observations provide further corroboration of this fact. While I have absolutely no intention of recommending unattended labor, this alternative history of childbirth is important for affirming women's ability to give birth—that is, their ability to cope with the extreme form of childbirth that developed in humans. Nor should we forget the existence of supporting attitudes and simple but effective birthing techniques.

Birthing pangs as a shared primate heritage

Similarities and continuities may also be observed between humans and chimpanzees in terms of birthing pangs. The term 'difficult births' has three senses; namely, the pain associated with childbirth, the extended duration of childbirth, and the frequency of problems resulting from childbirth. The pain of childbirth varies considerably from person to person, although has been described by one woman as 'painful, like cracking my hip bones' (Sakamoto 2011).[12] Since labor pains are mainly caused by uterine contractions, the pains during childbirth are experienced in common with other primates (Lefebvre and Carli 1985).[13] Labor pains on the part of chimpanzees can be recognized by the parturient female's restless posture changes, bodily contortions, and clenching of teeth (Matsubayashi 2002). Goodall and Athumani (1980: 546) vividly describe the distress exhibited by a multiparous female named Winkle during childbirth, as follows:

> There appeared to be 25 distinct 'bouts' of contractions. During a bout *Winkle* bore down between two and six times in succession (total 110 times, average median per bout, four times). In between these bouts she rested (for half a minute to 8 min, average median 1.5 min). Usually, *Winkle* grunted loudly as she bore down. Her face was contorted, eyes tightly closed, brow wrinkled, lips fully retracted from partially clenched teeth. Often her arms, legs, and whole body trembled with effort.

Given that this pain was enough to make even a chimpanzee—an animal with a reputation for a high pain threshold—squirm, we can imagine that this was significant. However, since other reports (e.g., Fujisawa and Matsuzawa 2012; Nishie and Nakamura 2017) found no evident signs of distress, there is presumably a broad spectrum of individual differences. A report by Kuze and Igarashi (2014) describes a chimpanzee pelvis that showed traces of what appeared to be a chip fracture sustained while giving birth, so it is possible that chimpanzees may feel the additional pain of the partial detachment of the sacroiliac ligament and pubic symphysis when the birth canal expands, as in humans. Matsubayashi (2002) also reports labor pains among Japanese macaques.

Next let us compare the time spent in childbirth by chimpanzees with that used for humans. In humans, the first stage of labor starts when contractions begin to occur regularly at a rate of about every ten minutes, or more than six times an hour. The second stage consists of the passage through the birth canal and the delivery, from the time the cervix is fully dilated until the crown of the baby's head becomes visible. Finally, the third stage is the postpartum period, when the placenta is delivered. Between the first and second stages, the descending fetal head breaks the amnion, releasing the amniotic fluid. The delivery time is up to twice as long for a primiparous (first-time) mother as for a multiparous mother, who has given birth before (Table 18.1).

Among chimpanzees, the first stage is defined as the time when labor pains are clearly observed, but because it is quite possible that the chimpanzee may have already been feeling labor pains before that, this first stage may be longer. The second stage lasts from the time the water breaks to the time of delivery, so the comparison seems more or less valid. So, while the first stage seems to be shorter for chimpanzees, the second stage seems consistent with that of multiparous human women. Fujisawa and Matsuzawa (2012) report that a pregnant female chimpanzee in the wild gave birth in 'the blink of an eye,' about twenty minutes after making up her bed. Since neither labor pains nor water breaking were observed in this case, we cannot know whether the first and second stage combined lasted a minimum or maximum of twenty minutes, or whether the second stage alone lasted twenty minutes at maximum. However, even in the human case, if we consider only cephalic-presentation (normal-presentation) deliveries, the second stage generally lasts at maximum one hour for primiparous mothers and less than fifteen minutes for multiparous mothers (Kitagawa and Uchiyama 2013). Let me quote some examples of short delivery, as reported by Sakamoto (2011):

Parturition category	1st stage	2nd stage	3rd stage	Reference
Primiparous humans	10–12 hours	1–2 hours	15–30 minutes	Kitagawa and Uchiyama 2013
Multiparous humans	5–6 hours	0.5–1 hour	10–20 minutes	Kitagawa and Uchiyama 2013
Chimpanzees	3.5–7 hours*	0.5–1 hour	no data	Matsubayashi 2002
Primiparous chimpanzee	no data	41 minutes	17 minutes	Kiwede 2000
Multiparous chimpanzee	52 minutes	1 hour	7 minutes	Goodall and Athumani 1980

Table 18.1: Comparison of parturition times

*N = 6; Time data for the first stage were estimated by subtracting 0.5–1 hour in the second stage from the total of 4–8 hours for the first and second stages together.

Case 1 (3rd child, birth weight 2932g): Labor pains started in earnest at 23:28. The cervix was fully dilated at 23:33. The baby was born at 23:56. The first and second stages combined lasted a total of 28 minutes. The second stage lasted 23 minutes.

Case 2 (2nd child, birth weight 3236g): Just after 8:00 in the morning, the mother was painful in every position. Sitz bath (to relieve pain) for about 30 minutes. At 11:31, the cervix was dilated 9 cm. Soon, the head emerged. The baby was born at 11:47. From just after 8:00, the first stage lasted 3 hours and 30 minutes, and the second stage lasted 16 minutes.

Out of 100 chimpanzee pregnancies observed at Sanwa Kagaku Kenkyusho (a Japanese pharmaceuticals company), the single CPD-related death was a stillbirth after a difficult birth resulting from a narrow pelvis. In addition, since there was another case in which a caesarean section was performed in the expectation that the mother's narrow pelvis would result in a difficult birth (Udono et al. 1999), there is a strong likelihood that the neonatal mortality would have been 2%. As noted earlier (in 'The obstetric dilemma: childbirth brought to the extreme'), neonatal mortality occurs frequently in smaller primates.

Even if the dataset is too small to be compared meaningfully with humans, we can nevertheless put a question mark over our tendency to think that birthing among apes is easier and shorter than is the case among humans. Nor can we confidently assert that humans have difficult births. More importantly, we find that the pain of childbirth did not begin with the obstetric dilemma, but predates even the ape stage, suggesting that the physiological system that suppresses the pain and fear of childbirth may have evolved alongside it.

Midwifery care and medical treatment for crises in childbirth

In humans, the gestation period ranges from thirty-seven to forty-two weeks. The period from twenty-two weeks to less than seven days after birth is called the perinatal period, and the period from six to eight weeks after delivery until the mother recovers is called the puerperium. During the perinatal period and the puerperium, the life of the mother and fetus (and then newborn) is likely to be threatened, and this risk is highest during the second stage of labor when the fetus passes through the birth canal.[14] According to one midwife in Hungary, 85% of childbirths do not require special care (Matsuoka 2014). However, since some of the remaining cases may result in the cessation of labor (a failure to progress), fetal paralysis, or major hemorrhaging in the expectant mother, these must not be underestimated. In addition to CPD, other possible risks include improper fetal position or posture that would prevent the fetus from entering the birth canal, and the length of the umbilical cord,[15] as well as the improper positioning of the placenta or its adhesion (placenta accreta) or premature separation.[16] The position and posture of the fetus can sometimes be corrected by waiting patiently or by the skill of the midwife, but most cases today are likely to be addressed by medical intervention with a caesarean section.

Loudon's (1986) analysis of maternal mortality in modern Britain cites two records of home births handled by the Westminster General Dispensary (dated 1781, N = 1897 and 1822, N = 687) as the first reliable statistics. Based on descriptions left by midwives, who were subject to strict rules, he estimated that 97%–97.5% of these births were unattended.[17] Furthermore, since these births took place in miserable conditions and were not elective, he deemed that this value would correspond to the expected value for the rate at which mother and child would have survived unattended labor, which can be used as an indicator to determine if a certain method of birth care is good or not. However, this was the maternal and child survival rate at the time of childbirth, and as such did not take into account subsequent deaths from puerperal fever or other postpartum illnesses. In situations where conditions were worse than in the slums of eighteenth- and nineteenth-century London, childbirth would be yet more dangerous. Until the establishment of methods of disinfection, postpartum fever was more dangerous than childbirth itself, so childbirth-related deaths would have been far higher than estimates from the value.

One example of the dangers of postpartum illness involves ancient Andean sites in northern Chile. Analyzing the remains of 128 women estimated to have died at the age of twelve to forty-five years old, Bernardo Arriaza and his colleagues determined that 14% of these women had died of childbirth-related causes and 83% of them died during the puerperium (Arriaza et al. 1988). Dating them by culture, it was found that the maternal mortality rate between 1300 BC and 600 AD was around 25%, dropping thereafter to between 8.8% and 4.7% until 1400 AD—this latter range being roughly consistent with the maternal mortality rate noted by a Spanish chronicler during the Indian-Spanish contact period (Arriaza et al. 1988). Two bodies were found to have died in childbirth, both belonging to women who were small in stature. A speculative explanation for the high mortality rate in the early period is that this may have been due to unbalanced nutrition resulting from farming, which led to shorter stature and more difficult births (Wells et al. 2012). Arriaza et al. (1988) also mention unattended labor as having been observed until the modern period among the peoples inhabiting the area near the ancient sites, as well as midwifery techniques performed by shamans in the case of difficult or abnormal childbirths. Such shamans would uncover a fetish called a Conopa (a piece of wood or pebble used in rituals) and place it on the woman's belly. The ritual setting reportedly 'produced a mystic, hypnotic state that not only provided more security to the mother in confinement or labor, but in some instances may also have resulted in muscular relaxation that contributed to correction of fetal malposition' (Arriaza et al. 1988: 39).

To get a true sense of at least some of the horrors of high maternal mortality, it may be necessary to consider the situation from the standpoint of those who faced it. In such times, high infant mortality was compounded by a high birth rate, so women repeatedly faced the cumulative fear of the probability of death in childbirth and the grief of losing a child. Leavitt (1986) cites the experience of a woman in eighteenth-century America who gave birth to twelve children, one every year or two between the ages of twenty-one and forty, of whom nine were lost. Leavitt (1986) states that networks of female companions supported women in such extreme situations of repeated suffering and grief.

Turning to Japan, the maternal mortality rate was 409.8 per 100,000 (0.4%) births in 1899, when statistics first began to be collected (NIPSSR 2021). In 2019, the rate had dropped to 3.3 per 100,000 births. No one would deny that this is due in large part to improvements in nutrition and hygiene, as well as the advent of modern midwifery and the contribution of medicine. However, in the eighteenth and nineteenth centuries in Europe and the United States, when male physicians began to intervene despite the fact that modern medicine had neither sufficient knowledge or proficiency in childbirth, nor consideration for the mother, the mortality rate of mothers and newborns actually increased, replacing the causes of maternal death by various diseases. Furthermore, even when the site of childbirth changed from the home to the hospital in the nineteenth century, the maternal mortality rate still increased to around 10% because medical personnel remained in their normal clothes and, in the most appalling cases, treated pregnant women in overcrowded settings directly after having performed pathological autopsies of those who died of puerperal fever (Leavitt 1986; Loudon 1986; Sewell 1993).

Michel Odent (2011), a French obstetrician who has been a pioneer in the field of natural childbirth, says that the most important thing is to create an environment that brings peace of mind to the mother, which requires a warm room and the supportive presence of someone who has experienced giving birth and whom the mother trusts. When a woman is anxious, adrenaline causes her muscles to stiffen. In some cases, the secretion of relaxin, which relaxes the pelvic ligament and the muscles of the soft birth canal, and of oxytocin, which promotes labor, can stop, which may lead to the cessation of labor (failure to progress). In addition, the embarrassment of the expectant mother in the presence of male obstetricians may likewise trigger birthing difficulties (Leavitt 1986; Matsuoka 1991, 2014; Odent 2011).

Fujie Sakamoto, who assisted with 4000 childbirths, says that carefree people tend to have easier births, while fearful and calculating people tend to have longer childbirths. She reassures expectant mothers by saying that things will be okay and encourages them to trust in the strength of newborns; even if the second stage of labor is a little prolonged, it is because the baby being born has chosen it. Midwives in England, Germany, and the Netherlands interviewed by Matsuoka (1991) also wait patiently for the second stage of labor to progress. Sakamoto (2011) says that if there is no problem with either the fetus or the mother, but the delivery is not progressing, then the posture and position of the fetus as it enters the birth canal is not appropriate, and she prompts the mother to help resume the delivery process. She asks the mother to lie down and raise her hips high, and then, after five contractions (about twenty minutes), corrects the woman's posture once the fetal head is separated from the pelvis. She says that this allows the delivery to proceed quickly.

Matsuoka (1991) reports watching a video of a birth scene in the Netherlands and being surprised to see that the midwife did nothing. It was explained that this was necessary to reinforce the mother's awareness that she had been the one to give birth, and that this recognition would give her confidence in raising the child. Midwives in Japan are more proactive about help. The autobiography of Sumi Nagasawa (1995), a midwife who assisted in more than 10,000 deliveries before and after the Second World War, details how she would relax the mother while reassuring those around her and getting them to help, such as by instructing them to heat the room and boil water. She would lead the birthing mother's breathing, protect the perineum, catch the newborn and promote the start of its respiration, then bathe the baby, deal with the umbilical cord, tend to any bleeding on the part of the mother, take care of the afterbirth, ensure cleanliness, and then ask the family to take care of the mother and her child. While most of these would seem to be things that could be done without one's needing to be a midwife, Nagasawa also relates her experiences of racing against time—using her fingers to peel away and remove an adherent placenta, of stopping sudden and massive hemorrhaging, of delivering babies with breech presentations, of teaching herself how to return a breech baby to a head-first presentation from outside the body, and uncoiling an umbilical cord. This wealth of experience and skill, combined with her friendly personality, could only have put expectant mothers at ease. On the other hand, when we read the accounts of women who have given birth alone, the reassurance of having someone they trust by their side may be enough for an expectant mother who is healthy and has a good environment.

Many doctors in the medical field, however, consider the second stage of labor to be a dangerous process, during which both mother and fetus are highly vulnerable. When the fetus continues to be compressed in the birth canal, its heart rate slows and a lack of oxygen places it into an asphyxia state. Hence, measures tend to be taken to ensure that the process finishes quickly. In several hospitals observed by Matsuoka (1991, 2014), a sequence of measures taken included the administration of stimulants, pushing on the mother's abdomen, using forceps to expand the birth canal and pull out the fetus, and using a vacuum to suck out the fetus. Intervening in childbirth too early without waiting patiently can produce tension in the expectant mother, disturbing her natural physiological response to childbirth, leading to a difficult birth that in turn necessitates more serious intervention. Consequently, such measures may be justified as measures dealing with difficult births (Matsuoka 1991, 2014).[18]

Buckley (2015) cautions that the effects of medical interventions on the hormone system of the expectant mother and fetus can be harmful. Odent (2011) warns that it is particularly dangerous to routinely give synthetic oxytocin to mothers, especially since studies have shown that synthetic oxytocin, a delivery stimulant, penetrates the placenta and reaches the fetus in concentrations dozens of times higher than are secreted by the fetus itself, and that exposure to high doses of oxytocin blocks receptor sensitivity (I discuss the effects of oxytocin more in the next section). Several large-scale follow-up studies found a significantly higher rate of delayed breastfeeding and early weaning within three months in newborns in the group of expectant mothers treated with oxytocin, and a significantly higher incidence of autism (Garcia-Fortea et al. 2014, Gregory et al. 2013). There are also reports on hospital deliveries in Canada, that cesarean section and instrumental labor increase the incidence of amniotic embolism, which is responsible for a quarter of all maternal deaths today (Kramer et al. 2006); further, questions about the lengths to which medical intervention should go and in which situations are far from settled. On the other hand, Odent (2011) cautions that people who choose to give birth at home or opt for natural childbirth do not always understand what this entails, and in some cases may increase the dangers of childbirth.

The significance of the narrow passage: a rereading of the obstetric dilemma to the labor hypothesis

From the foregoing, it does not necessarily seem inevitable that most pregnant women will suffer difficult births, unless they are quite young or old in terms of their maternal age, or obese or overweight, and as long as they follow precautions during pregnancy, continue to exercise moderately, and, at the time of childbirth, have a warm and clean environment in which they can adopt a comfortable posture with a dependable caregiver. However, so long as the fetus remains in the womb until it is almost too big to pass through the birth canal, it is inevitable that childbirth will be accompanied by some degree of difficulty and, in some few percent of cases, situations may occur that make vaginal childbirth difficult or impossible.

Veteran midwives, however, have a different idea. The important thing is not to say that the pain goes away, but to show sympathy for the mother's efforts to overcome the pain (Sakamoto 2011). This increases trust in both the caregiver and oneself and eases the childbirth itself. Also, as suggested by the expression in Japanese, '*hara wo itameta waga-ko*'—literally 'the child for whom I suffered pain in my belly', the travails of childbirth can foster women's attachment to their babies and a confidence in their ability to raise them, as well as a sense of general wellbeing. And also it can bring people together to watch over the suffering expectant mother (Matsuoka 2014; Nagasawa 1995). As for women's ability to give birth, from the fact that natural childbirth has been repeated time and again over 10,000 generations and over the 200,000 years since the emergence of anatomically modern humans, with no change in body structure, then as long as the conditions of nutrition and exercise are available, this ability should essentially be maintained even today.

So does the passage through the narrow birth canal have any significance for the newborn? To my knowledge, the only anthropologist to have considered this question is Ashley Montagu (1971), who hypothesized that the passage of the fetus through the birth canal would strengthen the bond between mother and child, on the basis that, in mammals, the bond between mother and child is formed when the mother licks and cleans the newborn. This hypothesis can be positively reinterpreted in light of subsequent research into the endocrine system as it relates to childbirth.

During childbirth, a variety of hormones with complementary or antagonistic effects are secreted by both mother and fetus to move the process of delivery forward (Buckley 2015). Oxytocin, a hormone secreted by both mother and fetus, and which increases rapidly during delivery, is particularly interesting. In addition to attachment formation and recognition of others, it also induces dopamine—which heightens arousal, pleasure, and happiness—and serotonin, which stimulates attachment to others. In childbirth, it induces powerful uterine contractions in the mother to promote delivery, encourages postnatal lactation, suppresses the mother's fear of her newborn, and stimulates attachment. It encourages newborns to suckle for milk, amplifying attachment to the mother, and furthermore protects the fetal cranial nervous system from intense stress responses during and immediately after delivery, suppresses pain sensation, and enhances immunity (Buckley 2015; Ben-Ari 2018; Odent 2011). Newborns secrete large amounts of stress hormones to maximize life-support functions in response to oxygen deficiency and pressure during labor, and to decreased body temperature and dehydration after birth. Oxytocin is what prevents these from damaging the brain, circulatory system, and immune system. Parturition—the delivery of the baby—is a key event in the development of these hormonal systems (Buckley 2015; Ben-Ari 2018).

Sakamoto (2011) asserts that the fact that a baby overcomes mortal suffering at the moment of birth instills the child with an absolute and unshakeable confidence with which to approach life's difficulties. I would agree with Sakamoto, though with the following caveat. The power of the fetus to overcome mortal suffering is undergirded by the hormonal system, which is the basis of the maintenance of life and of attachment to others, and which can be deployed even after growth. This is a power that has been acquired by humans as a result of evolution and, as such, is shared even by babies born by cesarean section.

I have long felt that among primates, humans have a particularly high stress tolerance. For example, gorillas suffer bouts of diarrhea just from being chased, and when they are captured, they can become dehydrated and die abruptly as a result. Japanese macaques and chimpanzees, while not so fragile, nevertheless lived shorter lives in captivity in the zoos of old than did their cousins in the wild. From 1956, when the Japan Monkey Center was established, until around 1960, many types of monkeys were brought to the zoo in cages. The monkeys were immediately moved to large cages and supplied with nutritious food, but within a month, many had dropped dead of organ failure or dehydration due to diarrhea.[19] Of course, many humans would have perished in the same situation, but monkeys and apes do seem to be much weaker in comparison to humans, who have endured slave ships and small, dark cells in more than a few instances.

The Japanese alpinist Kenji Kondō distinguishes between normal physical fitness, which he calls *kōdō tairyoku* (action stamina), and *bōei tairyoku* (defense stamina; namely, a resistance to physical, physiological, mental, and biological stresses), emphasizing the greater importance of the latter in the context of mountaineering (Yama Hack 2020). This corresponds to a resistance to cold, humidity, hunger, thirst, and disease, as well as to fear. When humans are exposed to stress, they deplete their store of vitamins and minerals and lose the ability to produce energy, which results in a loss of endurance and strength. Kondō asserts that the ability to prevent this from happening will also prevent people from going into distress. Oxytocin is thought to be involved in this, since it has the effect of ameliorating fear and depression and of increasing trust in others (which in turn leads to optimism and hope). It also suppresses excessive stress responses and moderates emotional reactions (Ben-Ari 2018; Buckley 2015; Neumann and Landgraf 2012).

Of course, not everyone's defenses can be strong at all times. Some might die of panic after getting lost in the woods,[20] while others have died of grief over the loss of a long-time companion. However, it is not uncommon for people to be able to withstand hunger, thirst, and cold, to be undeterred by illness, to endure loneliness while thinking of their lost loved ones, and to survive with a bright future ahead of them. I think there can be no doubt that humans as a whole have a higher stress tolerance than other mammals. Were this not the case, there could not have been so many slaves toiling in harsh environments to support the sweep of history from behind the scenes, so many warriors who devoted themselves to battle, or even so many workers performing what in the West is called 'labor' (from the Latin *labōrem,* meaning 'toil, distress, trouble'). Such tolerance can, of course, be improved by training, and it can be strengthened by having goals, hopes, devotion to one's fellows, or faith. However, that this is possible is surely due to the fact that we have a strong foundation of stress tolerance. Just as strong drinkers become alcoholics, thrill-seeking dependencies and stress addictions—often seen in soldiers, combat fanatics, martial artists, and celebrities—emerge when people go in search of further stress, precisely because there is an existing foundation of high stress tolerance. From the foregoing discussion, we can conclude that humans have a high stress tolerance and advance the hypothesis that this is a capacity that our species has acquired by dint of our repeated passage along a narrow path—that is, the extreme of birth.[21] In other words, we have increased our survival and social power by feeding on the extreme of birth: we are extremivora, all.

Finally, let us return to Washburn (1960). The 'obstetric dilemma' was solved by having the newborn emerge from childbirth in the fetal state, which in turn led to the emergence of the family. To this, we can now offer the following addendum. The resolution of the dilemma also represented the birth of children who were attached to their families and fellows, who were able to endure dangerous and arduous work, and who were capable of bringing home heavy game for the benefit of all. Although the act of bringing food back to camp from hunting and gathering excursions had existed since long before, the obstetric dilemma engendered people who engaged in these pursuits more deliberately and with joy. Thus the curse of childbirth labor imposed by God is transformed into a benediction that allows women to bear children and to endure labor and flourish, with bonds of affection and hope. In the end, more than either cognitive intelligence or the ability to wield weapons, it may be that this capacity, which has enriched human society in many ways, represents the sturdier evolutionary foundation.

Notes

1. Many of the most recently developed species of domestic livestock have higher rates of difficult births than humans. See note 18.
2. 'The midwives answered Pharaoh, "The Hebrew women are not like the Egyptian women. They are robust and give birth before the midwife arrives"' (Exodus 1:19). It should be noted that this was an excuse to disobey an order given by the Pharaoh that if a Hebrew woman were to bear a male child that the midwives should kill the baby on the spot.
3. A scheduled cesarean section takes about twenty minutes, has a good prognosis, and is less dangerous (Odent 2011).

4. Against this trend, the average weight of newborns in Japan has gradually decreased from 3.2 kilograms in 1980 to 3.0 kilograms in 2010, despite an increase in the rate of cesarean births to more than 20%. This regressive phenomenon is most likely due to Japanese women's pursuit of thinness as an aesthetic ideal (Choo et al. 2014). Moreover, obstetricians and the Ministry of Health, Labor and Welfare have warned that mothers who are too thin or too short are also susceptible to CPD, and that underweight infants born to thin mothers tend to rebound to obesity after birth, going on to develop lifestyle-related diseases in the future due to the fact of their adaptation to starvation in utero.
5. The survival rate of babies born by caesarean section in an extremely premature state of less than 1000 grams is about 90% in Japan, and about 50% for babies weighing less than 300 grams (BBC News 2019).
6. At the Second International Summit on Human Genome Editing (Hong Kong, November 2018), He Jiankui of China announced the birth of twins from genome-edited human fertilized eggs. The Japanese Society of Gene and Cell Therapy, the Japan Society of Human Genetics, the Japan Society of Obstetrics and Gynecology, and the Japan Society for Reproductive Medicine (2018), which have consistently taken the position that the clinical application of genome editing using human fertilized eggs should be prohibited, issued a new statement to that effect on December 4, 2018, in response to He's announcement.
7. The fertilized eggs were discarded after the experiment (Blakemore 2017).
8. The intent underlying such a subtle gesture is difficult to guess. My own idea is that she may have stopped the mother from pulling carelessly.
9. Chimpanzee females usually give birth alone, avoiding other individuals. This is associated with the prevention of infanticide by other individuals (Nishie and Nakamura 2017). Infanticide is unknown among bonobos. Females act in solidarity to repel male attacks, so infanticide by males is unlikely to take place.
10. This type of birth among chimpanzees was previously reported by Nissen and Yerkes in 1943 (Hirata et al. 2011). Accordingly, the observations made by Hirata's team were not unique.
11. We cannot conclude at this stage that 100% of the births are drop-births. However, since safety will differ depending on the posture of the newborn when it is pulled out, as well as the manner and timing of its withdrawal, data will need to be accumulated by further detailed observation.
12. Fujihara and Tsukisawa (2015) include a table by A. P. Nikolaev, summarized by Seizō Nagae, entitled 'Mechanisms of childbirth pain,' which states that 'a subset of parturients (7–14%) feel absolutely no pain during childbirth.' Nikolaev was a Soviet obstetrician who promoted painless childbirth, which became the basis of the Lamaze method (Fujihara and Tsukisawa 2015).
13. In their summary of reports of primate births, Lefebvre and Carli (1985) found that sixty-nine out of eighty-eight primates from twenty-nine species were recorded as appearing to suffer to some degree while giving birth. It might well be argued that it is meaningless to report the extent of pain and suffering, which are subjective sensations, felt by other people or other species. Here, however, I abide by the idea that it is possible to know at least whether other people or species that are close to us are in moderate or severe pain.

14. Pregnancy- and childbirth-related deaths include deaths attributable to pregnancy, deaths directly related to childbirth, and deaths due to postpartum causes such as infections or poor prognosis. Causes of death and mortality rates vary depending on retrospective determinations, the distinction drawn between the perinatal period and the puerperium, and whether to include stillbirths in childbirth statistics. There is, moreover, a history of such definitions changing over time. However, any minor changes in mortality rates caused by such differences do not present a problem for the discussion in this chapter, so statistics found in the literature for both maternal deaths (i.e., deaths during pregnancy, childbirth, and the puerperium) and childbirth-related deaths (i.e., deaths during childbirth and the puerperium) are cited without adjustment.
15. When the umbilical cord is too short, delivery may fail to progress or the placenta may be torn off prematurely, resulting in heavy bleeding. When it is too long, there is a risk of it becoming entangled with the fetus and causing suffocation during delivery.
16. The placenta is an organ that forms during pregnancy and attaches to the inner wall of the uterus. It nurtures the fetus by sending nutrients and oxygen from the mother to the fetus via the umbilical cord and carrying carbon dioxide and other waste products from the fetus back to the mother's body. Normally, the placenta detaches spontaneously from the uterus shortly after delivery and is expelled (as the 'afterbirth'). However, if detachment starts to take place before the placenta has fully separated from the uterus, this can cause heavy bleeding. In some cases, the placenta forms in a position that blocks the birth canal opening, which makes vaginal delivery impossible.
17. Judith Goldsmith (1984) quotes a statement by Grantly Dick-Read (1955) that 94%–98% of women in all of the ethnic groups he surveyed in Africa in 1953 had normal, natural childbirths and recovered after childbirth with only a few hours of rest, if they needed any at all. Note that this value is more or less consistent with that of Loudon's (1986) value for unattended births.
18. Difficult births in breeds of cattle and horses developed in recent years offer many hints for humans as well. The percentage of Holstein cows that cannot deliver a calf without human assistance is 10%–15% for cows' initial calving, which drops to 3%–5% for subsequent calving. Moreover, the mental stress arising from human monitoring and invasive assistance not only interferes with normal calving, but also causes difficult births when humans assist too early (Ishii 2014).
19. These examples are based on pathological autopsy records left by Sugio Hayama, at that time an assistant at the center. It is an astonishing record, even for a period when little was known about the mental health of primates. Hayama once remarked to me that the way primates were imported to and transported in Japan in the early days of the Monkey Center was not bad by the standards of the 1960s. He told me how he worked tirelessly to investigate the causes of the problems by performing autopsies, attempting improvements to rearing techniques, and keeping these records for subsequent research. His record constitutes a valuable resource that should be put to use in the fields of stress research and animal welfare.

20. The mechanisms of sudden death due to fear or severe stress include multiple circuits, such as stress hormones (for example, cortisol and adrenaline) causing damage to the circulatory, nervous, and digestive systems or intracellular magnesium being excreted, resulting in calcium entry and thereby causing circulatory disturbances (Nishimuta 2021).
21. This hypothesis could be corroborated by examining stress tolerance among smaller primates, which often have difficult births.

References

Arriaza, B., M. Allison and E. Gerszten (1988) 'Maternal mortality in pre-Columbian Indians of Arica, Chile'. *American Journal of Physical Anthropology*, 77: 35–41.

BBC News (2019, February 27) '"Tiniest baby boy" ever sent home leaves Tokyo hospital'. BBC, https://www.bbc.com/news/world-asia-47382058 (accessed January 27, 2022).

Ben-Ari, Y. (2018) 'Oxytocin and vasopressin, and the GABA developmental shift during labor and birth: Friends or foes?' *Frontiers in Cellular Neuroscience*, 12: 254.

Betti, L. (2017) 'Human variation in pelvic shape and the effects of climate and past population history'. *Anatomical Record*, 300: 687–697.

Blakemore, E. (2017, 2 August) 'First human embryos "edited" in U.S.: Get the facts'. National Geographic, http://news.nationalgeographic.com/2017/08/human-embryos-gene-editing-crispr-us-health-science (accessed January 11, 2022).

Boerma, T., C. Ronsmans, D. Y. Melesse, A. J. Barros, F. C. Barros, L. Juan, A.-B. Moller, L. Say, A. Hosseinpoor, M. Yi, D. L. R. Neto and M. Temmerman (2018) 'The global epidemiology of caesarean sections: Major increases and wide disparities'. *The Lancet Journal*, 392(10155): 1341–1348.

Buckley, S. J. (2015) *Hormonal Physiology of Childbearing: Evidence and Implications for Women, Babies, and Maternity Care*. Childbirth Connections Programs, National Partnership for Women & Families.

Choo, Dong-Mei, N. Sakamoto, N. Arata and Y. Oya (2014) 'Epidemiological study on maternal factors in low birthweight infants'. *Indicators of Health* 61(1): 1–8.

Demuru, E., P. F. Ferrari and E. Palagi (2018) 'Is birth attendance a uniquely human feature? New evidence suggests that Bonobo females protect and support the parturient'. *Evolution of Human Behavior*, 39(5): 502–510.

Dick-Read, G. (1955) *No Time for Fear*. Harper.

Douglas, P. H. (2014) 'Female sociality during the daytime birth of a wild bonobo at Luikotale, Democratic Republic of Congo'. *Primates*, 55(4): 533–542.

Engelmann, G. J. (1884) *Labor Among Primitive Peoples*. JH Chambers & Co. (available online,https://archive.org/details/laboramongprimi01engegoog; accessed January 13, 2022).

Fujihara, S. and M. Tsukisawa (2015) 'Seishin yobōsei mutsū bunben-hō o meguru sanfujinkai-tachi no ronsō: 1954–57-nen no sanfujinka gakujutsu zasshi-jō de tenkai sareta giron no bunseki to kentō' (An analysis of the controversy on the psychoprophylaxis method in the Japanese Medical Journal *Obstetrics and Gynecology*, 1954–1957). *Nihon ishigaku zasshi* (Journal of the Japanese Society for the History of Medicine), 61(2): 131–144.

Fujisawa, M. and T. Matsuzawa (2012) 'Yasei chinpanjī no shussan' (Chimpanzee birth in the wild). *Kagaku* (Science), 82(3).

Garcia-Fortea, P., G. E. Onzalez-Mesa, M. Blasco, O. Cazorla, M. Delgado-Rios and M. J. Gonzalez-Valenzuela (2014) 'Oxytocin administered during labor and breast-feeding: A retrospective cohort study'. *Maternal-Fetal & Neonatal Medicine*, 27(15): 1598–1603.

Goldsmith, J. (1984) *Childbirth Wisdom*. Boston Common Press.

Goodall, J. and J. Athumani (1980) 'An observed birth in a free-living chimpanzee (*Pan troglodytes schweinfurthii*) in Gombe National Park, Tanzania'. *Primates*, 21: 545–549.

Gregory, S. G., R. Anthopolos, C. E. Osgood, C. A. Grotegut and M. L. Miranda (2013) 'Association of autism with induced or augmented childbirth in North Carolina Birth Record (1990–1998) and education research (1997–2007) databases'. *JAMA Pediatrics*, 167(10): 959–966.

Hirata, S., K. Fuwa, K. Sugama, K. Kusunoki and H. Takeshita (2011) 'Mechanism of birth in chimpanzees: Humans are not unique among primates'. *Biology Letters*, 7(5): 686–688.

Ishii, M. (2014) 'Nyūgyū no nanzan: Gen'in, yobō, josan oyobi shitsu-i seifuku-hō' (Dystocia in dairy cattle: Causes, prevention, birth assistance, and dislocation recovery). *Kachiku shinryō* (Journal of livestock medicine), 61(3): 139–147.

Japanese Society of Gene and Cell Therapy, Japan Society of Human Genetics, Japan Society of Obstetrics and Gynecology, and Japan Society for Reproductive Medicine (2018, December 4) 'Statement by four relevant societies on the clinical application of genome editing of human fertilized eggs,' https://jshg.jp/news/2212/ (accessed May 13, 2022).

Kitagawa, M. and K. Uchiyama (eds.) (2013) *Kyō no josan: Mataniti saikuru no josan shindan jissen katei* (Midwifery today: Midwifery diagnosis and practice process in the maternity cycle). Nankodo.

Kiwede, Z. T. (2000) 'A live birth by a primiparous female chimpanzee at the Budongo Forest'. *Pan Africa News*, 7(2): 23–25.

Kramer, M. S., J. Rouleau, T. F. Baskett and K. S. Joseph (2006) 'Amniotic-fluid embolism and medical induction of labour: A retrospective, population-based cohort study'. *Lancet*, 368: 1444–8.

Kuze, N. and Y. Igarashi (2014) 'Ruijin'en ni mo ninshin shussankon ga aru no ka: Ruijin'en ni okeru kotsuban no jijōmen zenkō' (Do apes also have pregnancy birth marks?: Preauricular grooves on ape pelves). *Reichōrui kenkyū* (Primate research), Supplement 30(0): 58–59.

Leavitt, J. W. (1986) *Brought to Bed: Childbearing in America, 1750–1950*. Oxford University Press.

Lefebvre, L. and G. Carli (1985) 'Parturition in non-human primates: Pain and auditory concealment'. *Pain*, 21: 315–327.

Leutenneger, W. (1982) 'Encephalization and obstetrics in primates with particular reference to human evolution'. In E. Armstrong and D. Falk (eds), *Primate Brain Evolution: Methods and Concepts*. Plenum Press, 85–95.

Liston, W. A. (2003) 'Rising caesarean section rates: Can evolution and ecology explain some of the difficulties of modern childbirth?' *Journal of The Royal Society of Medicine*, 96: 559–561.

Loudon, I. (1986) 'Deaths in childbed from the eighteenth century to 1935'. *Medical History*, 30: 1–41.

Matsubayashi, K. (2002) 'Chinpanjī oyobi nihonzaru no shussan kōdō' (Childbirth behavior among chimpanzees and Japanese macaques). Oral presentation at the 7th Symposium of the Sectional Meeting on Human Evolution, Kyoto.

Matsuoka, E. (1991) *Shussan no bunka jinruigaku: Girei to sanba* (The cultural anthropology of childbirth: Rituals and midwives). Kaimeisha.

Matsuoka, E. (2014) *Ninshin to shussan no jinruigaku: Ripurodakushon o toinaosu* (The anthropology of pregnancy and childbirth: Rethinking reproduction). Sekai Shisosha.

Mitteroecker, P., S. M., Huttegger, B. Fischer and M. Pavlicev (2016) 'Cliff-edge model of obstetric selection in humans'. *Proceedings of National Academy of Sciences of the USA*, 113(51): 14680–14685.

Montagu, A. (1971) *Touching: The Human Significance of the Skin*. Columbia University Press.

National Academies of Sciences, Engineering, and Medicine (2019) *Second International Summit on Human Genome Editing: Continuing the Global Discussion: Proceedings of a Workshop in Brief*. The National Academies Press, https://doi.org/10.17226/25343

National Institute of Population and Social Security Research (NIPSSR) (2021) *Demographic Data Collection: Number and Rate of Maternal Deaths; 1899–2021*. NIPSSR.

Nagasawa, S. (1995) *Sanba no Osumi-chan ichidaiki* (Osumi-chan: The biography of a midwife). Soshisha.

Neumann, I. D. and R. Landgraf (2012) 'Balance of brain oxytocin and vasopressin: Implications for anxiety, depression, and social behaviors'. *Trends in Neurosciences*, 35(11): 649–59.

Nishie, H. and M. Nakamura (2017) 'A newborn infant chimpanzee snatched and cannibalized immediately after birth: Implications for "maternity leave" in wild chimpanzee'. *American Journal of Physical Anthropology*, 165(1): 194–199.

Nishikawa, M. (1997) *Aru kindai sanba no monogatari: Noto Takeshima Mii no katari yori* (The tale of a modern midwife: From the account of Mii Takeshima from Noto). Katsura Book.

Nishimuta, M. (2021) *Medical Note: Magnesium deficiency may be involved in heart disease due to stress overload*. https://medicalnote.jp/contents/170406-010-KM (Japanese URL accessed January 13, 2022).

Odent, M. (2011) *Childbirth in the Age of Plastics*. Pinter & Martin Ltd.

Portmann, A. (1956) *Zoologie und das neue Bild des Menschen. Biologische Fragmente zu einer Lehre vom Menschen* (Zoology and the new image of man. Biological fragments on a doctrine of man). Rowohlt.

Potter, J. E., E. Berquó, I. H. Perpétuo, O. F. Leal, K. Hopkins, M. R. Souza and M. C. Formiga (2001) 'Unwanted caesarean sections among public and private patients in Brazil: Prospective study'. *BMJ*, 323: 1155–1158.

Richmond, B. G., D. R. Begun and D. S. Strait (2001) 'Origin of human bipedalism: The knuckle-walking hypothesis revisited'. *American Journal of Physical Anthropology*, 44: 70–105.

Sakamoto, F. (2011) *Daijōbuya de* (It's okay). Sangyo Hensyu Center.

Sewell, J. E. (1993) *Cesarean Section: A Brief History*. National Library of Medicine.

Trevathan, W. (1988) 'Fetal emergence patterns in evolutionary perspective'. *American Anthropologist*, 90: 674–681.

Udono, T., M. Teramoto and I. Hayasaka (1999) 'Sanwa Kagaku Kenkyūsho ni okeru chinpanjī ninshin 100-rei no sōkatsu' (Summary of 100 chimpanzee pregnancies at Sanwa Kagaku Kenkyusho). *Reichōrui kenkyū* (Primate research), 15: 243–250.

Washburn, S. L. (1960) 'Tools and human evolution'. *Scientific American*, 203: 3–15.

Wells, J. C. K., J. M. DeSilva and J. T. Stock (2012) 'The obstetric dilemma: An ancient game of Russian roulette, or a variable dilemma sensitive to ecology?' *Yearbook of Physical Anthropology*, 55: 40–71.

WHO (World Health Organization) (2012) *Trends in Maternal Mortality: 1990 to 2010*. WHO, UNICEF, UNFPA and The World Bank Estimates.

Yama Hack (2020) *Thinking about Safety for the High Season of Mountain Climbing - From the 3rd National "Mountain Day" Forum*. https://yamahack.com/2547 (Japanese URL accessed January 13, 2022).

CHAPTER 19

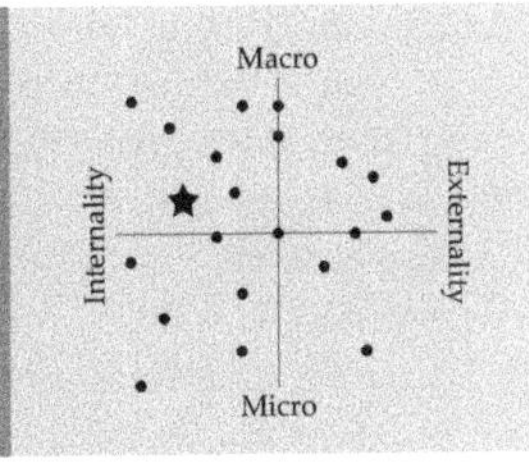

A Distinctively Human Aspect of Cooperative Food Preparation and Survival:

When Food Became a Community Resource

Yuko SUGIYAMA

Keywords: food, food processing and cooking, cooperation, cooperative work, semi-natural environment, food sharing, women, children, survival

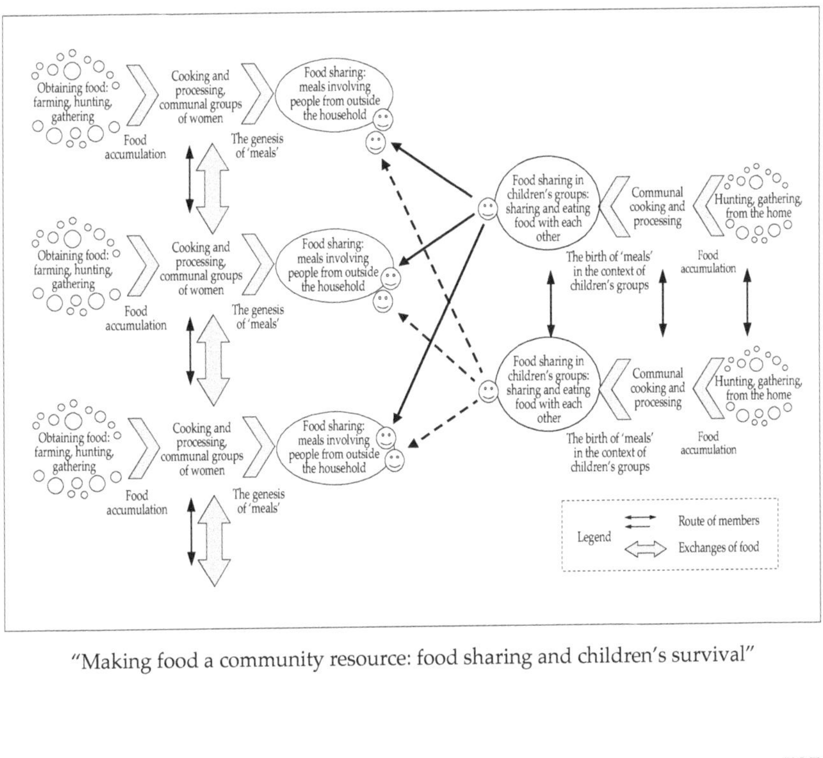

"Making food a community resource: food sharing and children's survival"

Human food systems are distinguished by the length of time between when food is harvested or hunted in the environment and when it is eaten, which is longer than among apes. During this time, meat is butchered, plant foods are processed, and both types of food are cooked—a series of events that involves the collection and distribution of foods acquired by multiple individuals. Food processing and cooking transforms raw materials into meals, producing social meaning and laying the groundwork for food sharing. Children, who are especially vulnerable group members, form multi-age groups with other children to eat together while also joining in other food-sharing groups at will, increasing their chances of surviving difficult conditions created by environmental changes.

The transportation and distribution of food laid the foundation for the development of distinctive human features such as bipedal locomotion and cooperative work. Likewise, the 'culinary leap' to using fire is believed to have encouraged human evolution. Having developed in this way, human food systems are distinguished by the length of time between when food is harvested or hunted in the environment and when it is eaten, which is longer than among apes. During this time, meat is butchered, plant foods are processed, and both types of food are cooked—a series of events that involves the collection and distribution of foods acquired by multiple individuals. Food processing and cooking changes raw materials into meals, producing social meaning and laying the groundwork for food sharing. Indeed, as Richard Wrangham (2009: 123) writes, quoting Catherine Perlès, 'cooking ends individual self-sufficiency.' In this way, food is transformed into a substance with great social meaning.

This chapter considers the 'cooking hypothesis,' which posits that cooking stimulated hominization, and builds on Kazuhiro Uzawa's (2012) discussion of the role of fire from an archaeological perspective to focus on the work of processing and cooking food (a subject that has thus far received little attention), as well as on gatherings of people involved in that work. This leads to an examination of environmental and food-habit change (i.e., the transition to cereal foods) and the nature of cooperative work, as well as the accumulation of human behavior patterns.

Ethnographers describe striking gatherings of women during the food processing stages, before cooking over fire, among both the Bemba, a swidden farming people, and the San, a hunter-gatherer people (Imamura 2010; Sugiyama 1987a). Especially in farming societies where grains are the staple food, more time and effort are spent on processing tasks such as threshing and refining grains and detoxifying cassava than on farming itself, and this work occupies the majority of occasions on which women gather together. I focus on the way in which this weakens the association between food and its original owner, easing the way for it to be distributed to and eaten by those who helped process it. I also note that cooperative food processing widens the variety of foods eaten and increases opportunities for children to obtain the food they need.

I am assuming here that the function of cooking in making food available to the wider community is a shared feature of human societies, regardless of differences in modes of subsistence. My particular focus is on hunter-gatherer and swidden farming societies that have adopted sedentary settlement patterns.

The adoption of a sedentary life causes humans to exert more constant pressure on the environment and deliberately change it. This happens because people view these changes as making the environment better suited to the particular lifestyle that settlement entails. However, lifestyles molded to a particular environment do not adapt easily to large changes in environmental conditions caused by climate change or other factors, and ultimately risk leading to extreme situations such as population decline or the disappearance of the group. What do cooperative food preparation and the survival of individuals within the group look like when a particular lifestyle emerges, and when it is placed at risk by environmental change?

In the following sections, I review the research on fire and food in human history and present a framework for considering food and environment from the perspective of cooking. I then examine how people incorporated cooking into their lives during the more than 20,000 years after food habits shifted to cereals and other starchy foods. I use the example of a contemporary African farming village to discuss foods acquired from a semi-natural environment that is shaped by agricultural modes of subsistence, as well as cooperative cooking and survival.

The cooking hypothesis, changing food habits, and the environment

The use of fire and human food systems

Richard Wrangham (2009) argues that humans began using fire far earlier than scholars previously believed, proposing the 'cooking hypothesis' that cooking was a crucial factor in hominization. According to this hypothesis, humans began eating meals cooked with fire when *Homo erectus* emerged between 1.8 and 1.9 million years ago, and this spurred the evolution of *Homo erectus*. Wrangham argues that cooking with fire increased the nutritional value of food and made it easier to digest, causing the masticatory organ and digestive tract to shrink and increasing energy supply to the brain due to an increase in meat consumption and other factors. In addition, because cooking enabled the detoxification of foods that are poisonous when raw, it expanded the range of edible foods. As Toshisada Nishida (2007) points out, Wrangham's linkage of cooking with fire to the emergence of male–female pairs and a decrease in sexual dimorphism in humans is difficult to accept. However, it is important to recognize that if food is raw, the sexual division of labor is unworkable (Wrangham 2009: 112), and that the division of labor in food processing and cooking ultimately led to the sharing of food with others.

Uzawa (2012) evaluates Wrangham's cooking hypothesis from an archaeological perspective, surveying the connection between fire and food at various stages of human evolution. Even taking into account the lack of a scholarly consensus as to when humans began using fire (based on the archaeological record), Uzawa cautions skepticism, given the apparent gap of almost 300,000 years between archaeological evidence and the timing assumed by the cooking hypothesis.[1] Nevertheless, he notes that the archaeological evidence suggests that archanthropine or early paleoanthropine humans with the skills to control fire dispersed widely and adapted to the natural environments they ended up in, and that the use of fire and cooking likely expanded food habits, allowing humans to disperse to diverse environments, including cold regions.

The earliest clear archaeological evidence of the use of fire, discovered at the Wonderwerk Cave in South Africa, dates back one million years, but the earliest discovery believed to show humans starting fires themselves instead of relying on natural phenomena is from the 790,000-year-old Daughters of Jacob Bridge (Gesher Bnot Ya'akov) site in Israel. The breakthrough allowing full control of fire occurred 500,000 years ago, when archanthropine humans emerged. The remains of what can be clearly identified as fireplaces appear 250,000 years later in the middle paleolithic age (at the Hayom site in Israel and Sodmein Cave in Egypt), and based on 70,000-year-old evidence at Blombos Cave in South Africa, which includes engraved stones and shell beads, as well as similar finds, 'it can be assumed that various features of modern humans existed, including the division of labor between those who acquired food and those who stayed in the dwelling to tend the fire (most likely men and women respectively), societal cooperation, and the distribution of resources' (Uzawa 2012: 54). A look at modern humans shows that many adaptations to climate change and shifts in food habits have since occurred. In particular, after grains became the major plant staples 10,000 years ago and sedentary settlement patterns were adopted, the human impact on the environment grew, leading to the coevolution of crops and cooking, according to Uzawa.

Thus, as cooking was incorporated into the steps carried out before eating, individuals came to depend heavily on others for their own survival. In that sense, it is accurate to say that food became a community resource long before the advent of agriculture. I propose that the repetition of daily activities under these conditions increased people's ability to cooperate with and recognize the intentions of others, forging social systems in which altruism served selfish ends. However, as Uzawa points out, the period during which these trends accelerated overlaps with the period when cereal grains and other starchy seeds became the main staple foods. This occurred during a period of climate and other environmental change, as well as increasing human pressure on the environment and increasing population density.

Food diversification and the shift to cereal foods as major plant staples

Environmental transformation caused by climate change and other factors led to changes in human food habits and spurred the development of tools for processing and cooking food. Mithen (2011) examines human evolution and environmental change caused by climate change beginning around 20,000 BCE.

As the planet warmed following the Last Glacial Period, the plants in the forest steppes of western Asia became more diverse, and the diets of hunter-gatherers in this region came to include wild-type barley, peas, lentils, flax, and similar wild plants.

By 14,500 BCE, the flora and fauna had become even more diverse, and around 15,000 years ago, oak, pistachio, and almond trees began to appear in woodlands. In the steppes, colonies of plants with edible rhizomes and wild-type barley, emmer, oats, goat-grass, and rye appeared, and eventually, as nutrition improved, the human population grew, expanding its living area to new woodlands and steppes. Even in the cold highlands, hunting flourished and the shapes and production methods of microliths changed.

Signs appeared that the grains used by humans were shifting from wild to cultivated varieties. The period of Early Natufian culture in western Asia between 12,300 and 10,800 BCE was an era of semi-domestication. Although people lived as hunter-gatherers in this period, over time they began to adopt sedentary settlement patterns. Villages formed on the edges between woodlands and steppes, and by gathering edible plants and hunting in these two different plant communities, villagers were able to attain a stable food supply. The presence at Tell Abu Hureyra (a representative archaeological site of this type) of sophisticated processing tools such as mortars carved from large, round rocks suggests that many people engaged in food processing. A greater diversity of plants was used, including acorns, almonds, lupin beans, hawthorn, peas, and wild-type emmer, oats, and oat grass. Before they were eaten, these latter three plants[2] required several hours of work using hand scythes for harvest, bowls,[3] mortars and other tools, bark serving dishes, and stones for cooking (Mithen 2011).

When the quantity of grains being processed and cooked is small, individual work may be more efficient than cooperative work. However, considering that fewer tools than dwellings have been discovered, it is conceivable that grains were prepared through some type of cooperative work. In addition, the use of grains such as wheat that ripen all at once requires timely harvest, likely further increasing the need for cooperation.

The use of fire not only increased the variety of plants people could eat but also led to the simultaneous use of multiple different environments, which stabilized food acquisition. At this stage, the work of processing, cooking, and preserving food became more cooperative. In addition, during the Last Glacial Maximum and the period of warming that followed, food habits shifted to cereal foods, and for most of the population grains became 'the staple food' (Uzawa 2012: 60).

Reviewing the history of environmental and food-habit change in this way reaffirms that the shift to cereal foods, the last major change in the history of food habits, was a significant turning point in human evolution. The fact that grains, which cannot be digested without cooking, became the staple food means that humans had reached a point at which they could not stay alive without fire, after which 'cooking really did make us: human' (Uzawa 2012: 60). I suggest that this both enabled the adoption of sedentary settlement patterns and was premised on the development of cooperative harvesting, processing, and cooking methods.

Climax environments and change: formation of semi-natural environments

The shift to starchy seeds as staples and adoption of sedentary settlement patterns

Grains are not the only foods that became more important as people began to process and cook food. Through this same process, the starchy seeds of trees in the beech family, such as chestnuts and acorns, became core foods for people in regions where they grew. The use of nuts by indigenous people on the west coast of North America and the use of beech-family seeds during the Jōmon period on the Japanese archipelago fall into this category. Staples, the foods which supply the bulk of dietary calories, emerged. As the use of these nuts became more central, people became more sedentary, and populations increased and became denser. This quantitatively increased pressure on the environment.

As David Sprague (2015) describes in his overview of land-use change on the timescale of human history, untouched natural environments were reorganized and reproduced as 'semi-natural environments' as humanity developed. To acquire and maintain set food habits, it became important to interact with the environment in ways that yielded particular resources. Even if unintentional, this continued use of particular resources altered the environment, producing distinctive landscapes. In addition, synchronized cycles emerged between the landscape and human technology and knowledge, labor allocation, cooperative work, settlement patterns, and residential groups, optimizing particular uses of the environment; that is, the relationship between humans and nature reached a climax—or, in other words, an extreme—state.

Survival in changing climax environments

The adoption of sedentary settlement patterns and semi-domestication led to a rapid development of techniques and tools for processing and cooking food. However, the technologies and human groups refined in response to particular environments, as well as the climax human–environment relationships described above, likely had difficulty responding to climate change and other major changes. In the absence of such changes, climax human–environment relationships lead to the development of skills for effectively and reliably obtaining food, but in the face of major change, they risk drawing the entire group into extreme situations that put its survival at risk.

Masaki Nishida (1997) discusses the elaborate relationships between people and the environment, centered around the use of nuts, that developed in a large Jōmon village. This village had gathering sites for food processing and cooking, and there is evidence of an alignment between human activities and the use-regeneration cycle of the surrounding secondary woodland. During the long, temperate Jōmon period, tools, skills, settlement locations, cooperative work systems, and ceremonies existed in balance with the natural environment—or, to borrow Sprague's (2015) term, 'semi-natural environments' developed—and in a sense it is probably accurate to say the environment reached a climax state.

However, these elaborately developed relationships between humans and the environment were unable to adapt to the climatic cooling that occurred in the late Jōmon period. Nobuhiko Kamijō (2016) sorts through an immense quantity of Jōmon artifacts to examine the food processing techniques of this period and concludes that this was a turning point for tools. According to Kamijō, the scale of villages and the distribution of the human population also changed during this period. For example, the population of western Japan suddenly decreased and village size in northeastern Japan shrank, suggesting that people dispersed into small groups to search for new ways of life. Eventually, as livelihoods based on hunting, fishing, and gathering semi-domesticated plants gave way to different types of livelihood based on agriculture, the Jōmon culture vanished (Kamijō 2016).

In that changes to livelihood and settlement patterns are unavoidable during periods such as this, when food habits change due to climate changes and other factors, the survival of residential group members is put at risk. For example, the transition to cereal foods is efficient in terms of caloric intake, but nutrition may become unbalanced because people feel full if grains are the only food they eat. At Tell Abu Hureyra, research indicates that the population did, in fact, decline for a period (Uzawa 2012). And while the timescale differs, nutrition also worsened temporarily for the contemporary Bemba, as discussed below. These examples demonstrate that during periods of change in food habits, people must increase food variety on a trial-and-error basis and obtain new combinations of foods.

When food habits change and people are searching for new food combinations, securing a stable supply of the nutrients and calories necessary for the survival of children, a particularly vulnerable type of group member, becomes an essential condition for the continuation of the group. How does the group ensure the survival of these highly vulnerable members who are directly impacted by critical situations during transition periods? Cooperative modes of food processing and cooking can be effective in these situations. Even when a given household is unable to secure adequate food, by cooperatively processing and cooking food, it is possible to supplement the food supply and acquire the necessary food for children, who must eat several meals each day.

Today, 500,000 years after the breakthrough that allowed humans to control fire and more than 70,000 years after remains appear suggesting 'that various features of modern humans already existed, including the division of labor, societal cooperation, and the division of resources' (Uzawa 2012), how do humans incorporate cooperative food preparation into society? How is this related to the environment and the survival of group members? Beginning in the next section, I draw on an example of contemporary farmers to consider these questions.

Cooperative food processing and cooking among contemporary farmers in Africa: insights from a Bemba farming village

Differences in processing and cooking various foods

Beginning in this section, my focus shifts to the lives of contemporary African farmers. First, however, I would like to review some of the differences in processing and cooking different foods.

Differences in foods are reflected in the ways they are processed and cooked, the cooperative work involved in those processes, and the ways people gather to do that work. Even considering only agricultural foods, cooperation with others is necessary when harvesting and preparing grains and nuts that ripen all at once, as well as for cassava, Japanese horse chestnuts, and other foods requiring detoxification. In addition, processing and cooking acorns, walnuts, and similar foods is generally labor intensive. By contrast, foods such as root crops and bananas, which do not need to be detoxified or threshed, require little preparation and can be cooked and eaten immediately after harvest. Furthermore, since they can be harvested little by little, farmers can harvest and prepare them individually (see Table 19.1). The need for cooperative work thus varies depending on food type. Viewed in this context, the lives of contemporary Bemba farmers are fascinating in that even when foods can be individually harvested and prepared, farmers nevertheless cook them cooperatively, or at least gather to help one another with tasks other than cooking. They appear to prioritize cooperative work over efficiency.

Crop	Harvesting method	Cooking processes						
		1st	2nd	3rd	4th	5th	6th	7th
cassava	digging	peeling and bleaching	leaving/ drying	grinding/ grilling	cooking *ubwali*			
taro	digging	steaming/ boiling/ grilling						
yam	digging	steaming/ boiling/ grilling						
banana	picking	steaming/ grilling/ cooking						
maize	fruit picking	drying	threshing	immersion	removal of skin (in a mortar)/ wind winnowing	partial polishing (2–3 times)	milling	cooking
pearl millet	root cutting/ ear harvesting	drying	threshing/ wind winnowing	grinding	cooking			
sorghum	ear harvesting	drying	threshing/ wind winnowing	grinding	cooking			
finger millet	ear harvesting	drying	threshing/ wind winnowing	grinding	cooking			
teff	root cutting	threshing/ stomping	threshing/ wind winnowing	wind winnowing several times	grinding	cooking		
barley	root cutting	drying	threshing/ wind winnowing	grinding	cooking			
wheat	root cutting	drying	threshing/ wind winnowing	grinding	cooking			
acorn (non-astringent)	gathering	shelling	crushing	cooking				
chestnut	gathering	shelling	crushing	cooking				
horse chestnut	gathering	shelling	astringency removal	drying	crushing	cooking		

The above list includes non-African crops and plants (created by the author).

Table 19.1: Processes involved in processing and cooking various staple foods

Bemba modes of subsistence and the development of semi-natural Miombo woodlands

In this section, I examine how the Bemba, a swidden farming society in Zambia's Northern Province, use the environment and work cooperatively to process and cook food.

The Bemba are known for their development of the *chitemene* (also *citemene*) system, a unique form of swidden agriculture that they use in the Miombo woodlands where they live (Allan 1965; Richards 1939; Kakeya and Sugiyama 1987). The region is seemingly unsuited to agriculture, with poor soil and scattered vegetation. However, Bemba farmers have developed a method of supplementing organic matter in the soil and attaining adequate harvests through the *chitemene* system of cultivation, which involves cutting, piling, and burning tree branches, as described by soil scientists (e.g., Stromgaard 1985; Araki 1996). After several years of cultivation, the land on the outer edge of the deforested area that is not affected by fire is fallowed for a period of sixteen years, and for more than thirty years on cultivated land that is burned, allowing the vegetation to recover so that the land can be cultivated again (Oyama 2013).

Juichi Itani (2003) has categorized indigenous farming systems in the Miombo woodlands based on restoration of soil fertility. According to his study, these systems can be broadly divided into those that rely on an external nutrient supply and those that restore fertility through fallowing. After converting forest to grassland, fallowing systems can be further divided into grassland fallow systems, which alternate between using the land as grassland and cultivated land, and woodland fallow systems, which wait for the forest to recover before cultivating the land again.

The Bemba's *chitemene* system is a woodland fallow system, but in the interval while the Miombo woodland is regrowing, cutting and burning alter the composition of species. For example, certain medicinal plants and preferred edible wild plants are found on cultivated land immediately after fallowing. In addition, because many field mice inhabit these fields, they are good hunting grounds for children. Several years after fallowing, a type of tree locally called *mutondo* emerges thickly on cultivated land. This tree is a host plant for the edible caterpillars that are a preferred food of the Bemba, and during the caterpillar season, these fields are the best place to gather them.

Through varied livelihood activities, including hunting and gathering, the Bemba make complex use of the diverse species that emerge as a result of their swidden cultivation system. By interacting over time with the natural environment of the Miombo woodland through farming, hunting, and gathering, they continuously reproduce a semi-natural environment suited to their own survival, creating a unique landscape (Sugiyama 1998). This has enabled a livelihood strategy that maintains the cyclical use of the surrounding Miombo woodland and ensures a stable food supply.

It is also worth noting that the Bemba's regenerative cycle in the Miombo woodlands is synchronized with generational cycles and the movement of people. Indigenous farming systems such as the *chitemene* system are not simply technologies for productivity but, instead, are deeply linked with the formation of values and worldviews shared by the people who live in these places, and they exist as systems that integrate people's lives into local material cycles (Sugiyama 1998).

Accordingly, when this system is abruptly disturbed by population growth, agricultural policy changes, or land tenure changes, the negative impacts appear first in the weakest members of residential groups. In the Bemba village of M—, strong government policies aimed at commercializing agriculture in the mid-1980s triggered explosive growth in the cultivation of maize as a cash crop. This caused hardship for female heads of households who did not have access to male labor, and the impacts were felt by children in these households.[4] However, because women gathered regularly to cooperatively process and cook food, a type of sharing arose and was observed to have a mitigating effect on the impact of these changes.

The meaning of cooperation in food processing and cooking

The Bemba have had a traditional monarchy since the seventeenth century, and today many people continue to speak Bemba as their native language. Residential-group villages are small, made up of about ten to seventy households. The basic livelihood units are nuclear-family households, with livelihood activities clearly divided according to gender. Men cut the trees for swidden cultivation while women pile the branches and perform most of the other agricultural work. Women also do the daily food processing and cooking. Villages are made up primarily of matrilineal relatives, and women gather daily to process and cook the harvest.

Audrey Richards (1939: 132) described cooperative cooking groups (joint cooking groups) in a Bemba village in Northern Rhodesia (now Zambia) in the 1930s. According to Richards, cooperative cooking greatly lightened female labor. During the peak of the farming season, she observed women cooperating by dividing labor between those who brought ingredients for side dishes back from the fields and those who prepared the staple foods. She emphasized the significant contribution that cooperative cooking made to the stability of household modes of subsistence, as well as the role collaborative work played in transmitting knowledge.

In the 1930s, the imposition of a poll tax by the colonial government prompted many men of working age to travel to copper mining regions to earn money. This was a period of dramatic change in everyday life due to the resulting labor shortages in farming villages and the penetration of the money economy originating from men who returned from the mines. These changes caused the size of cooperative cooking groups to shrink and food processing and cooking to become individualized, depriving people of the advantages mentioned above, and Richards reported that some households suffered hardship during food shortages.

I observed a similar chain of events while living in a Bemba village in the late 1980s and 1990s. During this period, cultivation of hybrid maize as a cash crop rapidly expanded, and small-scale flour mills were introduced around the same time. This enabled women to do most maize processing individually, from removing kernels to grinding them, so they had fewer opportunities to gather. This significantly impacted the activities of women in households that were beginning to focus on maize production. In households where women gathered less often with others to process and cook food, they had fewer opportunities to engage in the minor exchanges of food that take place during cooperative work, or to eat together, which follows from these exchanges. Studies have reported a reduction in food variety in households shifting to hybrid maize production (Bolt et al. 1987).

Cooperative food processing and cooking is meaningful not only because it makes work more efficient, but also because the act of gathering itself is deeply significant. As the examples above make clear, women's cooperative work encourages food exchange between households, which supports survival by stabilizing modes of subsistence and ensuring a more diverse diet. In the situation I observed, the transition from finger millet to maize as a staple food and from hand grinding to machine grinding reduced opportunities for women to gather. In this way, even when cereal foods remain the staple, changing crops and processing methods leads to changes in the nature of cooperative work, strongly impacting food acquisition for individual households and members.

The communal nature of food processing and cooking also helps to create an image of communal cooperativity through the cooked food. An important societal responsibility of adult Bemba women is to ceremonially maintain the purity of the fire. The 'dirt'[5] in fires that are not appropriately managed is believed to 'pollute' the fires and food of other households through ember sharing and cooperative food processing and cooking, causing respiratory illnesses that can spread to all the children in the village and, in the worst cases, cause death. This belief is connected to ancestor worship, reproducing in daily life general images of the village as a 'community of consumption' (Sugimura 2004).

Preparing for and carrying out cooperative food processing and cooking

This section uses examples from a survey of all thirteen households in the Bemba village of M— conducted in 1984 and 1985 to examine the contemporary state of food processing and cooking (Sugiyama 1987a).

Swidden cultivation among the Bemba utilizes a several-year crop rotation. In the first year, edible gourds and winter squash are planted on the periphery of cultivated fields, while native greens and tomatoes, cassava cuttings, and finger millet (the Bemba's staple food) are planted in the fields. In the second year, after harvesting the finger millet, peanuts and Bambara groundnuts are planted and harvested. During the third and fourth years, when the cassava roots have grown large enough to eat, they are harvested on an ongoing basis. In the fifth year, when all the cassava has been eaten, the soil is mounded into round ridges to plant green beans and a plant used as fish poison. After these are harvested, the land is fallowed.

Among the crops produced in the *chitemene* system, routine cooperative processing and cooking take place to thresh, refine, and grind finger millet, to make cassava flour, to pound cassava leaves, and to grind sprouted finger millet to brew alcohol. Table 19.2 lists the processing and cooking steps and tools for finger millet and cassava root.

As for the ownership of tools, women in each household own winnows, brooms, pots, and wooden spatulas, but there is a clear division between households that do and do not own the mortars and grinding stones essential for threshing finger millet and grinding finger millet and cassava. Mortars and grinding stones are owned by the older generation of women, who own them

Crop	Processing and cooking	Tools
Finger millet	Threshing	Mortar and pestle, winnowing basket
	Wind winnowing	Winnowing basket, metal basin
	Milling	Grindstone, winnowing basket, broom
	Cooking	Pot, wooden spatula
Cassava	Peeling	Knife, basket
	Drying	Reed mat or waterproof cloth
	Grinding 1	Mortar and pestle, sieve, winnowing basket
	Grinding 2	Grindstone, broom
	Cooking	Pot, wooden spatula

Table 19.2: Processing and cooking staple food crops and tools used

as a set (Figure 19.1). For Bemba women, the ownership of tools is synchronized with advancement through life stages and accompanied by an increase in social authority. In this way, it is deeply connected to sociocultural context (Sugiyama 1987a).

The sociocultural context of limited tools for processing and cooking foods essential to the local diet increases the need for women in M— to cooperate when processing and cooking food and encourages gatherings centered on the frequent borrowing and lending of tools.

Generally, finger millet is processed and cooked as follows:

- Transportation and threshing: once a week, women retrieve a basket worth (about 30 kilograms) of finger millet from granaries near their fields and bring it back to the village to process. Each woman is very familiar with her household's consumption rhythm and adjusts the amount she prepares according to the size of her household's harvest that year. Women who have harvested more than they need will fill their baskets with more finger millet than usual or go to the granary more often.

  The finger millet in the granaries still has a small portion of stem attached to each ear. Back in the village, the ears are spread on a mat to dry, then pounded in mortars in small batches to remove the hulls. After

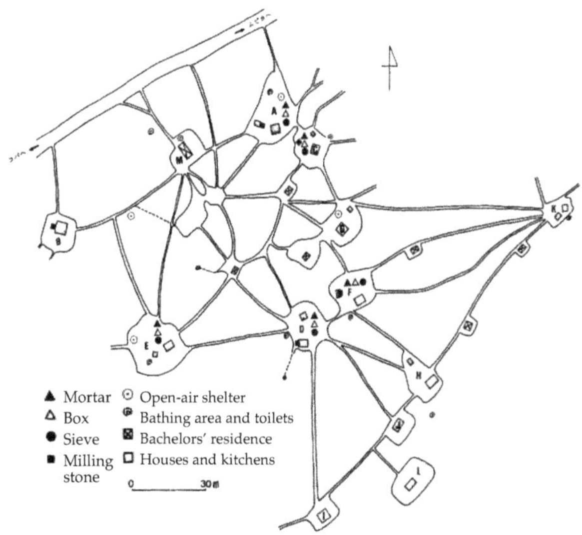

Figure 19.1: Arrangement of tools used for processing and cooking food in M— village (created by the author. A—, B—, C—, D—, E—, and F— are older women, called *bakalamba,* elders, in Bemba language)

pounding, they are winnowed several times to remove chaff before grinding. One 30 kilogram basket of finger millet yields about 27 or 28 kilograms of threshed, winnowed grain. This quantity takes three to four hours to process. The day's work ends with this step.

- Grinding: grinding is done by hand using a built-in grinding stone. Speed varies from person to person, but on average 3 to 4 kilograms of finger millet take one-and-a-half to two hours to grind into flour. At this pace, one basket of finger millet takes fourteen to fifteen hours to grind. Typically, women at first grind only enough flour for two days of meals. The remainder of the finger millet is ground in several early morning or evening sessions and stored in cylindrical metal containers. If for any reason a woman has not ground enough flour and runs out, it is usual for other women to lend or give her some.

Cassava is processed and cooked as follows:

- Peeling, soaking, and drying: cassava is peeled in the field where it is harvested. Slightly more than 30 kilograms of peeled cassava roots are placed in a basket and transported to a water source for soaking. After two or three days, the softened cassava is removed from the water and spread on a nearby field for about a day to remove moisture. At this stage, the cassava is not fully dried.

- Grinding 1 and 2: after the initial drying, the cassava is brought back to the village and pounded in a mortar until it is semi-ground. The pounded cassava is divided into small portions and set on reed mats or tarpaulins to dry. Once dry, it is pounded in a mortar, sifted, and pounded again into flour. These stages of cassava processing take several days.

As Sahlins (1972) describes in his theory of 'underproduction,' swidden farmers spend little time on farming. The Bemba spend an average of two hours per day on agricultural work (Sugiyama 1987b) but more than twice as long preparing harvested foods for meals. Of the time women spend working each day, more than half is spent on processing and cooking food.

Processing the staple foods finger millet and cassava alone requires several hours per day. This involves continuous repetition of the simple tasks of pounding in a mortar and grinding with a stone. Women prefer to do this work cooperatively. This is because pounding with two or three people using the same mortar progresses much faster and makes the simple work enjoyable. A grinding stone cannot be used by more than one person at a time, but if women ask for help, others will take over, or several women will take turns grinding relay-style, or a woman who is not participating in the day's work in the field will ask to borrow a grinding stone to make flour for another woman. Young girls volunteer to help as well. Various kinds of information and goods other than those related to processing and cooking food are also exchanged: a woman may say, 'Today I don't have enough firewood' and borrow a piece or two from another woman, or she may borrow or lend seasonings, or ask for some snuff, or exchange medicinal plants or herbs with which to treat a sick child.

Cooperative food processing and cooking is thus closely linked to tool ownership, efficiency, allotment of time to various tasks throughout the day, and food acquisition, but, in addition, the gatherings provide opportunities to exchange goods and information not related to food and, in doing so, village women form and maintain friendly relationships with one another.

Cooperative food preparation and food sharing among the Bemba

In this section, I discuss field observations on gatherings of women in M— in 1984 and the results of a meal survey conducted there. The field observations took place over the course of three-and-a-half arbitrarily chosen days in May and early June 1984, during which I recorded the locations and activities of people throughout the village each hour between 07:00 and 20:00. Of 224 observations of people processing and cooking food, 184 involved gatherings of two or more women Thus gatherings of women account for more than 80% of the total observations. Considering that at the time of my fieldwork, the elderly women B— and E— were avoiding gatherings with other women for personal reasons, this figure would likely be even higher at other times.

What were these gathered women doing? I was able to confirm the activities of women gathered around houses in a total of seventy-three instances. Of these, the largest number were observed chatting (thirty-eight instances) and the next largest were observed processing or cooking food (twenty-seven instances), making up the bulk of observations (Table 19.3). A considerable number of

Activity	Total people (in all instances)
Conversation	38
Processing and cooking food	27
Meals	5
Sleeping and resting	3
Total	73

Table 19.3: Activities of women of M— village when gathered together

the women observed chatting were waiting for their turn with the mortar or grinding stone, and I frequently observed them taking over jobs or helping other women. A small number of women were sleeping or resting in front of the other women. It is likely accurate to call these sites of food processing and cooking not simply worksites, but places where women gather as friends.

When women gather, they also perform tasks such as pounding casava leaves in a mortar to make an important side dish, shelling beans, or removing fibrous strings from greens. The woman who owns the house where this is done will begin cooking side dishes right away, placing cassava leaves together in a pot without asking whose leaves she is using. When preparing greens, if she notices there are not enough, she will add leaves picked by other women.

Once the owner of the house begins cooking her household meal after processing food, she will invite everyone who is there to share the meal. Those who want to eat will join without hesitation, but some will simply thank the hostess without touching the food or take just a few bites before returning to their conversations. This latter category of women will often receive some of the prepared side dish before the meal starts to take home for their own household meals, setting the food aside in a pot or other container they have brought with them.

Among the most enthusiastic participants in these meals are children aged three to six who are accompanying female family members at the gathering, as well as other children not yet in school who quickly gather after noticing that a meal has begun. When many children are present, dishes of *ubwali* (a staple food) and side dishes will be prepared specially for them, and they will eat in a separate circle from the women. If the children in the hostess's household are coming home from school, food will be set aside for them before the meal begins.

I observed meals in three households in this village for two weeks each and found that people other than household members shared in 70% of meals. Most of these non-household members were children under ten years old. In the households of older women, these children were their grandchildren, the grandchildren of their siblings, the children of relatives from other villages who were living with them, or neighboring children.[6]

It is typical for village children to eat meals not only at their own homes but also at the homes of older women in the village until they are teenagers. Particularly when their parents are not home, children join the households of older women for meals. When too many people gather and there is not enough food, older children in the household will sometimes receive permission from their mother to take out staple-food flour and cook additional food to eat with the other children, along with leftover side dishes. When there are no leftover side dishes, they will ask neighbors to share some, salt-roast some peanuts, or cook field mice, mole rats, or insects that they have caught themselves.

Children's meals and cooperative food processing and cooking

Reproductive histories of Bemba women and child groups

I have already noted that as a result of processing and cooking food together, the people who have gathered will join in the meal afterwards. In particular, young children accompanying women and neighborhood children enthusiastically participate in these meals. Because meals typically take place outside, children can eat at any house in the village.[7] It is not unusual for Bemba children and even babies to be cared for by people other than their mother. As children grow older, they separate from their parents and form child groups that eat at multiple places using multiple methods.

Figure 19.2 shows the reproductive histories of A—, a woman born in 1930, and her eldest and third daughters, who lived in M— at the time of the study. The information is based on life histories of the three women. A— married immediately after her first menstrual period and gave birth to her eldest son at age sixteen. She subsequently gave birth to nine more children, of which her second, fifth, and sixth daughters died when young. The youngest child, her third boy, was born in 1974. A—'s eldest daughter, born in 1953, gave birth to a daughter in 1968, the first of nine children each born two or three years apart. A—'s third daughter, born in 1958, gave birth to a boy in 1974, the first of four children born about three years apart. At the time of my fieldwork in 1984, a twelve-year-old child and a ten-year-old child were living in A—'s household; a seven-year-old, five-year-old, and zero-year-old were living in her eldest

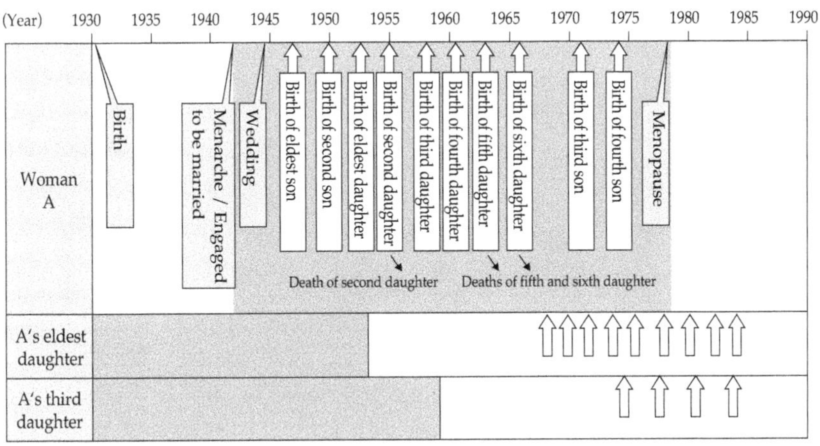

Figure 19.2: Reproductive history of A— and her daughters (until 1984)

daughter's household; and a ten-year-old, seven-year-old, four-year-old, and one-year-old were living in her third daughter's household.[8]

Bemba women spend most of the thirty-something years between giving birth to their first child and approaching menopause either pregnant or nursing. It is not uncommon for a mother and her daughter to both nurse babies at the same time, or for the children of a mother and her daughter to form a group and take care of one another.

In the above example, even excluding A—'s third son (age twelve) who was beginning to distance himself from the children's group as he became a teenager, and A—'s fourth daughter's toddler and her third daughter's nursing baby, A— and her two daughters collectively had six children between the ages of four and ten in 1980.[9] They were closely related, but they were also friends who spent their days together, eating mainly with A—'s household but also often joining in meals with their parents' households, or the households of A—'s siblings and their daughters.

Multi-age groups of children

How are these child groups structured? They are notable for the fact that while they are made up of children between certain ages, they typically include children of different ages within those limits, with the older children tending to maintain group order. Figure 19.3 shows a model of children's developmental stages in Bemba farming villages. Among the Bemba, infants are kept constantly with their mothers only until the age of about three months. They begin to walk independently around age one, and at age two at the earliest and age three in typical cases, they become the youngest members of child groups. These groups, however, do not always have the same members, but shift with time and place.

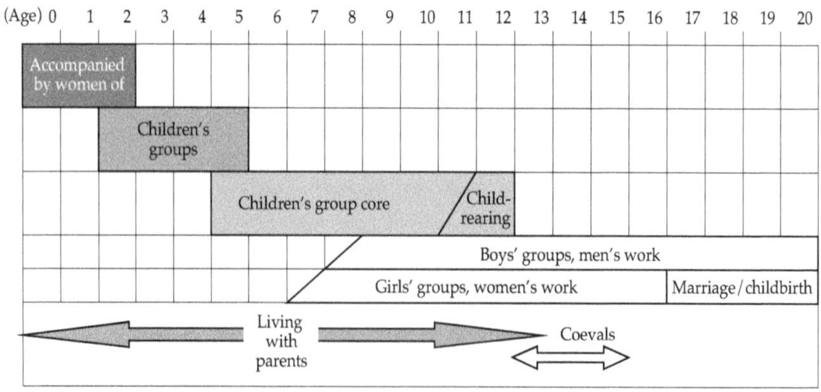

Figure 19.3: Child's growth process and changes in companions

Depending on the scope and content of activities, child groups tend to split into gatherings of children aged around two to five and four to ten. When children reach age eleven or twelve, they usually begin separating from the group but still care for the youngest children. The two- to five-year-olds play or gather food near the village or where the older children can see them. The four- to ten-year-olds, regardless of gender, go to the Miombo woodlands around the village to play, hunt, and gather, bringing their harvest back home to cook and share with younger children who happen to be nearby.

When children reach five or six, they begin to spend more time with children of the same gender and age. Boys preform 'male work,' hunting for field mice or mole rats, while girls gather wild fruit in the Miombo woodland or play at 'female work' at water sources or in the village. Girls are also in charge of fetching water. In this way, the scope of girls' and boys' activities changes as they grow older. Eventually they spend more time with children their own age, and when they become teenagers, they leave their parents' house and find a shed to sleep in with friends of the same gender.

From a young age, children are taught to express their own wishes about where to sleep and eat, and these wishes are respected. In child groups, too, members express their own will, and when disagreements arise, older children usually mediate. Children from households without enough food are able to take part in meals at other households, as well as to obtain food within the child group. They also learn how and where to hunt and gather edible wild foods. It is probably accurate to say that by forming groups with other children—through which they acquire food, work cooperatively to cook it, share it among themselves, and join in meals at various houses in the village—children get the food they need. In addition, these activities teach them skills and knowledge that instill habits of dividing food and eating together as an intuitive behavioral logic.[10]

In the example of the Bemba, each household carries out livelihood activities independently, but because village women share a limited supply of tools and (as a result) process and cook food cooperatively, they not only accomplish their work more efficiently but also have more opportunities to share food and eat together. Also, because meals take place at different times in each household, children can participate in multiple meals.

Children also cook cooperatively among themselves. These multi-age child groups secure the food they need by acquiring ingredients as part of their play, cooking side dishes cooperatively, and receiving flour from their parents to make their own *ubwali*.

Processing and cooking food cooperatively lessens the sense that a given food belongs to the household or individual that originally acquired it, which encourages exchanges of food and eating together. In this way, cooperative food processing and cooking turns food into a community resource.

However, a widespread value of sharing food and eating together does not on its own guarantee that children will have opportunities to get the food they need. Rather, these opportunities are ensured by regular gatherings of women to process and cook food, a scale of residential groups that allows for children to choose between multiple meal places, and the shifting membership in groups centered on matrilineal relatives.

The composition and scale of child-friendly residential groups

Kumiko Sakamoto (2015) conducted a questionnaire survey on infant mortality rates in three of the 'poorest' regions in Tanzania. The study reaffirms that, as past research has indicated, the place where a birth occurs, the education level and knowledge regarding maternal and child health of the caregiver or mother, and diet-related factors all affect infant mortality rates. The study also focuses on regional differences.

The survey results show that when household meals were insufficient, survey participants who said children could eat at the homes of other people or relatives reported rates of child survival that were higher by a statistically significant degree. The number of female household members was also correlated with child mortality numbers. Mutual aid, between women in different households, related to food clearly contributed to infant survival. However, Sakamoto reports that the benefits flow only to people with relatives or friends to help them and who are included in systems of 'food mutual aid,' leaving 'a significant number of women and children not involved in mutual aid.' Of the three villages surveyed, mutual aid related to food functioned only in the village on Zanzibar; in the other two villages, located in southeast and central Tanzania, it would be inaccurate to say that mutual aid functioned village-wide to ensure children received food. In all three villages, behavior related to sharing food and eating together was valued, but that did not necessarily mean the safety net of mutual aid encompassed the entire village.

A likely reason is that farming villages in Tanzania are large, with many households in them. Due to a policy beginning in the late 1960s promoting clustered settlements, it is not unusual for villages in Tanzania to have 1000 households. The villages Sakamoto studied were of a similarly large scale, and mutual aid was limited to close relatives and friends. People with scanty networks naturally were left out of mutual aid, and it was difficult for their children to receive support in obtaining the food they needed.

Another likely factor in the limited reach of mutual aid networks is the smaller number of women gathering to process and cook food, and the smaller number of gatherings themselves. Unlike in Bemba farming villages, maize (which is ground using machines) became the staple food in the 1970s in the Tanzanian villages Sakamoto studied. The work of stripping corn kernels from cobs, too, is performed by household members or small groups of relatives. Because cooking takes place out of public sight in individual household kitchens, it is difficult for non-household members to join in meals.

By comparison, cooperative work is easier in Bemba farming villages due to village scale and member composition, as well as food processing and cooking methods and sites. The Bemba village I studied had thirteen households and about 100 residents and was organized around matrilineal kin groups. This scale allows each resident to remain aware of the activities of all other members, and due to the layout, most houses can be seen at a single glance. Because food processing and cooking take place outside (or in open-air shelters), it is readily apparent where and when a meal can be had. The open design of houses and their arrangement in the village are important factors enabling food to become a community resource.

Significance of reclaiming 'externalized' food in the context of human history

Harvesting, processing, cooking, and eating food

Because the act of processing and cooking food has been interposed between gathering and catching food, on the one hand, and eating it, on the other, these two behaviors have been distinguished from each other as distinct spheres of social activity. Food processing and cooking in contemporary human society is both a means of preparing resources and a process of intervention by others in the food that is essential for individual survival.

This characteristic, unique among animals, gives rise to a situation in which people who cooperatively process and cook food within residential groups gain the ability to eat foods that they did not acquire themselves, an arrangement that enables stabilization of the group's food intake. The tendency toward 'averaging out of food' among the Tongwe, a swidden farming people in Tanzania (Kakeya 1974), and the leveling mechanism in Bemba society (Kakeya and Sugiyama 1987) tend to equalize inter-household imbalances in food. I believe I have shown in this chapter that the group members who benefit most strongly from this tendency are children.

Food processing and cooking, in particular, are performed at sites that are visible to many members of the residential group, and in places like Bemba villages, where meals are eaten at different times depending on household circumstances, children who miss a meal with their own household or do not get enough to eat can participate in a meal at another house when they need to. Even if their own household is suffering from a food shortage, they can eat multiple times with other households to get the food they need, and if the whole village is suffering from a food shortage, children can stay and eat with relatives in other areas to get through the crisis.

The multi-age groups that children routinely spend time in provide opportunities to acquire hunting, gathering, and cooking skills, as well as to obtain substantial, animal-protein-centered food on their own. In this sense, women's cooperative food processing and cooking not only has the effect of somewhat leveling household differences within the village, but also of creating flexible sites of consumption where children can obtain the food they need for survival even when the food supply worsens.

The shift in food habits to cereal grains and other starchy foods, together with the adoption of sedentary settlement patterns, increased the time and labor required to process and cook foods, increasing the need for cooperative work. This cooperative work not only serves to pool food harvested by individual members and integrate it into the sociocultural product that is a meal, but also to create sites of casual food sharing and eating together. In addition, gathering itself has important societal meaning. The nature of cooperative work changes easily in response to the characteristics and processing methods of the foods being prepared, and changes in cooperative work are closely tied to the selection of food crops and the nature of sharing.

In addition, when food brought by individual members of the cooperative group is pounded together in the same mortar or cooked in the same pot, people pay less attention to which food belongs to which person by the time the meal is ready. (The same phenomenon is observed at men's drinking parties. Alcohol purchased by individual men is mixed in a large bucket and poured into a cup that is passed around, leading participants to focus less on who paid for the alcohol.) It is also important to note the ability of women, who take primary responsibility for processing and cooking food, to understand and adjust the household consumption rhythm and to distribute food.

Cooking becomes invisible and industrial: outcomes of the enclosure of food

Turning to the highly modernized and industrial food system of contemporary Japan, the industrialization of food processing and the transition to cooking by individuals or nuclear families is striking. The chief aim of the modernization of food processing and cooking has been to shorten the amount of time spent on these activities at home. The acts of harvesting and eating food occupy different societal spheres of activity, and while the basic human condition of needing to cook in order to eat has not changed, modernization has caused the bulk of food processing and cooking to be industrialized, nearly eliminating the need to engage in cooperative work before eating. Traditional foods have been replaced with those which can be cooked and eaten right away, and various activities focused on eating have been enclosed within the home as a private sector.

Furthermore, the generally accepted notion that cooking is an expression of affection toward the family constricts women. Meals in contemporary Japan have changed from being open to the community to being confined to the family or household, or extremely individualized. Contemporary urban ways of eating and the problems associated with them are deeply linked to the severing and large-scale industrialization of food production, processing, and consumption, and particularly to the rendering invisible of food processing and cooking. The processes that come before placing food into one's mouth, tasting it, and swallowing it are today left to invisible others, an extreme situation that might well be called an anomaly among animals.

Against this context, numerous initiatives are underway to recreate community food systems. I will not describe these efforts in detail here, but examples include women's cooperative cooking initiatives and efforts to make food processing and cooking more visible in contemporary rural Japan; in cities, the creation of children's dining halls make meals (that were formerly enclosed within the home) public, so that even poor children can reliably access meals, and create related gathering places for adults. These efforts connect cooperative cooking to the pleasure of gathering and lessen resistance to receiving food from others, making it easier for people to share food and eat together.

In contemporary food systems, there is a strong tendency to focus only on the act of putting food in one's mouth. This devalues the time and effort put into processing and cooking food, a result of externalizing and reducing the labor needed for these steps. However, movements to reconsider the industrialization and externalization of food and to reconstruct 'food as a sequence' that includes processing and cooking are gaining strength. These movements are attempting to visualize the steps of processing and cooking food before we eat, by organizing new styles of cooperative work. People are enjoying themselves through these activities of processing, cooking, and eating food together with others, regardless of gender and age. These phenomena, too, are interesting in the context of human history.

Notes

1. While Uzawa (2012) acknowledges recent research developments, to avoid complicating his argument he uses the old categories of Archanthropinae, Paleoanthropinae, and Neoanthropinae.
2. These are shattering varieties, but non-shattering traits appear to have been mixed among them. It took between 200 and 250 years for them to become non-shattering.
3. Processing and cooking are believed to have taken place as follows. An ear of emmer, oats, or goat-grass was placed in a bowl. A heated stone was added to char the grains and then removed. The grains were then crushed in a mortar to extract the kernels. The crushed grains were transferred to a bark dish where the chaff was separated. The kernels were ground to flour in the mortar. The flour was then mixed with water and cooked like a pancake on a stone.
4. In addition, government land expropriation occurred in the 1990s, and revision of land laws in 1995 destabilized the very foundations of people's livelihoods. Today, the continuation of the *chitemene* system in M— appears difficult (Oyama 2011).
5. Because the methods for managing the hearth fire are a secret taught only to girls who have undergone the initiation ceremony, I cannot provide details here. Proper bodily care before and after sexual intercourse is directly connected to management of the fire. Fire that has been touched by an improperly cared-for body is said to be polluted with 'dirt' that will spread to other fireplaces through the borrowing or lending of embers.
6. At the time of my fieldwork, the children of relatives in other areas were staying and eating at three of the thirteen households in M—, and children from one other household were staying with relatives in another village.
7. By contrast, when hosting important guests, they alone will be invited into the house, and a meal will be served with the door closed. While the male head of the household may also be present, children are not permitted to join on such occasions.
8. The age of first marriage rose in the 1980s, with more girls marrying at sixteen or seventeen and giving birth to their first child at eighteen. In addition, some women have complied with family planning initiatives that encourage them to have only three or four children. A—'s third daughter was among those who had planned pregnancies.
9. Two of A—'s eldest daughter's children were staying with relatives elsewhere and attending school, and therefore were not in M— at the time of my fieldwork in 1984.
10. Food sharing in children's groups is much more rigid than in adult groups.

References

Allan, W. (1965) *The African Husbandman*. Oliver and Boyd.

Araki, S. (1996) 'Tsuchi to Mionbo-rin' (The soil and the Miombo woodland). In J. Tanaka, M. Kakeya and M. Ichikawa (eds), *Zoku shizen shakai no jinruigaku* (The further anthropology of natural society). Academia Publishing, 305–338.

Bolt, R., M. Silavwe and M. Bolt (1987) *Food Availability and Consumption Patterns in the Northern Region of Zambia ARPT Trial Areas: Implications for Research Extension and Policy*. ARPT.

Imamura, K. (2010) *Sabaku ni ikiru onna-tachi* (Women of the desert). Dōbutsusha.

Itani, J. (2003) 'Afurika Mionbo-rintai to sono shūhen chiiki no zairai nōhō' (Indigenous farming systems in Miombo woodlands and surrounding areas in East Africa). *Ajia Afurika chiiki kenkyū* (Asian and African area studies), 2: 88–104.

Kakeya, M. (1974) 'Tongwe-zoku no seikei iji kikō: Seikatsu kankyō nariwai shokuseikatsu' (Tongwe livelihood maintenance mechanisms: Living environment, livelihood, and dietary habits). *Kikan jinruigaku* (Anthropology quarterly), 5(3): 3–90.

Kakeya, M. and Y. Sugiyama (1987) 'Chū nanbu Afurika sorin-tai ni okeru Benba-zoku no yakibata nōkō' (Swidden cultivation by the Bemba people in the sparse woodlands of central and southern Africa). In I. Ushijima (ed.), *Shōchō to shakai no minzokugaku* (The ethnological study of symbols and society). Yuzankaku, 111–140.

Kamijō, N. (2016) *Jōmon jidai ni okeru dakkoku funsai gijutsu no kenkyū* (A study of Jōmon-era threshing and grinding techniques). Rokuichi Shobo.

Mithen, S. (2011) *After the Ice: A Global Human History, 20,000–5,000 BC*. Harvard University Press.

Nishida, M. (1997) 'Saibai to nōkō, shutsugen katei no seitaigaku' (Origins of cultivation from ecological standpoints). *Reichōrui kenkyū* (Primate research), 13: 173–181.

Nishida, T. (2007) *Ningensei wa doko kara kita ka: Sarugaku kara no apurochi* (Where has human nature come from? An approach from primatology). Kyoto University Press.

Oyama, S. (2011) 'Zanbia ni okeru shin tochi-hō no seitei to benba nōson no konkyū-ka' (Zambia's new land law and the impoverishment of rural Bemba communities). In M. Kakeya and J. Itani (eds), *Afurika chiiki kenkyū to nōson Kaihatsu* (African area studies and rural development). Showado, 246–280.

Oyama, S. (2013) 'Zanbia hokubu ni okeru chitemene kōsaku no kankyō riyō to kyūkan kikan no sanshutsu: Saiteki kyūkan kikan to iu gainen no teiji' (Miombo woodland and *citemene* shifting cultivation of Bemba people in northern Zambia: Specifying and calculating a concept of 'optimum fallow period'). *Ekumēne kenkyū* (Ökumene research), 2: 21–37.

Richards, A. I. (1939) *Land, Labour and Diet in Northern Rhodesia*. International African Institute, Oxford University Press.

Sahlins, M. (1972) *Stone Age Economics*. Aldine.

Sakamoto, K. (2015) 'Influencing factors on children's mortality and morbidity: Comparative analysis of case studies in central and southeast Tanzania'. *Utsunomiya Daigaku kokusai gakubu kenkyū ronshū* (Journal of the Faculty of International Studies, Utsunomiya University), 40: 13–33.

Sprague, D. (2015) 'Ningen katsudō no rekishi' (A history of human activity). In K. Morita and H. Ikeda (eds), *Ningen katsudō to seitaikei* (Human activity and ecosystems). Kyoritsu Shuppan, 1–20.

Stromgaard, P. (1985) 'A subsistence society under pressure: The Bemba of northern Zambia'. *Africa*, 55 (1): 39–58.

Sugimura, K. (2004) *Afurika nōmin no keizai* (African peasant economies). Sekai Shisosha.

Sugiyama, Y. (1987a) 'Usu o kashite kudasai: Seikatsu yōgu no shoyū to shiyō o meguru Benba josei no maikuro poritikusu' ('May I use your mortar?' Micro-politics of the Bemba women on the ownership and usage of daily utensils). *Afurika kenkyū* (Journal of African studies), 30: 9–39.

Sugiyama, Y. (1987b) 'Kiru koto to yaku koto' (Cutting and burning). *Afurika kenkyū* (Journal of African studies), 53: 1–19.

Sugiyama, Y. (1998) '"Kiru koto" to "yaku koto": *Chitemene* no kaikon hōhō ni kan suru benba no setsumei ronri to "gijutsu" ni kan suru kōsatsu' (Cutting trees and piling branches: A folk explanation of the *citemene* cultivation of the Bemba of Northern Zambia). *Afurika kenkyū* (Journal of African studies), 53: 1–19.

Uzawa, K. (2012) 'Jinrui no shinka to hi' (Fire and human evolution). In T. Asakura (ed.), *Hi to shoku* (Fire and food). Domesu Publishers, 44–66.

Wrangham, R. (2009) *Catching Fire: How Cooking Made Us Human*. Basic Books.

Zeder, M. A. (2011) 'The origins of agriculture in the Near East'. *Current Anthropology*, 52(S4): S221–S235.

CHAPTER 20

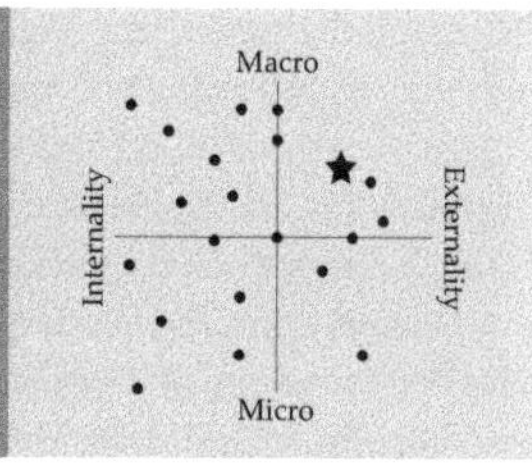

Exploring the Habitat of *Ardipithecus ramidus*:
by Modeling the Survival Environment for Extant Primate Groups

Naofumi NAKAGAWA

Keywords: groups, environment, activity time budget, time constraints, extant primates, fossil hominids

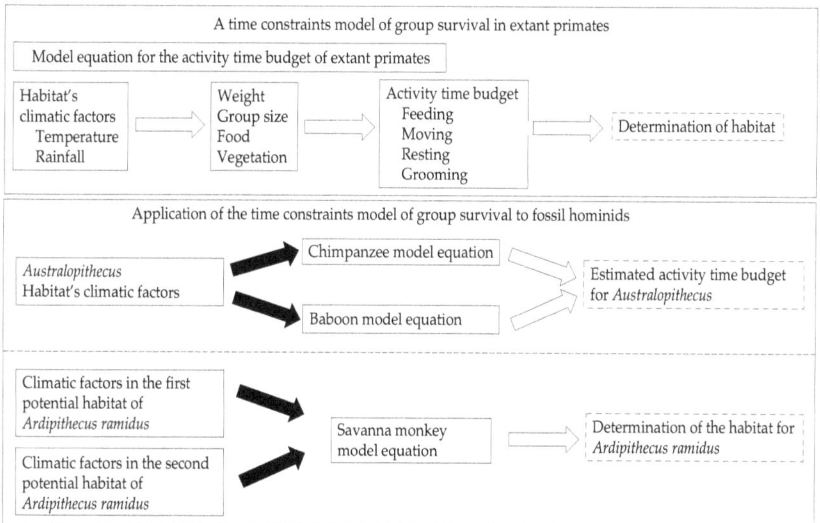

Although the time spent engaging in activities necessary for survival, such as feeding and moving, is dependent on meteorological factors such as temperature and precipitation, time is finite. Based on this idea, the activity time budget model was derived to predict the habitat in which extant primates can survive as groups. I tried applying the equations from this model to fossil hominids. First, on the assumption that our understanding of the paleoenvironment of the *Australopithecus* habitat is correct, I estimated their activity time budget from a combination of the activity time budgets of extant chimpanzees and baboons. Next, considering similarities in morphology and ecology, I inferred the paleoenvironment of the habitat of *Ardipithecus ramidus* on the assumption that their activity time budget could be approximated by that of savanna monkeys.

Introduction: group survival and Dunbar's idea of time constraints

Just as it is often said that 'humans are social animals' and that 'humans cannot survive on their own,' implicit within the statement that humans 'live' is the understanding that they do so 'as social beings.' This is precisely why, in this book as well, the social isolation of individuals within group contexts has been discussed as an example of an extreme (see Soga's discussion in Chapter 12). Furthermore, groups such as regional communities in depopulated areas are also at risk of disappearing, and this too has been discussed as an example of an extreme (see Kitamura's discussion in Chapter 14). Since nonhuman primates are also gregarious, with the exception of orangutans (*Pongo* spp.) and nocturnal prosimians, for these creatures, 'living' is *almost* synonymous with living as a member of the 'groups' that constitute the basic social unit[1]—so, naturally, social isolation could be an extreme here, as well (see Nishie's discussion in Chapter 8). Therefore, when considering the most extreme environments in which they can survive, we need to consider environments in which they can do so *as groups*. Yet when we think in terms of survival, we often think about the survival of individuals or populations (i.e., groups of a single species of organisms living within a defined area). And while this does seem to be a natural approach, if we think about it a little, primate habitats have rarely been conceptualized on the basis of such an idea.

Another issue that has only rarely been thought about in the context of the study of wild primates is time constraints. Humans know that time is finite. This is precisely why, bemoaning their perpetual lack of time, so many people cut back on hours spent sleeping, spending their days racing against time, begrudging even time spent eating and socializing. However, when we observe primates in the wild, we rarely get the sense that they are in any hurry. With diurnal primates, most observers surmise that they sleep from sunset until dawn, but even if they were cutting back on their sleeping time, the observers have no way of sensing this. Also, even during daytime observation, they mostly appear to be fully diverted with resting and grooming, which may be a reason that primatologists have so rarely treated time as a constraint on survival.

In this context, Robin Dunbar, a British primatologist and anthropologist, who was conducting a survey of gelada baboons (*Theropithecus gelada*) in Ethiopia at the same time as a team of researchers from Kyoto University led by Masao Kawai (Kawai 1984), published a paper entitled, 'Time: A hidden constraint on the behavioral ecology of baboons' (Dunbar 1992a).[2] From the fact that the amount of time that primates spend grooming shows a positive correlation with group size, Dunbar had considered that grooming plays a role as a so-called 'glue' for maintaining group bonds. Moreover, this paper was written based on the idea that since the time for activities other than grooming, such as feeding, moving, and resting, is determined by meteorological factors such as temperature and rainfall, then, depending on habitat, time would become a limiting factor for maintaining a constant group size. So, using baboons (*Papio* spp.) as a case study, he derived model equations that sought to explain the proportion of total activity time spent on each activity. He then attempted to predict whether survival was possible in environments where this total did not exceed 100%, and, if so, the maximum size of such a group, and also verified the validity of this prediction.

In the first half of this chapter, I detail the prediction of group survival based on the model equations for the baboon activity time budget developed by Dunbar (1992a), then verify the validity of this prediction. In addition, by applying the model equations for baboons and chimpanzees (*Pan* spp.) to fossil excavation sites that constituted the habitat of australopithecine fossil hominids (*Australopithecus* spp.), I present a study by Dunbar and his colleagues (Bettridge et al. 2010) estimating their activity time budget. In the second half, after presenting my personal view that the subject of my own research, namely patas monkeys (*Erythrocebus patas*) and savanna monkeys (*Chlorocebus* spp.), can serve as a model for fossil hominids, I attempt to infer the habitat of the early hominin *Ardipithecus ramidus*—a matter of no small debate—by applying the model equations for the activity time budget of the savanna monkey.

In this chapter, I refer to the prediction of group survival using the activity time budget model equations described above as the 'time constraints model of group survival.' Moreover, in the context of this chapter, which was written on the basis of this model, the smaller the maximum viable group size (hereinafter, 'maximum ecologically tolerable group size') in a given environment, the closer that environment is to a limiting extreme.

A time constraints model of baboon group survival

Dunbar's old model

Dunbar (1992a) investigated the activity time budget, group size, and habitat-related meteorological factors for anubis baboons, yellow baboons, chacma baboons, guinea baboons, and hamadryas baboons, which make up five species of savanna baboon (*Papio cynocephalus*),[3] in search of the optimal equation for expressing their respective activity time budgets in terms of meteorological factors and group size. The activity time budget is expressed by broadly dividing the activities into four categories (feeding, moving, grooming, and resting) and then finding the ratio of time spent on each of these activities out of the total amount of activity time, which is roughly equivalent to the length of a day. Dunbar derived model equations that summarized the feeding time budget and moving time budget based solely on group size and the meteorological factors of rainfall and temperature. Also, in terms of the grooming time budget, he obtained the following equation (Dunbar 1992a) expressing a relationship with group size on the basis of the idea that grooming facilitates the formation of group bonding:

$$\text{grooming time (\%)} = 4.533 + 0.0764\ \text{group size}$$

After estimating feeding time and moving time based on rainfall, temperature, and group size, and setting aside only 5% for resting time, he hypothesized that the remaining time is diverted to grooming time, thereby obtaining the maximum ecologically tolerable group size for that area. As a result, he found that places where the annual average temperature was 10°C or less or 35°C or more had a maximum ecologically tolerable group size of one or less, regardless of annual rainfall. In other words, this result showed that these environments were beyond the limiting extreme for baboons. Moreover, in cases where the annual rainfall was 2700 mm or more, although groups could form where temperatures ranged between 30°C and 39°C, environments with temperatures below 30°C or reaching 40°C or higher were also beyond the extreme. The resulting inference was that in environments where groups could form, large groups would form at moderate levels of temperature and precipitation. I should reiterate here that I will refer to the prediction of group survival using the activity time budget model equations as the 'time constraints model of group survival.'

This baboon model has only been applied to the baboon habitats of the African continent, but let us attempt to apply it to the Japanese archipelago so that we can illustrate it more easily. According to data made publicly available by the Japan Meteorological Agency (2018), the annual average temperature of the Japanese town of Ohma is around 10°C, so we would calculate that baboons would not form groups here. On the other hand, on the island of Yakushima, which represents the southern limit of the range of Japanese macaques, the annual average temperature is approximately 26°C. Accordingly, given a maximum annual rainfall of around 2500 mm, we would estimate a habitat that could narrowly support groups of up to thirty individuals. In fact, however, since this is actually Japan's rainiest region, with annual precipitation sometimes in excess of 5000 mm, the model posits that baboons would not actually be able to live here. Incidentally, Tokyo, with an annual average temperature of approximately 16°C and annual rainfall in the neighborhood of 1500 mm, would be able to support the formation of baboon troops of up to seventy-nine individuals.

Next, as part of his validation of the model, Dunbar first of all calculated the ecological stress using the following formula:

$$\text{ecological stress} = \frac{\text{(group size (observed value))}}{\text{(maximum ecologically tolerable group size for the region (predicted value))}}$$

Then, by comparing feeding time and the total time devoted to resting and grooming between groups with an ecological stress factor of 1 or more and groups with a factor of less than 1, he found that although there was no significant difference in feeding time, the total time devoted to resting and grooming was much shorter among groups with a stress factor of 1 or more than among those with a stress factor of less than 1. This result indicates that when the stress factor is 1 or more, the total time devoted to resting and grooming is reduced in order to supplement feeding time. Further, a comparison of ecological stress among groups that frequently fragmented while feeding and among those that did not do so found that stress levels were higher among the former. This result suggests that with the reduction in grooming time, group bonds are no longer sustainable, leading to group fragmentation.

A new model by Bettridge, Lehmann, and Dunbar

Following his publication of the aforementioned old model, in collaboration with colleagues, Dunbar attempted to apply the model to primate species other than baboons. Time constraints models of group survival were derived by similar methodologies for chimpanzees (Lehmann et al. 2007), spider monkeys (Korstjens et al. 2006), red and black-and-white colobus monkeys (Korstjens and Dunbar 2007), and gorillas (Lehmann et al. 2008). Moreover, while minimum group sizes could not be introduced with any degree of precision, unlike in the new baboon model (Bettridge et al. 2010) described below, the distributions predicted by the models showed a high degree of conformity with actual distributions, proving the versatility and robustness of the model.

The new baboon model makes the following three assumptions:

- feeding and moving time are each independently determined by climatic factors, diet, and group size
- grooming time (%) = 1.55 + 0.23 group size
- grooming time was constrained to a minimum of 1%, and all other time budget components to a minimum of 5%.

On this basis, the maximum ecologically tolerable group size in each region is calculated by starting with a group size of one individual, and then increasing the size of the group by one individual at a time until the sum of all activity time budgets reaches 100%. As a result, in broad terms, the model posits that feeding time decreases as the climate gets drier (though there is also a factor whereby feeding time increases in completely desertified environments where there is little fluctuation in rainfall), that resting time increases with more drastic temperature variation, and that with higher annual average temperatures, resting time increases at the expense of moving time in order to avoid increases in body temperatures. Although the variables have become more complex and difficult to interpret as compared with the old model, the accuracy of the model is improved.

Although conceptually distinct from the idea of time constraints, the new model introduces the notion of minimum permissible group size in an effort to refine predictions of group survival. Although the hypothesis that the minimum permissible group size in a given area is determined by predation risk is quite self-evident, its novelty lies in its comparison of the actual minimum group size depending on the type and presence or absence of predators. The results obtained in support of this hypothesis include the fact that minimum group size is larger where predators are present, with the effect that groups are larger where there are lions than where there are leopards, and that minimum group size is inversely related to the body mass of adult females.

This was followed by a verification of the model's validity. When the predicted value of the maximum ecologically tolerable group size calculated by the new model was compared with the observed value of average group size, the observed value was generally smaller. Also, when compared with the observed values for maximum group size, they largely tended to be in agreement. The final verification task was to investigate the degree of conformity between the area predicted to be within the realizable niche space (a group size between the minimum permissible group size and maximum ecologically tolerable group size) and the actual distribution of baboons. The results of a similar analysis of savanna monkeys are detailed below (see 'Willems and Hill's time constraints model for group survival among savanna monkeys'), so for now I will only note that they were also generally in agreement. Moreover, as a further analysis, a comparison of the time budgets for each activity predicted from meteorological factors in areas where there were no baboons versus those in which baboons were actually present found a large difference only in terms of feeding time, which would have been much longer in the former areas. This result indicates that baboons are unable to live in habitats that would require an extremely high investment in feeding time.

Applying the time constraints model of group survival to *Australopithecus* fossil hominids

Australopithecus

Phylogenetically speaking, the closest extant primates to modern humans like ourselves (*Homo sapiens sapiens*) are two species of the genus *Pan* that live in Africa—namely, the common chimpanzee (*P. troglodytes*) and bonobo (*P. paniscus*)—followed by two species of gorilla, the eastern gorilla (*Gorilla beringei*) and western gorilla (*G. gorilla*). Together with three species of orangutan living in Southeast Asia—the Bornean orangutan (*Pongo pygmaeus*), Sumatran orangutan (*P. abelii*), and a newly described species called the Tapanuli orangutan (*P. tapanuliensis*; see Nater et al. 2017)—these species constitute what are commonly known as the great apes, and which, together with humans, are classified as the taxonomic family Hominidae. After the divergence by the common ancestor of modern humans and extant chimpanzees, many species belonging to multiple genera in the lineage of hominids, which includes modern humans—officially known as the Hominini in taxonomic terms using tribe (which is the classificatory category directly above genus)—have been excavated as fossils.

These species include the fossils of individuals attributed to the genus *Australopithecus*, whose habitat is considered to have spanned the territory from East Africa to South Africa from approximately 4.2 million to 2 million years before the emergence of *Homo*. The first fossil of *Australopithecus africanus* was discovered in South Africa in 1924 and was known as the Taung Child because it was the fossil of an infant child. Although it took time to be accepted as the oldest human then known—which was partly due to the fact that, prior to this discovery, humans had been conventionally believed to have originated in Asia—it was indisputably a human ancestor (now thought to date to approximately 2.5 million years ago). In 1974, fifty years after the discovery of *A. africanus*, came the discovery of *Australopithecus afarensis*. This species was estimated to be between 3 and 3.75 million years old—predating *A. africanus* by between 500,000 and one million years, and for a long time after its discovery this fossil hominid was regarded as humanity's oldest ancestor. The female, whose fossilized skeletal remains (approximately 40% of the entire body) were discovered in Hadar, Ethiopia, in East Africa, was named Lucy (after the Beatles song 'Lucy in the Sky with Diamonds,' which was playing at the time of her excavation). Fossilized footprints discovered in 1978 in Laetoli in northern Tanzania, which are often attributed to two adults and a child and indisputably belong to creatures with an upright bipedal walk, have suggested the existence of a family. Other known species include *A. anamensis* (considered to be a forerunner of *A. afarensis*), which is regarded as the oldest *Australopithecus* species and is thought to have lived in the environs of Kenya's Lake Turkana as early as 4.2 million years ago, and *A. garhi*, which is considered to be the most recent *Australopithecus* species and is thought to have lived in Ethiopia's Middle Awash River Basin around 2.5 million years ago. *A. garhi* is regarded as a species that links *A. afarensis* with the genus *Homo*.

Meanwhile, quite apart from *Homo*, another line that evolved from *Australopithecus* is *Paranthropus*. This genus originated with *P. robustus*, which was discovered at the South African site of Kromdraai in a stratum dating to

2 million years ago. In East Africa, *P. boisei* from Olduvai Gorge in Tanzania is known as a roughly contemporaneous species, while its ancestor *P. aethiopicus*, from between 2.7 and 2.3 million years ago, is known from the shores of Lake Turkana in Kenya, as well as southern Ethiopia. *Paranthropus* are also known as 'robust australopithecines,' since, in comparison with *Australopithecus*, they are characterized by larger mandibles, jutting cheekbones (zygomatic arches), and a larger sagittal crest on their crown, conferring them with an overall rugged impression.

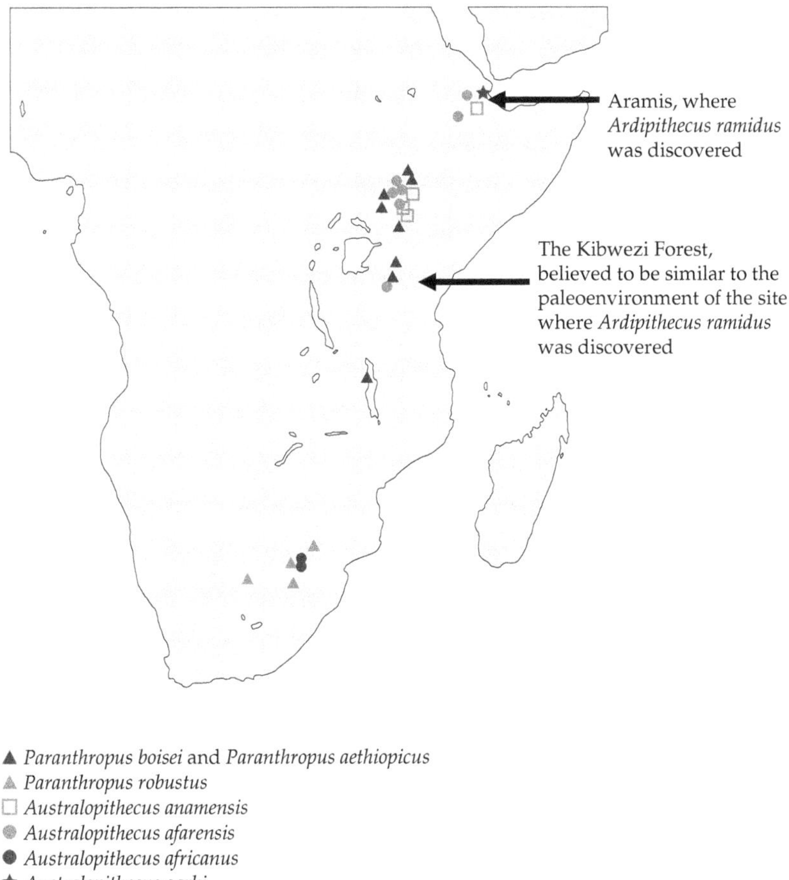

Figure 20.1: Locations of the discovery sites of *Australopithecus* and *Paranthropus*, along with the discovery site of *Ardipithecus ramidus*, at Aramis and Kibwezi Forest, which is considered to resemble its paleoenvironment (adapted from Dunbar 2016: Figure 4-1)

Applying the model equations for chimpanzee and baboon activity time budgets to the paleoenvironment inhabited by *Australopithecus* to estimate their activity time budget

Dunbar's student Caroline Bettridge attempted to apply a model obtained from chimpanzees, which are the most closely related species to humans (Lehmann et al. 2007), and activity time budget model equations obtained from baboons, as extant primates living in the savanna where human ancestors lived (Bettridge et al. 2010), to paleoenvironment sites where australopithecine fossils (in the broad sense including *Paranthropus*) were found and to sites where they were not found. First, when the chimpanzee model was applied, the largest ecologically tolerable groups of *Australopithecus* predicted for most sites where fossils were found were less than about fifteen individuals. This is well below the minimum value of approximately forty individuals for extant chimpanzee group size. In other words, this result suggested that *Australopithecus* would not have been able to survive in the habitats that they were thought to have actually inhabited, indicating that it was not feasible to apply the chimpanzee model equations to *Australopithecus* without modification. The source of this difficulty lay in the chimpanzees' moving time budget. In the chimpanzee model, the moving time budget is expressed as follows:

$$\text{moving time }(\%) = 102 + 2.6\text{ party size} - 0.1\text{ (annual rainfall)} + 0.000003\text{ (annual rainfall)}^2$$

Chimpanzee groups do not always gather all members of the group together to range and forage, as do baboons and many other types of primates; rather, they split up into several subgroups called small parties. Since these parties repeatedly gather together and exchange members before separating again, chimpanzee groups are said to have a fission-fusion society (see the detailed discussion by Nishie in Chapter 8). The amount of time that chimpanzees spend moving increases as the party size grows larger, and then increases rapidly once the number of chimpanzees exceeds ten individuals. This is because the within-group scramble for foods exerts a powerful influence, since chimpanzees feed on ripe fruits, of which there are limited resources. In other words, when a large party is formed, the per capita share of ripe fruit will decrease. As a result, the party must head for the next fruit-bearing site to make up for the reduced amount, but since the number of ripe fruits is limited, this will require moving over a longer distance.[4] Moving time also increases with decreased annual rainfall. This is because annual rainfall generally shows a positive correlation with the availability of food resources, so that food requirements can be met in sites where food is available without requiring much movement. Thus, by forming a fission-fusion society that feeds by breaking up into small parties, chimpanzees are able to inhabit marginal areas close to the tropical lowland forests near the equator, which receive heavy amounts of rainfall (see Itoh's discussion in Chapter 2).

However, even if *Australopithecus* were to have reduced their party size to ten individuals or fewer, in the manner of chimpanzees, it was calculated that they would still experience a dearth of time in most of their habitats (Bettridge et al. 2010). I expect that most who read this far will be feeling a fundamental concern, or perhaps be feeling that something is amiss. Whereas chimpanzees and baboons alike both locomote with a quadruped walk, *Australopithecus* did so with an upright bipedal walk. This is precisely why they are regarded as a species of human. Naturally, Dunbar and Bettridge both thought about this point and offered a supplementary explanation, which I summarize here. The first consideration was the improved efficiency of movement afforded by the upright bipedal walk. Comparing the energy efficiency of the bipedal walk and quadruped walk in the trees and on the ground, they introduced the view that a bipedal walk is superior when spending at least 65% of the time moving on the ground, noting also that the shift to the bipedal walk may have been triggered by an increased dependence on the ground. However, since *Australopithecus* was not as competent at the bipedal walk as modern humans, it seems reasonable to think that they benefited not from the bipedal walk itself, but solely from the increase in stride length. According to their calculations, whereas the lower limbs of female chimpanzees were 44 cm, those of *Australopithecus* females were about 52 cm. The resulting increase in stride length would have reduced moving time for *Australopithecus* by 2.5%, but not much more than that. The next consideration was the cooling effect of the upright bipedal walk. Since in an upright posture, as opposed to a four-legged posture, only the head and shoulders are exposed to sunlight, it would seem that resting time can be economized during the hours of the day when the sunlight is strong. To that extent, it would become possible to be active in the sunshine. However, it was pointed out that activity itself leads to the generation of body heat, so that any advantage conferred by the bipedal walk would thereby be reduced. Thus the residual advantage would mainly be the cooling effect owing to the loss of hair and the accompanying perspiration. However, since it has been generally accepted that the loss of hair did not take place until after the emergence of *Homo*, it would seem that the cooling effect of the bipedal walk for *Australopithecus* would have been negligible.

Therefore, Bettridge adjusted the model equations for the moving time budget of baboons to account for the larger body mass of *Australopithecus* and used this to replace the model equations for the moving time budget of chimpanzees. In doing so, the goodness of fit for the time constraints model for chimpanzee group survival predicting the range of *Australopithecus* for a minimum group size of ten individuals, as is already known for chimpanzees, increased sharply from 26% to 76%. In particular, it became possible to accurately predict that *Australopithecus* could inhabit all of the sites in which they were thought to have actually lived. This suggests that *Australopithecus,* at least in terms of foraging, were similar to baboons, whose diet extends to a variety of foodstuffs such as fruits (including unripe fruits), seeds, leaves, and roots and bulbs, as well as insects. In other words, *Australopithecus* appears to have been a type of human resembling a chimpanzee that foraged in the manner of a baboon (see Dunbar 2016 for details).

A revision of my own theory: savanna monkeys as an evolutionary model for *Ardipithecus*

The old evolutionary model: patas monkeys as an evolutionary model for *Homo*

In my book, *'Futsū' no saru ga kataru hito no kigen to shinka* (Monkeys tell us human origin and evolution) (Nakagawa 2015), I developed an argument through a field survey of patas monkeys and tantalus monkeys (*Chlorocebus tantalus*)—two of the most closely related guenons (classified in the tribe Cercopithecini) living sympatrically in the savanna of northern Cameroon—that the patas monkey is an optimal species for exploring the evolution of *Homo*. On the other hand, although I did not emphasize the point, having positioned the tantalus monkey as a comparator, I also wrote that 'for the time being, we might say that it is a model for *Australopithecus afarensis*.' The origins of this personal theory lay in the length of the patas monkey's limbs and its affinity with the savanna. These relatively long limbs are consistent with those of *Homo erectus*, which emerged around the time that the environment dried and the savanna expanded. The tantalus monkey is one of six species of the genus *Chlorocebus*, and although

Photograph 20.1: Although their species diverged only recently, the common patas monkey (foreground right) has long limbs, while the tantalus monkey (foreground left) has short limbs; both animals in the foreground are adult males

it lives in the savanna, as suggested by the term 'savanna monkeys' by which these species are also collectively known, it primarily makes use of the riverine forests and woodlands adjoining the grasslands, with shorter limbs and a stronger arboreality than the patas monkey (Photograph 20.1, 20.2).

Studies have found that by virtue of their superior mobility, which capitalizes on the length of their limbs, patas monkeys, while making use of highly nutritious but small animal matter (such as grasshoppers) that can be found scattered about the grasslands, feed selectively on high-quality plant-based foods such as acacia pods and gum. These foods can only be reached by ranging far and wide across the vast savanna but, when reached, can be accessed in large quantities at one time, and the patas monkeys make extensive use of sites where they are concentrated. The shorter-limbed tantalus monkeys have established a narrow home range that covers less than a square kilometer and centers on the riverine forest, and while subsisting primarily on the fruits found therein, they also feed indiscriminately on various plant foods such as leaves, seeds, flowers, gum, and insects, including low-quality high-fiber materials such as grasses in the Gramineae family (for a more detailed discussion, see Nakagawa 2007).

Based on these findings, the scenario I envision for evolution from *Australopithecus afarensis* to *Homo* is as follows. Some *A. afarensis* that had been living in the riverine forest began to advance into the grasslands, which were expanding as the environment dried. Longer-legged individuals were then able to cross the vast grasslands faster and even more efficiently, allowing them to feed in the acacia groves and other riverine forests where high-quality food was plentiful. Further, moving through these vast grasslands afforded the opportunity to encounter and make use of progressively higher-quality animal carcasses. Thus, with the increased ease of acquiring large amounts of high-quality food, the burden on the digestive tract was lessened, which in turn provided latitude

Photograph 20.2: Tantalus monkeys, which live predominantly in riverine forests (left), and patas monkeys, which also venture far from there into the distant grasslands (right)

for the brain to expand, leading to the advent of *Homo* with its long legs and large cerebrum.

Savanna monkeys as an evolutionary model for *Ardipithecus*

In Nakagawa (2015), after having painstakingly developed my hypothesis that patas monkeys can be seen as an evolutionary model of *Homo*, when the time came to write the postscript, I found that the conventional view—central to my theory—that the lengthening of limbs had occurred in *Homo* had been cast into doubt, and that a new finding (Pontzer 2012) had been submitted that longer limbs were already present among *Australopithecus*. Figure 20.2 plots a comparison of the estimated body mass and estimated hindlimb length of modern hominids, fossil hominids, and two genera of extant great apes (*Pan* and *Gorilla*). While the *Homo* fossil hominids certainly have longer limbs relative to their weight as compared with the great apes, *Australopithecus* falls entirely within the range of a standard deviation from modern humans. Previously, the specimens regarded as representative of *A. afarensis* and *A. africanus* were considered to have short limbs, owing to their small bodies, but the subsequent discovery of a large-bodied specimen brought *Australopithecus* within a standard deviation of *Homo*, which includes modern humans. On the other hand, the early hominin

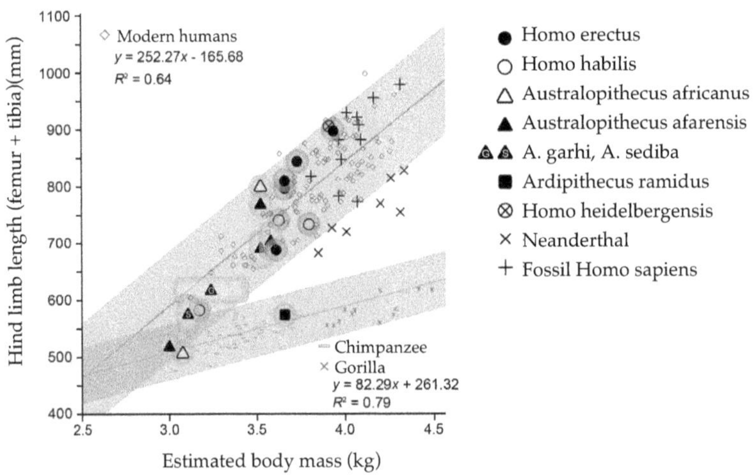

Figure 20.2: Estimated hindlimb length versus body mass for fossil hominids; the shaded area is ±2 times the standard deviation of the regression obtained from modern humans and extant great ape species, respectively (reproduced with permission from Figure 2 in Pontzer (2012: S351); permission no. 4476220872297)

Ardipithecus ramidus, which lived 4.4 million years before *Australopithecus*, did have short hindlimbs, and these specimens fell within one standard deviation of the great apes (Pontzer 2012). In my book, I described these facts in the postscript, with the addendum that 'if this new theory is correct, then it may be that patas monkeys should serve as an evolutionary model of *Australopithecus*... and savanna monkeys as a model for *Ardipithecus*' (Nakagawa 2015).

While a follow-up article has been published in connection with this new finding (Anton et al. 2014), there nevertheless seem to be a few issues. The most basic analytical error concerns the regression equation for the great apes, in that it is found without any consideration given to the difference in the number of samples for each genus (*Pan*, sixty-six; *Gorilla*, twenty-two), or to intra-genus species differences and sexual differences. Among *Gorilla* species, which are characterized by considerable sexual dimorphism in terms of body mass, a regression equation with a slightly steeper slope than that of the regression equation for great apes in general can be seen, as though males and females were lined up in parallel. In addition, the larger of the two *A. africanus* specimens is sometimes considered to be an early *Homo* (Grabowski et al. 2015). Furthermore, there is ample possibility that the body masses of the specimens, other than that of Lucy (the smallest of the four *A. afarensis* specimens), were actually slightly heavier (Suwa, personal communication). Given the foregoing, Gen Suwa (see the '*Ardipithecus*' section below), one of the discoverers of *Ardipithecus ramidus*, has suggested that while it is certainly possible that the lengthening of the limbs took place at some point between *Ardipithecus* and *Australopithecus*, there is also a strong likelihood that it took place between *Australopithecus* and *Homo*, as argued by the conventional theory (Suwa, personal communication).

For the rest of this chapter, I will now restrict my focus to *Ardipithecus*. Thus, in view of the superior arboreality of the savanna monkey, while I conclude that they are more suitable as a model for *Ardipithecus* than for *Australopithecus*, for the purposes of this chapter, at least, I will leave aside the question of whether patas monkeys can be regarded as an *Australopithecus* model. That said, for whichever consideration, I wish to note that the following modifications are inevitable. Patas monkeys and savanna monkeys are estimated to have diverged approximately 5 million years ago (Page et al. 1999). This means that, when considering patas monkeys as a model for *Homo*, there is a considerable gap between the emergence of patas monkeys 2.4 million years ago and that of *Homo erectus* 1.8 million years ago. However, when considered as a model for *Australopithecus*, this date brings us much closer to the era of their emergence. In other words, this means that the lengthening of the limbs would have taken place slightly earlier than the expansion of the savanna among both Hominidae and Cercopithecini. However, because the drying of the environment had already begun 8 million years ago, leading to the formation of small grasslands, there is little doubt that an environment was emerging in which longer limbs would work effectively, so this fact poses no major contradiction. Also, while we can assume for *Homo* that animal carcasses would have served as high-quality foodstuffs found scattered across the grasslands, corresponding to the insects consumed by patas monkeys, the equivalent for *Australopithecus* would have been tubers. Tubers, which represent a high-quality food that is rich in starch, with less difficult-to-digest fiber in comparison with the herbaceous piths that serve as a fallback food for chimpanzees to eat when fruits are not in season, have been posited as a possible fallback food for *Australopithecus* (Wrangham 2010). Moreover, in terms of animal-based food sources, notwithstanding the fine scratches on the surface of fossil bones from 3.39 million years ago that are regarded as the oldest instance of the use of stone tools and the oldest evidence of the consumption of animal matter (McPherron et al. 2010), it is also appropriate to assume that termites were eaten as part of the australopithecine diet (Backwell and d'Errico 2001; Lesnik 2014).

Finally, for the purposes of this chapter, I do not consider *Australopithecus* to include *Paranthropus*, sometimes known as the 'robust' *Australopithecus*. Investigation of the stable isotope composition ratio of tooth enamel carbonate among both *Paranthropus boisei* from East Africa (Cerling et al. 2011) and *Paranthropus robustus* from South Africa (Codron et al. 2008) has shown a composition ratio close to that of Gramineae and Cyperaceae (sedges), indicating a diet similar to that of extant *Papio* species, and therefore suggesting that the baboon should be regarded as a *Paranthropus* model (Codron et al. 2008; Macho 2014).

Ardipithecus

Following his discovery of a maxillary third molar from Aramis in Ethiopia's Middle Awash Basin, Gen Suwa, an anthropologist at the University of Tokyo, was one of the first to take part in the subsequent excavation of *A. ramidus* (Kawai 2010). A joint team of American, Japanese, and Ethiopian researchers led by the American anthropologist Tim White excavated loose teeth, the mandibular fragments of an infant, and adult skull and humerus fragments, all estimated to be 4.4 million years old. In 1994, the discovery was initially classified as a new species of australopithecine, *Australopithecus ramidus*. However, owing to the fact that it was considerably more primitive than *Australopithecus*, in 1995 it was announced as a new genus and named *Ardipithecus ramidus* (Leakey and Walker 1997). The discovery of an almost intact full female skeleton, dubbed 'Ardi,' followed, and the results of an analysis of Ardi from various angles were published in eleven papers published in the October 2, 2009 issue of *Science* (e.g., White et al. 2009a). Ardi was estimated to weigh approximately 50 kg and to be about 120 cm tall, with a small brain on the scale of that of a chimpanzee (White et al. 2009a). The absence of any significant size difference between males and females, judging from the trunk and canines (Suwa et al.

2009; White et al. 2009a), was suggestive of a monogamous mating system with weak male competition for females (Lovejoy 2009). *Ardipithecus ramidus* locomoted using an upright bipedal walk rather than a knuckle walk (which constitutes a quadruped walk with palms curled into a fist so that the backs of the fingers make contact with the ground, as in the case of chimpanzees and gorillas on the ground). On the other hand, interestingly, the presence of what seemed to be a grasping big toe that faced the other toes (Lovejoy et al. 2009a) suggested that they were strongly arboreal, and it was thought that rather than using the forelimbs to hang from branches and brachiate in the manner of chimpanzees and gorillas, they employed a quadruped walk that involved placing their palms on branches (Lovejoy et al. 2009b). Moreover, in terms of diet, they were generally omnivorous and frugivorous, without being heavily dependent on ripe fruits (as chimpanzees are), leaves (as gorillas are), or hard foods (as *Australopithecus* were) (Suwa et al. 2009).

In 1997, fossils older than *A. ramidus* were uncovered in the same Middle Awash Basin, in a stratum dating to 5.6 million years ago. Although this finding was initially regarded as a subspecies of *A. ramidus* at the time of its announcement in 2001 (Haile-Selassie 2001), it was subsequently reclassified as a distinct species of *Ardipithecus*, known as *A. kadabba* (Haile-Selassie et al. 2004).

Applying the time constraints model of group survival to the fossil hominid *Ardipithecus*

My idea and the background to beginning my analysis

Wishing to argue that the patas monkey represents a suitable evolutionary model for *Homo* or *Australopithecus*, I wanted to try something similar to Bettridge's attempt (detailed earlier in this chapter) to apply the model equations for chimpanzee and baboon activity time budgets to the paleoenvironment inhabited by *Australopithecus*. Unfortunately, however, patas monkeys did not feature in the series of time constraints models for group survival developed by Dunbar and his colleagues. Given the dearth of even the most basic ecological data for an activity time budget for patas monkeys—the only data available being from my own study of Kala Maloue National Park in Cameroon (Nakagawa 2000) and a study of Kenya's Laikipia Plateau by a team of American researchers (Isbell et al. 1998)—there was no way to put together a model.

Therefore, I resolved to try to apply the model equations for the activity time budget of savanna monkeys (Willems and Hill 2009)—which I had begun to think of as suitable as an evolutionary model for *Ardipithecus*—to the paleo-environment that would have been inhabited by *Ardipithecus*. However, because the habitat of *Ardipithecus* (both *A. ramidus* and *A. kadabba*) was limited to the

Middle Awash Basin, it was not possible to perform an analysis estimating the activity time budget of *A. ramidus* by combining the model equations for the chimpanzee activity time budget to compare sites where fossils were found with those where they were not, as Bettridge attempted for *Australopithecus*.

This is when I noticed the habitat that *Ardipithecus ramidus* was thought to have inhabited. Whereas, based on paleontological evidence, the researchers who discovered *A. ramidus* consider this to have been a forest-like environment (White et al. 2009b; Suwa and Ambrose 2014; White 2014), their interlocutors, based in part on evidence from stable carbon isotopes, consider this to have been a mosaic environment of savanna and riverine forest (Cerling et al. 2010, 2014; Gani and Gani 2011, 2016; Dominguez-Rodrigo 2014), prompting a fierce debate. Kibwezi Forest in southern Kenya is regarded by the discoverers as a modern environment approximating the paleoenvironment of *A. ramidus* (Gibbons 2009), while the Ethiopian site of Aramis, where the discovery took place, is regarded by their interlocutors as a closer approximation of the habitat (Gani and Gani 2011). Using model equations for the activity time budget of savanna monkeys adjusted for the larger bodies of *A. ramidus*, I decided that I would attempt to compare the activity time budget in each environment to investigate which of them seemed more realistic and thereby determine which habitat would have been more suitable as a native paleoenvironment. This is a reversal of Bettridge's idea. She estimated the *Australopithecus* activity time budget as a combination of the activity time budgets of chimpanzees and baboons, assuming that her prior inference of the paleoenvironment of the *Australopithecus* habitat was correct. In contrast, I am attempting to infer the paleoenvironment of the habitat on the assumption that the activity time budget of *Ardipithecus ramidus* can be approximated by that of the savanna monkey, in view of their similarities in terms of their affinity with the savanna, short limbs, arboreality, and omnivorous diet.

Willems and Hill's time constraints model for group survival among savanna monkeys

The time constraints model for group survival among savanna monkeys I employ was developed by Dunbar's collaborators Russell Hill (e.g., Hill and Dunbar 1998) and Erik Willems (Willems and Hill 2009). Currently, there are generally considered to be six species of savanna monkeys but, as noted earlier, these authors treat them in the conventional fashion as one species with six subspecies. Of these six subspecies, activity time budget data could be obtained for fourteen populations of four of the subspecies; namely, vervet monkeys (*Chlorocebus pygerythrus*), green monkeys (*C. sabaeus*), tantalus monkeys, and djamdjam monkeys (*C. djamdjamensis*).[5] As with a series of models developed by Dunbar and his colleagues (e.g., Lehmann et al. 2007, 2008; Bettridge et al.

2010), they investigated the activity time budget, group size, and meteorological factors in the habitat to find the optimal formula for expressing each activity time budget in terms of group size and various meteorological factors. However, in addition to temperature and rainfall, as one of these meteorological factors they also made use of the Normalized Difference Vegetation Index (NDVI). The NDVI expresses the density of vegetation as calculated by satellite data measuring the differential reflection of visible red and near-infrared light.[6] Another major difference with the model equations developed by Dunbar and his colleagues is the estimation of grooming time. Whereas grooming time was determined solely as a function of group size in Dunbar and his colleagues' model equations, Willems and Hill's model equations treated it just like any other activity. As a result of their analysis, they obtained the following relational expressions:

(1) feeding time (%) = $-78.76 + 130.60P_{H'} - 104.32\ NDVI_{min} + 0.96\ T_{SD} + 0.03\ P_{annual}$

(2) moving time (%) = 10.22 + 0.76 feeding time occupied by eating leaves

(3) grooming time (%) = $-\ 24.86 + 1.87\ P > 2T + 1.65\ T_{daily\ range} + 2.09\ Day\ length_{range}$

(4) resting time (%) = 84.60 – 1.24 feeding time – 0.52 group size

(5) feeding time on leaves (%) = $51.74 - 111.96\ NDVI_{min} + 0.81\ P_{dry\ month}$

Here, $P_{H'}$ is an index expressing monthly rainfall variations, $NDVI_{min}$ is the value for the NDVI of the least productive month, T_{SD} is the standard deviation for mean monthly temperature (°C), P_{annual} is annual precipitation (mm), $P > 2T$ is the number of months in which the precipitation exceeds twice the value of the average monthly temperature, which corresponds to the length of the growing season for plants, $T_{daily\ range}$ is the mean of the range of temperature on a given day, Day length $_{range}$ is the annual range of day length (hours), and $P_{dry\ month}$ is the precipitation (mm) for the driest month.

Places where the sum of each of the activity time budgets in (1) to (4) exceeds 100% are predicted by the model to be environments beyond the limiting extreme that savanna monkeys would find habitable.

Next is the verification of the validity of the time constraints model for group survival predicted thus. Compared with the actual distribution of savanna monkeys, we find that the proportion of accurate predictions—in other words, the sum of the proportion of places predicted to be uninhabitable and where monkeys are not actually found (45.92%) and the proportion of places predicted to be habitable and where they actually are found (41.51%)—was as high as 87.43%. Of the remaining 12.57%, where the predictions were not correct, 5.15% consisted of sites predicted to be uninhabitable in which monkeys were actually found and 7.42% of sites that were predicted to be habitable but where they were not actually found (Figure 20.3).

In addition, moving group size to the left side of the expression in (4) derives the following equation:

$$\text{group size} = \frac{84.60 - \text{resting time} - 1.24\ \text{feeding time}}{0.52}$$

By keeping only 10% of resting time as the minimum necessary time and substituting feeding time for each location into this equation, the maximum ecologically tolerable group size at a given location can be obtained with the following equation:

$$\text{maximum ecologically tolerable group size} = \frac{74.60 - 1.24\ \text{feeding time}}{0.52}$$

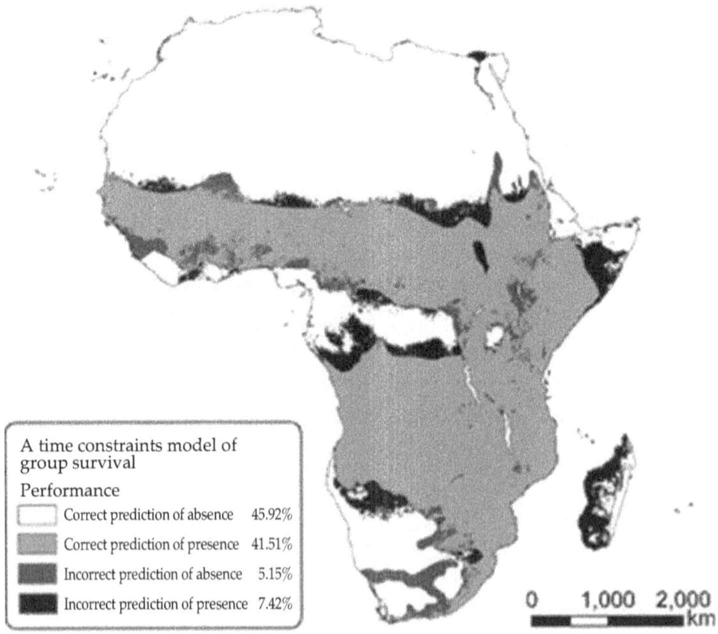

Figure 20.3: Goodness of fit of the group survival time budget model for savanna monkeys (reproduced with permission from Figure 5(a) in Willems and Hill (2009: 2307); permission no. 447621120909)

Willems and Hill then compared actual group sizes by adding the twenty-two populations for which we have group size data to the fourteen populations used to create the model. It was found that the predicted maximum group size exceeded the actual maximum group size in thirty-one of these populations (86.1% of the total). This result also attests the model's validity.

Kibwezi Forest (Kenya) and Aramis (Ethiopia)

Kibwezi Forest in southern Kenya, considered by the researchers who discovered *Ardipithecus* to be a close analogue for its habitat (Gibbons 2009), is a forest reserve situated in the foothills bordering Chyulu Hills National Park, located at 2° 26′S and 37° 55′E (see Figure 20.1). The annual average temperature is 21.7°C, with a total annual rainfall of 689 mm. The dominant tree species consist of the leguminous Lebombo wattle (*Newtonia hildebrandtii*), several species of fig (*Ficus* spp.), the umbrella thorn (*Acacia tortilis*) and other *Acacia* species, and *Terminalia brownii*, a tall tree in the Combretaceae or white mangrove family (Photograph 20.3).

The Ethiopian site of Aramis, where the current discoveries took place and which the interlocutors consider to be a better candidate as a habitat, is located at latitude 10° 30′N and longitude 40° 30′E (see Figure 20.1). Its annual average temperature is 25.2°C, with a total annual rainfall of 555 mm. While closed- and open-canopy forests extend 1 to 2 km and 0.5 to 1 km, respectively, along the Awash River and its tributary the Jara River, the locale is also characterized by the open grasslands dotted with acacias that spread away from these stands (Photograph 20.4).

Photograph 20.3: Kibwezi Forest in southern Kenya, which is considered to approximate the habitat of *Ardipithecus ramidus* (photograph reproduced courtesy of the photographer, Dr. Tim D. White)

Inferring the paleoenvironment habitat of *Ardipithecus ramidus* by applying model equations for the activity time budget of savanna monkeys

Finally, let us try to apply these model equations to the paleoenvironment of the *Ardipithecus ramidus* habitat. To do so, however, we need to correct for body mass, as when Bettridge's baboon model was applied to *Australopithecus*. Specifically, we will correct feeding time and moving time as follows (cf. Bettridge and Dunbar 2012):

$$\textit{Ardipithecus}\text{ feeding time (\%)} = \text{savanna monkey feeding time} \times \left(\frac{\textit{Ardipithecus}\text{ body mass}}{\text{savanna monkey body mass}}\right)^{0.404}$$

$$\textit{Ardipithecus}\text{ moving time (\%)} = \text{savanna monkey feeding time} \times \left(\frac{\textit{Ardipithecus}\text{ body mass}}{\text{savanna monkey body mass}}\right)^{0.333}$$

Feeding time is corrected by the body mass ratio to the power of 0.404 for the following reasons. For one thing, Kleiber's Law tells us that energy requirements scale to the power of 0.75 of an animal's body mass. On the other hand, since the retention time in the intestinal tract, which represents the efficiency with which nutrients are absorbed from food, scales with exponent 0.346, the effect of both would be proportional to the power of 0.75 – 0.346 = 0.404.

Moving time entails a correction for stride length or, in other words, a correction for limb length. Since the body mass ratio approximates a volume ratio, and volume is the cube of length, the requisite correction is to the 1/3 power of the body mass ratio (i.e., the exponent 0.333).

Also, in terms of the body mass of *Ardipithecus ramidus*, as described earlier, Ardi was estimated to have a body mass of about 50 kg. However, according to a

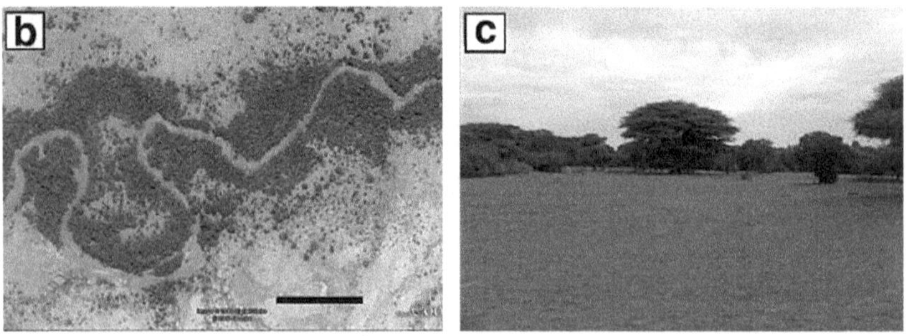

Photograph 20.4: Near Aramis in contemporary Ethiopia; (b) satellite photograph of the Jara River and (c) Jara River floodplain (reproduced with permission from Figure 4 in Gani and Gani (2011: 4); permission no. 4445730567742)

recent paper published by Grabowski et al. (2015) arguing that the body mass of fossil hominids is generally overestimated, I have performed calculations based on both the 32.1 kg value posited in this paper and the conventional 50 kg value. Meanwhile, in terms of the body mass of the savanna monkey, I used 4.38 kg as the average value for the four (sub)species for which there are data for at least ten males and ten females, based on the second volume (on primates) of the *Mammals of Africa* series of field guides (Butynski et al. 2013).[7]

Table 20.1 shows values for the various activity time budgets and the maximum ecologically tolerable group size for *Ardipithecus ramidus* as predicted for Kibwezi Forest and Aramis, respectively, together with values for the meteorological factor variable values called for by the prediction equation. For reference, values for the savanna monkey are also listed. Assuming a body mass of 32.1 kg for *A. ramidus*, the results predict that in Kibwezi Forest *Ardipithecus* would spend 29.9% of its time feeding, 50.1% moving, and 3.2% grooming, with the resulting calculation that this habitat could support groups of up to 72.3 individuals. In contrast, at Aramis, the predicted feeding time on its own was as high as 73.0%, which, added to moving time alone, calculated at 77.6%, would far exceed 100%. Since we cannot here set aside the minimum requirement of 10% for resting time, it is impossible to calculate the maximum ecologically tolerable group size, meaning that for *A. ramidus*, Aramis would be uninhabitable. Assuming a body mass of 50 kg, the values predicted for *A. ramidus* are 35.7%

	Climate and other environmental variables								Savanna monkey					*Ardipithecus ramidus*			
Location	$NDVI_{min}$	$P_{dry\ month}$	P>2T	P_{annual}	$P_{H'}$	$T_{daily\ range}$	T_{SD}	Day $length_{range}$	Feeding time (%)	Feeding time on leaves (%)	Moving time (%)	Grooming time (%)	Max. group size	Feeding time (%)*	Moving time (%)*	Grooming time (%)*	Max. group size*
Kibwezi Forest	0.286	1	4	636	0.778	12.1	1.389	0.282	13.36	20.53	25.82	3.17	111.61	29.86 35.72	50.11 58.10	3.17 3.17	72.25 –**
Aramis	0.163	7	2	479	0.856	14	2.319	1.23	32.67	39.18	40	4.55	65.56	73.04 87.36	77.63 89.97	4.55 4.55	–** –**

Table 20.1: Predicted maximum ecologically tolerable group size and percentage of time spent engaging in each activity for savanna monkey and *Ardipithecus ramidus* in Kibwezi Forest, Kenya, and Aramis, Ethiopia, respectively, with values for climate and other environmental variables required in the prediction formula

$NDVI_{min}$: value for the Normalized Difference Vegetation Index of the least productive month; $P_{dry\ month}$: precipitation (mm) for the driest month; P>2T: number of months in which the precipitation value exceeds twice the value of the average monthly temperature (°C) (this corresponds to the length of the growing season for plants); P_{annual}: annual precipitation; PH′: an index of monthly variation in precipitation; $T_{daily\ range}$: the mean of the range of temperature on a given day; T_{SD}: standard deviation of the monthly mean temperature; $Day_{length\ range}$: annual range in day length (hours).

* The upper row of values for *Ardipithecus ramidus* assumes a body weight of 32.1 kg and the lower row assumes a body weight of 50 kg.

** Unable to secure 10% as the necessary minimum rest period.

for feeding time, 58.1% for moving time, and 3.2% for moving time, leaving only 3% to be set aside for resting time, indicating that *A. ramidus* would not have been able to inhabit Kibwezi Forest either. Calculating backwards to find a body mass for which we would be able to secure a resting time of exactly 10% yields a body mas value of about 40 kg, for which the maximum ecologically tolerable group size would have been 65.6 individuals. In other words, as long as the body mass for *A. ramidus* was in the range of 32.1 to 40 kg, it appears that an environment like the present-day Kibwezi Forest could support groups of between 65.6 and 72.3 individuals.

Let us now revisit the meteorological factors to see why there should be such a significant difference between the two areas. In the savanna monkey model, the most influential environmental variable is $NDVI_{min}$, which represents the density of vegetation. As is evident from the magnitude of the coefficients in the model equations, this is the most influential factor in determining moving time and the second most influential factor in determining feeding time. Since lower values for $NDVI_{min}$ work in both cases to increase the time budget, activity time in Aramis (where the 0.163 value for $NDVI_{min}$ 0.123 is less than the 0.286 value for Kibwezi Forest) will be greatly reduced. Moreover, high values for $P_{H'}$—which, as an index of monthly fluctuation in the volume of rainfall, is the most influential factor in determining feeding time—lead to increases in the time budget. Accordingly, activity time will be less available in Aramis, where the value is 0.856, than in Kibwezi Forest, with a value of 0.778.

In the savanna monkey case, the same trends are present given the use of the identical model. Here, however, even in Aramis, although moving time is still quite high (40%), it represents an environment that can support groups of up to 66.6 individuals. Since *Ardipithecus ramidus*, which weighed at least seven times more than a savanna monkey, would not have had longer limbs relative to its body mass like *Australopithecus*, its moving time would have been extended accordingly, leading to the depletion of its activity time.

Conclusion: in lieu of a summary

There is no doubt that the early hominid *Ardipithecus ramidus* lived 4.4 million years ago in what is now Aramis in Ethiopia. However, as compared to Aramis today, which consists of open-canopy forests along a river surrounded by open grassland, its home would have been a forest environment spreading out from spring-fed pools, with a lower annual average temperature and higher annual rainfall. Although it would have moved with an upright bipedal walk over land, it would have employed a quadruped walk in the trees. Like extant savanna monkeys, which have short limbs and split their time equally between the ground and the treetops, it was likely omnivorous, consuming a variety of foods while relying on fruits as a staple. Although it would have been possible for several monogamous families to congregate in groups of up to seventy individuals, given that their bodies were more than seven times larger than those of a savanna monkey, they would presumably necessarily have had to spend a considerable amount of time feeding and moving. Later, the progressive drying of the African continent would likely have pushed their livelihoods to the brink of collapse in terms of time, plunging them into extremis and eventually beyond that limit. Then, with the emergence of *Australopithecus* from 4.2 million years ago, longer limbs would likely have economized on moving time, allowing them to thrive until the advent of *Homo*. However, after giving rise to many species, *Australopithecus* too met with extinction, as did many later species of *Homo*. We are now aware of considerable evidence for multiple species of hominins living sympatrically during the same period, in which case extinction may have resulted not only from meteorological factors but also interspecies competition (see Terashima's discussion in Chapter 21). In any case, there is no doubt that in the history of human evolution, the overall habitat of individual species has repeatedly exceeded the extremes representing the limits of group survival.

Acknowledgments

Dr. Erik Willems of the University of Zurich, who co-authored the model with Russel Hill, sought out and provided me with environmental data, including NDVI values for Kibwezi Forest and Aramis. Dr. Tim D. White of the University of California, Berkeley, contributed a photograph of Kibwezi Forest. Dr. Gen Suwa of the University of Tokyo was kind enough to read and comment on my manuscript for this chapter, and allow me to quote his personal communications. Spring Nature Group, John Wiley and Sons, and the University of Chicago Press have allowed me to reprint the figures. For all of the above, I would like to express my deepest gratitude. Part of this research was also supported by the Japan Society for the Promotion of Science, a Grant-in-Aid for Scientific Research (B) (17H02436, Naofumi Nakagawa).

Notes

1. Even among primates that live in groups, some males that have attained sexual maturity live alone as solitary 'floaters.'
2. As may be inferred from the fact that this paper was published in the same year as 'Neocortex size as a constraint on group size in primates' (Dunbar 1992b), which provided the basis of the well-known 'Social Brain Hypothesis' (Dunbar 1998a), these two papers are closely linked. The Social Brain Hypothesis holds that the complexity of society is what drove the evolution of the brain based on the fact that the neocortex ratio of primates positively correlates with the average group size for each species. Based on this relationship, an estimation of the optimal size of human groups from the human neocortex ratio yields a

figure of 150 people. This number, called 'Dunbar's Number,' is reputedly the maximum number of people whom a person is able to recognize properly and recall their mutual relationship on an individual basis. Furthermore, by applying this human group size of 150 people to a relational equation derived from nonhuman primates expressing grouping time as a function of group size, Dunbar calculated backwards to estimate grooming time for humans. In doing so, he calculated that there would not be enough time for grooming, which basically can only be accomplished by two individuals at the same time, which in turn inspired his theory that humans devised language as a means of interacting with many individuals simultaneously to maintain group bonds (Dunbar 2003; for more detailed accounts, see Dunbar 1998b and 2011).

3. It is now more common to view these as separate species (Butynski et al. 2013).
4. Competition arises when the dominant individual in the group drives out inferiors or when inferior individuals avoid the dominant individual, even when there is no direct competition for food.
5. I have listed the mainstream scientific names for each species when regarded as an independent species (Butynski et al. 2013).
6. This index shows a value between –1.0 and 1.0, primarily representing greenery. Negative values generally represent clouds, water, and snow, while values below 0.1, which approximate 0, primarily represent rocks and the Earth's surface. Values between 0.2 and 0.3 indicate shrubs and grasslands, while values between 0.6 and 0.8 indicate temperate forests and tropical rainforests.
7. While I have also written on the subject of tantalus monkeys (Nakagawa 2013), they are not included in my calculations because I only had body mass data for one individual.

References

Anton, S. C., R. Potts and L. C. Aiello (2014) 'Evolution of early *Homo*: An integrated biological perspective'. *Science*, 345: 1236828-1–1236828-13.

Backwell, L. R. and F. d'Errico (2001) 'Evidence of termite foraging by Swartkrans early hominids'. *Proceedings of the National Academy of Sciences of the United States of America*, 98: 1358–1363.

Bettridge, C., J. Lehmann and R. I. M. Dunbar (2010) 'Trade-offs between time, predation risk and life history, and their implications for biogeography: A systems modelling approach with a primate case study'. *Ecological Modelling*, 221: 777–790.

Bettridge, C. M. and R. I. M. Dunbar (2012) 'Modeling the biogeography of fossil baboons'. *International Journal of Primatology*, 33: 1278–1308.

Butynski, T. M., J. Kingdon and J. Kalina (eds) (2013) *Mammals of Africa, Volume II: Primates*. Bloomsbury Publishing.

Cerling, T. E., E. Mbua, F. M. Kirera, F. K. Manthi, F. E. Grine, M. G Leakey, M. Sponheimer and K. T. Uno (2011) 'Diet of *Paranthropus boisei* in the early Pleistocene of East Africa'. *Proceedings of the National Academy of Sciences*, 108: 9337–9341.

Cerling, T. E., F. H. Brown and J. G. Wynn (2014) 'On the environment of Aramis: A comment on White in Dominguez-Rodrigo'. *Current Anthropology*, 55: 469–470.

Cerling, T. E., N. E. Levin, J. Quade, J. G. Wynn, D. L. Fox, J. D. Kingston, R. G. Klein and F. H. Brown (2010) 'Comment on the paleoenvironment of *Ardipithecus ramidus*'. *Science*, 328, 1105d1–1105d4.

Codron, D., J. A. Lee-Thorp, M. Sponheimer, D. de Ruiter and J. Codron (2008) 'What insights can baboon feeding ecology provide for early hominin niche differentiation?' *International Journal of Primatology*, 29: 757–772.

Dominguez-Rodrigo, M. (2014) 'Is the "Savanna hypothesis" a dead concept for explaining the emergence of the earliest hominins?' *Current Anthropology*, 55: 59–81.

Dunbar, R. I. M. (1992a) 'Time: A hidden constraint on the behavioral ecology of baboons'. *Behavioral Ecology and Sociobiology*, 31: 35–49.

Dunbar, R. I. M. (1992b) 'Neocortex size as a constraint on group size in primates'. *Journal of Human Evolution*, 20: 469–493.

Dunbar, R. I. M. (1998a) 'The social brain hypothesis'. *Evolutionary Anthropology*, 6: 178–190.

Dunbar, R. I. M. (1998b) *Grooming, Gossip, and the Evolution of Language*. Harvard University Press.

Dunbar, R. I. M. (2003) 'The social brain: Mind, language, and society in evolutionary perspective'. *Annual Review of Anthropology*, 32: 163–181.

Dunbar, R. I. M. (2011) *How Many Friends Does One Person Need?* Harvard University Press.

Dunbar, R. I. M. (2016) *Human Evolution: Our Brains and Behavior*. Oxford University Press.

Gani, M. R. and N. D. Gani (2011) 'River-margin habitat of *Ardipithecus ramidus* at Aramis, Ethiopia 4.4 million years ago'. *Nature Communications*, 2: 1–5.

Gani, N. D. and M. R. Gani (2016) 'On the environment of Aramis: Concerning reply of White to Cerling et al. in August 2014'. *Current Anthropology*, 57: 219.

Gibbons, A. (2009) 'A new kind of ancestor: *Ardipithecus* unveiled'. *Science*, 326: 36–40.

Grabowski, M., K. G. Hatala, W. L. Jungers and B. G. Richmond (2015) 'Body mass estimates of hominin fossils and the evolution of human body size'. *Journal of Human Evolution*, 85: 75–93.

Haile-Selassie, Y. (2001) 'Late Miocene hominids from the Middle Awash, Ethiopia'. *Nature*, 412: 178–181.

Haile-Selassie, Y., G. Suwa and T. D. White (2004) 'Late Miocene teeth from Middle Awash, Ethiopia, and early hominid dental evolution'. *Science*, 303: 1503–1505.

Hill, R. A. and R. I. M. Dunbar (1998) 'An evaluation of the roles of predation rate and predation risk as selective pressures on primate grouping behaviour'. *Behaviour*, 135: 411–430.

Isbell, L. A., J. D. Pruetz, M. Lewis and T. P. Young (1998) 'Locomotor activity differences between sympatric patas monkeys (*Erythrocebus patas*) and vervet monkeys (*Cercopithecus aethiops*): Implications for the evolution of long hindlimb length in *Homo*'. *American Journal of Physical Anthropology*, 105: 199–207.

Japan Meteorological Agency (2018) 'Past meteorological data search', http: / / www.data.jma.go.jp / obd / stats / etrn / index.php (accessed November 18, 2018).

Kawai, M. (1984) *Jinrui shinka no kakurezato: Gerada hihi no shakai* (A hidden village of human evolution: Gelada baboon society). Heibonsha.

Kawai, N. (2010) *Hito no shinka 700 man'nen-shi* (Seven million years of human history). Chikuma Shobō.

Korstjens, A. H. and R. I. M. Dunbar (2007) 'Time constraints limit group sizes and distribution in red and black-and-white Colobus'. *International Journal of Primatology*, 28: 551–575.

Korstjens, A. H., I. L. Verhoeckx and R. I. M. Dunbar (2006) 'Time as a constraint on group size in spider monkeys'. *Behavioral Ecology and Sociobiology*, 60: 683–694.

Leakey, M. and A. Walker (1997) 'Early hominid fossils from Africa'. *Scientific American*, 276 (6): 74–79.

Lehmann, J., A. H. Korstjens and R. I. M. Dunbar (2007) 'Fission-fusion social systems as a strategy for coping with ecological constraints: A primate case'. *Evolutionary Ecology*, 21: 613–634.

Lehmann, J., A. H. Korstjens and R. I. M. Dunbar (2008) 'Time management in great apes: Implications for gorilla biogeography'. *Evolutionary Ecology Research*, 10: 517–536.

Lesnik, J. J. (2014) 'Termites in the hominin diet: A meta-analysis of termite genera, species and castes as a dietary supplement for South African robust australopithecines'. *Journal of Human Evolution*, 71: 94–104.

Lovejoy, C. O. (2009) 'Reexamining human origins in light of *Ardipithecus ramidus*'. *Science*, 326: 74e1–74e8.

Lovejoy, C. O., B. Latimer, G. Suwa, B. Asfaw and T. D. White (2009a) 'Combining prehension and propulsion: The foot of *Ardipithecus ramidus*'. *Science*, 326: 72e1–72e8.

Lovejoy, C. O., S. W. Simpson, T. D. White, B. Asfaw and G. Suwa (2009b) 'Careful climbing in the Miocene: The forelimbs of *Ardipithecus ramidus* and humans are primitive'. *Science*, 326: 70e1–70e8.

Macho, G. A. (2014) 'Baboon feeding ecology informs the dietary niche of *Paranthropus boisei*'. *PLoS One*, 9: e84942.

McPherron, S. P., Z. Alemseged, C. W. Marean, J. G. Wynn, D. Reed, D. Geraads, R. Bobe and H. A. Bearat (2010) 'Evidence for stone-tool-assisted consumption of animal tissues before 3.39 million years ago at Dikika, Ethiopia'. *Nature*, 466: 857–860.

Nakagawa, N. (2000) 'Seasonal, sex, and interspecific differences in activity time budgets and diets of patas monkeys (*Erythrocebus patas*) and tantalus monkeys (*Cercopithecus aethiops tantalus*), living sympatrically in Northern Cameroon'. *Primates*, 41: 161–174.

Nakagawa, N. (2007) *Saban'na o kakeru saru: Patasu monkī no seitai to shakai* (The monkeys running in the savanna: The ecology and society of patas monkeys). Kyoto University Press.

Nakagawa, N. (2013) '*Chlolocebus tantalus*'. In T. M. Butynski, J. Kingdon and J. Kalina (eds), *Mammals of Africa, Volume II: Primates*. Bloomsbury Publishing, 271–273.

Nakagawa, N. (2015). *'Futsū' no saru ga kataru hito no kigen to shinka* (Monkeys tell us human origins and evolution). Pneuma-sha.

Nater, A., M. P. Mattle-Greminger, A. Nurcahyo, M. G. Nowak, M. de Manuel, T. Desai, C. Groves, M. Pybus, T. B. Sonay, C. Roos, A. R. Lameira, S. A. Wich, J. Askew, M. Davila-Ross, G. Fredriksson, G. de Valles, F. Casals, J. Prado-Martinez, B. Goossens, E. J. Verschoor et al. (2017) 'Morphometric, behavioral, and genomic evidence for a new orangutan species'. *Current Biology*, 27: 3487–3498 e10.

Page, S. L., C. H. Chiu and M. Goodman (1999) 'Molecular phylogeny of Old World Monkeys (Cercopithecidae) as inferred from gamma-globin DNA sequences'. *Molecular Phylogenetics and Evolution*, 13: 348–359.

Pontzer, H. (2012) 'Ecological energetics in early *Homo*'. *Current Anthropology*, 53: S346–S358.

Suwa, G. and S. H. Ambrose (2014) 'On the environment of Aramis: A comment on White in Dominguez-Rodrigo reply'. *Current Anthropology*, 55: 473–474.

Suwa, G., R. T. Kono, S. W. Simpson, B. Asfaw, C. O. Lovejoy and T. D. White (2009) 'Paleobiological implications of the *Ardipithecus ramidus* dentition'. *Science*, 326: 94–99.

White, T. D. (2014) 'On the environment of Aramis: A comment on White in Dominguez-Rodrigo reply'. *Current Anthropology*, 55: 471–472.

White, T. D., B. Asfaw, Y. Beyene, Y. Haile-Selassie, C. O. Lovejoy, G. Suwa and G. WoldeGabriel (2009a) '*Ardipithecus ramidus* and the paleobiology of early Hominids'. *Science*, 326: 64–64, 75–86.

White, T. D., S. H. Ambrose, G. Suwa, D. F. Su, D. DeGusta, R. L. Bernor, J. R. Boisserie, M. Brunet, E. Delson, S. Frost, N. Garcia, I. X. Giaourtsakis, Y. Haile-Selassie, F. C. Howell, T. Lehmann, A. Likius, C. Pehlevan, H. Saegusa, G. Semprebon, M. Teaford and E. Vrba (2009b) 'Macrovertebrate paleontology and the Pliocene habitat of *Ardipithecus ramidus*'. *Science*, 326: 87–93.

Willems, E. P. and R. A. Hill (2009) 'A critical assessment of two species distribution models: A case study of the vervet monkey (*Cercopithecus aethiops*)'. *Journal of Biogeography*, 36: 2300–2312.

Wrangham, R. W. (2010) *Catching Fire: How Cooking Made Us Human*. Basic Books.

CHAPTER 21

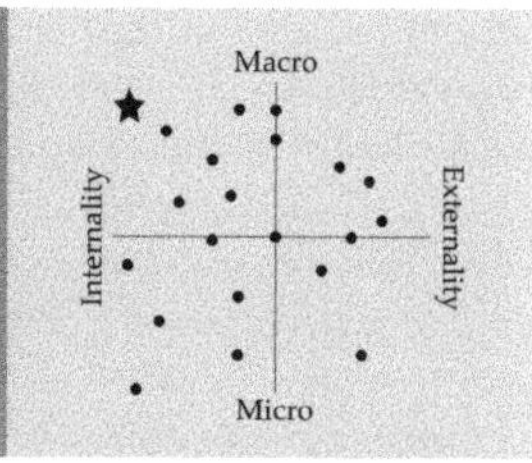

Evolutionary and Cultural 'Swing-by' through Encounters with Others:

A Comparison of Archaeological and Ethnographic Data

Hideaki TERASHIMA

Keywords: *Homo sapiens*, replacement of old human species by new ones, African tropical rainforest, hunter-gatherers and farmers, intermarriage

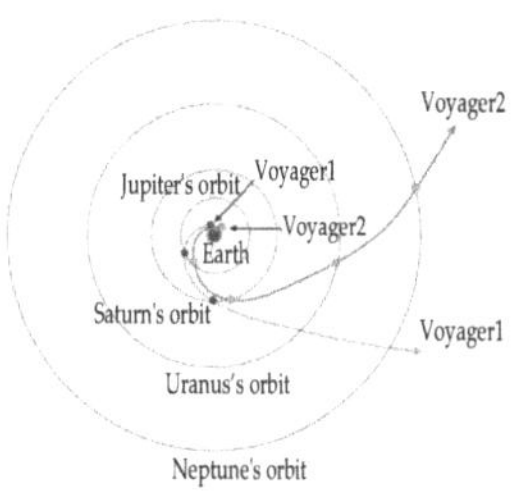

Voyager 1 and 2 space probes, which made use of swing-by

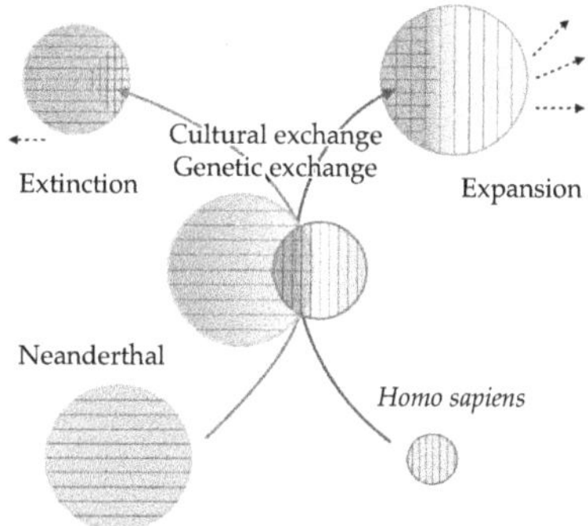

Homo sapiens galvanized to expand after interaction with Neanderthals

Genomic studies of fossilized bones have revealed that modern humans (*Homo sapiens*) engaged in social interactions (including sexual contact) with Neanderthals (*Homo neanderthalensis*) and other ancient human species after leaving their African homeland. What DNA does not tell us, however, is how *Homo sapiens* dealt with such encounters, which in some cases led to cross-species hybridization. This chapter attempts to decipher the interactions between prehistoric *Homo sapiens* and Neanderthals by tracing the history of interactions between indigenous Pygmy hunter-gatherers and later-arriving Bantu and other agriculturalists in the tropical rainforests of Central Africa. It argues that the construction of equal social relationships, including intermarriage with an indigenous population, may have been the cultural and physical impetus behind the geographical expansion of *Homo sapiens*.

The beginnings of travel

Homo mobilitas and *Homo communicans*

As explained by German biologist Jakob von Uexküll (1864–1944), for people, the environment is a world that consists of various things and phenomena that people perceive, interact with, and occasionally compete with or avoid (Uexküll 2010). In addition to things that belong to the natural world, things in the imaginary world—if they have an effect on human behavior—could also be said to be part of the environment. On the other hand, things that we have little to do with, either physically or mentally, do not constitute a meaningful environment, no matter how present they are around us. The way that the environment manifests is intimately related to the skills and knowledge of a human population, its way of life, and its consciousness. The environment is not fixed but, rather, is always in (sometimes violent) flux. In this chapter, I consider such situations—in which an environment undergoes a massive change and reorganization—as extremes of the living environment.

However, I am not concerned here with extremes in the sense of the limits of biological survival. However harsh nature is in objective terms, it makes little sense to call it extreme for people who have already established a stable livelihood there. The same is true for natural disasters and catastrophes, which have an enormous impact on people but leave little room for mutual negotiation. On the other hand, encounters with groups of strangers, which are the subject of this chapter, although they may sometimes pose an existential threat, can also be approached in a variety of ways. As a result, new patterns of behavior and lifestyles may emerge and bring about major social changes, which can sometimes turn out to be fortuitous for both groups.

Here I would like to point out two capacities that people have that have relevance to environmental change. The first is their exceptional mobility. After the emergence of modern humans (*Homo sapiens*) in Africa between 200,000 and 300,000 years ago, some of them left Africa to expand into Asia and Europe more than 100,000 years ago, so that by around 10,000 years ago, they had made their mark on all of the world's continents save Antarctica. This characteristic mobility has led some to refer to humans as *Homo mobilitas* (man of mobility) (Akazawa 2012). It seems clear that the history of human mobility has been one of repeated encounters and leave-takings between human groups that are culturally and physically different from each other.

The other capacity I wish to draw attention to is our ability to communicate—the ability to negotiate with the various entities and phenomena that comprise our world. Hitoshi Imamura (2000), a social philosopher, observing that living while interacting with others (in various senses) lies at the root of our existence as social animals, proposes the term *Homo communicans* (man of communication).

Our daily lives are mediated by the exchange of various words and goods with others. This is not a mere economic act, however; Imamura finds a deeper meaning. Namely, humans are animals that live while engaging physically and mentally in exchanges not only with their own compatriots, who share the same culture, but also with strangers from faraway lands, gods, and spirits, as well as all manner of beings and phenomena around them, including the earth, animals, and plants. Imamura uses the word 'communication' in its broadest sense to encompass the entirety of human social life.

From the time of our inception, humans as *Homo communicans* have repeatedly encountered various others unknown to them. Conceivably, this would have entailed some instances of collision upon meeting, but there must also have been processes of mutual recognition, mutual negotiation, and the establishment of trade relations. The powerful vectors of movement, encounter, and reciprocal action have created the diverse yet mutually comprehensible world we inhabit today. An important task of ecological anthropology, which is concerned with the broad exploration of human lifeways, is to investigate the subtleties and consequences of the movements and encounters that have colored human history.

In this chapter, I discuss two cases of such encounters between separate groups. The first is the encounter that took place between the Neanderthals and *Homo sapiens* during the Upper Paleolithic, while the other is the encounter between hunter-gatherers and farming peoples in the rainforests of Central Africa in the mid-Holocene. Although these two events, disconnected in space and time, might seem to be unrelated, they share one major point in common. Namely, as a result of these encounters, the various groups who were party to them began to have mixed families. Using this point as a touchstone, I would like to consider the role of migration and encounters, and the establishment of intermarriage as a sequence in the context of human history, based on data from anthropology, archaeology, genetics, history, and ethnography.

Neanderthals and *Homo sapiens*

It has recently been discovered that *Homo sapiens* (modern humans), who began their exodus from their homeland on the African continent more than 100,000 years ago, often subsequently had encounters in various parts of Eurasia with archaic humans similar to themselves. A representative case is the encounter between *Homo sapiens* and Neanderthals (*Homo neanderthalensis*) in Europe during the Upper Paleolithic.

Neanderthals, who are considered to have evolved in Europe from a population of earlier humans called Heidelberg man (*Homo heidelbergensis*), which came out of Africa, appeared on the prehistorical stage around 300,000 years ago. They were a species of human with stocky bodies who survived the long and harsh Ice Age by hunting large animals with their outstanding physical strength and athletic ability. The size of the Neanderthal brain surpassed that of *Homo sapiens*, and they would also have been highly intelligent. In recent years, evidence of intellectual and cultural activities once thought to have been unique to *Homo sapiens*, such as cave painting and the production of ornaments, have been discovered in a succession of Neanderthal sites (Zilhão et al. 2010; Hoffmann et al. 2018).

Homo sapiens, in turn, evolved about 200,000 to 300,000 years ago from the *Homo heidelbergensis* that stayed in Africa, and some of them left Africa more than 100,000 years ago. Estimates in recent studies have suggested that settlement in western Europe began about 47,000 years ago. There they encountered the Neanderthals, with whom they coexisted for up to 5000 years or more. Then, around 40,000 years ago, the Neanderthals disappeared. *Homo sapiens*, on the other hand, survived to spread quickly throughout Eurasia. This mysterious prehistorical event has been called the 'replacement of Neanderthals by modern humans' (Higham et al. 2014; Nishiaki 2013, 2014, 2015; Sano and Omori 2015; Callaway 2017).

Many hypotheses have been advanced as to the cause of Neanderthals' demise. Some have suggested that *Homo sapiens* destroyed the Neanderthals by virtue of their intelligence and military might. As yet, however, little archaeological evidence of such violence has been uncovered. Based on changes in the number of sites, the Neanderthal population was already on the wane by about 48,000 years ago, before the arrival and settlement of *Homo sapiens*. It has been speculated that even the Neanderthals could not adapt to the coldest conditions of the last glacial period and died out naturally without engaging in major conflicts with *Homo sapiens* (Sano and Omori 2015). So it was that, by 40,000 years ago, Neanderthal sites had finally disappeared from Europe. No definitive cause has yet been established as to why the Neanderthals disappeared. However, the members of *Homo sapiens*, who were making their way into Europe with lithic tool technology that remained on almost the same level as that of the Neanderthals, seem to have been somewhat superior to the Neanderthals in terms of trade networks, adaptability to diverse environments, and creative learning ability. It may be that these abilities were accelerated by the advance of *Homo sapiens* into the harsh environment of the European Ice Age.

However, little concrete evidence has been found of negotiations between Neanderthals and *Homo sapiens*, and the realities of specific contacts between the two populations has remained shrouded in a deep fog. Molecular genetic research that makes use of DNA, which has developed rapidly since the end of the twentieth century, has opened possibilities for approaching this vexing situation. In 1997, a team from the Max Planck Institute for Evolutionary Anthropology in Germany succeeded in extracting mitochondrial DNA from fossilized Neanderthal bone tissue. Then, in 2010, the news that a complete Neanderthal nuclear genome had been sequenced took the world by surprise (Pääbo 2014; Green et al. 2010). As a result of these developments, it was found that a few percent of the genes of all living humans were of Neanderthal origin.

Although *Homo sapiens* and Neanderthals are often treated as separate species, their divergence from a common ancestor took place no more than several hundred thousand years ago, and to that degree it was still possible for them to interbreed and reproduce. The recent discovery of the mixture of *Homo sapiens* with Neanderthal and Denisovan peoples in the Altai Mountains of Central Asia has had an even greater impact (Kuhlwilm et al. 2016; Viola and Pääbo 2013; Reich 2018).

Although Neanderthal-derived genes account for only a few percent of the *Homo sapiens* genome, it is surprising that they have been passed on to most humans existing today, from Eurasia to the New World, including Africa (Chen et al. 2020). Nor is their function insignificant. The whiteness of the skin of humans inhabiting the north, for example, is reportedly a trait inherited from Neanderthals. At the time of the initial migration into Europe, it is thought that *Homo sapiens* had dark skin derived from an African heritage, but in the high latitudes of northern Europe, where sunlight is weaker, lighter skin proved more advantageous for the body's synthesis of vitamin D. The immune system against viruses unique to Eurasia also seems to be a function of Neanderthal-derived genes. Moreover, adaptation to highlands, such as the Tibetan plateau, has been suggested to be a Denisovan-derived trait (Huerta-Sánchez et al. 2014).

The discovery that *Homo sapiens* interacted with Neanderthals and Denisovans, including sexually, and that this interaction has had an impact on human evolution that has lasted into the present day, has prompted a major rethinking of our understanding of human evolution. In the face of these facts, Svante Pääbo, a leading researcher on Neanderthal DNA, arrived at the conclusion that 'low levels of mixing with earlier humans seemed to have been the rule rather than the exception when modern humans spread across the world' (Pääbo 2014: Chapter 23).

By what process did these encounters take place, and what kind of mutual negotiation went on in such contexts? Unfortunately, there is a limit to how much genetic research is able to reveal. Archaeological information is also scarce. In this chapter, I propose to address these questions using historical and ethnographic data from the African rainforest. These two narratives, in Europe and Africa, separated by tens of thousands of years and thousands of kilometers, may seem at a tremendous remove from each other both historically and spatially. On an evolutionary scale, however, the significance of their mutual remove is much less significant.

Specifically, I focus on the encounter in the middle of the Holocene between two groups of people with different origins, distinct cultures, and different ecologies: namely, hunter-gatherers, who were the first inhabitants of the Central African rainforests, and agriculturalist farmers, who arrived in the forests later. Happily, their descendants have survived to the present day, and a variety of information is available, including not only ethnographic data from field surveys but also the legendary and anecdotal information that comprises their oral traditions. In recent years, the molecular genetic analysis of existing ethnic groups has opened possibilities for understanding how these groups interacted with each other in the past. Lexical analysis and glottochronology (the study of the rate of vocabulary changes) also represent powerful tools (Klieman 2003; Bahuchet 2012; Vansina 1990).

Hunter-gatherers and farmers in the African tropical rainforest

In the beginning, a loose fellowship

The tropical rainforests of the massive Congo Basin, which stretches from the Great Rift Valley in Central Africa to the Atlantic Ocean, were inhabited by hunter-gatherers from about 300,000 years ago (which corresponds to the Middle Stone Age in the context of African archaeology) into the Late Stone Age (Taylor 2011; Mercader et al. 2000; Mercader 2002). Molecular genetic studies suggest that the ancestors of the current Pygmy[1] population diverged from the non-Pygmy population, settling in the forest somewhere between around 60,000 and 90,000 years ago. Later, they split again into two populations—one in the west and one in the east—between about 20,000 and 30,000 years ago (Quintana-Murci et al. 2008; Verdu et al. 2009; Batini et al. 2011; Patin et al. 2014).

In the savanna areas of southwestern Nigeria and western Cameroon, Bantu-speaking peoples began farming about 5000 years ago, after the end of the last glacial period. They then began to migrate eastward and southward, also spreading into the forests and into the savanna areas to the north (Vansina 1990; Klieman 2003; Bahuchet 2012). The beginnings of the Bantu migration are said to date somewhere between 4000 and 5000 years ago, though there are some indications that it dates back even earlier to 6000 to 8000 years ago (Klieman 2003; Montano et al. 2011; Moïse 2014). Those farmers who entered into the forests would have come into contact with the indigenous Pygmy hunter-gatherers wherever they went during their migration.

Extant African Pygmy populations are distributed across ten countries (Angola, Cameroon, Equatorial Guinea, Gabon, Central African Republic, Republic of the Congo, Democratic Republic of the Congo (DRC; formerly Zaire), Uganda, Rwanda, and Burundi). They are divided into more than twenty distinct ethnic groups according to linguistic and regional differences, and their total population has been estimated between 200,000 and 400,000 (Bahuchet 2012; Robillard and Bahuchet 2012). While there are certainly physical, cultural, and social variations among the various peoples known as Pygmies, they also share strong commonalities. They form small groups called bands, which consist of a few families to a dozen families, and until a few decades ago they lived a nomadic life of hunting and gathering in the forest. Even in areas where they have long since settled, many people still spend several weeks or even months of the year hunting and gathering in the African forests (Yasuoka 2011; Hattori 2012). Their common values include the freedom and autonomy of the individual and band, gender equality and egalitarianism among young and old, a commitment to sharing within the band, and a powerful sense of belonging to the forest. It is also important to note that all extant Pygmy groups seem to have formed social relationships, including those of economic exchange, with neighboring farmers in their respective territories.

Despite the absence of empirical data concerning contact that took place between hunter-gatherers and farmers in prehistoric times, there is much that can (to some extent) be inferred from other sources, including a wealth of folklore, folktales, and linguistic studies (Gusinde 1956; Turnbull 1976 [1965]; Schebesta 1977 [1933], 1978 [1936]; Bahuchet and Guillaume 1982; Bahuchet 2012; Kazadi 1981; Vansina 1990; Klieman 2003; Moïse 2014). Based on a lexical analysis of the Proto-Bantu language, farming at the time of contact appears to have been concentrated on the cultivation of millet in the east, root vegetables and fruit trees in the west, and oil palm and yam in both the east and west (Vansina 1990). Early farmers seem to have settled first along the banks of rivers, since these crops prefer sunny areas such as woodland edges. These people used stone tools and earthenware, but metalwork had not yet appeared on the scene.

How would those farmers who first moved into the forests have behaved? First and foremost, the forest would have been an extraordinary place, rife with danger for outsiders. If someone were to lose track of the time or the direction in which they were headed, they could find themselves in mortal danger. Biological resources, such as those used for food and natural drugs for medical treatment, were also quite different from those of the savanna or woodland. At such times, the presence of Pygmies as the indigenous forest people must have been quite reassuring for the farmers. Even today, the knowledge and abilities of Pygmy peoples regarding every aspect of their natural forest universe are highly valued by farmers, who acclaim them as 'guides of the forest,' 'our compass,' and 'our teachers.' Living freely in the forest, Pygmies are accorded a special status and sometimes play special roles in farmers' ceremonies. In many farmers' folktales and folklore, Pygmies appear as benefactors who impart various cultural elements to farmers, such as fire, iron, agriculture, and even the incest taboo.

One thing that farmers had to rely on the Pygmies for was their relationship with the spirits of the land. Originally, the Bantu settlers practiced a philosophy of respecting the indigenous people of the land. This was because they believed that the indigenous people possessed access to the various autochthonous spirits and deities, and that having these supernatural forces under control was essential for continuing to live in the forest (Vansina 1990; Klieman 2003). For the Pygmies, too, the ceramics and polished stone tools that the farmers had brought with them would have held considerable allure. And they would also likely have had an interest in people who used nature in a different way than they did. Conceivably, as *Homo communicans*, they would have been receptive to each other as they gradually began visiting each other's places of residence and exchanging gifts, eventually developing a relationship that included fictive kinship, as I describe below.

In the earliest days of settlement, farmers' livelihoods would not have depended solely on agricultural crops but would have made use of multiple modes of subsistence that included hunting, fishing, and gathering foods from the wilderness. Such activities were carried out in the vicinity of the village, however, and never took them deep into the forests. The impact of incipient farming on the environment would have been minimal, their respective niches did not overlap, and their conflicts of interest were minimal; socially speaking, they appear to have been more or less equal (Vansina 1990; Bahuchet 2012; Sato 2010). It is quite likely that they coexisted peacefully and amicably for thousands of years until the coming of the next era. During this time, linguistic consolidations took place in which each Pygmy group adopted the language of the adjacent farming populations. As I discuss below, interethnic marriages are also likely to have occurred with considerable frequency during this period.

Iron and bananas: waxing and waning relationships

The power balance between the two groups was tipped by the advent of iron. The production of iron began in the forested regions of Central Africa sometime around the first century AD, and iron products became quite popular in the region by around the tenth century. Some of the iron implements that were fashioned include axes (still in common use today), large machetes, spears, and arrowheads. Use of these tools greatly augmented the efficiency of farming, making it easier to clear and expand fields for cultivation. Pygmies also obtained these iron products via neighboring farmers to use in their own subsistence practices, and provided the farmers with meat and honey from the forest in exchange for iron goods. Accordingly, as suppliers of iron products, the farmers' position was gradually strengthened (Vansina 1990).

Another development was the introduction of bananas. These would bring about a major change in the agriculturalist lifestyle, which in turn would affect the relationship between the farmers and the Pygmies. While the details are opaque, there is a likelihood that bananas appeared in equatorial Africa at the beginning of the first millennium AD and spread throughout the region by the middle of the millennium. Bananas proved to be a crop perfectly suited to the tropical rainforest. Unlike with yams, there is no need to clear the forest completely for planting. They are easy to harvest and prepare. But the most outstanding characteristic was their ability to produce. With a yield per unit area more than ten times greater than that of yams, the only comparable crop is said to be manioc (cassava), which was not introduced from South America until the nineteenth century (Vansina 1990). The cultivation of bananas meant that fields and villages no longer had to be confined to the margins of the forest. This resulted in a dramatic increase in the farmers' mobility. Moreover, this increase in production led to a burgeoning population, which in turn further increased productivity.

Bananas also presented a very attractive food source for Pygmies. The fruit is easy to transport, easy to store, and easy to prepare—ideally suited to the nomadic life. Thus, by virtue of their bananas and ironware implements, farmers were able to obtain surplus crops and facilitate their exchanges with the Pygmies. For their part, by securing surplus crops from the farmers, the Pygmies were now able to devote themselves to acquiring wilderness goods such as game meat and honey. The groundwork was now laid for farmers and hunter-gatherers to specialize in the products of the fields and forests, respectively.

However, a problem had also arisen. Lacking plantations of their own, the Pygmies' only choices were to obtain bananas from the farmers or to take them without permission. The banana plantations were located within a few kilometers of the villages and the trees were laden with bananas. Next to these plantations were roads that the Pygmies traversed every day. For the Pygmies, bananas growing in the unsupervised plantations were a source of great temptation. Although it was an easy matter to poach bananas from such plantations, it was also obvious that this would incense the farmers. Still, it would have been impossible for the farmers to keep a constant watch on their fields, and there could be no effective way to prevent theft. The only thing they could do would be to resort to witchcraft. Thus the management of their crops became a headache for the farmers.

With the further development of agriculture, there was greater expansion of villages and cultivated land. As more and more forest land was cleared, conflicts of interest between the farmers and hunter-gatherers gradually mounted. Meanwhile, the farmers continued to have high expectations about the game meat and honey they received from the Pygmies, and the Pygmies would be unable to manage without the iron implements and crops supplied by farmers. The farmers now had to produce crops for the Pygmies, and to that extent this meant they had to carry out more work, and the assistance of the Pygmies was now crucial at the height of the busy farming season. Farmers and Pygmies continued to search for a way to meet their respective needs, so that ultimately the relationship of coexistence between the two parties was strengthened, and a division of labor was established by which they were able to satisfy each other's needs and desires (Sato 2010).

However, during this period, farmers still lived in small groups of a few dozen people at most that were dispersed through the forest. Employing the slash-and-burn methods of swidden cultivation, they would also have had to move their residences every few years. Thus, even here, the Pygmies and farmers still lived in their own niches and had not been assimilated into a single system. In that farmers' economic strength increased on the basis of their crops, their dominance also increased to some extent, but their impact on subsistence livelihoods in the forest would likely have remained quite limited.

The world outside and the forest people: amplifying disparity

In the modern era, the forest world changed drastically with the arrival of traders from the world outside the forest, who came seeking slaves and ivory. This section focuses on the situation that unfolded in the Ituri forest, in the easternmost part of the Congo Basin. In the mid-nineteenth century, Arab traders known as the Bangwana appeared in Ituri bearing guns and started trading for slaves and ivory. With the arrival of this outside force, the balance of power in Ituri, which had previously been made up solely of farmers and Pygmies, now took on the aspect of a tripolar structure (Figure 21.1). The Bangwana, by virtue of their military and economic might and the influence of their Islamic religion, established their dominance over the native farmers and ruled all of Ituri until the army of the Congo Free State crushed them in the 1890s. From the early twentieth century, the Belgian colonial government assumed control of the Congo (Schebesta 1978 [1936]; Wilkie and Curran 1993).

For the ivory traders, attentions were focused on the Pygmies, who had long enjoyed a reputation for bravery in hunting elephants armed only with spears—an extremely dangerous prospect. In addition to skill with the spear, successfully bringing down an elephant required a knowledge of elephant ecology and behavior, as well as other qualities such as physical strength, courage, and determination. Even among Pygmies who are master hunters with the bow and arrow or with the net, few are good at elephant hunting. A good elephant hunter is called a *tuma* and commands great respect.

But the ivory traders never negotiated directly with Pygmies. First, rather, they visited the farming villages, where they entered into negotiations with prominent villagers. Beyond the fact that Pygmies shied away from negotiating with strangers, the ivory traders also despised the hunter-gathers, whom they had no interest in negotiating with directly. Acting as intermediaries, villagers would engage Pygmies with whom they were on good terms to go hunting.

When the Pygmies successfully obtained ivory, the villagers would sell it to the ivory traders, giving the Pygmies a small reward for their trouble. Since the ivory traders were only interested in the ivory, the Pygmies also ended up with a large amount of elephant meat, and the entire community—including not only the Pygmies who took part in the hunt but the farmers, as well—enjoyed days of celebration (Harako 1976; Bailey 1991).

Elephant hunting predates the introduction of iron tools, but the advent of iron spears and ivory traders undoubtedly augmented its efficiency, as well as motivation on the part of the Pygmies. However, the exclusive control that the farmers had over negotiations with the outside world further weakened the Pygmies' social status in negotiations with outsiders relative to the farmers' own status.

In 1885, the Congo Free State (geographically congruent with the current DRC) was recognized as the private domain of King Leopold II of Belgium. This was accompanied by the conquest of the Ituri region by the Congo Free State's armed forces in the 1890s, which brought an end to the rule of the Bangwana. Soon, however, the Congo Free State faltered in the face of international criticism against its inhumane policies regarding the collection of wild rubber

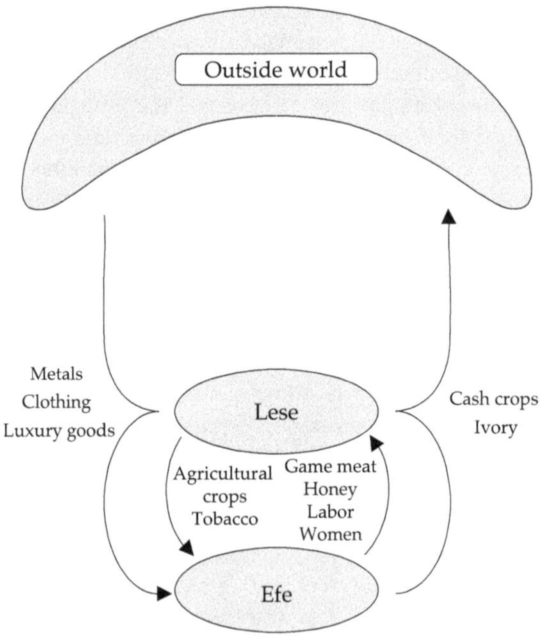

Figure 21.1: Traditional exchange relationships among the forest inhabitants and with the outside world

for export, and the territory became a colony of Belgium in 1908. In the 1930s and 1940s, the Belgian Congo's colonial government built highways through the Ituri forest to promote the region's industrial development, compelling local farmers—who until then had been practicing swidden cultivation (slash-and-burn agriculture) in the forests—to relocate to settlements along the highways. Many of the Pygmies who had befriended the farmers followed their lead. In the new settlements, the government imposed agricultural policies that forced farmers to grow cotton and groundnuts for export. Farmers' agency was reduced to meeting their quotas, and the production of subsistence crops began to rely more heavily on the labor of Pygmies (Wilkie and Curran 1993).

The farmers and Pygmies who had once lived in small groups scattered throughout the forest were now forced to congregate in a limited space along the road and to live in an environment with which they had no previous experience. The settled (or something approximating settled) lifestyle proved quite damaging to the Pygmies' nomadic lifestyles. As it became more difficult to secure wild plants in the vicinity for food, rather than deliberately traveling deeper into the forest, Pygmy women increasingly turned to helping farmers in their fields and obtaining crops in payment. Meanwhile, it was also becoming difficult for farmers to freely cultivate new land every year as they once had. Lacking sufficient land that would allow them to rotate their crops and fallow their fields, they struggled as their crop yields declined (Wilkie and Curran 1993; Wæhle 1989).

The colonial government's policy toward those who lived in the forests served to foster discrimination and conflict rather than equality and harmony among the inhabitants. Government officials discriminated against Pygmies, whom they considered more uncivilized than farmers. They concentrated local power in the hands of farmers and allowed the Pygmies to be forcibly conscripted to work on plantations. Such policies further reinforced discriminatory attitudes toward Pygmies on the part of the farmers.

On the heels of the Congo gaining its independence from Belgium in 1960, the Congo Crisis (1960~1965, a series of civil wars involving various dimensions of army revolts, conflicts between political parties as well as ethnic groups, and also of proxy conflict in the Cold War) plunged the whole country into a prolonged period of unrest. In the Ituri forest, various groups that included stragglers from the Simba rebellion (1963~1965, that once captured much of eastern Congo) caused a great deal of mischief. When Mobutu Sese Seko, a former military officer, seized control of the government in a 1965 coup d'état, the disturbances seemed to have finally settled. In the 1970s, however, the new government's policy of shutting out foreign companies and the deterioration of infrastructure such as roads and public transportation led to economic collapse, as well as rampant political corruption. With the rapid decline in the market economy, many villagers suffered as they found themselves increasingly persecuted and unjustly dispossessed by the government and military forces (Wilkie and Curran 2003; Wæhle 1989; Grinker 1994; Mwanyimi-Mbomba 1985; Ichikawa 1991; Sawada 2010).

A world inhabited by hunter-gatherers and farmers

The foregoing represents a sketch of the history of contact between hunter-gatherers and farmers in the tropical rainforests of Central Africa, from a primordial era that began more than 5000 years ago and was characterized by gentle equality, or perhaps dominance by the Pygmies as the indigenous peoples (Bahuchet 2012), to an era in which the Pygmies gradually became more dependent on the iron tools and crops supplied by farmers (which led to a decline in their social and economic status) and finally to the era that came after the mid-nineteenth century, when incipient disparities were amplified by factors such as the intrusion of outside forces, the market economy, and discriminatory government policies. The relationship between the two populations in the modern era is frequently described as a patron–client relationship, a social institution based on a dominant–subordinate relationship (Wilkie and Curran 2003; Grinker 1990, 1994). However, even if it appears superficially stratified, the reality is far from being what the 'patrons' (as the dominant party) might prefer (Turnbull 1976 [1965]; Wæhle 1989; Bahuchet 2012; Terashima 1997). The surface relationships, which are colored by political and social disparities, appear markedly different in the everyday sociocultural contexts that underlie the surface. As a core consideration of this, I would next like to discuss intermarriage and the construction of affinal relations between ethnic groups through fictive kinship.

The reality and process of interethnic marriage

Although the occurrence of interbreeding between Pygmies and farmers has long been discussed (Schebesta 1977 [1933], 1978 [1936]), genetic studies since the mid-twentieth century have revealed a history of widespread mixing between Pygmy and farmer populations in Cameroon and Central Africa (Cavalli-Sforza 1986). More recent studies using DNA analysis have found that intermarriage has taken place with considerable frequency throughout the African tropical rainforest, from the past up to the present day.

After examining the DNA of nine extant Pygmy groups and nine Bantu populations spanning a territory from West Africa to Kenya, Patin et al. (2014) noted a bidirectional gene flow in almost all Pygmies and farmers. Among the Babongo in Gabon and the Bezan in Cameroon, in particular, DNA flow from farmers to Pygmies averaged 40%–50%. In the other populations, this value averaged around 10%. Overall, it was found that the flow of Pygmy

DNA into farmer populations averaged around 10% (0.7%–15.7%). In a similar study, Verdu et al. (2009) point to the historical existence of a situation strongly characterized by bidirectional mixture.

Ethnographic studies have also been conducted in this regard, although only to a limited degree. Table 21.1 shows the state of intermarriage with neighboring farmers for ten Pygmy populations found in ethnographic studies from the past few decades (Takeuchi 2014). In two Aka populations in the Republic of the Congo and the Central African Republic, and one Mbuti population in the DRC, almost no intermarriage was found to be taking place with neighboring farmers. Meanwhile, in the three Baka populations in Cameroon, the rate of intermarriage with farmers ranged between 2% and 9%. Conversely, among three Efe populations in the DRC, intermarriage with Lese agriculturalists ranged between 17% and 28%, while the frequency of intermarriage with the Babongo in Gabon was even higher (42%). In addition, in Gabon, 11% of such marriages took place between female farmers and male Pygmies, which is rarely seen elsewhere (Matsuura 2011, 2012). Today it is not uncommon for farmers to express discriminatory and disdainful attitudes toward the Pygmies that prompt them to reject the prospect of intermarriage out of hand. In these villages, however, intermarriage was a common practice historically and, in some cases, remains quite widespread today.

Pygmy ethnic group	Farmers	Village	Intermarriage index		Source
			Female	Male	
Aka	Minputu	Ibenga	0† (0, 72)††	0† (0, 151)††	Takeuchi 2014
Aka	Ngandu	Bakoka	rare	0	Hewlett 1991
Mbuti	Kaheku, Nande, and Pakombe	Southern Ituri	a few	0	Hart 1979
Baka	Bangando	Koumela	0.02 (3, 140)	0 (0, 120)	Matsuura 2012
Baka	No data	Lomie	0.045	no data	Hewlett 1996
Baka	No data	Abongmbang	0.086	no data	Hewlett 1996
Efe	Lese	Malembi	0.17 (21, 123)	0 (0, 97)	Bailey 1985
Efe	Lese	Mangala	0.25 (14, 55)	0	Terashima 1987
Efe	Lese	Andiri	0.28 (17, 60)	0	Terashima 1987
Babongo	Massango	Ogoulou	0.42 (18, 43)	0.11 (3, 28)	Matsuura 2012

Table 21.1: Interethnic marriage between Pygmy hunter-gatherers and farmers (modified from Takeuchi 2014)

† Ratio of interethnic marriage

†† First values in parentheses indicate the number of Pygmies married to farmers, while the second value is the total number of female or male Pygmies, respectively

Now I would like to offer a description of the process of intermarriage between different ethnic groups based on my own field notes from research conducted in the late 1970s and early 1980s among a group of Lese farmers and Efe Pygmies living in the vicinity of the village of Andiri, which is located in the northeastern part of the Ituri forest, which stretches to the northeastern edge of the DRC. The Efe and Lese have maintained a mutual relationship for at least several hundred years—as long as two millennia by some accounts (Wilkie and Curran 2003). The relationship between the two groups is not merely economic, but also has ritual elements, and the crux of the arrangement was a social relationship that I dubbed the '*Efe-maia—muto-maia*' (EM) relationship (Terashima 1987). The term '*Efe-maia*' means 'my Efe' in Lese, while '*muto-maia*' means 'my villager.' These are used as mutual terms of address by a specific Efe male and a Lese male who act as though they are true brothers. Such customs of mutual recognition and mutual obligation between people who do not have a consanguineous relationship is known as fictive kinship and is also found in Pygmy societies in other regions (Joiris 2003).

The EM relationship extends from the principals concerned to their families. If the parents are in an EM relationship, their children will have the same relationship to each other. An Efe will regard the child of his or her *muto-maia* as his or her own, and the same is true of a Lese for the child of an *Efe-maia*. Those party to an EM relationship are mutually obliged to provide each other with their free and full support. People in EM relationships are considered to be like family members and may not marry. However, there is an important link between the EM relationship and intermarriage between Efe and Lese.

The most desired form of marriage in Efe society is what is known locally as 'head and head.' A bride-giving group can demand a woman from the bride-taking group as a bride for its own group. In a hunter-gatherer society, women's work is universally important. The loss of a woman's labor is a great blow, but the loss could be immediately compensated if the bride-giving group were to receive a bride for itself as well. In anthropological terms, this method is known as marriage by sister exchange (here, the term 'sister' is classificatory, and extends to include female patrilateral first and second cousins).

In the absence of a suitable woman in the groom's group, alternative measures will be taken. One of these is marriage by bride service, a matrilocal pattern in which the newlywed couple remains with the bride's group for a few years to assist her parents. The other is marriage by bride price, in which the groom supplies the bride's family with gifts such as dishes, clothes, and ornaments, or cash or livestock. For an Efe, however, preparing the necessary items on his own, even if he opts for bride price, is far from an easy matter, and in many cases, his *muto-maia* will shoulder this responsibility. In such a case, the *muto-maia* would be able to adopt a daughter born from the resulting marriage. While the daughter is young, she will live with her parents, but when the time comes, she will move to the village of the *muto-maia* where, once she is of age, she will usually marry a villager. In such a case, the bride price paid by the groom will belong to the *muto-maia* (Terashima 1987). Consequently, the *muto-maia* will have recouped the bride price that he paid on behalf of his *Efe-maia* (Figure 21.2). I should add that this custom is not a special case of the EM relationship and is held to be a general practice in Lese society with regard to bride price exchange.

A young Efe woman does not necessarily have to be a Lese's adopted daughter to marry a Lese man. However, as a prerequisite for marriage, it is essential that she is able to adapt to life in the village, and this is where everyday interactions based on EM relationships have much to contribute. When an Efe and Lese are in an EM relationship, their entire families will be on friendly terms. The Efe's daughters will visit the village often from a young age to spend time in the *muto-maia* household doing housework, helping in the fields, and minding children. When they are the right age, they will also undergo the female initiation ritual known as *ima* together with young Lese women of their own age. In this way, as they spend more time in the village and become accustomed to the rhythms of village life, they may begin to think about marrying a Lese. 'You might be born an Efe,' the Lese proclaim, 'but when you grow up in a village, you are a villager.'

As far as I have been able to determine, couples consisting of an Efe male and Lese female are unknown. The Lese explain that this is because 'village women cannot live in the forest.' Indeed, it would be impossible for a woman who grew up in a village to suddenly become a hunter-gatherer living a nomadic life. And while I cannot say definitively that discrimination toward the Efe might not be part of the story, it is nevertheless clear that differences in

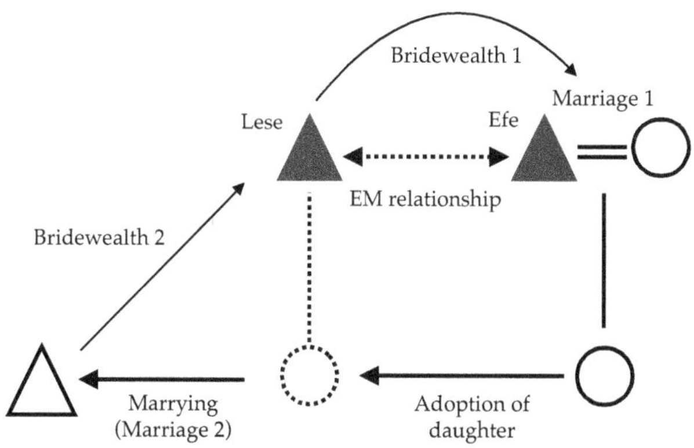

Figure 21.2: Marriage and bridewealth transfers through the *Efe-maia—muto-maia* relationship

lifestyle are a major factor in the unidirectionality of intermarriage. In Andiri, there was little evident disdain or discrimination against Efe brides and their children who lived in the village. As observed by the farmers interviewed long ago by the Austrian ethnologist Paul Schebesta (1977 [1933], 1978 [1936]), the Efe are prolific and hard working, and are therefore more desirable as wives. However, I should note that an Efe girl's marriage to a villager does not serve to raise the status of her parents or benefit them in any economic sense.

Among the Massango farmers of Gabon, conversely, there is no objection to their daughters marrying local Babongo Pygmies. One possible reason for this is also the similarity of their lifestyles. According to Matsuura (2011), by the mid-twentieth century, the Babongo had already adopted a settled lifestyle and were making their living by farming. While they also engaged in hunting and gathering, these activities were confined to the vicinity of their villages, and their subsistence patterns were more or less consistent with those of the Massango. Their villages and houses were also almost identical, and both had matrilineal clans and practiced virilocal residence patterns. The Babongo regularly spoke the Massango language, and the two groups celebrated coming-of-age ceremonies and other important rituals together. Hence there were few physical barriers that might hinder intermarriage.

Equality and the beginnings of intermarriage

Cavalli-Sforza (1986), who sought to use genetic analyses to elucidate the history of intermarriage between Pygmies and farmers, speculated that the early period of contact was the most active period of intermarriage, judging by the extensive scale of intermarriage that his survey revealed between the two groups. The reason for this, paradoxically, is the lack of intermarriage between female farmers and male hunter-gatherers in modern times. Cavalli-Sforza felt that this was due to the discriminatory attitudes of farmers toward hunter-gatherers.

Hence, if there had been a time in the past when intermarriage had been more frequent, it would likely have been in the early period, when such discriminatory feelings were not as common.[2] It is certainly the case that prior to the introduction of iron implements and bananas, the agriculturalists, like the Pygmies, had also engaged in hunting, fishing, and gathering activities, and there would not have been any differences in their respective lifestyles that would have stood in the way of intermarriage. In addition, there was little influence from outside society, and thus no social disparities that would have stemmed from this. Rather, as mentioned earlier, the Pygmies had the advantage of being indigenous to the area, and thus familiar with all aspects of the forest, including the spirits of the land.

This early period of equal coexistence continued for a prolonged period of time. The beginning of the Bantu migration was once assumed to be somewhere between around 1000 and 2000 BC (Vansina 1990). More recently, however, using the technique of glottochronological analysis, Klieman (2003) has estimated the migration to have begun around 4000 BC. Meanwhile, a study by Montano et al. (2011) using molecular genetics has suggested a possible date as far back as 6000 BC. If contact between Pygmies and farmers began thousands of years earlier than previously assumed, this would suggest that their loose and equal relationship would have lasted that much longer, increasing the likelihood of their having established close social relationships and bilateral intermarriage in various places during this period.

The relationship between equality or equivalency and intermarriage is an important one. Presumably, before love across ethnic boundaries could be consummated as a matter of course in the form of marriage, social relationships suitable for intermarriage would have to have been established to some extent. As mentioned above, where intermarriage is now commonplace, there exists a background that facilitates it—in other words, interdependent and more or less equal and amicable social relations.

The relationship between the Pygmies and the villagers has sometimes been perceived in terms of a patron–client relationship. Seen thus as the relationship between a protector and a ward, there would seem to be some overlap with the EM relationship found in Andiri. In fact, however, there is a fundamental difference. That is, the framework of the former is a hierarchy in which the farmers are clearly superior and seek to control the conduct of the subordinate Pygmies. The client is regarded as an instrument to support the patron's wealth and power base (Wilkie and Curran 1993). Such a relationship would be completely unacceptable to the Pygmies, who value individual freedom and equality.

In the context of EM relationships, the relationship between parents is inherited by the children, but from the point of view of the Efe, the villagers who serve as *muto-maia* are not always good people. Tolerant and generous *muto-maia* are viewed approvingly by the Efe, but in some cases *muto-maia* can also be arrogant and demanding. Such villagers are disliked and kept at a distance. This does not necessarily erase the fact of an EM relationship, but the relationship is not necessarily binding. In the event that a *muto-maia* is deemed unsatisfactory, the Efe can cause the attenuation of the EM relationship by various means, such as by keeping his distance from the *muto-maia* in question or establishing a new trade partnership with another farmer (Wæhle 1989; Terashima 1997). The option of absconding beyond the reach of avaricious and high-handed villagers or officials was always open to the Pygmies. The forest was their ultimate refuge, and as long as it was there to escape to, they could be free and secure for the time being.

Farmers with strongly discriminatory attitudes would repeatedly vent their frustrations with diatribes against Pygmies who were unable or unwilling to behave as the farmers wanted, accusing them of being inconstant, capricious, sloppy in their work, lazy, brutish, and worse. Yet any designs they might have had on enslaving the Pygmies by force never achieved any result. Turnbull (1976 [1965]) saw their behavior as what he called 'wayward servants' as the Pygmies' optimal strategy against any farmers who sought to constrain their freedom and independence.

In the end, no relationship can last very long if either party feels that it is a one-sided exploitation. Since the period of their earliest contact, Pygmies had been accorded respect by the farmers as the forest's original inhabitants. The Pygmies, meanwhile, had no objection to living in the forest with farmers for as long as the latter did not undermine the Pygmies' dignity or their values, such as with regard to their independence and equality, and they greatly appreciated the crops and metal products that the farmers provided.

Furthermore, the intimate association between the two groups enabled them to move freely between the two worlds of the village and the forest, to enjoy each space as a site of recreation and a refuge in which to seek solace from the stresses of their own society (Terashima 1991). In this way, a relationship of symbiosis or 'consortism' (Terashima 2001; Wæhle 1989) was developed between people of broadly different origins and ecologies. It is not at all unusual that such contexts should engender instances of love and produce new romantic partnerships. And with the proliferation of intermarriages, the social relationship between the two groups becomes that much more intimate.

Encounters and departures: swing-by to the new world

Here I would like to return once again to the encounters and interactions between *Homo sapiens* and archaic humans such as Neanderthals that took place in prehistoric times. DNA research on *Homo sapiens* and archaic humans by Svante Pääbo and others has generated completely new insights into the relationship between these different human species, which until recently had often been viewed as hostile. As Pääbo (2014: Chapter 23) writes:

> I had come to believe that although the big picture of modern human spread was one where the replacement crowd pushed other groups into extinction, this was not a total replacement. Rather, some DNA seemed to leak over into the groups that lived on, so much so that I started using a term I had picked up from somewhere to describe this process: 'leaky replacement.'

On the recent discoveries of interspecific reproduction among some humans, Chris Stringer (quoted in Morelle 2016) commented that 'anywhere in southern Asia could theoretically have been the location of this early interbreeding, since we really don't know how widespread Neanderthals and early modern humans might have been in the regions between Arabia and China…' He went on to mentioned three possibilities of such encounter and happening among them: (1) relatively peaceful exchanges of partners, (2) one group raiding another and stealing females (which happens in chimps and some modern hunter-gatherers), and (3) adopting abandoned or orphaned babies (Morelle 2016).

The first of these theories—that of peaceful negotiation—seems to be the only one that is clearly plausible, while the second and third theories, if they could have taken place at all, would have been sudden phenomena at most, and thus unlikely to have left traces over such a wide swath of the earth's extant human population. We can only assume that some process must have taken place that made sense to both *Homo sapiens* and Neanderthals, and which allowed them to reach the point of interbreeding.

Considering the size of their brains, the *Homo sapiens* and Neanderthals must have been equipped with a high degree of sociality and communicative ability. One recent study of Neanderthal hyoid bones (the bones situated at the root of the tongue) excavated from Kebara Cave in Israel suggests that they would have possessed vocal abilities more or less similar to those of *Homo sapiens* (D'Anastatio et al. 2013). Nevertheless, communication does not necessarily entail the use of spoken language to that end. Even in the historical period, it must frequently have occurred that people of different ethnicities, encountering each other for the first time, found their words mutually incomprehensible. Nevertheless, *Homo sapiens*, as *Homo communicans*, has been able to communicate its ideas sufficiently by various means other than words. With this also in mind, and in light of the interactions presented in this chapter between hunter-gatherers and farmers that have taken place in the tropical African rainforest from prehistoric times to the present day, Stringer's first theory of interbreeding through peaceful negotiation offers a convincing story.

An encounter with an unknown stranger is a process that begins with exploratory communication between the two parties, and which proceeds to gradually deepen their mutual empathy, recognition, and appreciation through mutual visits and exchanges of gifts with each other. Where people are hesitant to meet in person, they can communicate through other means, such as silent trade. As one party discovers tools, techniques, and knowledge that the other lacks, the two will increase the density of their interactions and the level of their trade through access to these discoveries. Interactions like rituals, which take place in the symbolic realm, will gradually increase, and groups that were once 'others' will come to be recognized as firmly identifiable neighbors in their respective daily lives. Institutionally, as well, they will learn to maintain intimate and stable relationships with each other, such as, for example, by likening each other to kin.

The establishment of fictive kinship is a common means of exchange found in many human societies, and presumably has ancient roots in the evolution of human society. The creation of such social relationships allows people of widely different origins to be socially recognized as members of a marriageable group. Once marriages between the two groups can be solemnized without difficulty, even stronger forms of cooperation will emerge.

Although the above scenario remains speculative, if we consider it alongside the 5000 years of coexistence between *Homo sapiens* and the Neanderthals, and the predispositions of humans as *Homo communicans*, it is perhaps not such an unreasonable assumption. It is not uncommon in human history for those belonging to different ethnic groups to have bonded with each other, leaving behind offspring and creating new groups in the process. Strangers who differ from ourselves are typically feared as potential enemies, but it is also often the case that it is by virtue of their being 'aliens' that they are eagerly awaited. That a belief in deities from the outer world who bring happiness and good fortune is ubiquitous throughout the world indicates a strong interest in and expectations for odd people from distant worlds, who may have abilities that we do not possess.

It is said that nature abhors inbreeding. In the higher animals, this is avoided by means of a variety of ecological and behavioral processes. Jun'ichirō Itani (1987), one of the founders of Japanese primatology, once elucidated the evolution of primate societies from primates to apes and humans by plotting this development along two axes; namely, innate behavior that leads to the avoidance of inbreeding and herd inheritance. In addition, while the pair-bond between males and females is a fundamental structure of human society (Chapais 2008), this pair-bond is buttressed by the inbreeding avoidance system. Claude Lévi-Strauss (1949) clearly stated that the intentional prohibition of incest is the dividing line between nature and culture in human society. The avoidance of inbreeding, which functioned as an adaptive natural behavior among nonhuman primates, functions as the intentional social institution of exogamy in human society, engendering social solidarity and cultural exchange between different human groups. Our keen interest in people who are different from ourselves, without going as far as to mention the legends of visiting gods, is simply a trait that has been biologically and socially embedded in our primate DNA since time beyond memory.

There exists an astrophysical technique called 'swing-by' that makes use of universal gravitation. According to the Japan Aerospace Exploration Agency (JAXA 2007), swing-by is a method by which planetary probes, which can carry only a small amount of fuel (i.e., propellant), accelerate themselves to travel over vast distances by exploiting the gravity of each planet they approach along the way (see the figure at the beginning of this chapter). It is a space navigation technology that gained popular attention when it was used by NASA's twin planetary probes Voyager 1 and Voyager 2 and JAXA's Hayabusa asteroid probe. I would like to conclude by drawing an analogy between the encounters and interactions among different peoples that we have discussed in this chapter and stories of departures setting out in search of new worlds.

Homo sapiens, which emerged in Africa when the Neanderthals were beginning to be active in Europe, was an ordinary human species with little to recommend it in the way of physical or technological prowess. Even so, encouraged by characteristics as *Homo mobilitas,* its members eventually left Africa in search of new lands. Quite unexpectedly, they encountered indigenous inhabitants here and there that we call the 'archaic humans,' and each time they began to engage in communication with each other as *Homo communicans.* Almost before they knew it, they found themselves enjoying friendly relations with these people, and in addition to conveying and exchanging knowledge and skills, they also began to engage in sexual relations. As a result, through genetic exchange with the archaic humans, the *Homo sapiens* newcomers were able to acquire physical abilities such as natural immunities and adaptive traits suited to their new natural environment. Their interactions with indigenous populations during their travels unintentionally nourished the cultural and physical development of *Homo sapiens.* As the days and months passed, *Homo sapiens* once again surrendered to the pull of the unseen world that lay over the horizon and embarked on a new journey as *Homo mobilitas.* Thus it was that, through repeated swing-by encounters and farewells with strangers, *Homo sapiens* grew mentally and physically, until it found itself in a new world far from home.

The various steps of the natural world occasionally exhibit violent aspects in the form of great leaps, transformations, and ruptures. The same is true of the course of human evolution. The process described in this chapter of chance encounters with strangers and subsequent interactions in archaeological and prehistoric contexts must certainly have been events such as these. And, where those involved were able to make the most of it, this could be said to have been a wonderful way of taking advantage of an extreme situation.

Notes

1. In this chapter, I follow popular usage in employing the term 'Pygmy' to refer broadly to the hunting and gathering peoples that have inhabited the tropical rainforests from prehistoric times to the present day in territories from West Africa to the Great Rift Valley.
2. Patin et al. (2014), reporting on a genetic analysis of present-day rainforest hunter-gatherers (i.e., Pygmy) populations, found that while these groups presented as much as 50% of farmer genomic ancestry, large-scale admixture only began about 1000 years ago. On the other hand, Quintana-Murci et al. (2008) say that asymmetric maternal gene flow from Pygmies to the ancestors of the farming populations began no more than 40,000 years ago and persisted until a few thousand years ago.

References

Akazawa, T. (2012) 'Seven million years of *Homo mobilitas*. In M. Into (ed.), *Human Great Journey: From Africa to Easter Island*. Asahi Shinbun Shuppan, 7–32. [in Japanese]

Bahuchet, S. (2012) 'Changing language, remaining Pygmy'. *Human Biology*, 84(1): 11–43.

Bahuchet, S. and H. Guillaume (1982) 'Aka-farmer relations in the northwest Congo Basin'. In E. Leacock and R. Lee (eds), *Politics and History in Band Societies*. Cambridge University Press: Cambridge/Maison des Sciences de l'Homme, 189–211.

Bailey, R. C. (1991) *The Behavioral Ecology of Efe Pygmy Men in the Ituri Forest, Zaire*. Anthropological Papers, Museum of Anthropology, University of Michigan, No. 86.

Batini, C., J. Lopes, D. M. Behar, F. Calafell, L. B. Jorde, L. van der Veen, L. Quintana-Murci, G. Spedini, G. Destro-Bisol and D. Comas (2011) 'Insights into the demographic history of African Pygmies from complete mitochondrial genomes'. *Molecular Biology and Evolution*, 28(2): 1099–1110.

Callaway, E. (2017) 'Oldest *Homo sapiens* fossil claim rewrites our species' history'. *Nature News*, 8 June.

Cavalli-Sforza, L. L. (ed.) (1986) *African Pygmies*. Academic Press Inc.

Chapais, B. (2008) *Primeval Kinship: How Pair-Bonding Gave Birth to Human Society*. Harvard University Press.

Chen, L., A. B. Wolf, W. Fu, L. Li and J. M. Akey (2020) 'Identifying and interpreting apparent Neanderthal ancestry in African individuals'. *Cell*, 180: 677–687.

D'Anastatio, R., S. Wroe, C. Tuniz, L. Mancini, D. T. Cesana, D. Dreossi, M. Ravichandiran, M. Attard, W. C. Parr, A. Agur and L. Capasso (2013) 'Micro-biomechanics of the Kebara 2 hyoid and its implications for speech in Neanderthals'. *PLOS ONE*, 8(12): 1–7.

Green, R. E., J. Krause, A. W. Briggs, T. Maricic, U. Stenzel, M. Kircher, N. Patterson, H. Li, W. Zhai, M. H. Fritz, N. F. Hansen, E. Y. Durand, A. S. Malaspinas, J. D. Jensen, T. Marques-Bonet, C. Alkan, K. Prüfer, M. Meyer, H. A. Burbano, J. M. Good et al. (2010) 'A draft sequence of the Neandertal genome'. *Science*, 328: 710–722.

Grinker, R. R. (1990) 'Images of denigration: Structuring inequality between foragers and farmers in the Ituri forest, Zaire'. *American Ethnologist*, 17(1): 111–130.

Grinker, R. R. (1994) *Houses in the Rain Forest: Ethnicity and Inequality among Farmers and Foragers in Central Africa.* University of California Press.

Gusinde, M. (1956) *Die Twiden: Pygmäen und Pygmoide im tropischen Afrika.* W. Braumüller.

Harako, R. (1976) 'The Mbuti as hunters: A study of ecological anthropology of the Mbuti Pygmies (1)'. *Kyoto University African Studies*, 10: 37–99.

Hart, J. A. (1979) *Nomadic Hunters and Village Cultivators: A Study of Subsistence Interdependence in the Ituri Forest of Zaire.* A thesis submitted to Michigan State University.

Hattori, S. (2012) *Challenge for Coexistence of Forests and People: Conservation of Tropical Rainforests and the Culture of Hunter-gatherers in Cameroon.* The Center for African Area Studies, Kyoto University. [in Japanese]

Hewlett, B. S. (1991) *Intimate Fathers: The Nature and Context of Aka Pygmy Paternal Infant Care.* The University of Michigan Press.

Higham, T., K. Douka, R. Wood, C. B. Ramsey, F. Brock, L. Basell, M. Camps, A. Arrizabalaga, J. Baena, C. Barroso-Ruíz, C. Bergman, C. Boitard, P. Boscato, M. Caparrós, N. J. Conard, C. Draily, A. Froment, B. Galván, P. Gambassini, A. Garcia-Moreno, S. Grimaldi et al. (2014) 'The timing and spatiotemporal patterning of Neanderthal disappearance'. *Nature*, 512: 306–309.

Hoffmann, D. L., C. D. Standish, M. García-Diez, P. B. Pettitt, J. A. Milton, J. Zilhão, J. J. Alcolea-González, P. Cantalejo-Duarte, H. Collado, R. de Balbín, M. Lorblanchet, J. Ramos-Muñoz, G. C. Weniger and A. W. G. Pike (2018) 'U-Th dating of carbonate crusts reveals Neandertal origin of Iberian cave art'. *Science*, 359: 912–915.

Huerta-Sánchez, E., X. Jin, Asan, Z. Bianba, B. M. Peter, N. Vinckenbosch, Y. Liang, X. Yi, M. He, M. Somel, P. Ni, B. Wang, X. Ou, Huasang, J. Luosang, Z. X. Cuo, K. Li, G. Gao, Y. Yin, W. Wang et al. (2014) 'Altitude adaptation in Tibetans caused by introgression of Denisovan-like DNA'. *Nature*, 512: 194–197.

Ichikawa, M. (1991) 'Barter and cash trade in the Ituri Forest of Zaire: On the discrepancy of exchange system'. In Y. Tani (ed.), *Reading Culture: Between Fields and Texts.* Jinbun Shoin, 48–77. [in Japanese]

Imamura, H. (2000) *Homo communicans: Anthropology of Gift and Exchange.* Kodansha. [in Japanese]

Itani, J. (1987) *Evolution of Primates Society.* Heibonsha. [in Japanese]

JAXA (Japan Aerospace Exploration Agency) (2007) 'A story of "Hayabusa," a small inter-planet asteroid prover', http: //spaceinfo.jaxa.jp/hayabusa/about/principle1.html [in Japanese]

Joiris, D. V. (2003) 'The framework of Central African hunter-gatherers and neighbouring societies'. *African Study Monographs*, Supplementary Issue, 28: 57–79.

Kazadi, N. (1981) 'Méprises et admires: L'ambivalence des relations entre les Bacwa (Pygmées) et les Bahemba (Bantu)'. *Africa*, 51(4): 836–847.

Klieman, K. A. (2003) *The Pygmies were Our Compass: Bantu and Batwa in the History of West Central Africa, Early Time to c. 1900 C.E.* Heinemann.

Kuhlwilm, M., I. Gronau, M. J. Hubisz, C. de Filippo, J. Prado-Martinez, M. Kircher, Q. Fu, H. A. Burbano, C. Lalueza-Fox, M. de la Rasilla, A. Rosas, P. Rudan, D. Brajkovic, Ž. Kucan, I. Gušic, T. Marques-Bonet, A. M. Andrés, B. Viola, S. Pääbo, M. Meyer et al. (2016) 'Ancient gene flow from early modern humans into Eastern Neanderthals'. *Nature*, 530: 429–433.

Levi-Strauss, C. (1949) *Les Structures Élémentaires de la Parenté*. Presses Universitaires de France.

Matsuura, N. (2011) 'Historical changes in land use and interethnic relations of the Babongo in southern Gabon. *African Study Monographs*, 32(4): 157–176.

Matsuura, N. (2012) *The Modern 'Forest People': An Ethnography on Babongo Pygmies in Central Africa*. Showado. [in Japanese]

Mercader, J., M. Garcia-Heras and I. Gonzalez-Alvarez (2000) 'Ceramic tradition in the African forest: Characterisation analysis of ancient and modern pottery from Ituri'. *Journal of Archaeological Science*, 27: 163–182.

Mercader, J. (2002) 'Forest people: The role of African rainforests in human evolution and dispersal'. *Evolutionary Anthropology*, 11: 117–124.

Moïse, R. E. (2014) '"Do Pygmies have a history?" revisited: The autochthonous tradition in the history of Equatorial Africa'. In B. S. Hewlett (ed.), *Hunter-Gatherers of the Congo Basin: Cultures, Histories, and Biology of African Pygmies*. Transaction Publishers, 85–116.

Montano, V., G. Ferri, V. Marcari, C. Batini, O. Anyaele, G. Destro-Bisol and D. Comas (2011) 'The Bantu expansion revisited: A new analysis of Y chromosome variation in Central Western Africa'. *Molecular Ecology*, 20: 2693–2708.

Morelle, R. (2016) 'Neanderthals and humans interbred "100,000 years ago"'. BBC News, 17 February, https://www.bbc.com/news/science-environment-35595661

Mwanyimi-Mbomba, M. (1985) *Chronologie Générale de l'Histoire du Zaïre (Des Origines à mon Jours)*. Centre de Recherches Pedagogiques.

Nishiaki, Y. (ed.) (2013) *Homo sapiens and Neanderthals: The Replacement Viewed from Palaeolithic Archaeology*. Rokuichi Shobo. [in Japanese]

Nishiaki, Y. (ed.) (2014) *Homo sapiens and Neanderthals 2: Learning Viewed from Archaeology*. Rokuichi Shobo. [in Japanese]

Nishiaki, Y. (ed.) (2015) *Homo sapiens and Neanderthals 3: The Replacement of Humans and Culture*. Rokuichi Shobo. [in Japanese]

Pääbo, S. (2014) *Neanderthal Man: In Search of Lost Genomes*. Basic Books. [Kindle version]

Patin E., K. J. Siddle, G. Laval, H. Quach, C. Harmant, N. Becker, A. Froment, B. Régnault, L. Lemée, S. Gravel, J. M. Hombert, L. Van der Veen, N. J. Dominy, G. H. Perry, L. B. Barreiro, P. Verdu, E. Heyer and L. Quintana-Murci (2014) 'The impact of agricultural emergence on the genetic history of African rainforest hunter-gatherers and agriculturalists'. *Nature Communications*, doi: 10.1038/ncomms4163

Reich, D. (2018) *Who We Are and How We Got Here: Ancient DNA and the New Science of the Human Past*. Oxford University Press.

Robillard, M. and S. Bahuchet (2012) 'Les Pygmées et les autres: Terminologie, catégorisation et politique'. *Journal des Africanistes*, 82–1/2: 15–51.

Sano, K. and T. Omori (2015) 'The process of the replacement of *Homo sapiens* and Neanderthals'. In Y. Nishiaki (ed.), *Homo sapiens and Neanderthals 3: The Replacement of Humans and Culture*. Rokuichi Shobo, 20–35. [in Japanese]

Sato, H. (2010) 'The reasons that hunter-gatherers did not change to farmers'. In D. Kimura and K. Kitanishi (eds), *Ecological Studies of Forest Dwellers: People, Nature and History, Vol. I*. Kyoto University Press, 119–139. [in Japanese]

Sawada, M. (2010) 'Change of the image of death in Efe hunter-gatherers'. In D. Kimura and K. Kitanishi (eds), *Sociological Studies of Forest Dwellers: Humans, Nature and History, Vol. II*. Kyoto University Press, 347–366. [in Japanese]

Schebesta, P. (1977 [1933]) *Among Congo Pigmies*, Gerald Griffin (trans.). AMS Press Inc.

Schebesta, P. (1978 [1936]) *My Pygmy and Negro Hosts*, Gerald Griffin (trans.). Hutchinson & Co.

Takeuchi, K. (2014) 'Interethnic relationships between Pygmies and farmers'. In B. S. Hewlett (ed.), *Hunter-Gatherers of the Congo Basin: Cultures, Histories, and Biology of African Pygmies*. Transaction Publishers, 299–320.

Taylor, N. (2011) 'The origins of hunting & gathering in the Congo basin: A perspective on the Middle Stone Age Lupemban industry'. *Before Farming*, 2011/1 article 6: 1–20.

Terashima, H. (1987) 'Why Efe girls marry farmers?: Socio-ecological backgrounds of inter-ethnic marriage in the Ituri forest of central Africa'. *African Study Monographs*, Supplementary Issue, 6: 65–83.

Terashima, H. (1991) 'Honey and holidays: The interactions mediated by honey between Efe hunter-gatherers and Lese farmers in the Ituri Forest'. *African Study Monographs*, Supplementary Issue, 25: 123–134.

Terashima, H. (1997) *Forest of Symbiosis (World of Tropical Forest 6)*. University of Tokyo Press. [in Japanese]

Terashima, H. (2001) 'The logic of symbiosis: A history of interaction between the people of tropical forest and the outsiders'. In S. Wada (ed.), *Ethnic Relationships in Modern Africa*. Akashi Shoten, 223–243. [in Japanese]

Turnbull, C. M. (1976 [1965]) *Wayward Servants: The Two Worlds of the African Pygmies*. Greenwood Press.

Uexküll, J. von (2010) *A Foray into the Worlds of Animals and Humans: With a Theory of Meaning (Posthumanities)*, J. D. O'Neil (trans.). University of Minnesota Press.

Vansina, J. (1990) *Paths in the Rainforest: Toward a History of Political Tradition in Equatorial Africa*. The University of Wisconsin Press.

Verdu, P., F. Austerlitz, A. Estoup, R. Vitalis, M. Georges, S. Théry, A. Froment, S. Le Bomin, A. Gessain, J.-M. Hombert, L. Van der Veen, L. Quintana-Murci, S. Bahuchet and E. Heyer (2009) 'Origins and genetic diversity of Pygmy hunter-gatherers from Western Central Africa'. *Current Biology*, 19: 312–318.

Viola, B. and S. Pääbo (2013) 'What's new in Central Asia?' In *Basic Issues in Archaeology, Anthropology and Ethnography of Eurasia*. Institute of Archaeology and Ethnography, SB RAS Press, 555–565.

Wæhle, E. (1989) *Elusive Persistence: Efe (Mbuti Pygmy) Autonomy Strategies in the Ituri Forest, Zaïre*. Department of Social Anthropology, University of Oslo.

Wilkie, D. S. and B. Curran (1993) 'Historical trends in forager and farmer exchange in the Ituri rainforest of northeastern Zaïre'. *Human Ecology*, 21(4): 389–417.

Yasuoka, H. (2011) *Ecological Anthropology of the Baka Hunter-gatherers: Rethinking Hunter-gatherer Subsistence in the African Rainforests*. Kyoto University African Study Series 004, The Center for African Area Studies, Kyoto University. [in Japanese]

Zilhão, J., D. E. Angelucci, E. Badal-García, F. d'Errico, F. Daniel, L. Dayet, K. Douka, T. F. G. Higham, M. J. Martínez-Sánchez, R. Montes-Bernárdez, S. Murcia-Mascarós, C. Pérez-Sirvent, C. Roldán-García, M. Vanhaeren, V. Villaverde, R. Wood and J. Zapata (2010) 'Symbolic use of marine shells and mineral pigments by Iberian Neandertals'. *PNAS*, 107(3): 1023–1028.

Index

Subjects

Persons

Ethnic groups and indigenous peoples

Species and fossil hominids

Geography and ancient civilization